Zeolites and Clay Minerals as Sorbents and Molecular Sieves

ACADEMIC PRESS INC. (LONDON) LTD.
24/28 Oval Road,
London NW1

United States Edition published by
ACADEMIC PRESS INC.
111 Fifth Avenue
New York, New York 10003

Library of Congress Catalog Card Number: 77 79300
ISBN: 0 12 079350 4

Printed by Unwin Brothers Limited
The Gresham Press,
Old Woking, Surrey

Zeolites and Clay Minerals as Sorbents and Molecular Sieves

R. M. BARRER, FRS
Chemistry Department,
Imperial College, London

1978

ACADEMIC PRESS
London · New York · San Francisco
A Subsidiary of Harcourt Brace Jovanovich, P

Preface

To mineralogists and earth scientists the zeolites and clay minerals have been of great interest for a very long time. Chemists and engineers became interested primarily in the years following the Second World War, and their attention has for the most part been concentrated upon the zeolites. So great has been the effort expended that the volume of research on these minerals is now too large to digest and summarize in a single, reasonably critical volume. As with other technical and scientific areas of rapid growth a point has been reached where monographs are required, each of which reviews a chosen part of the broad field. It was on this basis that the present book was planned. A number of monographs, either with a single author, or (more commonly) cooperative efforts by a number of authors, have appeared which deal with clay minerals or with zeolites. Nevertheless the coverage in certain areas is inadequate, and the present book will, it is hoped, fill one of these gaps. It provides an account of sorption and intercalation by zeolites and clay minerals, emphasizing the structural, physicochemical and theoretical bases of the phenomena observed. In some respects at least, clay minerals and zeolites provide model sorbents, which also have a growing range of applications both in laboratory practice and in industry.

Because of the volume of research in this subject, the choice of reference material has been restricted. Other papers could equally well have been selected to illustrate the topics reviewed and discussed. To those many researchers whose excellent work has not been quoted I can only plead the need for reasonable brevity. For the same reason I also thought it advisable to limit the survey to well known crystalline silicate sorbents with zeolites as the major group but with (it is hoped) an adequate account of clay mineral sorbents. The similarities and differences between the sorbent properties of these two groups are particularly interesting. In the zeolites the rigid frameworks emphasize the molecule sieving function; on the other hand with the clay minerals, frequent one-dimensional swelling introduces a new factor, controlled by variables not evident in sorption by zeolites. Nevertheless it does not appear to have been generally realized that, as first shown in 1955, permanent intracrystalline porosity can readily be introduced into clay minerals. Such porous forms can then duplicate in many directions the molecule sieving function of the zeolites. Like the zeolites, clay minerals can either be synthesized or taken from natural deposits. At present the mined products are used industrially but not the synthetics. With zeolites the reverse has tended to be the case.

In the text the recommendations of the Zeolite Nomenclature Committee of IUPAC have been employed, as presented to the Third Zeolite Symposium held in Zurich in 1973. It should also be noted that the original units of the authors have usually been retained when numerical data have been quoted. It was felt that readers will be familiar with the older units such as calories as well as with the corresponding SI units, because the literature is now a good mixture of each. I wish to thank many authors for permission to reproduce diagrams; and also to thank in particular those colleagues who have individually read sections of the manuscript, made constructive suggestions and pointed out errors or ambiguities. They include Mr. R. Ash and Drs. J. A. Barrie, I. S. Kerr, D. Nicholson, N. G. Parsonage and L. V. C. Rees. My thanks are also due to the Leverhulme Trust for the award of an Emeritus Fellowship.

Richard M. Barrer

April 1978

Contents

1
The Zeolites: their Nature and some Uses

1. ZEOLITES

The zeolites form a large family of aluminosilicates which has been studied by mineralogists for more than 200 years. This is illustrated by the dates of discovery of individual members of the group, recorded by Breck[(1)] and given in serial order in Table 1. Fine crystals are often found lining cavities and cracks in basaltic rocks of volcanic origin, as exemplified in Figs 1 and 2. The cavities are considered to be formed by bubbles of fluids in the parent magma, and the zeolite crystals grow as a result of the chemical action on the magma of these fluids or of fluids which have subsequently displaced them. In general the bulk compositions of the zeolites tend to correlate with those of the parent rock—more aluminous zeolites are associated with rocks deficient in silica and more siliceous zeolites with rocks high in silica. There may be more than one zeolite species present and evidence is sometimes found of slow replacement of one zeolite by another rather than of co-crystallization. Zeolites of this volcanic origin are usually dispersed in occurrence and are not therefore suitable for industrial purposes.

More massive deposits occur as very small crystal grains in certain sediments and low grade metamorphic rocks. Some of these deposits are extremely large. Aqueous and normally alkaline solutions acting upon volcanic ash deposited in lakes account for much of the zeolitization of this kind. Phillipsite and clinoptilolite are found in ocean sediments and are again thought to have formed from volcanic debris. Zeolites are sometimes found in arid alkaline soils which may also be rich in salts. Beds rich in zeolites may be hundreds of metres thick; indeed beds of tuff converted largely or in part to zeolites can reach thicknesses of several kilometres. Some deposits can be primarily of one zeolite species and the deposits may contain little else than

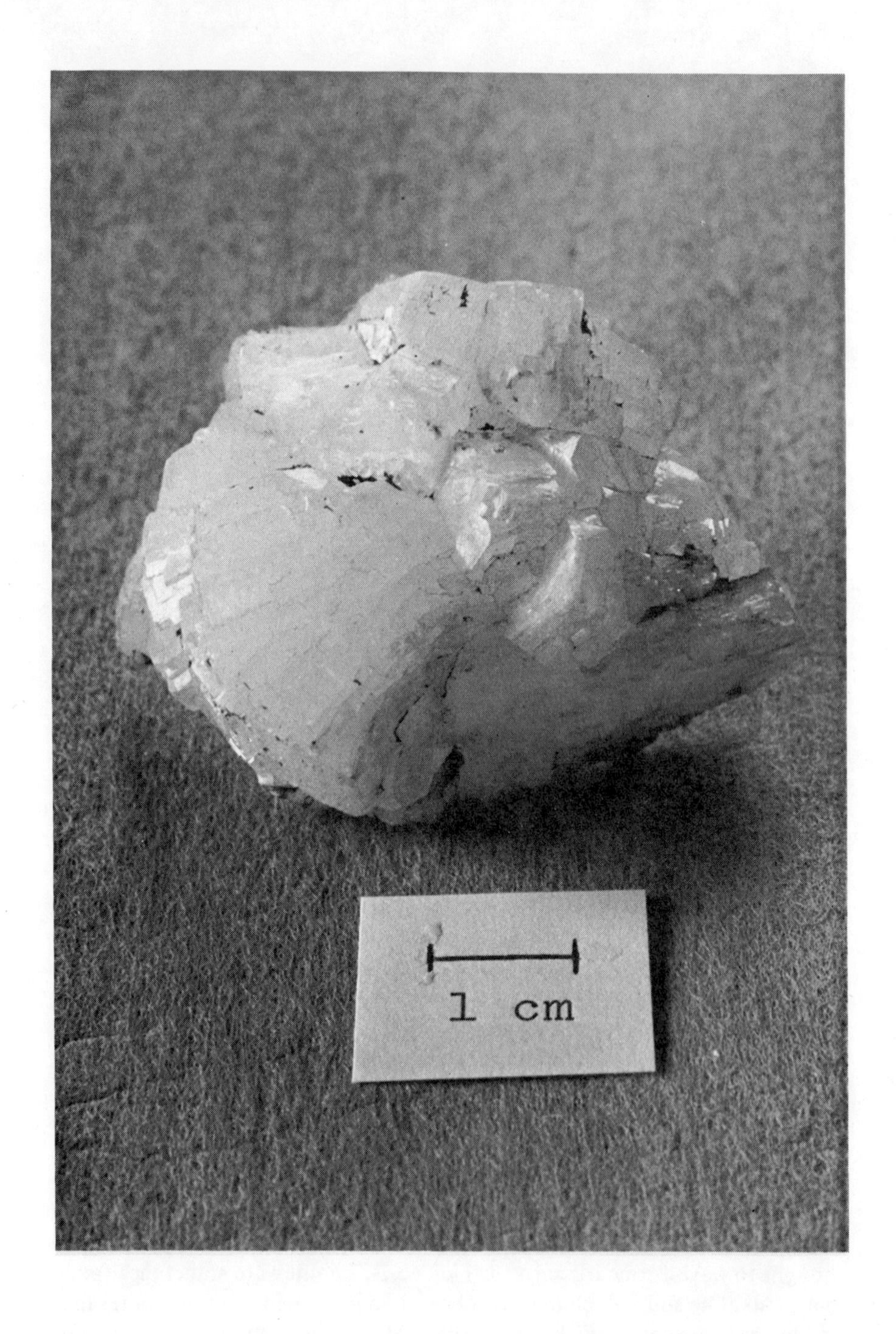

FIG. 1. Crystals of heulandite, showing the characteristic platy structure.

Fig. 2. Crystals of stilbite of typical form.

zeolite. Such occurrences should be particularly useful for commercial exploitation.

Other deposits rich in zeolites have been the result of low grade metamorphism and metasomatism of comparatively deep-seated rocks. It is not clear how far zeolitization represents a metastable condition, but once formed, the zeolites often resist subsequent transformation, at least for long periods of time. Their nucleation and growth can be controlled by kinetic factors rather than by the thermodynamic stability relations of mineral assemblages, especially when the temperature of formation is low. Sheppard[2] and Kossowskaya[3] have among others listed examples of large occurrences of zeolites of sedimentary origin. The zeolites so found are mainly analcime, chabazite, erionite, laumontite, clinoptilolite, mordenite and phillipsite.

The fundamental building units in natural zeolites are tetrahedra SiO_4 and AlO_4. Zeolites are tectosilicates, that is they are formed by the linking together of these tetrahedra to give three-dimensional anionic networks in which each oxygen of a given tetrahedron is shared between this tetrahedron and one of four others. Thus there are no unshared oxygens in the frameworks. This fact means that in all tectosilicates (zeolites, felspars and felspathoids) $(Al + Si) : O = 1 : 2$. For every Si^{IV} which is replaced in the

TABLE 1

Dates of discovery of some zeolites

Zeolite	Date	Zeolite	Date
Stilbite	1756	Mordenite	1864
Natrolite	1758	Clinoptilolite	1890
Chabazite	1772	Offretite	1890
Harmotome	1775	Erionite	1890
Analcime	1784	Kehoeite	1893
Laumontite	1785	Gonnardite	1896
Thomsonite	1801	Dachiardite	1905
Scolecite	1801	Stellerite	1909
Heulandite	1801	Ferrierite	1918
Gmelinite	1807	Viseite	1942
Mesolite	1813	Yugawaralite	1952
Gismondine	1816	Wairakite	1955
Brewsterite	1822	Bikitaite	1957
Epistilbite	1823	Paulingite	1960
Phillipsite	1824	Garronite	1962
Levynite	1825	Mazzite	1972
Herschelite	1825	Barrerite	1974
Edingtonite	1825	Merlinoite	1976
Faujasite	1842		

framework by Al^{III} a negative charge is created which is neutralized by an electrochemical equivalent of cations. Additionally in zeolites but not in felspars or some felspathoids the framework is sufficiently open to accommodate water molecules as well as cations. This openness imparts characteristic zeolite properties: the water molecules can move easily within the crystals and so can the cations. These ions therefore undergo ready exchange with other cations, and the water molecules can be removed or replaced in a continuous manner and often reversibly. Sorption isotherms do not normally show steps such as are found in crystallohydrates.

Table 1 shows that 18 naturally occurring zeolites had been found by 1825. In the remainder of the nineteenth century seven more were recorded, and another 12 in the twentieth century. In addition a considerable number of synthetic zeolites have recently been made which do not appear to have natural counterparts. The table includes only natural species, some of which are variants of one framework topology (e.g. stilbite, stellerite and barrerite; analcime, kehoeite, viseite and wairakite; and chabazite and herschelite). Usually however the topologies are different for each zeolite, and their diversity has made them, as Eitel[4] has remarked, "the pride of mineralogists". Until the early 1940's attempts to synthesize zeolites were largely made by mineralogists interested in the stability relationships of zeolites with each other and with other minerals. The reaction conditions were those thought likely to have arisen naturally and often involved autoclaving, always under hydrothermal conditions, at high pressures and several hundreds of degrees centigrade. Summaries, up to about 1937, of reported but not always proven hydrothermal syntheses of mineral-type compounds, were given by Niggli and Morey[5] and by Morey and Ingerson.[6] Ion exchange reactions of zeolites also received limited attention in qualitative or semi-quantitative experiments, and in the 1930's the term "zeolite" was extended to include gel exchangers, organic and inorganic. This confusing and erroneous use of the term has now been abandoned.

2. SORPTION

The three properties of zeolites which have industrial potential are their capacities to sorb gases, vapours and liquids; to catalyse reactions; and to act as cation exchangers. Some description of these three aspects follows.

The ability of several zeolites such as chabazite to take up large amounts of some guest molecules in place of water had been known for a considerable time before McBain[7] in 1932 wrote his monograph on physical sorption. This monograph devoted a chapter to sorption by chabazite, with some reference to the small amount of work then performed on other zeolite sorbents. It had been shown that the molecular volume of the sorbate had an influence upon the amounts which could be taken up by the zeolite.[8] There were indications therefore that, as with certain microporous carbons and silica gels, substantial selectivity based upon molecular dimensions could occur which could be thought of as a molecule sieving process, at least in part. Despite these indications there had been no drive to explore and quantify such possibilities until the early 1940's when the extraordinary versatility of molecule sieving by zeolites was well demonstrated in several papers[9, 10, 11] and patents.[12] About the same time, because massive deposits of appropriate zeolites were then unknown, synthesis of zeolites was also commenced with the specific objective of preparing molecular sieve sorbents from readily available materials. Early successes were the formation of Species P and Q[13, 14] with an aluminosilicate structure now known as the *ZK-5* framework; [15, 16] and also the synthesis of another useful molecular sieve, mordenite.[17] Species P and Q contained $BaCl_2$ and $BaBr_2$ respectively, replacing much zeolitic water. When the $BaCl_2$ was removed from Species P an excellent molecular sieve sorbent was obtained[18] which had the properties that had been demonstrated for Ca-rich chabazite[9, 10] and that were later realized also in the Ca-form of zeolite *A*,[19] at that time still to be synthesized. Thus the *ZK-5* structure was the first synthetic zeolite with no natural counterpart to be prepared and examined as a molecular sieve. Zeolite *ZK-5* (Species P freed of $BaCl_2$) and Ca-chabazite had the ability to separate *n*-paraffins from branched chain paraffins, aromatic hydrocarbons and cycloparaffins. The exciting aspect of the separations was their quantitative nature: they could be virtually complete in a single step.[10, 11]

This property of zeolites was considered to be due to the crystalline nature of the sorbents.[9] The porous crystals were thought to be permeated by channel networks, the free dimensions of which were determined by the positions of the framework oxygens just as exactly as the rest of the crystal lattice was defined by the positions of its constituent atoms. None of the irregularities characteristic of the pore structures of amorphous micropore carbons or silica gels could occur and therefore quantitative molecule sieving was possible. The zeolites were considered as rigid, crystalline sponges capable of imbibing large amounts of molecules small enough or of the right shapes to pass through their surfaces and enter the intracrystalline pores, but unable to sorb molecules having the wrong sizes or shapes. Zeolites could, it was shown, act in so clear-cut a way as to add a new order of selectivity to mixture separation by adsorption. The work led to the characterization of three[9] and later of five[21] classes of molecular sieve, in defining which certain molecules of known dimensions were used for calibration. These classes and examples of molecules which they would and would not sorb are given in Table 2. Instances of mixtures actually separated by three of these classes of sieve are given in Tables 3, 4 and 5[10, 11, 12] and the regular porous

TABLE 2

Classification of some molecular sieves

Molecular size increasing ⟶

	He, Ne, Ar, CO H_2, O_2, N_2, NH_3, H_2O	Kr, Xe CH_4 C_2H_6 CH_3OH CH_3CN CH_3NH_2 CH_3Cl CH_3Br CO_2 C_2H_2 CS_2	C_3H_8 n-C_4H_{10} n-C_7H_{16}	 C_2F_6 CF_2Cl_2[a]	SF_6 iso-C_4H_{10} iso-C_5H_{12}	$(CH_3)_3N$ $(C_2H_5)_3N$ $C(CH_3)_4$	C_6H_6 $C_6H_5CH_3$ $C_6H_4(CH_3)_2$	Naphthalene Quinoline, 6-decyl-1, 2, 3, 4-tetra- hydro- naphthalene, 2-butyl-1- hexyl indan $C_6F_{11}CF_3$	1, 3, 5-triethyl benzene	$(n$-$C_3F_9)_3N$
	Size limit for Ca- and Ba-mordenites and levynite about here ($\approx 3\cdot8$Å)		n-$C_{14}H_{30}$ etc. C_2H_5Cl C_2H_5Br C_2H_5OH $C_2H_5NH_2$ CH_2Cl_2 CH_2Br_2 CHF_2Cl CHF_3 $(CH_3)_2NH$ CH_3I B_2H_6	CF_3Cl $CHFCl_2$	iso-C_8H_{18} etc. $CHCl_3$ $CHBr_3$ CHI_3 $(CH_3)_2CHOH$ $(CH_3)_2CHCl$ n-C_3F_8 n-C_4F_{10} n-C_7F_{16} B_5H_9	$C(CH_3)_3Cl$ $C(CH_3)_3Br$ $C(CH_3)_3OH$ CCl_4 CBr_4 $C_2F_2Cl_4$	Cyclopentane Cyclohexane Thiophen Furan Pyridine Dioxane $B_{10}H_{14}$		1, 2, 3, 4, 5, 6, 7, 8, 13, 14, 15, 16-decahydro-chrysene	
Type 5										
	Size limit for Na-mordenite and Linde sieve 4A about here ($\approx 4\cdot0$Å)									
Type 4										
		Size limit for Ca-rich chabazite, Linde sieve 5A, Ba-zeolite and gmelinite about here ($\approx 4\cdot9$Å)								
Type 3										
							Size limit for Linde sieve 10X about here			
Type 2										
								Size limit for Linde sieve 13X about here (≈ 10Å)		
Type 1										

[a] Freon-type molecules provide interesting border-line cases and can differentiate between certain of the zeolites grouped as Type 3. Simple ketones and esters are also border-line cases.

TABLE 3
Resolutions of mixtures using chabazite

Mixture	Component(s) sorbed	Conditions and comments
$CH_3OH + (CH_3)_2CO$	CH_3OH	As liquid at ≈20°C Rapid and quantitative
$CH_3OH + C_2H_5Br$	CH_3OH	As above
$CH_3OH + CS_2 + CH_3CN + C_6H_6$	CH_3OH, CS_2 and CH_3CN	As above
$CH_3OH + H_2O + CH_3OCOCH_3$	CH_3OH, H_2O	As above
CH_3OH + $OCOCH_3$ / $OCOCH_3$	CH_3OH	Sorption via vapour in equilibrium with solution. Complete separation
$C_2H_5OH + C_6H_5CH_3$	C_2H_5OH	As liquid at ≈20°C Separation slow but complete
C_2H_5OH + CHCl:CCl_2	C_2H_5OH	As above
$C_2H_5OH + CH(CH_3)_2OH$	C_2H_5OH	As above
$C_2H_5OH + (CH_3)_3OH$	C_2H_5OH	As above
C_2H_5OH + n-C_7H_{16}	C_2H_5OH	As above
$C_2H_5OH + H_2O + (C_2H_5)_2O$	C_2H_5OH, H_2O	As above.
$C_2H_5OH + CH_3COC_2H_5$	C_2H_5OH	As liquid at 112°C Separation rapid and complete
$C_2H_5OH + CH_2Br_2$	C_2H_5OH	As liquid at ≈20°C Slow but complete separation
$CH_2O + H_2O + CH(CH_3)_2OH$	CH_2O, H_2O	As liquid at ≈20°C Rapid and quantitative
$CH_2O + H_2O + CH_3I$	CH_2O, H_2O	As above
$CO_2 + CH(CH_3)_2OH$	CO_2	As above, CO_2 initially in solution
$SO_2 + CHCl_3$	SO_2	As above, SO_2 initially in solution
$N_2O_3 + C_6H_6$	N_2O_3	As above, N_2O_3 in solution
$H_2S + C_6H_6$	H_2S	As above, H_2S in solution
$CS_2 + CH_3COCH_3$	CS_2	As liquid at ≈20°C Rapid and quantitative
$CS_2 + CHCl_3$	CS_2	As above
$CS_2 + C_6H_6$	CS_2	As above
CS_2 + Pyridine	CS_2	As above
$C_2H_5SH + C_6H_6$	C_2H_5SH[a]	As liquid at 50° Partial removal in a week
$CH_3NH_2 + C_2H_5OH + N(CH_3)_3$	CH_3NH_2, C_2H_5OH	As liquid at ≈20°C Complete removal of both constituents within 16 hours

[a] Slowly sorbed at room temperature.

TABLE 3—*continued*

Mixture	Components(s) sorbed	Conditions and comments
$C_2H_5NH_2$ + $(C_2H_5)_2NH$	$C_2H_5NH_2$[a]	As liquid at ≈20°. Slow but complete separation
$(C_2H_5)_2NH$ + C_8H_{18}	$(C_2H_5)_2NH$[a]	As liquid at 180°C. Separation only partial in five days
CH_3CN + Thiophen	CH_3CN	As liquid at ≈20°C. Rapid and quantitative
$CH_3CN + CH_2Br_2$	CH_3CN	As above
$CH_3CN + CH_2O$ + H_2O + n-C_7H_{16}	CH_3CH, CH_2O, H_2O	As above
C_2H_5CN + $CH(CH_3)_2OH$	C_2H_5CN[a]	As liquid at 100°C. Separation complete within two days
$HCl + CHCl_3$	HCl	As liquid at ≈20°C. Dissolved HCl removed quickly and completely
$Cl_2 + C_6H_6$	Cl_2	As above
$Br_2 + CCl_4$	Br_2	As above. Equilibrium separation nearly complete within 16 hours
I_2 + CHCl : CCl_2	I_2[a]	As liquid at ≈20°C Partial removal within four days
$CH_3I + C_6H_6$	CH_3I[a]	As liquid at ≈20°C Nearly complete in 12 days
CH_2Cl_2 + $CH(CH_3)_2OH$	CH_2Cl_2	As liquid at ≈20°C Rapid and complete
CH_2Cl_2 + $CH_3COC_2H_5$	CH_2Cl_2	As liquid at 112°C Rapid and complete
CH_2Cl_2 + Dioxane	CH_2Cl_2[a]	As liquid at 112°C Partial separation in three days
CH_2Cl_2 + sym-$C_2H_4Cl_2$	CH_2Cl_2	As liquid at ≈20°C Separation complete
C_2H_5Cl + $CH(CH_3)_2OH$	C_2H_5Cl[a]	As above
CH_2Br_2 + *iso*-C_8H_{18}	CH_2Br_2[a]	As liquid at ≈20°C Separation complete within 12 days
$CH_3Br_2 + C_6H_6$	CH_2Br_2[a]	As above
CH_2Br_2 + $CH(CH_3)_2OH$	CH_2Br_2[a]	As liquid at 112°C Separation nearly complete within 24 hours
$C_2H_5Br + C_6H_6$	C_2H_5Br[a]	As above. Separation complete
C_2H_5Br + $CH(CH_3)Cl_2$	C_2H_5Br[a]	As liquid at 97°C Separation complete within two days

surface of zeolite Ca-*A*, a class 3 sieve, is shown in Fig. 3.[22] Molecules are sieved by the 8-ring openings in the surface.

The possibilities thus revealed began to register with industry some years after the first disclosure. An important success came with the synthesis of zeolite *A* in the laboratories of Linde Air Products;[19] a synthetic zeolite with the ability, already noted, to separate *n*-paraffins from other hydrocarbons when in its Ca-exchanged form. Other major successes were the syntheses of variants of faujasite, an aluminous form termed zeolite X and later the more siliceous form, zeolite Y.[23] Much of the commercial side of current zeolite-based technology rests with zeolites *A*, X and Y, although there is growing interest in other zeolites such as mordenite, offretite, erionite, chabazite, clinoptilolite and the synthetic zeolites *L* and Ω.

Among the applications of zeolites one may mention hydrocarbon separations; the intensive drying of liquids, industrial gases, air and natural gas; removal of sulphur compounds from petroleum; and air separation. After the first measurements of sieving, the scale and range of selective sorption by zeolites has grown steadily, although only some of the separations depend upon molecule sieving. In others, all components of the mixture are sorbed, but at least one is more strongly sorbed than the others. This is the case for example in the separation of 2,6- and 2,7-dimethylnaphthalenes by the faujasite-type zeolite Na-Y;[24] or of *p*-xylene from a C_8 mixture containing

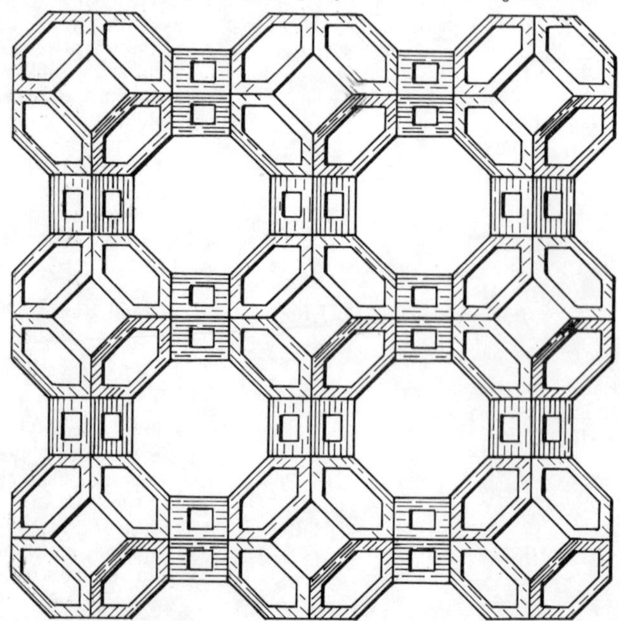

FIG. 3. A formal representation of the sieve-like structure of zeolite *A*[22]. There is an Al or Si atom centred at each corner and an O atom near to but not on the mid-point of each edge. Cations are not shown. For more detail see Chapter 2.

xylenes and ethylbenzene, also using faujasites with cations of Groups IA or IIA of the periodic table, or both.[25] Polar molecules are in general selectively sorbed in the presence of non-polar molecules, so that water, ammonia, sulphur dioxide and carbon dioxide may with suitable zeolites and conditions be removed from gas or liquid streams contaminated with them. The zeolites may be chosen to have a degree of openness of the intracrystalline pores which allows them to function as sieves, or they may have pores accessible to all components of the mixture, from which they preferentially remove the polar components. Zeolites select polar sorbates because they are themselves very polar. The anionic charges on the framework and the intracrystalline

TABLE 4

Resolution of mixtures using Na-mordenite

Mixture	Component(s) sorbed	Conditions and comments
$CH_3OH + CH_3COCH_3$	CH_3OH[a]	At 20°C. Slow but complete separation
$CH_3OH + C_2H_5OH$	CH_3OH[a]	As above
CH_3OH + Pyridine	CH_3OH[a]	As above
$CH_3OH + CH_2Cl_2$	CH_3OH[a]	As above
$CH_3NH_2 + C_2H_5OH + NH(C_2H_5)_2$	CH_3NH_2[a] CH_3NH_2[a]	As liquid at 73°C Separation nearly complete within 1 day
$CH_3CN + CH_2Br_2$	CH_3CN[a]	As liquid at 60°C Separation complete within 2 days
$CH_3CN + CH_3COCH_3$ + iso-C_8H_{18}	CH_3CN[a]	As liquid at ≈20°C Separation nearing completion in 14 days
$CS_2 + C_6H_6$	CS_2[a]	As liquid at ≈20°C Separation partial in 14 days
$Cl_2 + C_6H_6$	Cl_2	As liquid at ≈20°C Separation complete within 2 days
$HCl + CH_2Cl_2$	HCl	As liquid at ≈20°C Separation complete in less than 12 hours
$N_2O_3 + C_6H_6$	N_2O_3 ($NO + NO_2$)	As liquid at ≈20°C Partial separation in 1 day
$NH_3 + H_2O + C_2H_5OH$	NH_3, H_2O	As liquid at ≈20°C Separation complete within several hours

[a] Slowly sorbed at room temperature.

cations produce high local electrostatic fields and field gradients which interact with molecular dipoles and quadrupoles respectively and so augment the heat of sorption of polar guest molecules (Chapter 4).

In the pioneering work[9, 10, 11] it was established that the molecule sieving properties of a zeolite could be profoundly changed by ion exchange (see Table 2). Thus there were clear differences between Na- and Ca- or Ba-exchanged mordenites. This study was later extended to other cations in mordenite;[26] to Na- and Ca-exchanged chabazites;[27] to K-, Na- and Ca-exchanged forms of zeolite *A*[19] (respectively termed sieves 3-*A*, 4-*A* and 5-*A*); and to Na- and Ca-forms of zeolite X (respectively called 13-X and 10-X).[28] The numbers are supposed to indicate the van der Waals dimension in Å which must not be exceeded by the guest if it is to be imbibed by the crystals. Such figures are however crude indications not to be taken very literally. From the foregoing and subsequent work, ion exchange has been established as a standard way of tailoring molecular sieve zeolites so as to meet best the requirements of a particular separation. The reason is that cations are present in the same channels and voids as the guest species. The cations may sometimes occupy sites adjacent to the windows or apertures leading from one void to the next. Those cations so located may according to their sizes and numbers act as sentinels, effectively barring passage of a larger molecule while allowing passage of a smaller one, in situations where if there had been no cations in or near the windows, both molecules would have passed through. The effects of the cations on the molecule sieving can be very sensitive therefore to the size of the ions relative to the free dimensions of the windows, and also to the numbers of the ions in or near window sites. These numbers vary with ion valence ($2Na^+ \rightleftharpoons Ca^{2+}$) and with ion type. A zeolite usually provides several kinds of crystallographically distinct exchange site and the distributions of cations of a given type among these sites may change according to the ion (Chapter 2). Some quantitative aspects of the behaviour are described in Chapter 6, which considers intracrystalline diffusion of guest molecules.

The scale of hydrocarbon separation using molecular sieve zeolites may be seen from the following information, representing approximately the 1975 position.[29] World production of *n*-paraffins was about 1·6 million tonnes p.a., very largely obtained by using mole sieve sorbents, and with several variants of procedure. For example the Universal Oil Products (UOP) Molex process, introduced in 1964, accounts for about 500,000 tonnes p.a. with about 96% recovery of *n*-paraffins of more than 98% purity. The *n*-paraffins are used to make biodegradable detergents, chlorinated plasticizers, and single cell protein. For protein production they serve as food for bacteria in an appropriate fermentation process. A second application of zeolite sorbents is found in separation of *p*-xylene from C_8 reformate or isomerate mixtures. The 1972 estimate of world production was 2·32 million tonnes p.a. Universal

Oil Products' Parex zeolite process, from operational and planned units, will alone give 1·67 million tonnes p.a. with a *p*-xylene recovery of 99·7% and purity of 99·3%. A newer arrival is their Olex process with present capacity of about 282,000 tonnes p.a. for separating straight chain olefines from broad mixtures of paraffins and olefines. Product recovery and purity are respectively claimed as 93·4% and 93·7%, or better. As there is at present no other effective method of recovering straight chain olefines the mole sieve method has virtually all the market. Although the above remarks refer to the three UOP Sorbex processes (Molex, Parex and Olex) comparable processes using

TABLE 5

Resolution of mixtures using Ca- and Ba-mordenite

Mixture	Component(s) sorbed	Conditions and comments
$H_2O + C(CH_3)_3OH$	H_2O	As liquid at ≈20°C Separation rapid and complete
$H_2O + NH_3 + CH_3OH$	H_2O, NH_3	As above
$NH_3 + C_2H_5OH$	NH_3	As above
$NH_3 + C_2H_6$	NH_3	As gas at ≈20°C Separation quantitative
$NH_3 + C_3H_8$	NH_3	As above
$HCl + C_2H_6$	HCl	As above

zeolite sieves are operated or licensed at least for *n*-paraffin separation by other companies (e.g. British Petroleum, Shell, Union Carbide (IsoSiv process)).

Other hydrocarbon separation processes had by 1975 reached pilot plant stage. They include production of *p*-diethylbenzene (97·8% recovery; 99·1% purity); and the separation of *p*-cymene (isopropyl toluene) and *m*-cymene. The pilot plant results indicate 99·6% purity and 87% recovery for *p*-cymene; and 98·1% purity with 99% recovery for *m*-cymene.

Air separation with mole sieve sorbents is available for special purposes. Zeolites, including mordenite, chabazite, Ca-*A* and faujasite, sorb nitrogen preferentially to oxygen from dry CO_2-free air. This happens because nitrogen, but not oxygen, possesses a considerable molecular quadrupole moment; and a quadrupole-field gradient interaction between nitrogen and its intracrystalline polar environment enhances its sorption energy. The equilibrium gas phase is therefore enriched in oxygen,[30] and the sorbed phase in nitrogen. It is considered that for production of oxygen from air in quantities up to 25 or 30 tons per day, the process can be more economical than the cryogenic method. The Lindox and Unox[31] processes use a pressure swing to desorb the selectively sorbed nitrogen. The oxygen can serve for smaller scale uses such as secondary sewage treatment, with the air separation plant operating at the site and delivering the gas for more rapid improvement of BOD in the effluent. Pressure swing sorption and desorption from zeolite adsorbers is also used to prepare pure hydrogen.[32] In hydrogen formed by steam reforming CO_2, H_2O, CH_4, CO and N_2 are the impurities, all of which are much more strongly sorbed than H_2.

3. CATALYSIS

As noted earlier the systematic investigation of zeolite synthesis, selective sorption and ion exchange developed rapidly in the 1950's and the 1960's. However another area of great significance soon began to attract attention and has resulted in a major burst of activity, particularly in the petroleum industry. Zeolite-based catalysts became important when it was discovered that rare earth and hydrogen-forms of certain zeolites possessed cracking activity several orders of magnitude greater than that of conventional silica-alumina catalysts. This is illustrated in Table 6.[33] which compares temperatures required, for similar flow rates and pressures, to effect 5 to 20%

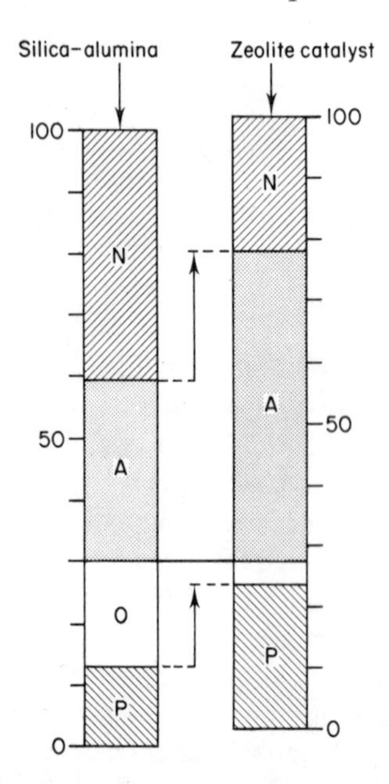

FIG. 4. The yields of different classes of hydrocarbon in the gasoline derived from cracking the same hydrotreated petroleum fraction with silica-alumina and zeolite cracking catalyst.[35] P = paraffin; O = olefine; N = naphthene; A = aromatic.

TABLE 6

Approximate temperatures for 5 to 20% conversion of *n*-hexane[33]

Catalyst	Main cation	Analysis (%) SiO_2	Al_2O_3	Na	Ca	RE[a]	T °C	α
Amorphous SiO_2, Al_2O_3 (standard)	—	—	—	—	—	—	540	1·0
Faujasite	Ca	47·8	31·5	7·7	12·3	—	530	1·0
Faujasite	NH_4	75·7	23·1	0·4	—	—	350	6400
Faujasite	La	—	—	0·4	—	29·0	270	7000
Faujasite	RE	—	—	0·39	—	28·8	<270	$>10{,}000$
Faujasite	RE, NH_4	40·0	33·0	0·22	—	26·5	<270	$>10{,}000$
Faujasite	RE, NH_4	—	—	—	—	—	420	20
Zeolite *A*	Ca	42·5	37·4	7·85	13·0	—	560	0·6
Zeolite *ZK–5*	—	—	—	—	—	—	400	38
Zeolite *ZK–5*	H	76·8	23·1	0·47	—	—	340	450
Mordenite	Ca	≈77	—	1·01	—	—	520	1·8
Mordenite	Ca, H	82·0	14·0	0·4	—	—	360 to 400	40 to 200
Mordenite	NH_4	—	—	—	—	—	<270	$>10{,}000$
Mordenite	H	80·1	13·4	0·3	1·54	—	300	2500
Mordenite	NH_4	—	—	0·1	—	—	<270	$>10{,}000$
Gmelinite	NH_4	—	—	—	—	—	<270	$>10{,}000$
							~200	$>10{,}000$
Chabazite	NH_4	—	—	—	—	—	<270	$>10{,}000$
Stilbite	NH_4	—	—	—	—	—	370	120

[a] RE = rare earth.

conversion of *n*-hexane to propane plus propene. The relative performances were estimated by finding the rate constant, k, for each catalyst at several temperatures and using the Arrhenius equation ($k = k_0 \exp -E/RT$) to extrapolate from the experimental to the reference temperature. The ratios α of the rate constants at the reference temperature for zeolite catalysts and a conventional alumina-silica cracking catalyst are given in the last column of Table 6. The activity of certain exchange forms of the faujasites, mordenites, gmelinite and chabazite is extraordinarily high. Comparable hydro-cracking activity has also been found with other zeolites such as erionite.(34)

Cracking catalysts based on zeolite Y were first introduced commercially in 1962 by the Mobil Company. Their performance won them rapid recognition and they have now displaced alumina-silica catalysts in nearly all US refineries and in many others throughout the world. They gave less coking, increased production capacity, and improved gasoline yields by as much as 25%. The different product patterns with silica-alumina and zeolite-based cracking catalysts are illustrated in Fig. 4.(35) There is a shift from olefines and naphthenes towards aromatics and paraffins with the zeolite catalyst. The transformations of olefines to paraffins and of naphthenes to aromatics nearly correspond in molar amounts. To account for this it has been suggested that olefins formed in the cracking reaction are hydrogenated to paraffins by hydrogen transfer from the naphthenes, which are thereby converted to aromatics.(35)

In hydrogen-zeolites it is believed that acid centres are formed of the type:(36)

$$\left[\begin{matrix} \diagdown & & HO & & \diagup \\ — Al & & & \diagdown Si & — \\ \diagup & & & & \diagdown \end{matrix} \right]$$

Various secondary reactions involving the Al and the –OH groups can occur, but need not be considered here. The Bronsted acid centres can by proton transfer result in the formation of carbonium ions as reactive intermediates, which can be involved in polymerization, isomerization and cracking. Since these acid centres occur throughout the body of the zeolite its molecule sieving and catalytic functions can be harnessed simultaneously. Thus in catalysts based upon erionite or zeolite *A* *n*-paraffins can enter the crystals and reach intracrystalline acid sites, but not branched chain or cycloparaffins, or aromatics. The *n*-paraffins can therefore be selectively cracked while the other hydrocarbons are relatively unchanged.(37) Zeolite-based isomerization catalysis and molecule sieving can also be combined, with zeolite Y in the isomerization units and with zeolite Ca-*A* in subsequent units to filter out unchanged *n*-paraffins. The *n*-paraffins may then be desorbed and returned to the reactant stream. In this way virtually complete isomerization of raffinates to branched chain paraffins, cycloparaffins and aromatics can be envisaged, with improved anti-knock properties without additions of lead tetraethyl. It is to be noted that while in gasoline one wishes to minimize the *n*-paraffin content, in diesel fuel one needs to maximize it. The longer *n*-paraffin chains required for diesel fuels can be separated from hydrocarbon oils by mole sieving as referred to in § 2.

Catalysis, selective for molecules of particular shapes, has been demonstrated for mixtures other than those referred to in the previous paragraph.[38, 39] Examples include the preferential combustion of *n*-butane (sorbed by the zeolite catalyst) in mixtures with isobutane (not sorbed); and the preferential hydrogenation of *n*-olefines (sorbed into the Pt-bearing

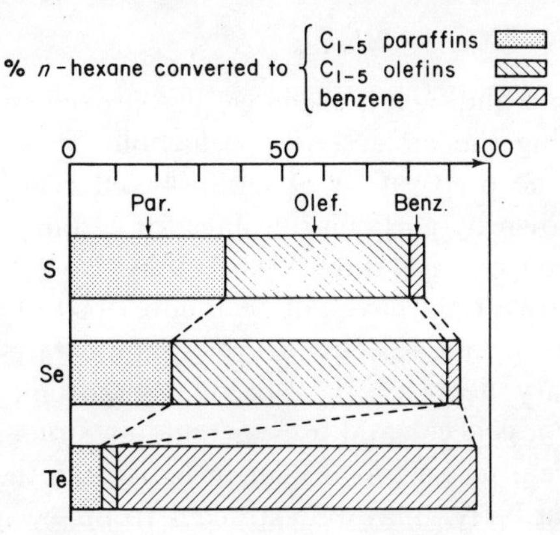

FIG. 5. The product distributions from *n*-hexane over S-, Se- and Te-complexed Na-faujasite.[35] (538°C, 1 atm, 9 s.)

zeolite) in mixtures with branched chain olefines (not sorbed). A mordenite catalyst containing both a Pt species and Na^+ ions was made which admitted ethylene or ethane but not propane. This material selectively hydrogenated ethylene in admixture with propylene.[35] No steady state hydrogenation of propylene is possible once some limit on intracrystalline propane concentration has been established, because this propane cannot either enter or leave the zeolite.

Elements like S, Se and Te may be sorbed by certain zeolites. Unexpected catalytic activity was found to be associated with zeolite Na-X containing a little of each of these elements.[40, 41] The sulphur-zeolite centres had a large hydrocarbon cracking activity resembling in product distribution a partly acid and partly radical catalysis mechanism. When S was replaced by Se and then by Te there were striking changes in the products with *n*-hexane as reactant, as shown in Fig. 5. With Se-Na-X there was a shift towards olefine formation and with Te-Na-X there was a highly selective dehydrocyclization to benzene.

Synthetic zeolites containing transition metal ions are active for oxidation of H_2, CO, C_2H_4 and NH_3.[42] With oxygen and Cr- or Ag-bearing zeolites, ammonia yielded N_2O and N_2; with Cu^{II}-Y propylene gave 2-propanol, acetaldehyde, acetone and acrolein in amounts depending upon the quantity of water also present in the system. With iron-bearing zeolites *A*, X or Y mixtures of propylene, ammonia and oxygen yielded $CH_2{=}CH{-}CN$ and water. In Ni- and Pd-faujasites benzene underwent hydrodimerization to phenylcyclohexane. These examples are a small selection of the many reactions effected with zeolite-based catalysts.[36, 43, 44, 45] The examples do however serve to indicate their versatility.

4. ION EXCHANGE

As indicated in §§ 2 and 3, ion exchange has proved to be of major significance in regulating the mole sieving behaviour and selectivity of zeolite sorbents; and in the preparation of zeolite-based catalysts. In addition the high selectivity shown by particular zeolites for certain cations[46] can assist in the concentration and isolation of such ions. Nevertheless ion exchange processes in their own right have not been developed so extensively as those based on molecule-sieving, selective sorption and catalysis. In this area, the massive sedimentary deposits[2, 3, 47] become important. Zeolite-rich rocks have had more or less accidental uses as building stone since ancient times. Large amounts are also used as a filler in paper. On the ion exchange side it is suggested that NH_4^+ may be extracted from sewage and agricultural wastes. There has already been considerable interest in the use of such materials for collecting and isolating radionuclides. Clinoptilolite for example is one of a number of zeolites selective for strontium and caesium,[48] two of the troublesome by-products of nuclear fission. The ultimate immobilization and disposal of radioactive nuclides is a matter of great concern to the atomic energy industry. Zeolites ion-exchanged with radionuclides could be compacted and sintered to amorphous or glassy phases which could then be stored underground.

Zeolites can exhibit ion sieving properties just as they possess molecule sieving characteristics. Clinoptilolite shows this behaviour with a range of alkylammonium ions.[49] While exchanging with simple monoalkylammonium ions zeolite *A* does not admit significant amounts of branched chain monoalkylammonium or of tetramethylammonium ions.[50] When the zeolite framework is more compact (analcime) or the intracrystalline voids less accessible (sodalite hydrate) ion sieving has been demonstrated for inorganic cations of different sizes. Thus Ag-analcime or Ag-sodalite hydrate will readily exchange Ag^+ by Na^+ but not by Cs^+. Accordingly CsCl solutions with NaCl impurity may easily be freed of the NaCl by the reaction:

$$NaCl + Ag\text{-Zeolite} \rightarrow Na\text{-Zeolite} + AgCl \downarrow.$$

If CsCl alone is contacted with the silver form of either analcime or sodalite hydrate the CsCl can be hydrolysed:

$$CsCl + 2H_2O + Ag\text{-Zeolite} \rightarrow CsOH + AgCl \downarrow + H_3O\text{-Zeolite}.$$

The hydrolysis can be demonstrated by the rise in pH of the aqueous solution. The same behaviour is shown with tetramethylammonium chloride and the silver forms of zeolite *A* or chabazite, or indeed of any zeolite from which the tetramethylammonium ion is excluded by ion sieve action.

The selectivities shown by a number of zeolites for Cs^+ or NH_4^+ (e.g. mordenite[52] and clinoptilolite,[48,49]) by chabazite for K^+ over Na^+,[53] by sodalite hydrate[54] or zeolite *A* for Ag^+, or by mordenite[55] for $[Cu(NH_3)_4]^{2+}$, $[Zn(NH_3)_4]^{2+}$ and $[Co(NH_3)_6]^{3+}$ illustrate a few of many ion separations which could be performed with zeolite exchangers.

An interesting separation of isotopes has also been effected using zeolites.[56] Lanthanide and actinide ions were introduced and fixed in the structures of zeolites X and Y by appropriate heat treatment, and the zeolites were then exposed to neutron radiation in a nuclear reactor. The (n, γ) reaction yielded an isotope one mass unit heavier than the parent ion. The γ-ray emission caused a recoil which dislodged a good proportion of the new isotope from the sites at which the ions were trapped to open sites from which they could be eluted by exchange solutions containing Li^+ or Ca^{2+}. The ions so removed in the first elution had the following proportions of new to parent isotope:

$$\frac{^{142}Pr}{^{141}Pr} = 44; \frac{^{140}La}{^{139}La} = 44; \frac{^{171}Er}{^{170}Er} = 24; \frac{^{241}Am}{^{240}Am} = 12 \text{ and } 13.$$

The separation factors are high and significant amounts of the recoil isotope can be concentrated in this way.

5. CONCLUSION

It has not been the purpose of this chapter to give any full account of the discovery, properties and uses of zeolites, but rather to introduce and illustrate them as materials of great diversity which have already found numerous applications. It seems surprising that so little attention was devoted to their exploitation until the 1940's although their unique properties were already foreshadowed in published researches dating from before the turn of the century. My attempts in the early and mid 1940's to interest several large companies in molecule sieving met with little response, perhaps in part because these attempts were ahead of their time, and in part because of wartime circumstances. However the receptiveness shown towards these novel materials subsequently changed dramatically, as this chapter indicates. In the field of sorption, the basic understanding of which is the objective of this monograph, a number of the actual or potential uses of zeolites in separation processes have been listed by Breck.[57] This list, already large, is constantly being augmented. An account of zeolite catalysts also forms part of a recent monograph.[58]

This introduction to the zeolite story may be ended by summarizing a number of the dates at which new directions or viewpoints emerged, several of which have been referred to earlier. One may start after 1756 when the first zeolite (stilbite) was named.

1850–52	Way and Thompson clarified the nature of ion exchange using soils as exchange media.[59]
1858	Action of salt solutions on silicates.[59a]
1862	The first synthesis of a zeolite was reported ("levynite", as optically uniaxial tablets, from aqueous K and Na silicates heated in glass tubes to 170°C.[60] The composition given corresponds with $Ca_{0\cdot25}Na_{2\cdot68}K_{2\cdot80}[(AlO_2)_{6\cdot5}(SiO_2)_{11\cdot5}]16\cdot8H_2O$).
1870–88	Various qualitative studies of ion exchange by zeolites.[61]
1898	Quantitative studies of water-zeolite equilibria.[62]
1899	Ion exchange via vapour phase using NH_4Cl.[63]
1902–5	Ion exchange via salt melts ($AgNO_3$, $TlNO_3$).[64]
1910	Sorption of heavy vapours in zeolites (I_2, Br_2, S, Hg, Hg_2, Cl_2, HgS).[65]
1928	Sedimentary deposit of analcime reported.[66]
1930–4	First zeolite structures determined (analcime and members of the natrolite group).[67]
ca 1932	Term "molecular sieve" coined to describe the behaviour of some microporous charcoals and zeolites.[7]
1942–45	Quantitative, single step separations by molecule sieving. Several classes of molecular sieve established. Ion exchange used for tailoring molecular sieves.[9, 10, 11, 12]
1948	First zeolite with no natural counterpart described and characterized as sorbent and sieve (Species P with *ZK*–5 framework). Mordenite synthesis reported (a zeolite now of industrial interest).[13, 14, 17, 68]
1949	Preparations of crystalline H-zeolites via NH_4-forms.[69]
1956–64	Synthesis and properties of industrially important zeolites *A*, X and Y revealed.[19, 23, 28]
1961	First direct syntheses of zeolites in absence of inorganic bases.[70]
1962	Mobil Co. introduced zeolite-based cracking catalysts.
1964	Crystalline dealuminated zeolitic SiO_2 made by acid extraction of clinoptilolite.[71]
1967	Preparation of ultrastable faujasite reported.[72]
1971–2	Very siliceous zeolites described (e.g. ZSM–5 and ZSM–8).[73, 74]

REFERENCES

1. D. W. Breck, "Zeolite Molecular Sieves", Wiley-Interscience (1974) p. 188.

2. R. A. Sheppard, *in* "Molecular Sieve Zeolites", Advances in Chemistry Series No. 101, Amer. Chem. Soc., Ed. R. F. Gould (1971) p. 279.
3. A. G. Kossowskaya, *in* "Molecular Sieves", Advances in Chemistry Series No. 121, Amer. Chem. Soc., Eds W. M. Meier and J. B. Uytterhoeven (1973) p. 200.
4. W. Eitel, private communication.
5. P. Niggli and G. W. Morey, *Z. Anorg. Chem.* (1913) **83**, 369.
6. G. W. Morey and E. Ingerson, *Econ. Geol.* (1937) **32**, 607.
7. J. W. McBain, "The Sorption of Gases and Vapours by Solids", Routledge and Sons (1932) Chapter 5.
8. T. Baba, *Bull. Chem. Soc. Japan* (1930) **5**, 190.
9. R. M. Barrer, *J. Soc. Chem. Ind.* (1945) **64**, 130.
10. R. M. Barrer and L. Belchetz, *J. Soc. Chem. Ind.* (1945) **64**, 131.
11. R. M. Barrer, *J. Soc. Chem. Ind.* (1945) **64**, 133.
12. R. M. Barrer, British Patent 548,905 (1942); US Patent 2,306,610.
13. R. M. Barrer, *J. Chem. Soc.* (1948) 127.
14. R. M. Barrer, British Patent 574,911 (1946).
15. R. M. Barrer and C. Marcilly, *J. Chem. Soc.*, A (1970) 2735.
16. R. M. Barrer and D. J. Robinson, *Z. Krist.* (1972) **135**, 374.
17. R. M. Barrer, *J. Chem. Soc.* (1948) 2158.
18. R. M. Barrer, *J. Chem. Soc.* (1948) 133.
19. D. W. Breck, W. G. Eversole, R. M. Milton, T. B. Reed and T. L. Thomas, *J. Amer. Chem. Soc.* (1956) **78**, 5963.
20. R. M. Barrer, *Quarterly Reviews* (1949) **3**, 293.
21. R. M. Barrer, *Brit. Chem. Eng.* (1959) May Issue, p. 1.
22. M. M. Dubinin, *in* "Zeolites, their Synthesis, Properties and Applications", 2nd All-Union Zeolite Conference, USSR (1965) Fig. 2, p. 8.
23. R. M. Milton, US Patent 2,882,244 (1959); D. W. Breck, US Patent 3,130,007 (1964).
24. J. A. Hedge, *in* "Molecular Sieve Zeolites", Advances in Chemistry Series, No. 102, Amer. Chem. Soc., Ed. R. F. Gould (1971) p. 239.
25. H. Lee *in* "Molecular Sieves", Advances in Chemistry Series No. 121, Amer. Chem. Soc., Eds W. M. Meier and J. B. Uytterhoeven (1973) p. 314.
26. R. M. Barrer, *Trans. Faraday Soc.* (1949) **45**, 358.
27. R. M. Barrer and J. W. Baynham, *J. Chem. Soc.* (1956) 2892.
28. D. W. Breck and E. M. Flanigen, *in* "Molecular Sieves", Soc. Chem. Ind. London (1968) p. 47.
29. *Chem. and Eng. News* (1975) October 6th, p. 18.
30. H. A. Stewart and J. L. Heck, *Chem. Eng. Prog.* (1969) **65**, 78.
31. D. W. Breck, "Zeolite Molecular Sieves", Wiley-Interscience (1974) pp. 709 *et seq.*
32. L. B. Batta, US Patent 3,564,816 (1968); and 3,636,679 (1972).
33. J. N. Miale, N.Y. Chen and P. B. Weisz, *J. Catalysis* (1966) **6**, 278.
34. cf H. E. Robson, G. P. Hamner and W. F. Arey, *in* "Molecular Sieve Zeolites", Advances in Chemistry Series, No. 102, Amer. Chem. Soc., Ed. R. F. Gould (1971) p. 417.
35. P. B. Weisz, *Chem. Tech* (1973) **3**, 498.
36. P. B. Venuto and P. S. Landis, *Advan. Catalysis* (1968) **18**, 259.
37. P. B. Weisz and V. J. Frilette, *J. Phys. Chem.* (1960) **64**, 382.
38. P. B. Weisz, P. B. Frilette, V. J. Maatman and E. B. Mower, *J. Catalysis*, (1962) **1**, 307.

39. N. Y. Chen and P. B. Weisz, *Chem. Eng. Prog. Symp. Ser.* (1967) **63**, 86.
40. J. N. Miale and P. B. Weisz, *J. Catalysis* (1971) **20**, 288.
41. W. H. Lang, R. J. Mikovsky and A. J. Silvestri, *J. Catalysis* (1971) **20**, 293.
42. K. M. Minachev and Y. I. Isakov, *in* "Molecular Sieves", Advances in Chemistry Series, No. 121, Amer. Chem. Soc., Eds W. M. Meier and J. B. Uytterhoeven (1973) p. 451.
43. P. B. Venuto, *in* "Molecular Sieve Zeolites", Advances in Chemistry Series, No. 102, Amer. Chem. Soc., Ed. R. F. Gould (1971) p. 261.
44. V. Penchev, N. Davidova, V. Kanazirev, H. Minchev and Y. Neinska, *in* "Molecular Sieves", Advances in Chemistry Series, No. 121, Amer. Chem. Soc., Eds W. M. Meier and J. B. Uytterhoeven (1973) p. 461.
45. J. Turkevich and Y. Ono, *Advan. Catalysis* (1969) **20**, 135.
46. H. S. Sherry, *in* "Molecular Sieve Zeolites", Advances in Chemistry Series, No. 101, Amer. Chem. Soc., Ed. R. F. Gould (1971) p. 350.
47. D. W. Breck, "Zeolite Molecular Sieves", Wiley-Interscience (1974) p. 192 *et seq.*
48. L. Ames, *Amer. Mineralog.* (1963) **48**, 868.
49. R. M. Barrer, R. Papadopoulos and L. V. C. Rees, *J. Inorg. nucl. Chem.* (1967) **29**, 2047.
50. R. M. Barrer and W. M. Meier, *Trans. Faraday Soc.* (1958) **54**, 1074.
51. R. M. Barrer and D. C. Sammon, *J. Chem. Soc.* (1956) 675.
52. R. M. Barrer and J. Klinowski, *J. Chem. Soc., Faraday I* (1974) **70**, 2362.
53. R. M. Barrer, J. A. Davies and L. V. C. Rees, *J. Inorg. Nucl. Chem.* (1969) **31**, 219.
54. R. M. Barrer and J. D. Falconer, *Proc. Roy. Soc.* (1956) A **236**, 227.
55. R. M. Barrer and R. P. Townsend, *J. Chem. Soc., Faraday Trans. I* (1976), **72**, 2650.
56. D. O. Campbell, *in* "Molecular Sieves", Advances in Chemistry Series, No. 121, Amer. Chem. Soc., Eds W. M. Meier and J. B. Uytterhoeven, (1973) p. 281.
57. D. W. Breck, "Zeolite Molecular Sieves", Wiley-Interscience (1974) p. 701.
58. Chapters 8 to 13 *in* "Zeolite Chemistry and Catalysis", Ed. J. A. Rabo, Amer. Chem. Soc. Monograph 171 (1976).
59. T. Way, *J. Roy. Soc.* (1850) **11**, 313; and (1852) **13**, 123. H. S. Thompson, *J. Roy. Agr. Soc., England* (1850) **11**, 68.
59a. H. Eichorn, *Ann. Phys.* Poggendorff (1858) **105**, 126.
60. H. de St Claire-Deville, *Compt. Rend.* (1862) **54**, 324.
61. J. Lemberg, *Z. Deutsch Geol. Ges.* (1876) **28**, 519.
62. G. Tammann, *Zeit. Phys. Chem.* (1898) **27**, 323.
63. F. Clark and G. Steiger, *Amer. J. Sci.* (1899) **8**, 245.
64. G. Steiger and F. Clark, *Zeit. Anorg. Chem.*, (1905) **46** 197.
65. F. Grandjean, *Bull. Soc. Franc. Mineralog.* (1910) **33**, 5.
66. C. S. Ross, *Amer. Mineralog.*, (1928) **13**, 195.
67. W. H. Taylor, *Zeit. Krist.* (1930) 74, 1; W. H. Taylor, C. A. Meek and W. W. Jackson, *Zeit. Krist.*, (1933) **74**, 373.
68. R. M. Barrer and D. W. Riley, *J. Chem. Soc.* (1948) 133.
69. R. M. Barrer, *Nature* (1949) **164**, 112.
70. R. M. Barrer and P. J. Denny, *J. Chem. Soc.* (1961) 971.
71. R. M. Barrer and M. B. Makki, *Canad. J. Chem.* (1964) **42**, 1481.
72. C. V. McDaniel and P. K. Maher, *in* Conference on Molecular Sieves, London (1967) *see* "Molecular Sieves", *Soc. Chem. Ind.* (1968) p. 186.
73. R. J. Argauer and G. R. Landolt, US Patent 3,702,886 (1972).
74. Mobil Co., Netherland Patent 7,014,807 (1971).

2

Zeolite Frameworks, Cations and Water Molecules

1. INTRODUCTION

The behaviour of a given zeolite as a diffusion medium, sorbent and molecular sieve is regulated in part by the topology of the anionic framework and the size, charge and locations of the exchange ions within the framework. Accordingly, structural studies are important in understanding zeolite sorbents. There have been various reviews of the structures of silicates in general[1, 2] and of zeolites in particular.[3, 4, 5, 6, 7, 8] Among other purposes these reviews have had as objectives the logical classification into groups related through the occurrence of shared structural features; and the prediction of novel frameworks belonging to a given group, but still awaiting synthesis in the laboratory or discovery in nature.

2. ZEOLITE CLASSIFICATION

On the basis of structure determinations, the zeolites and some related tectosilicates can be grouped as in Table 1. The second column gives idealized unit cell compositions which are somewhat arbitrary, but in which the Si/Al ratio falls within the range of analyses extant in the literature. The

TABLE 1

Groups of zeolites and tectosilicates with related topology

	Idealized composition per unit cell	Intracrystalline pore volume (as cm^3liq. H_2O/cm^3 of crystal)	Crystallographic data (a, b, and c in Å)	References
Analcime Group				
Analcime	$Na_{16}[(AlO_2)_{16}(SiO_2)_{32}]16H_2O$	0·18	Cubic, a=13·72	(9, 10)
Wairakite	$Ca_8[(AlO_2)_{16}(SiO_2)_{32}]16H_2O$	0·18	Monoclinic, pseudocubic, a=13·69, b=13·68, c=13·56, β=90·5°	(11, 12)
Leucite (felspathoid)	$K_{16}[(AlO_2)_{16}(SiO_2)_{32}]$	0	Tetragonal, a=12·98, c=13·68	
Pollucite (felspathoid)	$Cs_{16}[(AlO_2)_{16}(SiO_2)_{32}]$	0	Cubic, a=13·77	
Viseite	$Na_2Ca_{10}[(AlO_2)_{20}(SiO_2)_6(PO_2)_{10}(H_3O_2)_{12}]16H_2O$	–	Cubic, a=13·65	(13)
Kehoeite	$Zn_{5 \cdot 5}Ca_{2 \cdot 5}[(AlO_2)_{16}(PO_2)_{16}(H_3O_2)_{16}]32H_2O$	–	Cubic, a=13·7	(14)
Natrolite Group				
Natrolite	$Na_{16}[(AlO_2)_{16}(SiO_2)_{24}]16H_2O$	0·21	Orthorhombic, a=18·30, b=18·63, c=6·60	(15, 16)
Scolecite	$Ca_8[(AlO_2)_{16}(SiO_2)_{24}]24H_2O$	0·31	Monoclinic, a=9·848, b=19·978, c=6·522, β=110°6′	(15, 17)
Mesolite	$Na_{16}Ca_{16}[(AlO_2)_{48}(SiO_2)_{72}]64H_2O$	0·25	Orthorhombic, a=18·43, b=56·45, c=6·55	(18)
Thomsonite	$Na_4Ca_8[(AlO_2)_{20}(SiO_2)_{20}]24H_2O$	0·32	Orthorhombic, a=13·07, b=13·08, c=13·18	(15, 19)
Gonnardite	$Na_4Ca_2[(AlO_2)_8(SiO_2)_{12}]14H_2O$	0·35	Orthorhombic, a=13·19, b=13·32, c=6·55	(15, 20)

Edingtonite	$Ba_2[(AlO_2)_4(SiO_2)_6]8H_2O$	0·46	Orthorhombic, $a=9{\cdot}54$, $b=9{\cdot}65$, $c=6{\cdot}50$	(15)
Metanatrolite	$Na_{16}[(AlO_2)_{16}(SiO_2)_{24}]$	–	Pseudo-orthorhombic, $a=16{\cdot}3$, $b=17{\cdot}1$, $c=6{\cdot}6$	(16)
Heulandite Group				
Heulandite	$Ca_4[(AlO_2)_8(SiO_2)_{28}]24H_2O$	0·35	Monoclinic, $a=17{\cdot}72$, $b=17{\cdot}90$, $c=7{\cdot}43$, $\beta=116°\ 25'$	(21, 22)
Clinoptilolite	$Na_6[(AlO_2)_6(SiO_2)_{30}]24H_2O$	0·34	Monoclinic, $a=7{\cdot}64$, $b=17{\cdot}90$, $c=7{\cdot}40$, $\beta=116°\ 22'$	(23)
Brewsterite	$(Sr, Ba, Ca)_2[(AlO_2)_4(SiO_2)_{12}]10H_2O$	0·32	Monoclinic, $a=6{\cdot}77$, $b=17{\cdot}51$, $c=7{\cdot}74$, $\beta=94°\ 18'$	(24)
Stilbite	$Na_2Ca_4[(AlO_2)_{10}(SiO_2)_{26}]32H_2O$	0·38	Monoclinic, $a=13{\cdot}64$, $b=18{\cdot}24$, $c=11{\cdot}27$, $\beta=128°$	(25, 26)
Stellerite	$Ca_8[(AlO_2)_{16}(SiO_2)_{56}]56H_2O$	0·39	Orthorhombic, $a=13{\cdot}599$, $b=18{\cdot}222$, $c=17{\cdot}863$	(27)
Barrerite	$(Ca, Mg)_2(Na, K)_{12}[(AlO_2)_{16}(SiO_2)_{56}]52H_2O$	0·35	Orthorhombic, $a=13{\cdot}643$, $b=18{\cdot}200$, $c=17{\cdot}842$	(28)
Phillipsite Group				
Phillipsite	$(K, Na)_5[(AlO_2)_5(SiO_2)_{11}]10H_2O$	0·30	Monoclinic, $a=9{\cdot}87$, $b=14{\cdot}30$, $c=8{\cdot}67$, $\beta=124°\ 30'$	(29)
Harmotome	$Ba_2[(AlO_2)_4(SiO_2)_{12}]12H_2O$	0·36	Monoclinic, $a=9{\cdot}88$, $b=14{\cdot}14$, $c=8{\cdot}69$, $\beta=124°\ 81'$	(29)
Gismondine	$Ca_4[(AlO_2)_8(SiO_2)_8]16H_2O$	0·47	Monoclinic, $a=10{\cdot}02$, $b=10{\cdot}62$, $c=9{\cdot}84$, $\beta=92°\ 25'$	(30)
Zeolite Na-Pl	$Na_8[(AlO_2)_8(SiO_2)_8]16H_2O$	0·47	Pseudocubic, $a=10{\cdot}0$	(31)
Garronite	$NaCa_{2{\cdot}5}[(AlO_2)_6(SiO_2)_{10}]14H_2O$	0·41	Tetragonal, $a=10{\cdot}0$, $c=9{\cdot}9$	(32, 33)
Yugawaralite[a]	$Ca_4[(AlO_2)_8(SiO_2)_{20}]16H_2O$	0·30	Monoclinic, $a=6{\cdot}73$, $b=13{\cdot}95$, $c=10{\cdot}03$, $\beta=111°\ 31'$	(34)
Zeolite Li-*ABW*	$Li_4[(AlO_2)_4(SiO_2)_4]4H_2O$	0·28	Orthorhombic, $a=10{\cdot}31$, $b=8{\cdot}18$, $c=5{\cdot}00$	(35)

continued

	Idealized composition per unit cell	Intracrystalline pore volume (as cm^3liq. H_2O/cm^3 of crystal)	Crystallographic data (a, b, and c in Å)	References
Mordenite Group				
Mordenite	$Na_8[(AlO_2)_8(SiO_2)_{40}]24H_2O$	0·26	Orthorhombic, $a=18\cdot13$, $b=20\cdot49$, $c=7\cdot52$	(36)
Ferrierite	$Na_{1\cdot5}Mg_2[(AlO_2)_{5\cdot5}(SiO_2)_{30\cdot5}]18H_2O$	0·24	Orthorhombic, $a=19\cdot16$, $b=14\cdot13$, $c=7\cdot49$	(37)
Dachiardite	$Na_5[(AlO_2)_5(SiO_2)_{19}]12H_2O$	0·26	Monoclinic, $a=18\cdot73$, $b=7\cdot54$, $c=10\cdot30$, $\beta=107°\ 54'$	(38)
Epistilbite	$Ca_3[(AlO_2)_6(SiO_2)_{18}]16H_2O$	0·34	Monoclinic, $a=9\cdot08$, $b=17\cdot74$, $c=10\cdot25$, $\beta=124°\ 54'$	(39, 40)
Bikitaite	$Li_2[(AlO_2)_2(SiO_2)_4]2H_2O$	0·20	Monoclinic, $a=8\cdot61$, $b=4\cdot96$, $c=7\cdot61$, $\beta=114°\ 26'$	(41, 42)
Chabazite Group				
Chabazite	$Ca_2[(AlO_2)_4(SiO_2)_8]13H_2O$	0·48	Rhombohedral, $a=9\cdot42$, $\alpha=94°\ 28'$	(43, 44)
Gmelinite	$Na_8[(AlO_2)_8(SiO_2)_{16}]24H_2O$	0·43	Hexagonal, $a=13\cdot75$, $c=10\cdot05$	(42, 44)
Erionite	$(Ca, Mg, Na_2, K_2)_{4\cdot5}[(AlO_2)_9(SiO_2)_{27}]27H_2O$	0·36	Hexagonal, $a=13\cdot26$, $c=15\cdot12$	(45, 46)
Offretite	$(K_2, Ca, Mg)_{2\cdot5}[(AlO_2)_5(SiO_2)_{13}]15H_2O$	0·34	Hexagonal, $a=13\cdot29$, $c=7\cdot58$	(46, 47)
Levynite	$Ca_3[(AlO_2)_6(SiO_2)_{12}]18H_2O$	0·42	Hexagonal, $a=13\cdot338$, $c=20\cdot014$	(48, 49)
Mazzite (zeolite Ω)	$Na_{6\cdot03}K_{1\cdot91}Ca_{1\cdot35}Mg_{1\cdot99}[(AlO_2)_{9\cdot77}(SiO_2)_{26\cdot54}]28H_2O$	0·37	Hexagonal, $a=18\cdot392$, $c=7\cdot646$	(50, 51, 52)
Zeolite *L*	$K_6Na_3[(AlO_2)_9(SiO_2)_{27}]21H_2O$	0·28	Hexagonal, $a=18\cdot4$, $c=7\cdot5$	(53)
Sodalite Hydrate	$Na_6[(AlO_2)_6(SiO_2)_6]8H_2O$	0·34	Cubic, $a=8\cdot87$	(54, 55)
Cancrinite Hydrate	$Na_6[(AlO_2)_6(SiO_2)_6]8H_2O$	0·34	Hexagonal, $a=12\cdot72$, $c=5\cdot19$	(55, 56)
Zeolite Losod	$Na_{12}[(AlO_2)_{12}(SiO_2)_{12}]19H_2O$	0·37	Hexagonal, $a=12\cdot91$, $c=10\cdot54$	(57)

Faujasite Group				
Faujasite (Zeolites X and Y)	$Na_{12}Ca_{12}Mg_{11}[(AlO_2)_{59}(SiO_2)_{133}]260H_2O$	0·53	Cubic; $a=24\cdot67$	(58, 59)
Zeolite ZSM-3	$\{(Li, Na)_2[(AlO_2)_2(SiO_2)_{3\cdot2}]8H_2O\}_m$	≈0·53	Hexagonal; $a=17\cdot5$, $c=129$ (max)	(60)
Paulingite	$(K_2, Na_2, Ca, Ba)_{76}[(AlO_2)_{152}(SiO_2)_{525}]700H_2O$	0·48	Cubic; $a=35\cdot09$	(61, 62)
Zeolite *A*	$Na_{12}[(AlO_2)_{12}(SiO_2)_{12}]27H_2O$ (pseudo-cell)	0·47	Cubic; $a=12\cdot32$ (pseudo-cell)	(63, 64, 65)
Zeolite *RHO*	$(Na, Cs)_{12}[(AlO_2)_{12}(SiO_2)_{36}]mH_2O$	≈0·48	Cubic; $a=15\cdot0$	(66)
Zeolite *ZK-5*	$Na_{30}[(AlO_2)_{30}(SiO_2)_{66}]98H_2O$	0·45	Cubic; $a=18\cdot7$	(67)
Unclassified[b]				
Laumontite	$Ca_4[(AlO_2)_8(SiO_2)_{16}]16H_2O$	0·35	Monoclinic, $a=7\cdot55$, $b=14\cdot74$; $c=13\cdot07$, $\beta=111°\ 9'$	(68, 69)
Zeolite N (Z-21 and (Na, Me_4N)-V)	$\{Na[(AlO_2)(SiO_2)]1\cdot56H_2O\}_n$ where $n\approx176\times$density	≈0·35	Cubic, $a=36\cdot8$	(70)

[a] The assignment of yugawaralite to the phillipsite group is very tentative. It does not have the ribbons of 4-rings typical of this group, but like laumontite has 4-rings connected cornerwise. It could as well be placed with laumontite.

[b] Other unclassified zeolites include Li-H[(35)] ZSM-4[(70a)] reported as cubic with $a = 22\cdot2$Å; ZSM-5[(70b)] reported as tetragonal with $a = 23\cdot2$Å and $c = 19\cdot9$Å; ZSM-8[(70c)] and ZSM-10[(70d)] All the ZSM series appear to sorb cyclohexane and ZSM-5–8 are remarkably rich in silica.

framework composition is that part of the formula contained in the square brackets. It is a common experience with aluminosilicates that the Si/Al ratio can vary according to the composition of the parent magma and to the experimental conditions, such as the concentration of alkali present.[71, 72] Thus the zeolites will tolerate substitutions of the types

$$Na,Al \rightleftarrows Si$$
$$Ca,Al \rightleftarrows Na,Si.$$

Indeed zeolites have been made, or occur naturally, in which the ratios vary between extremes exemplified in Table 2. The lower limit to the compositions

TABLE 2

Compositional ranges observed in some tectosilicates

Tectosilicate	Low SiO_2/Al_2O_3	High SiO_2/Al_2O_3
Sodalite	2·0	10 (as $N(CH_3)_4$-sodalite)[73]
Chabazite	2·3 (as zeolite K-G)[74]	7·8 (in a natural chabazite)[75]
Faujasite	2·2 (as zeolite X)	6·8 (syntheses by Kacirek)[76]
Zeolite *A*	2·0 (as Na-*A*)	7·5 (as $N(CH_3)_4$-*A*)[77]

is seen to approach but not to fall below $SiO_2/Al_2O_3 = 2$, which is in accordance with Lowenstein's rule.[78] In tectosilicates, AlO_4-tetrahedra according to this rule are not linked directly to other AlO_4-tetrahedra, so that when $SiO_2/Al_2O_3 = 2$, Si and Al alternate upon the tetrahedral sites, giving ordered frameworks. Ordering of Si and Al is also possible for other SiO_2/Al_2O_3 ratios. It is found for example in natrolite and edingtonite when this ratio is three.

The upper limit to the ratio SiO_2/Al_2O_3 can in certain zeolites equal or exceed the value of ten recorded for $N(CH_3)_4$-sodalite. Such silica-rich zeolites include clinoptilolite, ferrierite and mordenite which are stable in acid solution.[79,80, 81] The concentration of Al is so low that virtually all can be extracted by treatment with mineral acid while still leaving the framework in its pristine configuration with nests of four OH groups replacing the AlO_4, M^+ of the parent structure. Such dealuminated zeolites remain crystalline to X-rays.

Germanium may also replace Si, and Ga may replace Al in both cases up to 100%[82,83,84] Fe^{III}, Be, B and P can also be considered as framework substituents, but borosilicates are not so far represented among zeolites,[85] while the evidence regarding substitution of framework atoms in zeolites by phosphorus is conflicting.[86,87, 88, 89, 90]

The water contents have been rounded to integral values, but since as a rule the water contents vary continuously with relative pressure and also with temperature, these values, though near, are not necessarily exact saturation uptakes. In column three the intracrystalline pore volumes are estimated as the volumes of liquid water which could be removed from each zeolite (by heat and evacuation) according to the formulae in column two. One can also estimate these pore volumes from the crystal structures; where this has been done there is reasonable correspondence between the two procedures. The remaining two columns give respectively, some crystallographic information, and references to selected researches on the zeolite structures.

The members of a given zeolite group fall into two categories:

(i) Those which have the same framework topology, but have different chemical compositions (cations, or Si : Al ratios) and in which as a result there may be minor framework adjustments, and also differences in chemical, crystallographic and physical properties. Thus, all the members of the analcime group are considered to have the same framework topology, but often differ widely in chemical composition. In the natrolite group, natrolite, scolecite and mesolite have one framework topology, and thomsonite and gonnardite another. Heulandite and clinoptilolite are also structural isotypes, and so are stilbite, stellerite and barrerite; phillipsite and harmotome; gismondine and zeolite Na-Pl; and faujasite, zeolite X and zeolite Y.

(ii) Those members of a group which have frameworks containing one or more common structural elements, linked together in different ways, so resulting in different topologies. Examples in the natrolite group are natrolite, thomsonite and edingtonite; in the heulandite group, heulandite, brewsterite and stilbite; all members of the chabazite group; phillipsite, gismondine and possibly garronite; mordenite, ferrierite and dachiardite; faujasite and zeolite *A*; zeolite *A*, zeolite *ZK*-5 and zeolite *RHO*.

In several other instances the structural resemblances are less clear, but were felt to be sufficient for tentative allocation to the groups of Table 1. These include yugawaralite and zeolite Li-*ABW* allocated to the phillipsite group; and bikitaite, allocated to the mordenite group. Although its structure has been determined, laumontite has not been classified since it does not have a clear enough structural relationship to members of any particular group. Finally there are cases where a given zeolite has structural relationships with zeolites in more than one group. For example sodalite hydrate has an equally close relationship with cancrinite hydrate, erionite or losod in the chabazite group and with faujasite or zeolite *A* in the faujasite group. With certain members of the chabazite group it has in common identical layers of 6-rings (rings of six (Al, Si)O_4 tetrahedra); with faujasite and zeolite *A* it has a common structural element in the form of 14-hedral cages of Type I (see Table 4).

The framework topologies of zeolites may also be repeated in non-zeolites. Thus in one of the least hydrated zeolites, analcime ($NaAlSi_2O_6.H_2O$) those forms containing large ions (leucite, $KAlSi_2O_6$; Rb-analcime, $RbAlSi_2O_6$; and pollucite, $CsAlSi_2O_6$) are anhydrous[91] and therefore are not zeolites but felspathoids. The leucite framework is tetragonal at ordinary temperatures and is somewhat shrunken compared with that of the cubic analcime structure; the Rb-form is also tetragonal and there is evidence that it may exist in two polymorphic varieties;[92] while the Cs-form (pollucite) is again cubic with a unit cell of edge equal to that of analcime. From the cell dimensions the space-filling capacity of ions for the framework is

$$K < Rb < Cs = (Na + H_2O).$$

The large Cs^+ ion in pollucite is almost immobile within the tight mesh of the framework so that exchange with Na^+ involves temperatures of $\approx 200°C$ or more.

The sodalite structure can in addition be made as the zeolite, sodalite hydrate,[93] $3(NaAlSiO_4).4H_2O$, and as the siliceous, water-free variant[73] $N(CH_3)_4.AlSi_5O_6$, in which the tetramethylammonium ions are locked during crystal growth into the 14-hedral cages of Type I (Table 4) of which the structure is composed. The $N(CH_3)_4^+$ ions fill all the intracrystalline free volume. The sodalite and cancrinite structures can moreover readily be synthesized in the presence of a wide variety of salts to form solid solutions of the salts within the tectosilicates.[55,56] As the salt content increases the amount of zeolitic water declines. The isotherms for salt uptake are continuous, smooth functions of the salt concentrations in the mixtures from which the tectosilicates crystallize.

Salt-bearing varieties of zeolite *ZK-5* can likewise be made[94,95] in which the salts are $BaCl_2$ or $BaBr_2$; and of the edingtonite-type zeolite K-F[95] in which the salts are KBr and KCl. In these cases the salt again displaces water and either cannot be removed without crystal breakdown (sodalite, cancrinite, K-F) or is removed with some difficulty (*ZK-5*). As a final example, the paracelsian and felspar structures contain the same double crankshaft chain (q.v.) as the phillipsite-type zeolites,[96] and so, although non-zeolites, they can be regarded as related to the phillipsite group. Thus there need be nothing about the framework topology alone which distinguishes zeolite from non-zeolite. Furthermore it is possible to have a continuous change from a zeolite to a non-zeolite by ion exchange (e.g. Na-form of analcime to K-form (leucite)) or by salt uptake (e.g. Na-sodalite hydrate to the felspathoid sodalite, $3(NaAlSiO_4)NaCl$).

Phases having the sodalite hydrate topology and which occur in nature are also chemically diverse and include the following non-zeolite variants, mostly cubic.

Sodalite	$Na_8[(AlO_2)_6(SiO_2)_6]Cl_2$	$P\bar{4}3m$	$a = 8\cdot88$Å
Nosean	$Na_8[(AlO_2)_6(SiO_2)_6]SO_4$	$P\bar{4}3m$	$a = 9\cdot07$
Haüyne	$Na_{5-8}Ca_{0-2}K_{0-1}$ $[(AlO_2)_6(SiO_2)_6](SO_4)_{1-2}$	$P\bar{4}3n$	$a = 9\cdot12$
Lapis lazuli	$(Na_2, Ca)_4[(AlO_2)_6(SiO_2)_6]$ $(SO_4, S, Cl_2)_2$	$P\bar{4}3n$	$a = 9\cdot08$
Tugtupite	$Na_8[(BeO_2)_2(AlO_2)_2(SiO_2)_8]Cl_2$	$I\bar{4}$	$a = 8\cdot58$; $c = 8\cdot82$
Danalite	$Fe_8[(BeO_2)_6(SiO_2)_6]S_2$	$P\bar{4}3n$	$a = 8\cdot20$
Helvine	$(Mn, Fe, Zn)_8[(BeO_2)_6(SiO_2)_6]S_2$	$P\bar{4}3n$	$a = 8\cdot40$
Genthelvine	$Zn_8[(BeO_2)_6(SiO_2)_6]S_2$	$P\bar{4}3n$	$a = 8\cdot12$

Similarly there are several non-zeolite hexagonal phases having the topology of cancrinite hydrate.

Cancrinite $(Na, Ca)_{7-8}[(AlO_2)_6(SiO_2)_6](CO_3, SO_4, Cl)_{1\cdot5-2}$, $1–5H_2O$
$P6_3$ $a = 12\cdot60$Å; $c = 5\cdot13$Å

Vishnevite $(Na, Ca, K)_{6-7}[(AlO_2)_6(SiO_2)_6](SO_4, CO_3, Cl)_{1-1\cdot5}$, $1–5H_2O$
$a = 12\cdot68$; $c = 5\cdot18$

Davyne $(Na, Ca, K)_{7\cdot5}[(AlO_2)_6(SiO_2)_6](Cl, SO_4, CO_3)_3$
$P6_3/m$ $a = 12\cdot70$; $c = 5\cdot33$

Microsommite $(Na, Ca, K)_{7\cdot5}[(AlO_2)_6(SiO_2)_6](Cl, SO_4)_{2\cdot5}$
$P6_322$ $a = 22\cdot08$; $c = 5\cdot33$

3. SECONDARY BUILDING UNITS

Isolated SiO_4^{4-} ions occur in orthosilicates such as olivine, Mg_2SiO_4. Individual tetrahedra may then link together by sharing an apical oxygen to form "island" secondary units.[1] Two linked groups form the ion $Si_2O_7^{6-}$ found, for example, in hemimorphite, $Zn_4Si_2O_7(OH)_2.H_2O$, and thorveitite $Sc_2Si_2O_7$. A triangular ion, $Si_3O_9^{6-}$, formed from three linked tetrahedra (i.e. a 3-ring) occurs in benitoite, $BaTiSi_3O_9$, and wadeite, $K_2ZrSi_3O_9$. In axinite, $Ca_2(Fe^{II}, Mn)Al_2(OH)BO_3Si_4O_{12}$, rings of four tetrahedra are found as separate groups. Six-ring anions, or these rings with some substitution by Al, are not uncommon. They are found in beryl, $Be_3Al_2Si_6O_{18}$, cordierite, $Al_3(Mg, Fe^{II})_2(Si_5Al)O_{18}$, and tourmaline, $(Na, Ca)(Li, MgAl)_3(Al, Fe, Mn)_6(OH)_4(BO_3)_3Si_6O_{18}$, as examples. Two 6-rings linked to form a hexagonal prism $(Si_{(1-x)}Al_x)_{12}O_{30}$ occur in milarite, $KCa_2Be_2AlSi_{12}O_{30}.\frac{1}{2}H_2O$ and osumilite, $(K, Na, Ca)(Mg, Fe^{II})_2(Al_2, Fe^{II}, Fe^{III})_3(Si, Al)_{12}O_{30}.H_2O$. In another area of silicon chemistry, the controlled hydrolysis of silicochloroform derivatives, $RSiCl_3$ (where R is a group such as C_6H_5- or CH_3-) yields among other products silsesquioxane species.[97, 98, 99, 100, 101] There is in these species, represented formally below, a Si at each corner with H or R attached and an oxygen centred near the mid-point of each edge.

With exceptions of the 3-ring and the pentagonal prism such secondary units often occur as elements in the continuous tectosilicate frameworks. The 4- and 6-rings appear very often and the hexagonal prism is found in chabazite, gmelinite and some other chabazite group zeolites, and also in faujasite and *ZK-5*. The double 4-ring (cubic unit) is found in the framework of zeolite *A*. In addition 5-, 8-, 10- and 12-rings and occasionally octagonal prisms are observed. The repeated occurrence of secondary

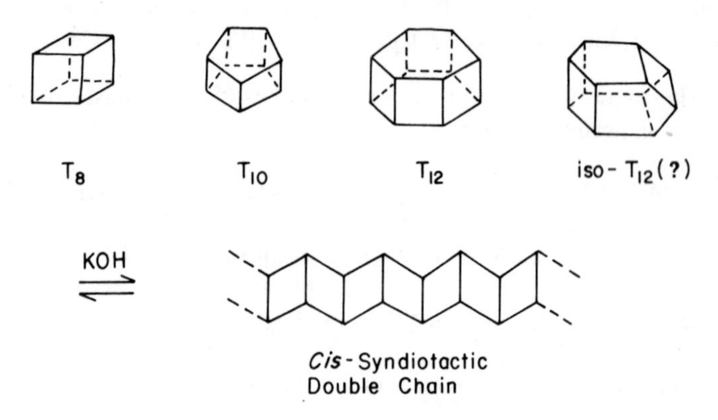

building units (SBU) led Meier[42] to identify in zeolites of then known structures the smallest number of SBU from which they could be constructed. In addition to the 4-, 6- and 8-rings, cubic units (4-4 unit) and hexagonal prisms (6-6 unit) already referred to, Meier identified the 4-ring with one extra attached tetrahedron; the 5-ring with attached tetrahedron and two linked 4-rings with an additional attached tetrahedron.

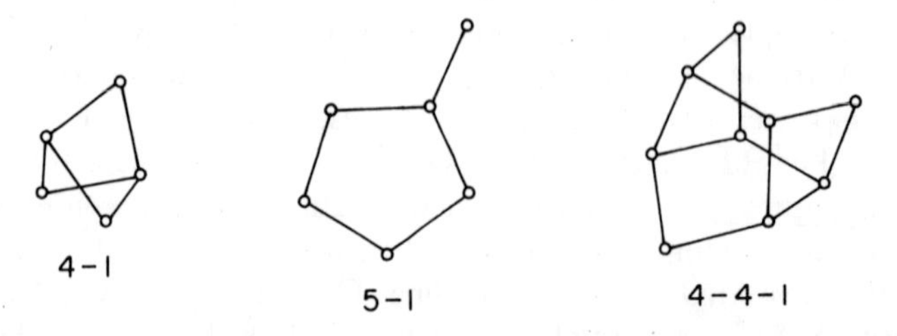

Some zeolites and the SBU present in them are indicated in Table 3. The appropriate linking of these can automatically create other features, for instance distorted 8-rings in analcime and erionite, or 12-rings in gmelinite,

TABLE 3

Secondary building units of zeolite frameworks[42]

Zeolite	Single rings			Double rings		Complex or multiple units		
	4	6	8	4–4	6–6	4–1	5–1	4–4–1
Analcime	+	+						
Natrolite						+		
Thomsonite						+		
Edingtonite						+		
Gmelinite	+	+	+		+			
Chabazite	+	+	(+)		+			
Offretite	+	+	+		+			
Erionite	+	+			+			
Zeolite *L*	+	+						
Levynite	(+)	+						
Cancrinite	+	+						
Sodalite	+	+						
Phillipsite	+		+					
Gismondine	+		+					
Brewsterite	+							
Heulandite								+
Stilbite								+
Mordenite							+	
Dachiardite							+	
Epistilbite							+	
Ferrierite							+	
Bikitaite							+	
Faujasite	+	+	+		+			
Zeolite *A*	+	+	+	+				
Zeolite *ZK-5*	+	+	+		+			
Paulingite	+		+					
Laumontite	+							

cancrinite, offretite, zeolite *L* and mordenite. Also in all zeolites various channel and cavity systems are formed which are of great importance in determining their sorption behaviour. One reason for considering zeolite topologies in terms of SBU was the surmise that one or more SBU anions might exist in the media from which the zeolites crystallized. If this was the case the growth of zeolite crystals could be more readily visualized.

However, as seen later, there are various other ways in which zeolite frameworks can be constructed and their formulation from the foregoing eight SBU need not indicate any actual mode of growth. For example the epistilbite framework in Fig. 1[4] can be made from 5-rings and 4-rings equally as well

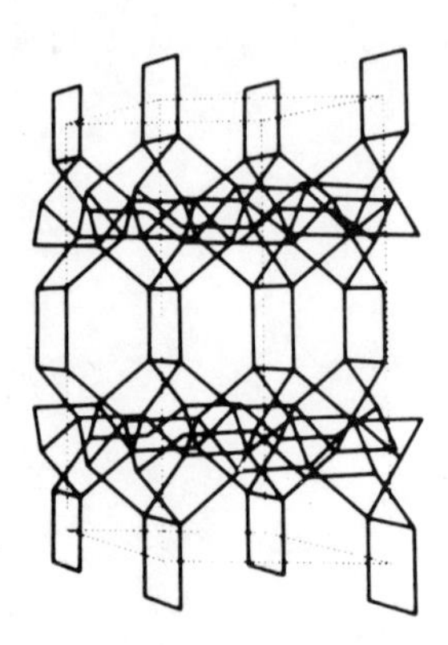
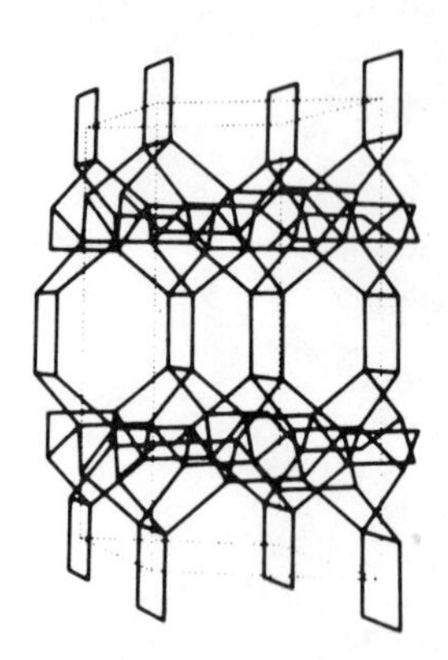

FIG. 1. A stereoscopic pair showing the epistilbite structure viewed along [001].[4] In this and similar framework representations there is Si or Al at each corner and oxygen near the mid-point of each edge.

as from the 5-1 SBU. In Fig. 1, and later formal representations of zeolite frameworks, Al or Si is centred at each vertex and oxygen near but not on the mid-point of each edge. Individual atoms are not shown.

A second reason for considering zeolite structures in terms of SBU is that these provide one basis for classifying zeolites.[5] Again, however, other structural blocks, chains and sheets exist which allow interrelationships to be more readily shown than do the SBU. The Breck[5] classification for instance combines as group 1 the analcime group of Table 1, the phillipsite group, and paulingite, laumontite and yugawaralite, on the basis that there are many 4-rings in each of these structures. However the structures and behaviour covered are then so diverse that the classification in terms of the 4-rings in each seems artificial.

4. POLYHEDRAL BUILDING BLOCKS IN ZEOLITES

As observed above, zeolite frameworks contain channels and cavities of many kinds. The voids are frequently polyhedral and many zeolite frameworks can be constructed by stacking one or more types of polyhedron in simple coordinations. There are only five polyhedra, which, stacked with others of the same kind in the same orientation can fill all space (Fig. 2).[102] Apart from two of the SBU (the 4-4 unit or cube and the 6-6 unit or hexagonal prism) there are two kinds of dodecahedron and the truncated octahedron (14-hedron of Type 1 in Table 4). The 4-4 or 6-6 units occur, as already noted, in some important zeolites (Table 4) and the truncated

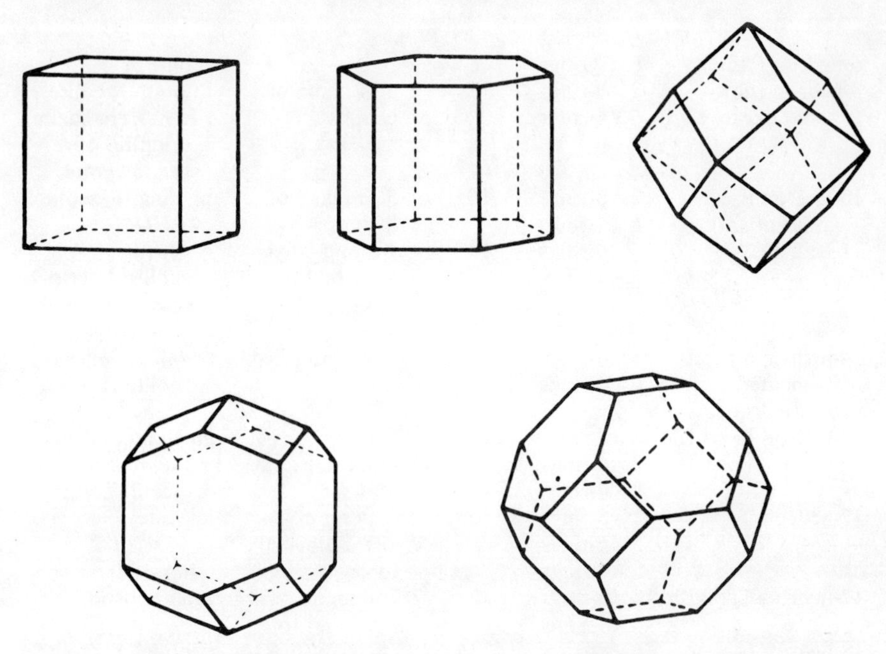

FIG. 2. Fedorov's five space-filling solids.[102]

octahedron is found in sodalite, faujasite and zeolite *A*. Sodalite (Fig. 3)[55] is the only zeolite composed only of one of Fedorov's space-filling polyhedra. A given 14-hedron is stacked in 8-fold coordination with respect to other polyhedra by sharing each of its eight 6-ring windows with one other 14-hedron. The free diameter of a 14-hedron is about 6·6Å and access to the 14-hedral voids is through 6-ring windows of free diameter $\approx 2 \cdot 3$Å. No zeolite framework has so far been found of which the dodecahedral cavities of Fig. 2 are building blocks. The linking of tetrahedra $(Al, Si)O_4$ always results in lining the insides of the polyhedra and of their polygonal "windows" with anionic oxygens, because Al and Si atoms are always buried among the four oxygens of each tetrahedron. A "free" distance means the distance between diagonally opposite points of the polyhedron or polygon which is not impinged upon by the oxygen atom linings.

Polyhedral cavities observed in zeolites are summarized in Table 4, with free dimensions, the numbers and kinds of polygonal faces of the polyhedra

TABLE 4

Some polyhedra in zeolites

Polyhedron	Faces	Vertices[a]	Approximate free dimension (Å)	Examples
6-hedron (cube)	6×4-rings	8	–	zeolite *A*
8-hedron (hexagonal prism)	2×6-rings 6×4-rings	12	2·3 in plane of 6-rings	faujasite, zeolite *ZK-5*, chabazite, erionite, offretite, levynite
10-hedron (octagonal prism)	2×8-rings 8×4-rings	16	4·5 in plane of 8-rings	paulingite, zeolite *RHO*
11-hedron	5×6-rings 6×4-rings	18	4·7 along *c* axis 3·5 normal to *c*	cancrinite, zeolite *L*, erionite, offretite, zeolite losod
14-hedron Type I (truncated octahedron)	8×6-rings 6×4-rings	24	6·6 for inscribed sphere	sodalite, faujasite, zeolite *A*
14-hedron Type II	3×8-rings 2×6-rings 9×4-rings	24	6·0 along *c* 7·4 normal to *c*	gmelinite, offretite, mazzite (zeolite Ω)
17-hedron Type I	3×8-rings 5×6-rings 9×4-rings	30	9·0 along *c* 7 to 7·3 normal to *c*	levynite
17-hedron Type II	11×6-rings 6×4-rings	30	7·7 along *c* 6·4 normal to *c*	zeolite losod
18-hedron (oblate spheroidal form)	6×8-rings 12×4-rings	32	10·8×6·6 (6·6 is measured between centre planes of opposite 8-rings)	paulingite, zeolite *ZK-5*
20-hedron	6×8-rings 2×6-rings 12×4-rings	36	11 along *c* 6·5 normal to *c*	chabazite
23-hedron	6×8-rings 5×6-rings 12×4-rings	42	15 along *c* 6·3 normal to *c*	erionite
26-hedron Type I (truncated cubo-octahedron)	6×8-rings 8×6-rings 12×4-rings	48	11·4 for inscribed sphere	paulingite, zeolite *ZK-5*, zeolite *A*, zeolite *RHO*
26-hedron Type II	4×12-rings 4×6-rings 18×4-rings	48	11·8 for inscribed sphere	faujasite (zeolites X and Y)

[a] If n denotes the number of faces, $2n-4$ gives the number of vertices.

and the number of vertices, which equals the number of (Al, Si)O_4 tetrahedra forming the polyhedron. Some of these polyhedra are illustrated in Fig. 4.[103, 57] From Table 4 one can see that the free dimensions of the polyhedra can be substantial and that they should often be able to hold clusters of small molecules, such as water, in addition to the appropriate

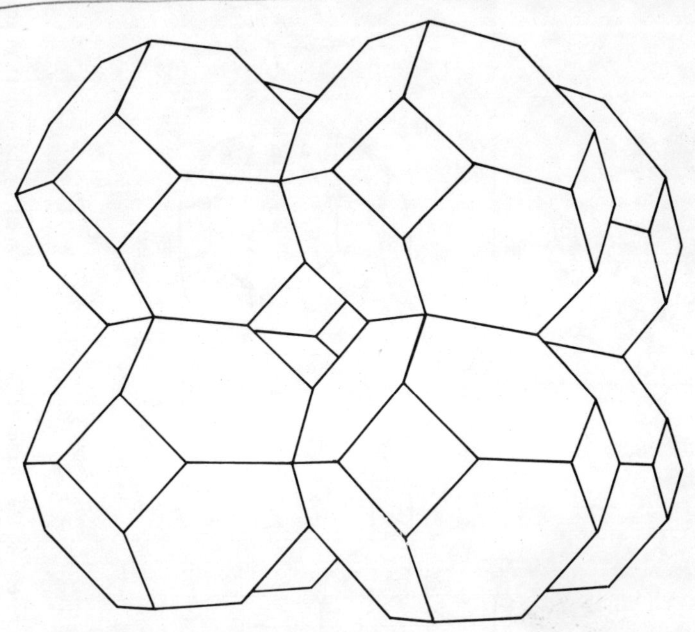

FIG. 3. The stacking of 14-hedra of Type I (sodalite cages) in 8-fold coordination, giving the sodalite structure.[55]

numbers of exchange cations. It is also seen that many of them have "windows" in the form of 8- or 12-ring apertures through which small molecules could readily pass. The maximum free diameters of the windows will be found when these are planar. Taking as 2·8Å the diameter of the oxygens* lining the inner peripheries of a window the free diameter of planar n-ring windows were estimated[103] as:

n	Free dimension (Å)
4	1·15
5	1·96
6	2·8
8	4·5
10	6·3
12	8·0

* 2·70Å would be a better estimate (see § 12). The reduction in free dimensions is however slight.

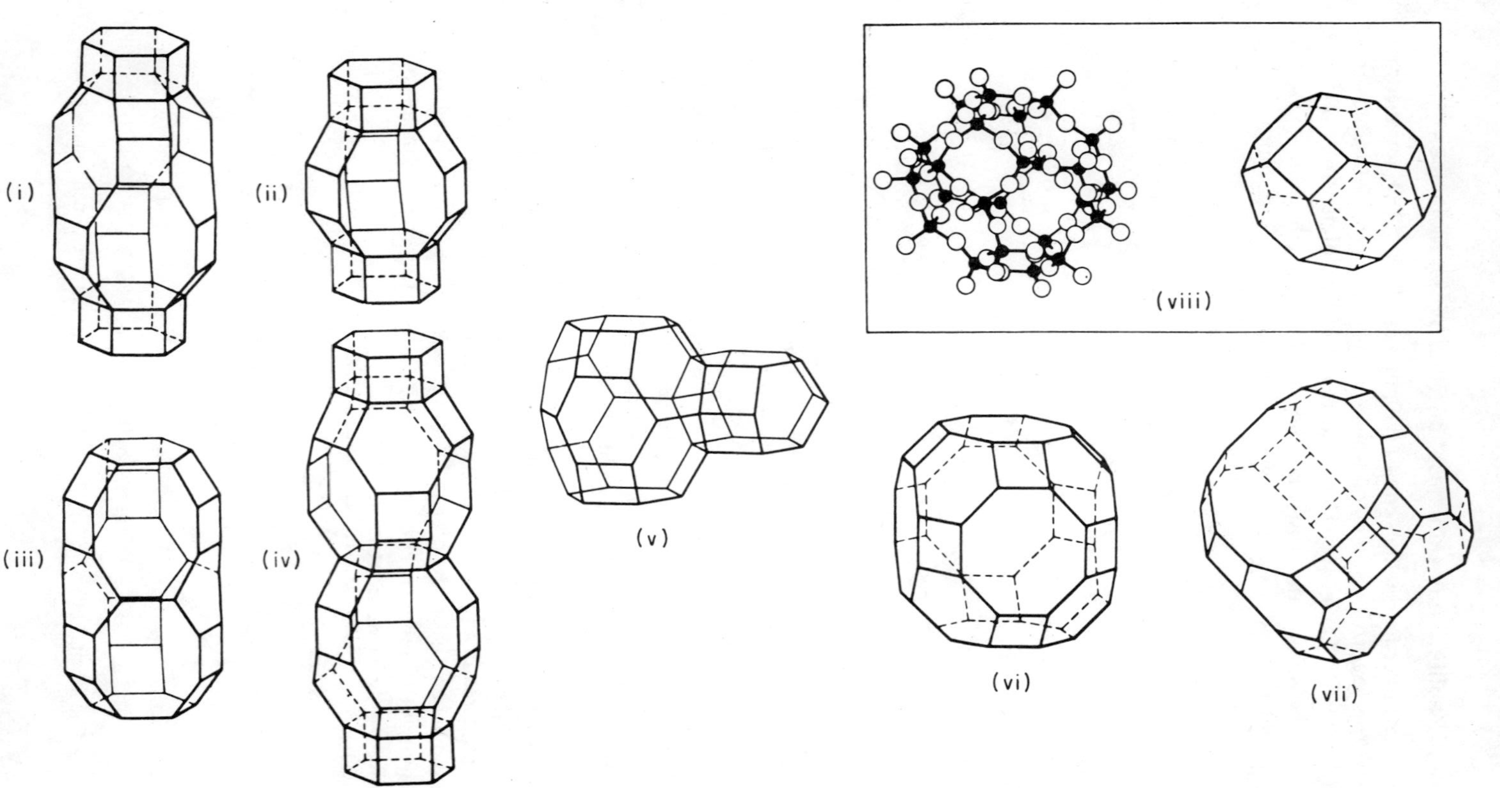

FIG. 4. Some of the polyhedral voids found in zeolite structures: (i) the chabazite 20-hedron, capped by hexagonal prisms; (ii) the gmelinite 14-hedron of Type II; (iii) the erionite 23-hedron; (iv) the levynite 17-hedron of Type I, with associated hexagonal prisms; (v) the losod 17-hedron of Type II, with associated 11-hedral canrinite cage; (vi) the zeolite *A* 26-hedron of Type I; (vii) the faujasite 26-hedron of Type II; (viii) the sodalite 14-hedron shown as in Fig. 3 and also as a "ball and stick" atomic model. Filled circles denote Al or Si and open circles denote oxygen. Part of the rear of the 14-hedron is, for clarity, omitted from the ball and stick model.

However, in the actual zeolite structures the polygonal windows are usually distorted to varying degrees. Thus estimated free dimensions of some 10- and 12-rings found in zeolites were as follows:

n	Zeolite	Approximate free dimensions in Å (maximum and minimum)
10	Heulandite	7·8 and 3·2
	Stilbite	6·2 and 4·1
	Dachiardite	6·7 and 3·7
	Epistilbite	5·3 and 3·2
	Ferrierite	5·5 and 4·3
12	Gmelinite	6·9
	Cancrinite hydrate	6·2
	Mordenite	7·0 and 6·7
	Faujasite	7·4

Unless blocked by the presence of exchange ions, or adventitious impurities there should be and often is little difficulty of access of small molecules through 8-, 10- and 12-ring openings to the voids and channels in zeolites. Figure 5[48] illustrates for various zeolites the degrees of distortion and free dimensions of 8-ring windows present in some zeolites and providing entry to the intracrystalline pores. The most nearly planar and regular 8-ring window is found in zeolite *A*. In other zeolites the 8-rings are boat- or chair-shaped.

The modes of linking of polyhedra and the relative numbers of these in zeolites of the chabazite and faujasite groups are presented in Table 5. The octagonal prism, 18-hedron and 26-hedron of Type I found in zeolite *ZK-5* are also found in paulingite. In paulingite, however, the arrangement of these three polyhedra shown in Fig. 6[4] does not fill all the unit cell and other structural features not based on simple stackings of polyhedra are present. Other members of the faujasite group are illustrated in Fig. 7[4, 6] and some important members of the chabazite group in Fig. 8[4].

In Table 5 an unknown zeolite closely resembling mazzite is referred to. This unknown was proposed as a possible structure[104] for zeolite Ω but the latter is most probably a synthetic variant of mazzite. The same columns of 14-hedra of Type II occur in both but the way in which these columns are linked around and form the wide channels is different. Such unknown structures are frequently apparent when frameworks are constructed from tetrahedra and various other examples will be given, the synthesis of which provides an interesting chemical challenge. Thus, Moore and Smith[105]

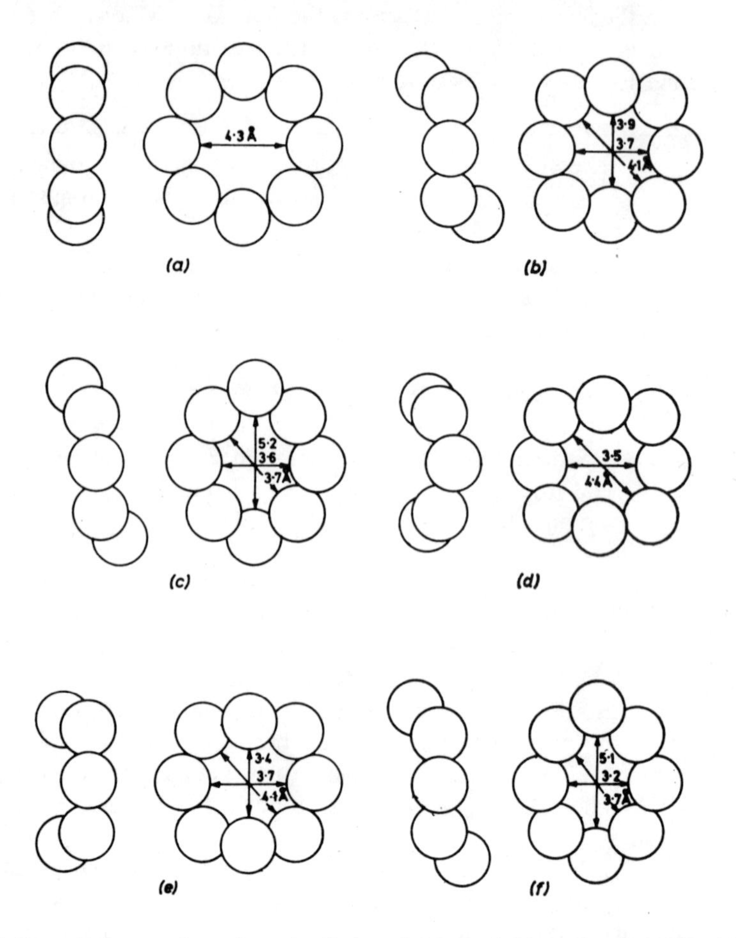

FIG. 5. Some 8-ring conformations for hydrated forms of (a) zeolite A; (b) chabazite; (c) erionite; (d) a hypothetical cubic zeolite; (e) gmelinite; and (f) levynite.[48]

considered the structures which could be formed by linking 14-hedra of Type I or 26-hedra of Type I. As a shorthand description of the possibilities they used the symbols: $H =$ hexagonal face; $H' =$ hexagonal prism; $S =$ square face; $S' =$ cubic unit; $O =$ octagonal face; $O' =$ octagonal prism. The type of contact between polyhedra was denoted by the appropriate one of these letters, and letters following, placed in parentheses, indicated the types of polygonal face opposing each other across the contact. In this way the possibilities of Table 6 were demonstrated in which of course sodalite, zeolite *A*, faujasite, zeolite *RHO* and zeolite *ZK-5* are included. Four of these structures are so far unknown. It is of interest that zeolite *RHO* was noted as a hypothetical structure[67] before it was synthesized.[66]

TABLE 5

Examples of combinations of polyhedra in zeolites

Zeolite	Polyhedra and other features	Proportions	Mode of combination
Cancrinite hydrate	11-hedra (cancrinite cages). Wide channels 11^l to c	–	11-hedra in columns 11^l to c. Six linked columns surround and create each wide channel. Alternate columns are displaced by $c/2$.
Losod	11-hedra in columns *A* 17-hedra of Type II in columns *B*	1 1	A column of *A* is surrounded by six of *B*, 11^l to c. *B*-columns are alternately displaced by $c/2$.
Chabazite	Hexagonal prisms 20-hedra	1 1	Prisms and 20-hedra alternate in columns 11^l to c. A given column is surrounded by six like columns, three displaced by $c/3$ and three by $2c/3$.
Gmelinite	Hexagonal prisms 14-hedra of Type II Wide channels 11^l to c	1 1	Prisms and 14-hedra alternate in columns 11^l to c. Six linked columns surround and create wide channels. Alternate columns displaced by $c/2$.
Erionite	Hexagonal prisms 11-hedra 23-hedra	1 1 1	Prisms and 11-hedra alternate in columns *A*, 23-hedra form columns *B*, both *A* and *B* 11^l to c. Each column *A* is surrounded by six of *B*, where columns *B* are alternately displaced by $c/2$.
Offretite	Hexagonal prisms 11-hedra 14-hedra of Type II Wide channels 11^l to c	1 1 1	Prisms and 11-hedra alternate in columns *A*; 14-hedra form columns *B*; both *A* and *B* are 11^l to c. Three *A* and three *B* surround and create each wide channel.
Mazzite	14-hedra of Type II Wide channels 11^l to c. Narrow channels 11^l to c	–	Six linked columns of 14-hedra surround and create wide and narrow channels 11^l to c. Alternate columns displaced by $c/2$.
Unknown	14-hedra of Type II. Wide channels 11^l to c	–	Six linked columns of 14-hedra surround and create wide channels. All at same height.
Faujasite	Hexagonal prisms 14-hedra of Type I 26-hedra of Type II	2 1 1	A given 14-hedron is linked by four prisms arranged tetrahedrally on four of its eight hexagonal faces, to its four nearest 14-hedron neighbours. Arrangement of 14-hedra is as are the atoms in diamond. This creates 26-hedral voids, also arranged like atoms in diamond.

TABLE 5—*continued*

Zeolite	Polyhedra and other features	Proportions	Mode of combination
Zeolite *A*	Cubic units 14-hedra of Type I 26-hedra of Type I	3 1 1	A given 14-hedron is linked by cubic units through its six 4-ring faces to a 4-ring face of each of six other 14-hedra. This arrangement creates 26-hedra.
Zeolite *RHO*	Octagonal prisms 26-hedra of Type I	3 1	A given 26-hedron is linked by octagonal prisms through each of its six 8-ring faces to an 8-ring face of one of six other 26-hedra.
Zeolite *ZK-5*	Hexagonal prisms 18-hedra 26-hedra of Type I	4 3 1	A given 26-hedron is linked by hexagonal prisms through its eight 6-ring faces to a 6-ring face of each of eight other 26-hedra. This creates the 18-hedral voids.

TABLE 6

Frameworks made by linking 14-hedra and/or 26-hedra, both of Type I

(a) From 14-hedra

Arrangement	Space group	a(Å)	c(Å)	Tetrahedra per unit cell	Example
$S(H\text{-}H)$	$Pm3m$	8·8	–	12	Sodalite
$S'(H\text{-}H)$	$Pm3m$	11·9	–	24	Zeolite *A*
$H(H\text{-}S)$	$Fd3m$	17·5	–	96	Unknown
$H'(H\text{-}S)$	$Fd3m$	24·7	–	192	Faujasite
$H(H\text{-}S)$ and $(H\text{-}H)$	$P6_3/mmc$	12·4	20·5	64	Unknown
$H'(H\text{-}S)$ and $(H\text{-}H)$	$P6_3/mmc$	17·5	28·5	128	Unknown

(b) From 26-hedra

Arrangement	Space group	a(Å)	c(Å)	Tetrahedra per unit cell	Example
$O(S\text{-}S) \equiv S'(H\text{-}H)$	$Pm3m$	11·9	–	24	Zeolite *A*
$O'(S\text{-}S)$	$Im3m$	15·1	–	48	Zeolite *RHO*
$H(O\text{-}S) \equiv O'(S\text{-}S)$					
$H'(O\text{-}S)$	$Im3m$	18·7	–	96	Zeolite *ZK-5*

(c) From 14-hedra and 26-hedra

Arrangement	Space group	a(Å)	c(Å)	Tetrahedra per unit cell	Example
$H(S\text{-}S) \equiv S'(H\text{-}H)$	$Pm3m$	11·9	–	24	Zeolite *A*
$H'(S\text{-}S)$	$Fm3m$	31·1	–	384	Unknown

Table 6 also indicates that zeolite *A* can be constructed in several different ways, and that the *O′*(*S*-*S*) and *H*(*O*-*S*) combinations of the 26-hedra are identical. The *H*(*H*-*S*) and *H′*(*H*-*S*) combinations of the 14-hedra have the centres of the polyhedra located as are atoms in diamond (or blende) while

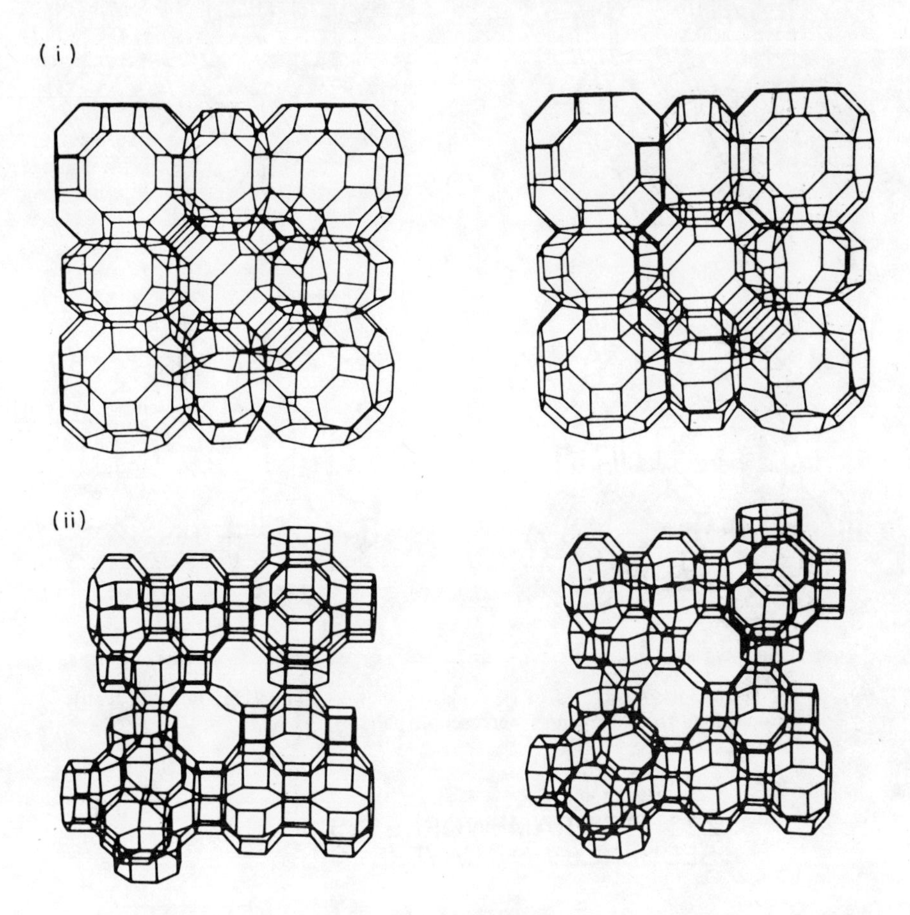

FIG. 6. The framework structures of (i) zeolite *ZK-5* viewed along [100] and (ii) paulingite viewed along [100], shown as stereoscopic pairs.[4]

the *H′*(*H*-*S*) + (*H*-*H*) variety has these polyhedra centred as are atoms in wurtzite. The *H′*(*S*-*S*) structure of linked 14-hedra and 26-hedra is analogous to the fluorite structure, with 14-hedra in fluorine atom positions and 26-hedra in the positions of the calciums.

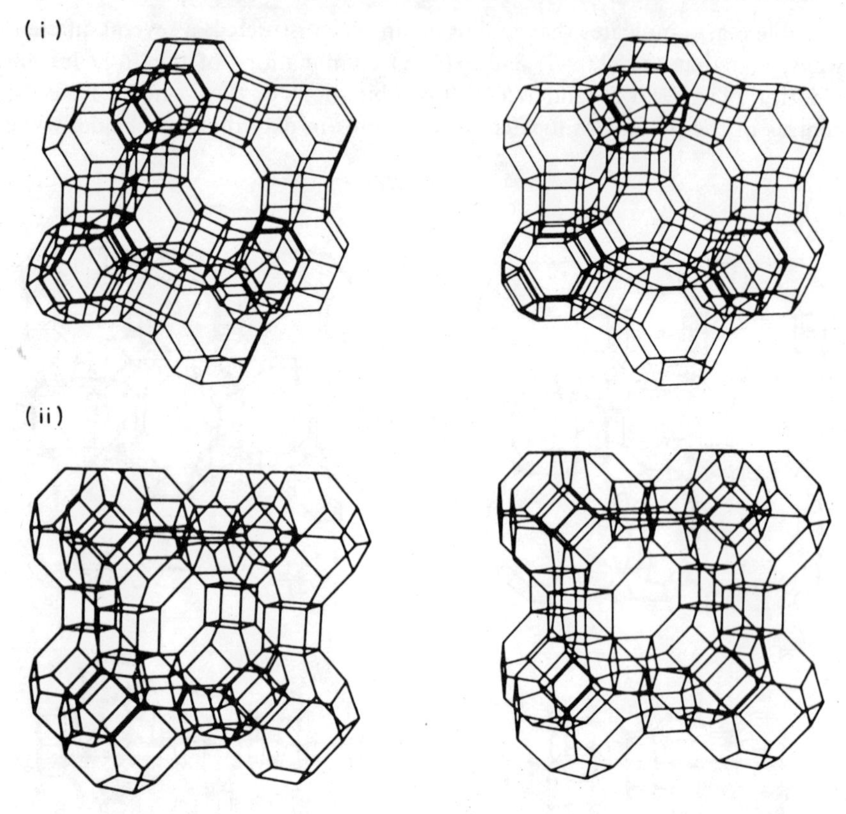

FIG. 7. The framework structures of (i) faujasite, viewed along [111] and (ii) zeolite *A* viewed along [100], shown as stereoscopic pairs.[4]

5. ZEOLITE FRAMEWORKS AS SEQUENCES OF LAYERS

Another fruitful way of considering zeolite structures is in terms of layers linked to other layers in different sequences. As already seen, in faujasite the 14-hedral (sodalite) cages have their centres at sites distributed as are the atoms in diamond or zinc blende, while in the structure $H'(H\text{-}S) + (H\text{-}H)$ of Table 6 these centres are distributed as in wurtzite. The blende and wurtzite structures are shown in Fig. 9,[106] where the blende structure has been drawn with one of the cube diagonals vertical and parallel to the principal axis of the wurtzite structure. Both structures so viewed can be visualized as formed by superposition of puckered layers. In blende, successive layers, denoted by *A*, are identical but are displaced with respect to each other to give a 1231 . . . sequence. In wurtzite, successive layers differ, being related by a rotation of 180° about the principal axis. If this second kind of layer is denoted by *B*, then the sequence is *ABA*. . . . Since *A* and *B* layers may be joined to each other, or with appropriate displacements, *A* may be joined to *A* and *B* to *B*, one can obtain an indefinite number of polytypes. The zeolite ZSM-3[60] appears to illustrate this. It has a unit cell with $a = 17{\cdot}5$Å, like the structure *H′*(*H-S*) + (*H-H*) of Table 6, but *c* was greater than 28·5Å and not more than 129Å. The distance between the layers of 14-hedral cages linked by hexagonal prisms is 14·3Å.

TABLE 7

Structures formed by linking layers containing single 6-rings (*a*, *b*, *c*) and layers containing double 6-rings (*A*, *B*, *C*)[8]

No. of layers in repeat unit	Structural type	Space group	*a*(Å)	*c*(Å)	Name
2	*ab*	$P6_3$	12·72	5·19	Cancrinite
3	*abc*	$P\bar{4}3m$	8·87	–	Sodalite
4	*abac*	$P6_3/mmc$	12·91	10·54	Losod
6	*ababac*	$P\bar{6}m2$	12·85	16·10	Liottite
8	*ababacac*	$P6_3mc$	12·77	21·35	Afghanite
10	*abcabcbacb*	$P\bar{3}ml$	12·88	26·76	Franzinite
2	*AB*	$P6_3/mmc$	13·75	10·05	Gmelinite
3	*ABC*	$R\bar{3}m$	13·78	15·06	Chabazite
2	*Ab*	$P\bar{6}m2$	13·29	7·58	Offretite
4	*AbAc*	$P6_3/mmc$	13·26	15·12	Erionite
4	*aBaC*	$P6_3/mmc$	13·26	15·12	–
6	*AbCaBc*	$R\bar{3}m$	13·34	23·01	Levynite

Some zeolites of the chabazite group and of related phases serve well to illustrate the kinds of structure which can be obtained from different sequences of two kinds of layer. One of these kinds contains 6-rings and the other double 6-rings (hexagonal prisms). The 6-rings or double 6-rings in a given layer are joined by 4-rings to 6-rings or double 6-rings in the layers above and below, the several layers being displaced with respect to each other. If layers containing 6-rings are called *a*, *b* and *c*, according to the displacement, and layers containing hexagonal prisms are correspondingly termed *A*, *B* and *C*, then the layer sequences in Table 7[8] refer to known species. In the 10-layer sequence of franzinite Merlino[8] additionally stated that the *c* layer in the sixth position was replaced by an a layer with probability one third. There are clearly possible new zeolites involving sequences of *A*, *B* and *C* layers analogous to those of *a*, *b* and *c* layers in zeolite losod,

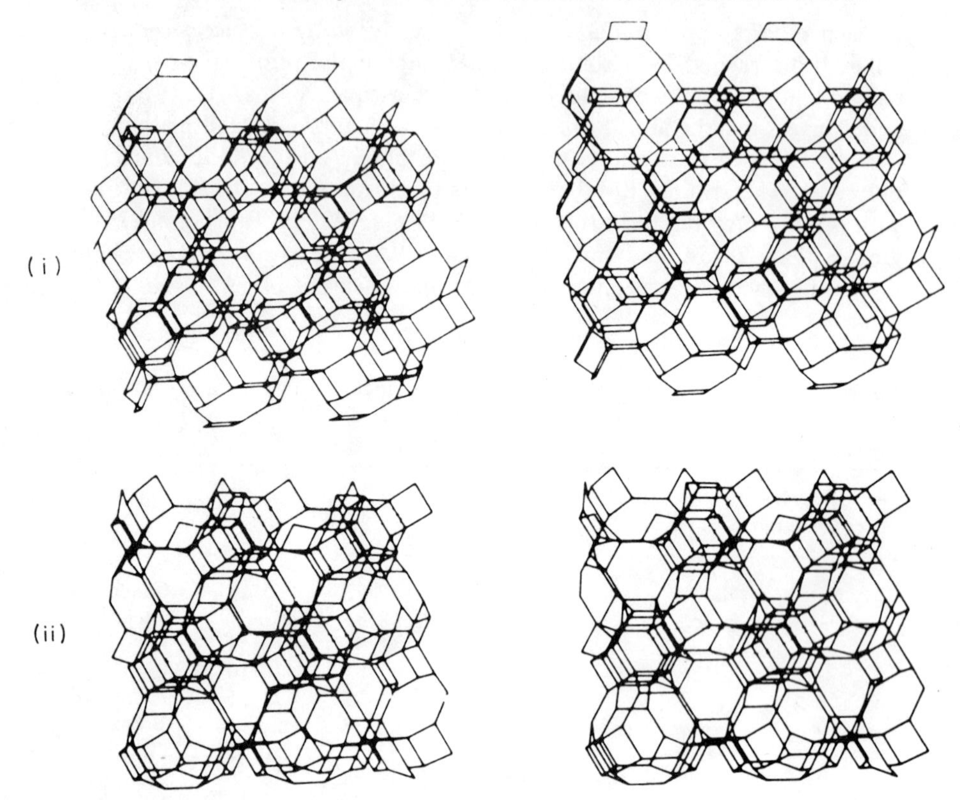

FIG. 8. Examples of frameworks chosen from the chabazite group all viewed along [001]: (i) chabazite; (ii) erionite; (iii) offretite; and (iv) zeolite *L*.[4]

and in liottite, afghanite or franzinite. Also sequences of layers of single 6-rings and of double 6-rings are possible in addition to those giving offretite, erionite and levynite. Two of these hypothetical structures belonging to the chabazite group were considered in some detail by Kokotailo and Lawton.[107] The layer sequences were *ABCB* . . . and *ABCACB*. . . . The hexagonal unit cells have $a = 13 \cdot 7$Å with $c = 20$ and 30Å respectively. Each contains elongated 26-hedral cavities of a third type (Type III). This cavity has nine non-planar 8-ring, two 6-ring and 15 4-ring faces, with a free length approaching 20Å and a free diameter of about 6·5Å, but varying somewhat along the length of the cavity. The *ABCB* structure also contains 14-hedral gmelinite cages, and the *ABCACB* structure both gmelinite and 20-hedral chabazite cages.

The mordenite group can also be described in terms of layers.[8, 108] Puckered sheets of 6-rings can be identified in these structures, joined with other sheets in several ways. In a given sheet the apices of the tetrahedra

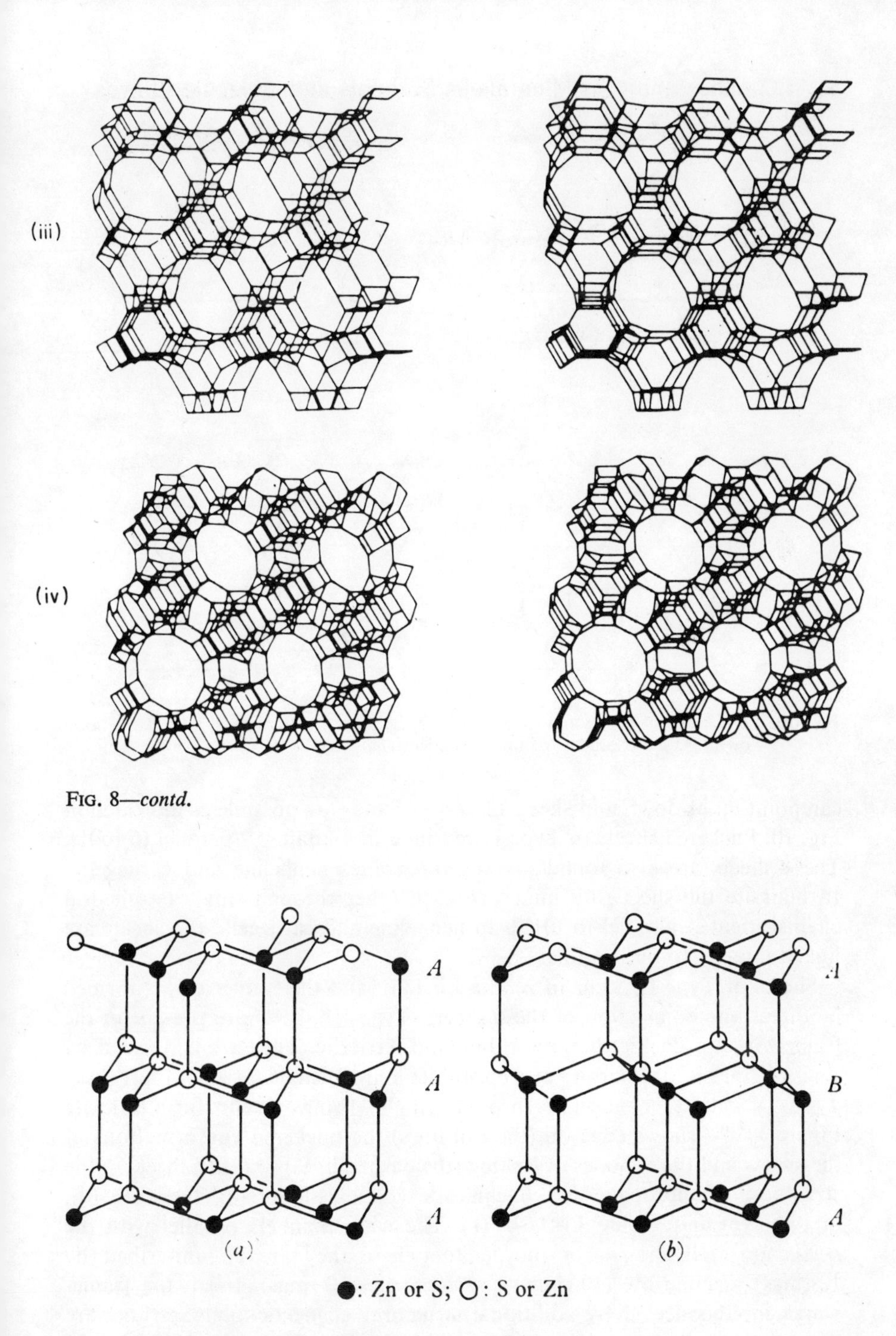

FIG. 8—*contd.*

FIG. 9. Clinographic projections of the structures of (a) zinc blende and (b) wurtzite showing the two layer sequences.[106]

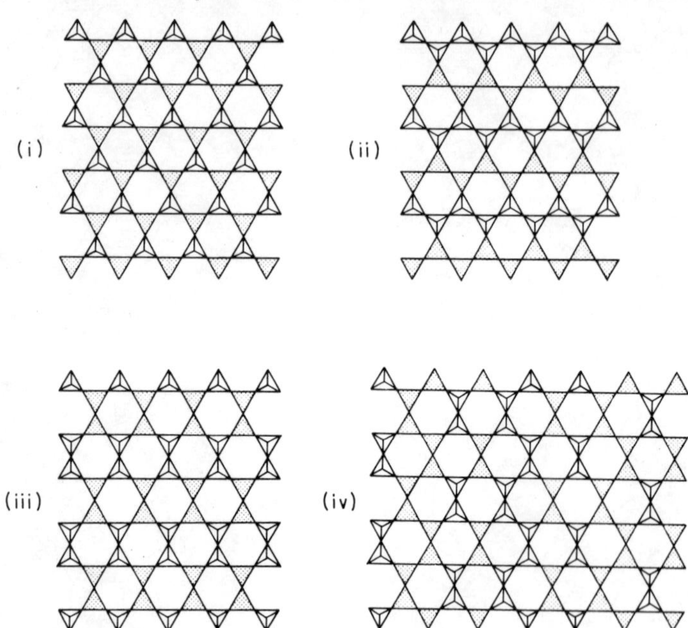

FIG. 10. Four kinds of hexagonal sheet in which the apices of the linked tetrahedra point down (shaded triangles) or up (unshaded triangles).[108]

can point up or down and sheets of various kinds are possible as indicated in Fig. 10. Puckered sheets of Type I are found in bikitaite[41] parallel to [001]. These sheets are also found in the non-zeolites nepheline and carnegeite. In bikitaite the sheets are linked to each other through single tetrahedral chains oriented parallel to [010]. In nepheline and carnegeite the sheets are linked directly to one another.

Sheets of Type II occur in zeolite Li-*ABW* and the framework is formed by direct interconnection of these sheets. Type III sheets are present in the frameworks of dachiardite, epistilbite and ferrierite. They are connected by single 4-rings in dachiardite and epistilbite and by single 6-rings in ferrierite. Type IV sheets, connected through single 4-rings occur in mordenite Figure 11[108] shows edge on (the full lines) the puckered conformations of the sheets and their modes of linking (the dashed lines) to other sheets in the structures of mordenite (a), dachiardite (b), epistilbite (c), ferrierite (d), bikitaite (e) and zeolite Li-*ABW* (f). The wide channels parallel with the *c*-axis are well shown for mordenite (where they are circumscribed by 12-rings), dachiardite (10-rings) and ferrierite (10-rings). In all the frameworks interconnected by additional structural elements many 5-rings are generated. By contrast, in Li-*ABW* where the sheets are joined directly no 5-rings appear. Figure 11f also shows that the framework can be made in

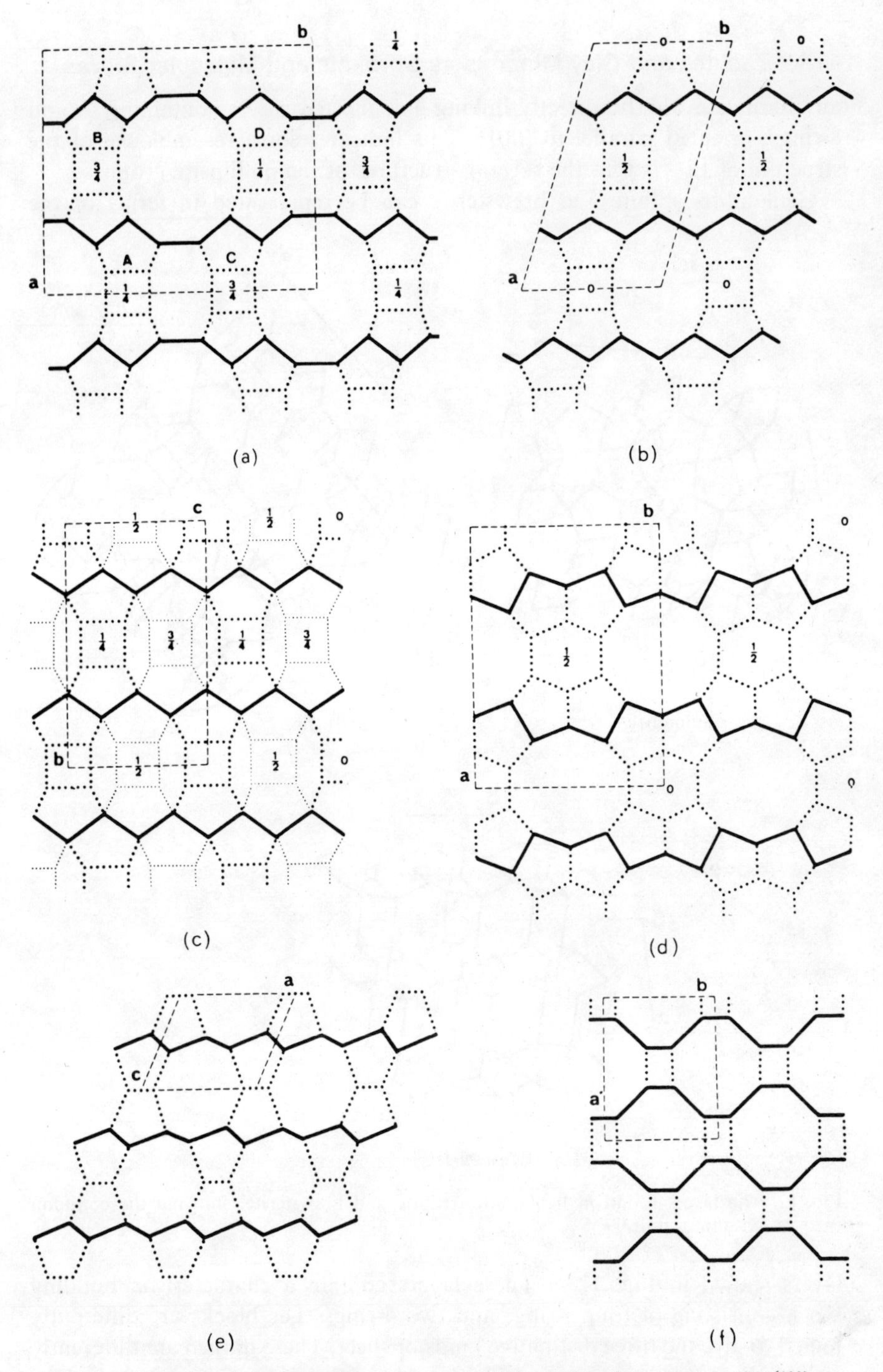

FIG. 11. The puckered conformations of hexagonal sheets found in some zeolites(108) are shown edge on (full lines). The interconnections between sheets are shown as dashed lines. The zeolites concerned are: (a) mordenite; (b) dachiardite; (c) epistilbite; (d) ferrierite; (e) bikitaite; and (f) zeolite Li-*ABW*.

an alternative way by directly linking 3-connected sheets containing 4- and 8-rings, oriented parallel to [001]. This indicates some resemblance of the structure of Li-*ABW* to the several structures of the phillipsite group.

Heulandite, stilbite and brewsterite can be represented in terms of the

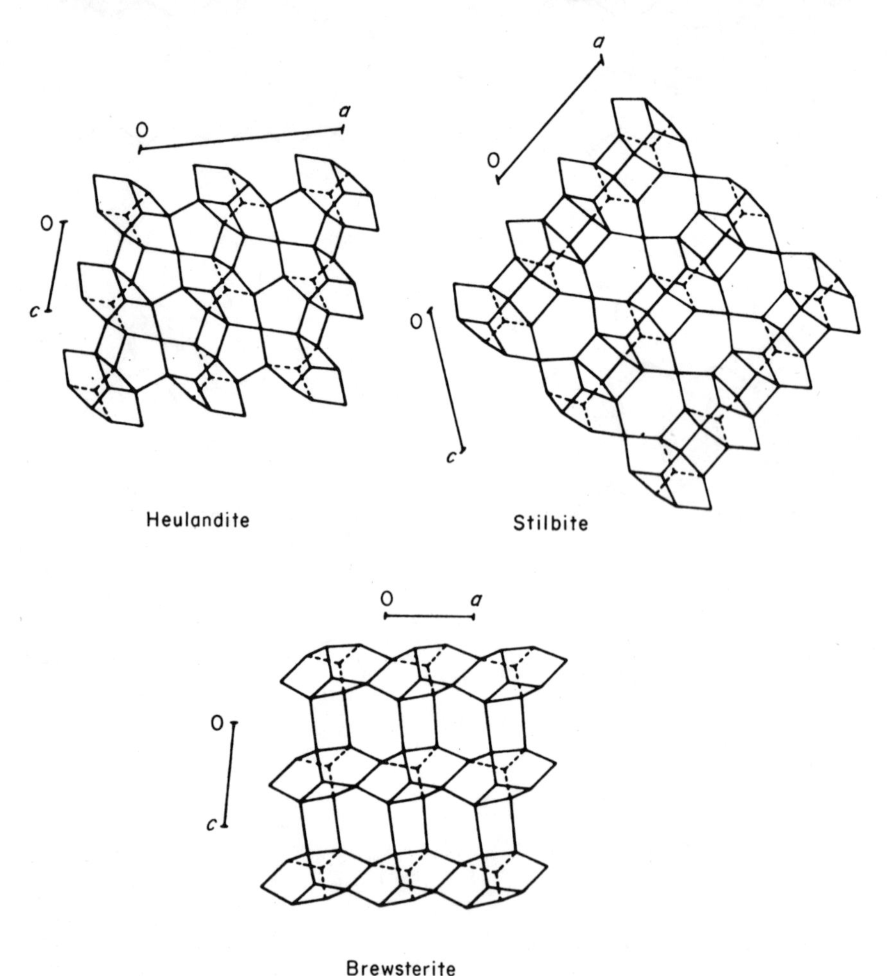

FIG. 12. The layers found in heulandite, stilbite and brewsterite, showing the common structural unit.[8]

layers shown in Fig. 12.[8] These layers contain a characteristic building block consisting of four 5-rings and two 4-rings. The blocks are differently joined to give the three distinctive kinds of sheet. These in turn are differently cross-linked to other sheets of like type in each of the three minerals. The bond density between layers is less than that in the layers, accounting for their platy character. The laumontite framework contains building blocks with four 6-rings and two 4-rings (Fig. 13).[8] In the plane shown in the figure, a given block is joined by four 4-rings to four other such blocks. This produces the rather open sheet of the figure, containing distorted 8-ring openings.

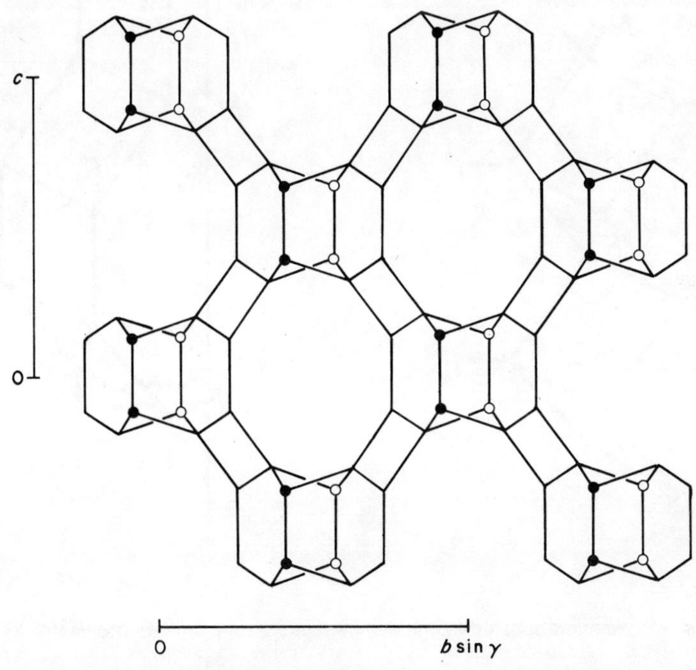

FIG. 13. The laumontite framework layer, showing the building blocks each composed of four 6-rings and two 4-rings.[8]

6. ZEOLITE FRAMEWORKS CONSTRUCTED FROM CHAINS

Frameworks of some zeolites of the chabazite, mordenite, natrolite and phillipsite groups can readily be constructed from several kinds of chain. In mazzite, zeolite *L*, gmelinite and cancrinite, for example, chains in the form of variously buckled ladders of 4-rings are identifiable,[53] and similar ladders are present in felspars and in the phillipsite group.[96] Figure 14[53] illustrates the conformations of two such ladders in relation to an operator axis. Each two unlinked oxygen atoms at the same height can be described as near (*N*) or far (*F*) from this axis. For generating frameworks from chains the axis can serve for *n*-fold, screw or inversion operations involving the ladders; at present one is considering it simply as a three-fold axis. The types

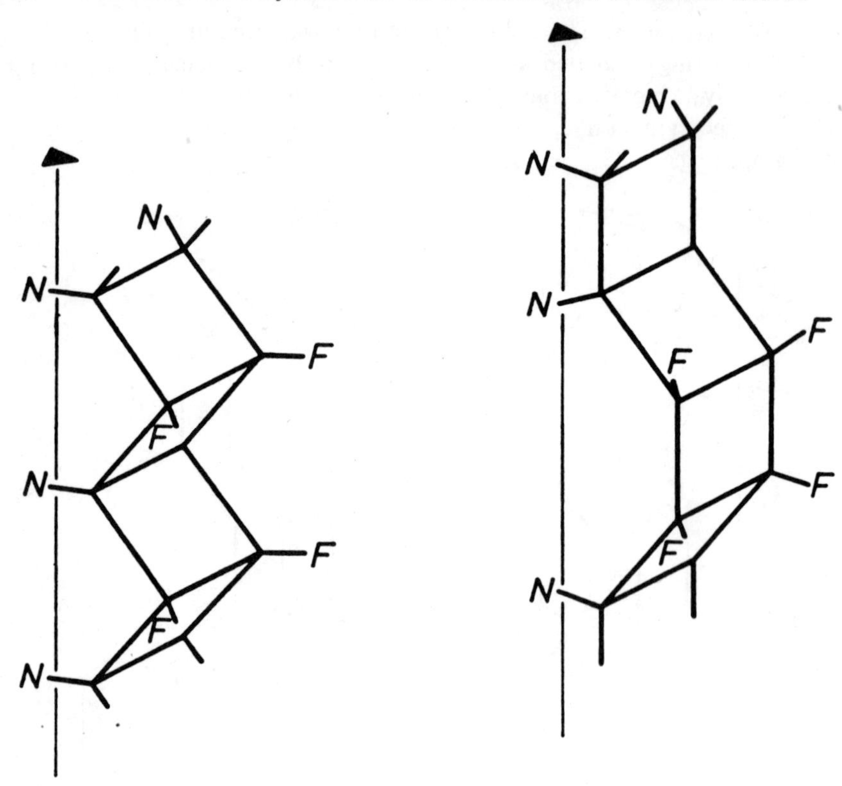

FIG. 14. Two configurations of 4-ring ladders relative to a three-fold operator axis.[53]

of chain indicated in Fig. 14 are designated *NF* and *NNFF* respectively in terms of the unlinked oxygens near and far from the axis. The numbers of these *N* and *F* atoms following one another in sequence may be only one or two because a three-fold repetition such as *FFF* would lead to unlikely configurations like double hexagonal prisms (6-6-6 units). The lengths of the identity periods are not restricted, but were limited to $n \leq 8$, the upper limit corresponding with $c \approx 20$Å, where $c = n \times 2{\cdot}5$.

The threefold operator of Fig. 14 serves to link the ladders through the unshared oxygens near the axis to produce the columns of polyhedra characteristic of chabazite group zeolites (cf. Table 5). There are then three ways of extending the network in the plane normal to the operator axis, using the oxygens far from the axis as links:

1. The chain considered is shared by two operators.
2. The chain considered is linked to a second chain belonging to another operator in such a way that

(a) the two connected chains do not face each other directly (Fig. 15a)
(b) the two connected chains face each other (Fig. 15c).

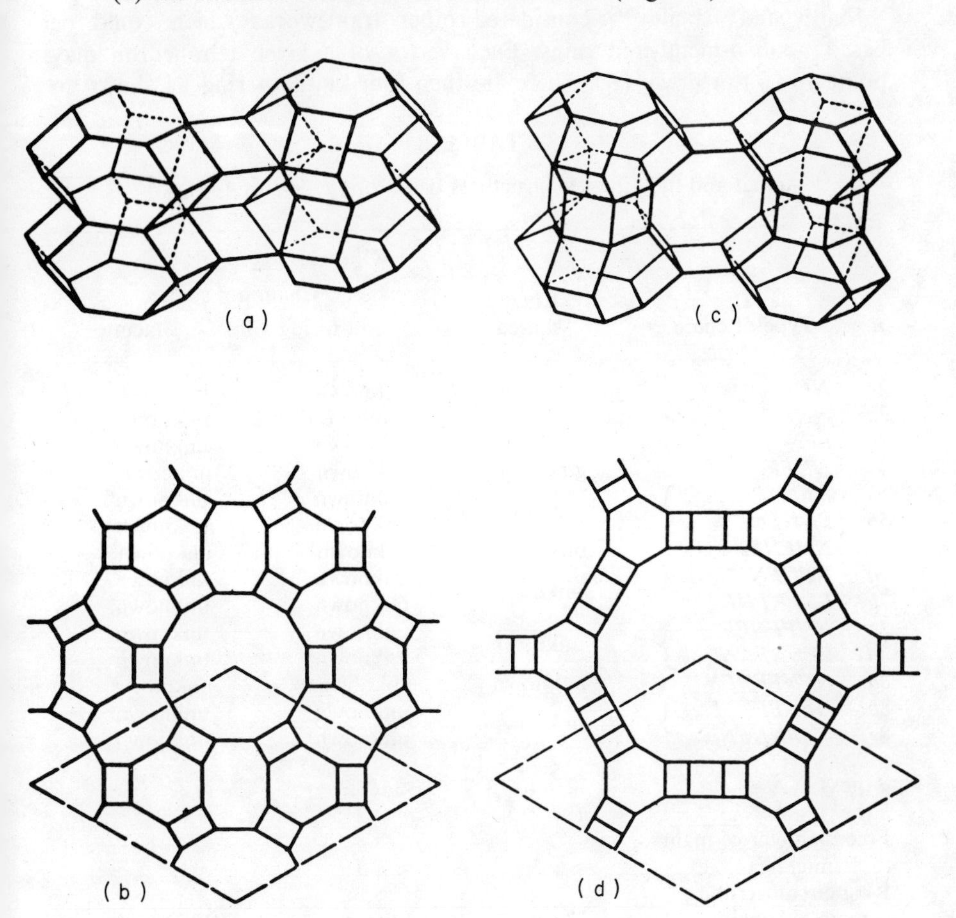

FIG. 15. In (a) the connected ladders of Fig. 14 do not face each other directly and generate the projection of Fig. 15(b), looking along the *c*-axis. In (c) the connected ladders of Fig. 14 face each other directly and generate the projection of Fig. 15(d), looking along the *c*-axis.(53)

One may then generate the frameworks illustrated in Table 8. Other members of the chabazite group such as sodalite, erionite, chabazite and levynite as well as additional hypothetical structures can be generated by applying 6_3 and $\bar{3}$ axes as operators. The projection looking down the *c*-axis shown in Fig. 15b is appropriate for all structures in column 4 of Table 8 and that in Fig. 15d for all structures of column 5. The remarkably large free diameters of all structures of this column and the fact that they are not blocked by stacking faults makes them of much interest. The large number of unknown channel structures of Table 8 is another interesting feature.

Smith and Rinaldi[96] considered other frameworks which could be based upon 4-membered rings. Each vertex of a given tetrahedron may point up (U) or down (D). There are then four kinds of ring as shown in

TABLE 8

Actual and hypothetical structures based on ladders of 4-rings.[53]

			Operator: ▲	
		Chain	Chain not shared	
n	Type of chain	shared	not facing	facing
2	*NF*	cancrinite	unknown	unknown
3*a*	*NNF*	offretite	zeolite *L*	unknown
3*b*	*FFN*		unknown	unknown[a]
4	*NNFF*	gmelinite	unknown	unknown[a]
5*a*	*NFFNF*	unknown	unknown	unknown[a]
5*b*	*FNNFN*		unknown	unknown
6	*NNFNFF*	unknown	unknown	unknown[a]
7*a*	*NFNFNFN*	unknown	unknown	unknown
7*b*	*FNFNFNF*		unknown	unknown
7*c*	*NNFFNFF*	unknown	unknown	unknown[a]
7*d*	*FFNNFNN*		unknown	unknown[a]
8*a*	*NFNFFNFN*	unknown	unknown	unknown[a]
8*b*	*NFNFFNFF*	unknown	unknown	unknown[a]
8*c*	*FNFNNFNN*		unknown	unknown
a (hex) in Å		12·8 to 13·7	≈18·5	≈22
c (hex) in Å		$n \times 2{\cdot}5$	$n \times 2{\cdot}5$	$n \times 2{\cdot}5$
Free diameter of main channel (Å)		≈6·5	≈7·5	≈15
Ring circumscribing main channel		12-ring	12-ring	18-ring
Stacking faults		Block main channel	Do not block main channel	Do not block main channel

[a] *FF* leads to very distorted 4-membered rings.

Fig. 16.[8] The first three of these may be joined to produce the three kinds of chain also shown in Fig. 16. The fourth kind of ring (*UUUU* or *DDDD*) forms the cubic (4-4) units found in zeolite *A*. The double crankshaft *UUDD* chain on the left of Fig 16 resembles the *NNFF* chain of Table 8 from which the threefold operator axis (Fig. 14) can generate the three channel structures of line 4 of the table, one of which is gmelinite. However Smith and Rinaldi considered other ways of cross-linking these chains and recorded 17 frame-

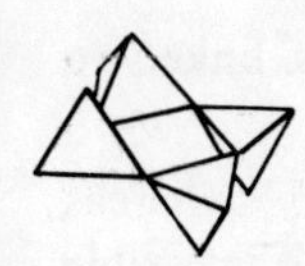
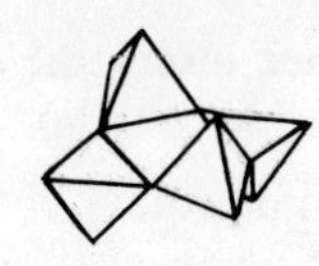
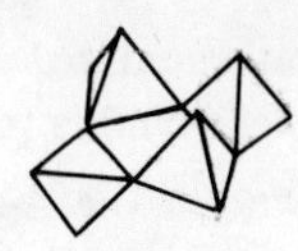
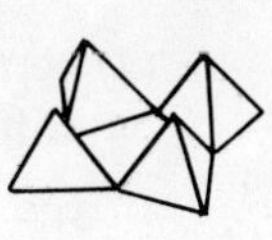

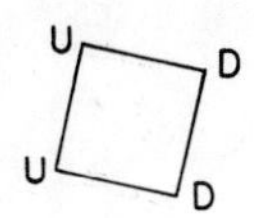

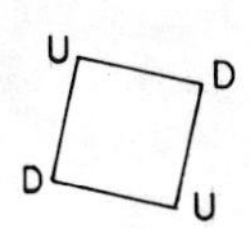

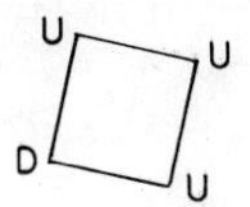

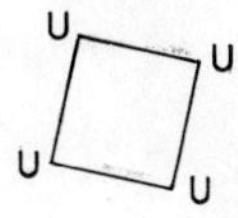

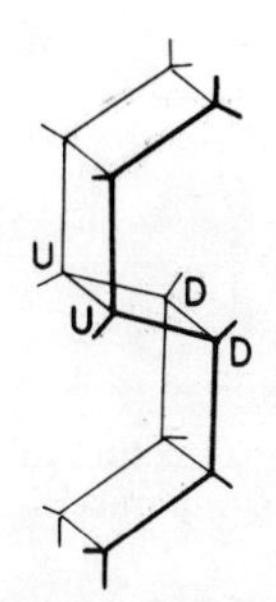

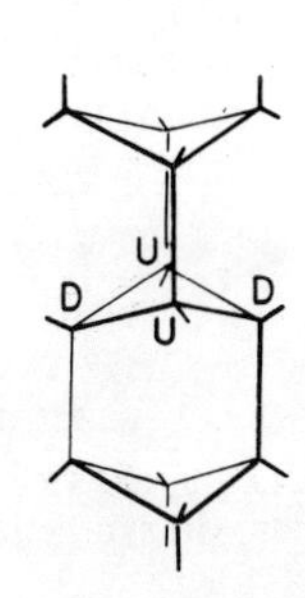

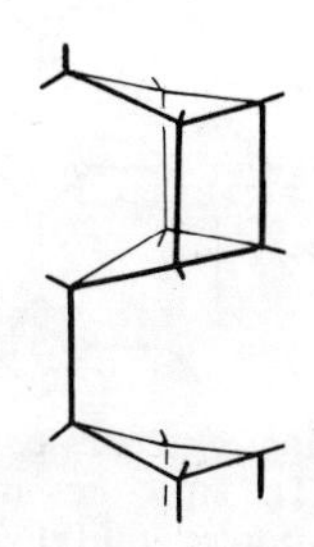
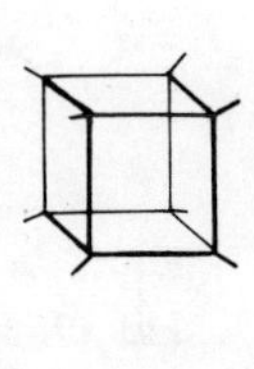

FIG. 16. Four types of 4-ring with apices pointing up (U) or down (D). At the bottom of the diagram three kinds of chain are shown, based on UUDD, UDUD and UDUU rings, together with the cubic unit which can be formed from two 4-rings of the fourth kind.[8]

works with repeat distances in the plane normal to the chain direction less than 15Å. Five were those of felspar, paracelsian, phillipsite, gismondine and the synthetic Phase A[109] ($BaAlSi_2O_6(Cl, OH)$, space group $I4/mmm$ with $a = 14 \cdot 194$Å and $c = 9 \cdot 934$Å).These patterns are shown in Fig. 17,[8] looking along the chain directions. For a real felspar this diagram is not exact because the rings do not superpose in projection along the chain axis, successive 4-rings being rotated relatively to one another. Chains can be linked to form frameworks in two ways:

(1) tetrahedra connected to one another along the chain are both connected to the same adjacent chain (the "flexible" mode of linking)

(2) tetrahedra connected to one another along the chain are linked to different adjacent chains (the "inflexible" mode).

In the inflexible mode the 4-rings cannot rotate cooperatively, whereas 4-rings of the flexible type can. Smith[(100)] reported thirteen of the inflexible types of structure corresponding with the flexible structures reported

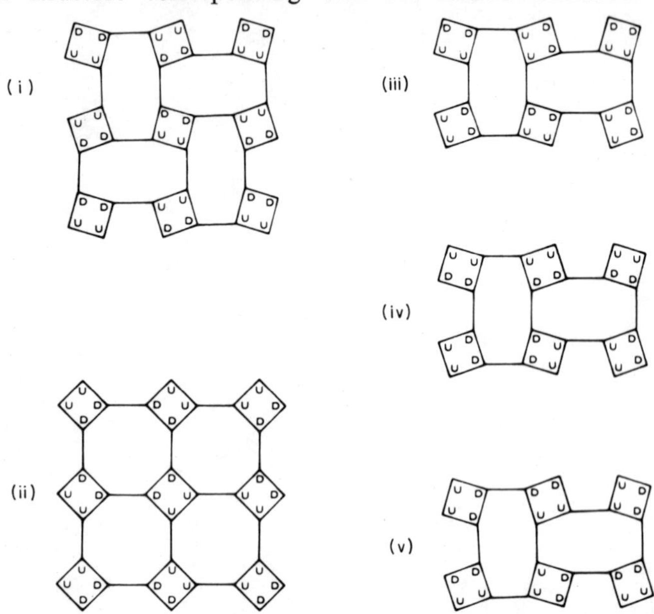

FIG. 17. Different linkings of chains based on UUD 4-rings, viewed along the direction of the chains.[(8)] The structures are: (i) felspar; (ii) synthetic Phase A; (iii) paracelsian; (iv) harmotome; and (v) gismondine.

earlier.[(96)] Paracelsian, phillipsite, gismondine and Phase A are all flexible types; felspar is an inflexible type. Many so far unknown structures are included of both flexible and inflexible types.

The second kind of chain (*UDUD* chains of Fig. 16) could be linked to others to give four structures,[(96)] then all unknown. Subsequently however one of these was identified in the non-zeolite, banalsite, $BaNa_2Al_4Si_4O_{16}$.[(111)] The third kind of chain in Fig. 16 (*UUUD* chains) can also be cross-linked to other chains to give frameworks none of which has so far been found in natural or synthetic compounds. Finally the cubic units of Fig. 16 (formed by *UUUU* and *DDDD* 4-rings) can be linked to other such units. One way of doing this connects a given cubic unit in 8-coordination via the unshared corner oxygens with each of eight other cubic units. This unknown cubic structure has an ≈ 10Å cell edge. It was originally suggested for zeolite Na-Pl,[(33)] which however has the topology of gismondine.[(31)]

A different kind of chain based on 5-rings has been identified in mordenite (Fig. 18[(36)]). Kerr[(112)] considered ways of cross-linking these chains and

TABLE 9

Structures obtainable by cross-linking mordenite-type chains[8]

Structure number	Space Group	Unit cell a(Å)	b(Å)	c(Å)	γ	Example
1	*B*2/*m*	18·73	10·30	7·52	107·9°	Dachiardite
2	*B*2/*m*	18·73	11·7	7·52	123°	Modified dachiardite
3	*Pmnm*	18·1	10·25	7·52		Unknown
4	*Amam*	18·1	20·5	7·52		Unknown
5	*Cmcm*	18·13	20·49	7·52		Mordenite
6	*Immm*	18·13	20·49	7·52		Modified mordenite
7	*Bbcm*	18·7	19·6	7·52		Unknown
8	*Bbmm*	18·7	19·6	7·52		Unknown

FIG. 18. The chain based on 5-rings which can be identified in mordenite.[36] The chain is shown as linked (Al. Si)O_4 tetrahedra.

constructed several unknown zeolite frameworks. Eight structures which could result are listed in Table 9 of which six are unknown. Modes of chain linkage are illustrated in Fig. 19. Figure 19(i) gives the *ab* section of structures 3 and 4 of the table, the latter structure having the 4-rings at the heights,

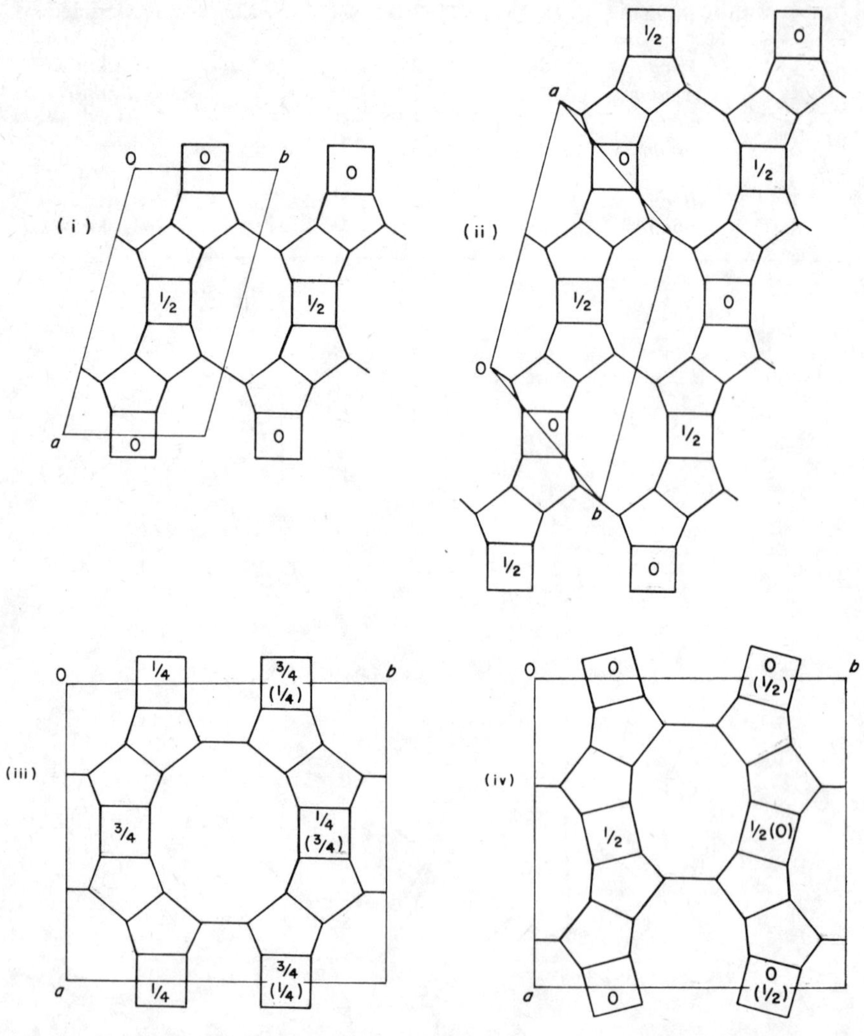

FIG. 19. Some known and hypothetical zeolites of the mordenite group, showing the projections on the ab plane.[8] The chains of Fig. 18 run parallel with the *c*-axis and are variously cross-linked as shown: (i) dachiardite; (ii) modified dachiardite; (iii) mordenite and modified mordenite; (iv) hypothetical structures 7 and 8 of Table 6. The numbers give the heights of the 4-rings as fractions of the cell *c*-dimension. Numbers in brackets in (iii) refer to modified mordenite and in (iv) to structure 8 of Table 6.

in fractions of c, given by the bracketed numbers. The "modified dachiardite" is shown in Fig. 19(ii), the relation between mordenite and "modified mordenite" is indicated in Fig. 19(iii), the latter with heights of 4-rings indicated by the bracketed numbers. The structures 7 and 8 are shown in Fig. 19(iv), the bracketed numbers again referring to the last of these. It is also to be noted that a modification of epistilbite can be constructed, as shown in Fig. 20, which can be compared with Fig. 11c.

A further illustration of the ready construction of zeolite frameworks by linking of chains is found in the natrolite group. The chains are composed of 4-rings of alternating U and D tetrahedra linked together by single

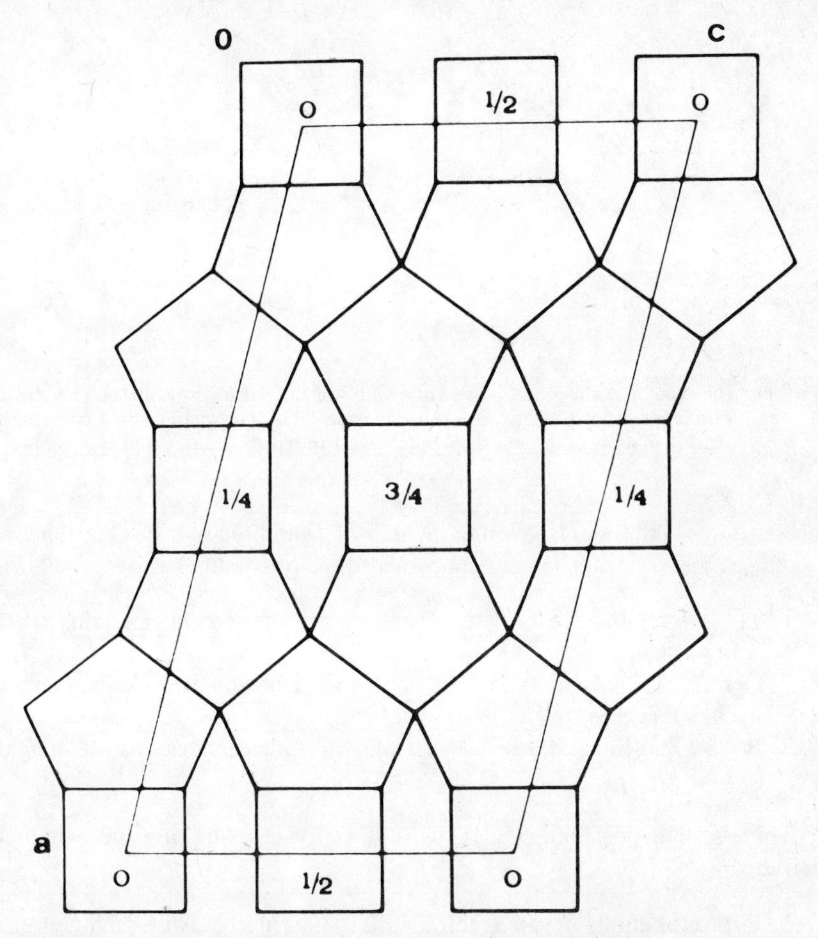

FIG. 20. A hypothetical structure related to epistilbite (modified epistilbite).[8] This structure may be compared with that of Fig. 11c. The first suggested unit cell (*ac* section, outlined in the figure) was later corrected.[8a] The true *ac* section is rectangular; $a = 35{\cdot}48$, $b = 15{\cdot}04$ and $c = 10{\cdot}3$ Å. The space group is *Fddd*.

tetrahedra as seen in Fig. 21.[3] The chains have a period of 6·6Å and the heights of the free vertices are multiples of $c/8$. Each chain is cross-linked to four other chains through its free vertices and each can be coded by the height, n, (i.e. $nc/8$) of the centre of its single tetrahedron[113]. The linkage of a given

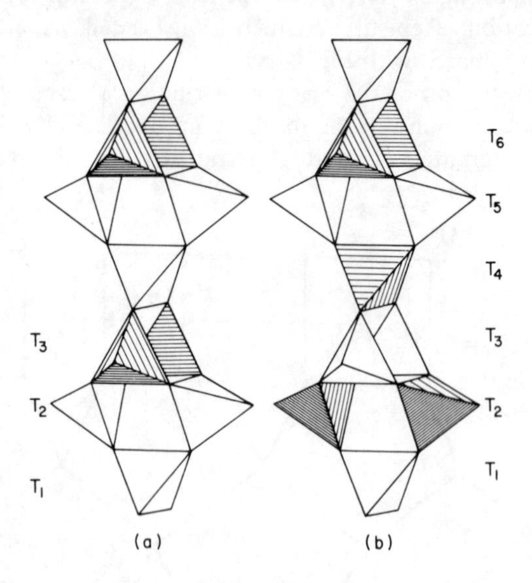

FIG. 21. The type of chain characteristic of zeolites of the natrolite group.[3] The shaded tetrahedra denote AlO_4, the unshaded area SiO_4. The chains show ordering in the distribution of the tetrahedra; (a) refers to natrolite and (b) to thomsonite.

chain to its neighbours is through the pair of *D* and the pair of *U* tetrahedra. As far as the *D* tetrahedra are concerned three possibilities arise:

1. The cross-links are both to *D* tetrahedra of two other chains (these being at height n).
2. One cross-link is to *D* and the other to *U* (the adjacent chains being at heights n and $(n - 2)$).
3. Both cross-links are to *U* tetrahedra (adjacent chains at heights $(n - 2)$).

Three similar possibilities arise as follows for cross-linking the pair of *U* tetrahedra:

1. Both cross-links are to *U*-tetrahedra (adjacent chains at height n).
2. One cross-link is *U-U* and one is *U-D* (adjacent chains at heights n and $(n + 2)$).
3. Both cross-links are *U-D* (adjacent chains at heights $(n + 2)$).

Alberti and Gottardi[113] therefore introduced the symbols

$n = n$ for a cross-link with a chain at the same height
$n \rightarrow n + 2$ for a cross-link with a chain at height $(n + 2)$
$n \leftarrow n - 2$ for a cross-link with a chain at height $(n - 2)$.

TABLE 10

Schemes for interconnecting chains of Fig. 21.[113]

Number	Direct	Unitary	Inverse	Rotational symmetry of scheme
1		$\Vert$ $= n =$ $\Vert$		4
2	$\Vert$ $\leftarrow n \rightarrow$ $\Vert$		$\Vert$ $\rightarrow n \leftarrow$ $\Vert$	2
3		$\downarrow$ $\leftarrow n \rightarrow$ $\uparrow$		2
4	$\Vert$ $= n \rightarrow$ $\Vert$		$\Vert$ $= n \leftarrow$ $\Vert$	1
5		$\downarrow$ $= n \rightarrow$ $\Vert$		1
6	$\uparrow$ $= n \leftarrow$ $\downarrow$		$\downarrow$ $= n \rightarrow$ $\uparrow$	1

TABLE 11

Space groups of natrolite group[a] zeolites[113]

Structural type	Disordered distribution	Natrolite-type order	Thomsonite-type order
1	$P\bar{4}2_1m$	$P2_12_12$ edingtonite	$P\bar{4}2_1c$
2	$Pman$ gonnardite (?)	$P2/b$	$Pcnn$ thomsonite
3	$I\bar{4}2d$ tetragonal natrolite	$Fdd2$ natrolite	not possible
4	$Pmma$	$P2/a$	$Pcca$
5	$Imma$	$B2/b$	not possible
6	$Pmna$	$P2_1/a$	not possible

[a] Mesolite and scolecite, which should be in the same box as natrolite, have lower symmetry.

There are accordingly nine ways of interconnecting a chain to each of four neighbouring chains which are given in Table 10. If corresponding direct and inverse schemes are considered as a single scheme there remain six possible schemes and therefore six different structures.

In fibrous zeolites two kinds of Si/Al ordering have been observed: the natrolite type for a ratio Si : Al = 3 : 2; and the thomsonite type for a

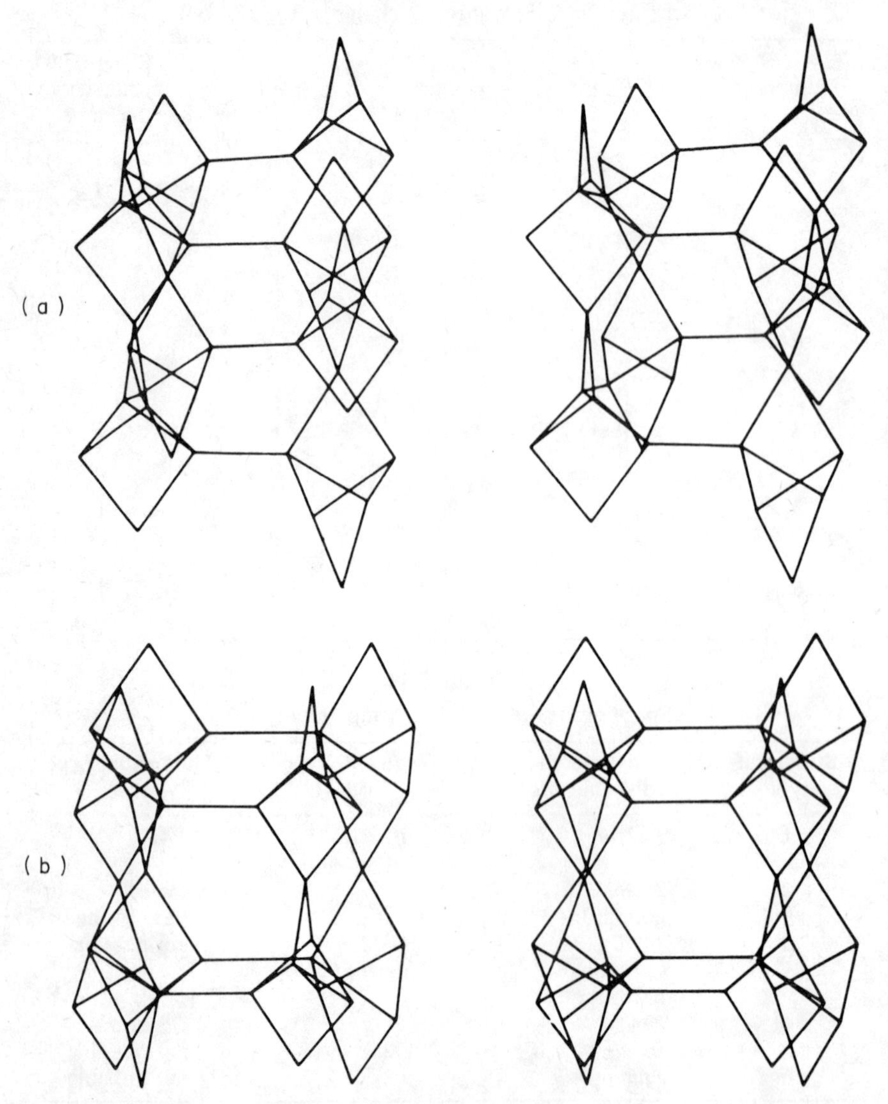

FIG. 22. Stereoscopic pairs showing the frameworks of (a) natrolite viewed along [110] and (b) thomsonite, viewed along [100].[4]

ratio Si : Al = 1 : 1 (Fig. 21). The latter implies that there must be a doubling of the chain period. If these possibilities of Si/Al ordering are combined with the six structures then 15 possibilities arise, which are summarized in Table 11. Ten refer to unknown phases, once more emphasizing areas for possible future discoveries, which will require single crystal measurements for proper identification of new fibrous zeolites. Stereoscopic pairs for natrolite, thomsonite and edingtonite are given in Fig. 22 (a and b) and Fig. 23 respectively.[4]

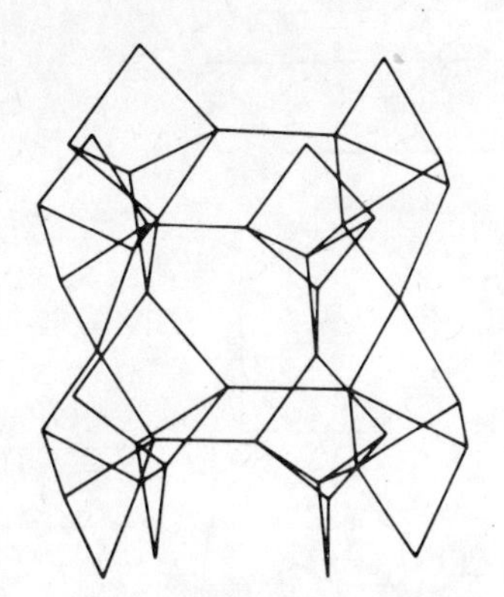
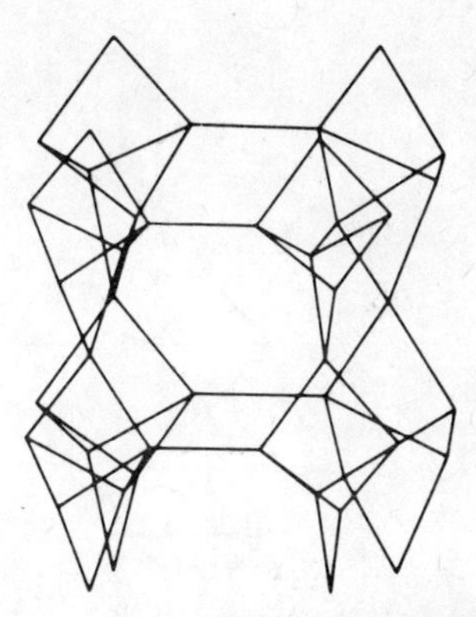

FIG. 23. Steoscopic pair showing the framework of edingtonite, viewed along [110].[4]

7. THE SIGMA TRANSFORMATION

Shoemaker *et al.*[114] have developed a further conceptual device for interrelating and building known and hypothetical zeolite structures, which they termed the "sigma transformation". This is an expansion of the tetrahedrally connected structure by imaginary fission of T (i.e. Si or Al) atoms lying on specified planes or surfaces running through the structure, and creation of new oxygen bridges connecting pairs resulting from the fission. The process is illustrated in Fig. 24, showing in the top half how a single tetrahedron by appropriate sigma transformations may yield 4-, 6- and 8-rings and 4-4-, 6-6- and 8-8-prisms. In the bottom half of the figure is shown the sigma transformation of the 14-hedron of Type I in three steps to give the 26-hedron of Type I. In addition to increasing the number of sides in a polygon and converting polygons to prisms (Fig. 24) the sigma transformation can convert strings to ladders. The fundamental requirement is that every T atom lying in the transformation plane must have two of its linkages lying in the plane and the other two emerging from opposite sides of the plane. An "inverse sigma transformation" can occur with a plane containing no T atoms if every oxygen bridge cut by the plane is a common edge of two 4-rings. This device may notionally reduce double rings (prisms) to single rings or ladders to strings.

The sigma transformation is a versatile method for generating new structures and for transforming one known structure into others. For example, starting with sodalite and with three successive transformations carried out on mirror planes, mutually at right angles and passing through the centre of the unit cell, zeolite *A* is generated. If these transformations are

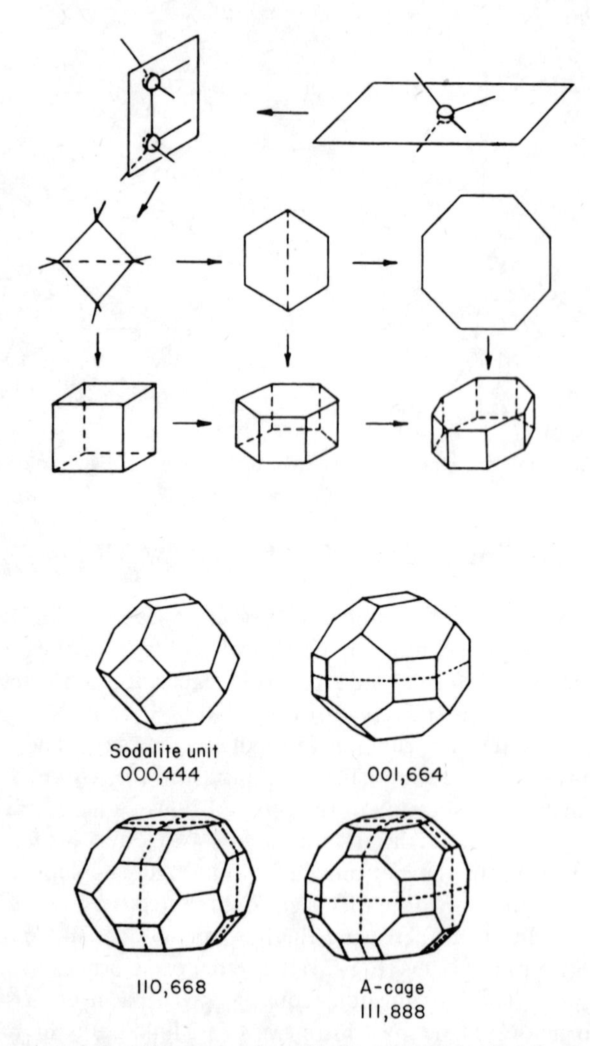

FIG. 24. Examples of the "sigma transformation". The top half of the figure illustrates the transformation of a single tetrahedron to yield 4-, 6- and 8-rings and 4-4-, 6-6- and 8-8- prisms. The bottom half shows stages in transforming a sodalite cage (14-hedron of Type I) to the 26-hedral cage of zeolite *A*.[114]

carried out one at a time, two intermediate structures are obtained: the first (after one such transformation) has a one-dimensional parallel channel system with 8-ring windows; and the second (after two transformations) has a two-dimensional rectangular net of such channels. Three more transformations, this time with three mirror planes passing through the unit cell corners, yield zeolite *RHO*, with two intermediate structures if the transformations occur one at a time.

If one considers in sodalite a set of sodalite cages (half those present) that are linked together in a diamond-like structure with shared 6-rings, and if, keeping these cages intact, sigma transformations are carried out on sets of planes containing these 6-rings for one diagonal direction at a time, then after four transformations involving three intermediate unknown structures one arrives at faujasite. The remaining half of the sodalite cages have been transformed by these operations into 26-hedra of Type II. Since the four sets of transformation planes are not mutually at right angles the "planes" for the second and third transformations must be mildly pleated.

Again starting with sodalite a series of no less than 12 other structures was generated. Still further transformations of sodalite yielded chabazite and two intermediate structures, while suitable transformations of cancrinite yielded offretite and then gmelinite. From tridymite the paracelsian and phillipsite topologies could be generated and from cristobalite the frameworks of Li-*ABW* and gismondine were obtained, as well as two unknown structures.[114]

8. CHANNEL PATTERNS AND DIMENSIONS

Sections 3 to 7 have indicated a number of the ways in which zeolite frameworks can be constructed and compared, which have shown large numbers of so far unknown framework topologies. From the viewpoint of sorption and catalysis other important aspects include the kinds and free dimensions of the channel systems which guest molecules must traverse in permeating the intracrystalline free volume.

If, starting from a given place in the crystal, the guest molecule may, through the channel system reach any other place within the crystal, then the channel system can be termed three-dimensional (3-D). If the guest at a given starting point is constrained to move only in a particular plane through the crystal the channel system can be termed two-dimensional (2-D); while if the guest starting from a given point can move only in one direction, the channel system is one-dimensional (1-D). Exactly the same definition may be adopted for diffusion of the exchangeable cations, although because these cations are often much smaller than guest molecules they can migrate in

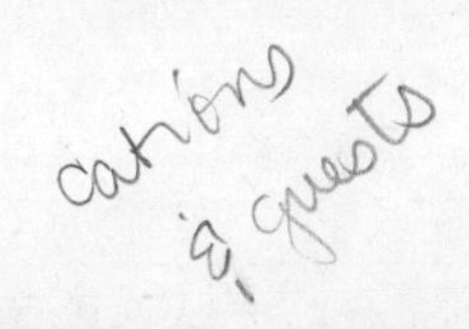

channel systems more restricted than those along which the guests can diffuse. Moreover, differences in molecular dimensions of guests may be such that for large guest species diffusion may be 1-D while for smaller ones it could be 2-D or 3-D in character. In cage structures such as chabazite, erionite, sodalite hydrate, zeolite *A* and faujasite the channel networks

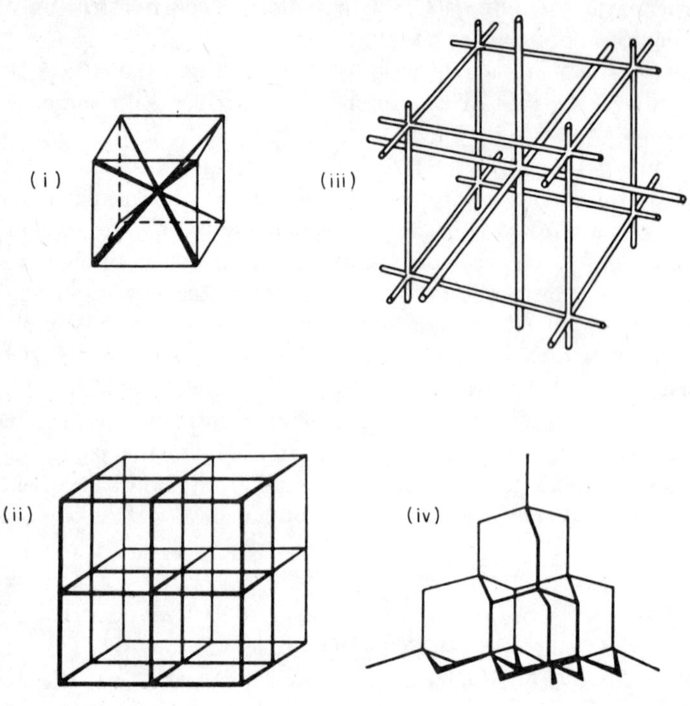

FIG. 25. Formal representations of the channel patterns in each of four different cubic zeolites: (i) Sodalite—from the centre of each 14-hedral sodalite cage eight directions give access to the centres of one each of eight other sodalite cages; (ii) Zeolite *A*—from the centre of each 26-hedron of Type I six directions lead to the centres of one each of six other identical 26-hedra; (iii) Zeolite *RHO*—there are two channel systems each of which has the symmetry of that in zeolite *A*. The two channel systems interpentrate but are not interconnected; (iv) Faujasite—from the centre of each 26-hedron of Type II four directions lead through 12-rings to the centres of one each of four other identical 26-hedra. The channel pattern has the symmetry of the bond pattern in the lattice of diamond.

allow 3-D diffusion of all guest molecules small enough to enter the main cavities. Channel patterns are illustrated formally in Figs 25 and 26 for a number of the above zeolites. In sodalite the access to the cavity is controlled by 6-rings; in chabazite, erionite and zeolites *A* and *ZK-5* the access to main cavities is through 8-rings; in faujasite it is through 12-rings; and in zeolite *RHO* there are two interpenetrating but non-intersecting 3-D channel systems. A guest molecule can diffuse in x, y or z directions in each but, if initially located in one of the two systems, cannot pass directly to the other. To do so it would have to leave the crystal and enter the other from the gas phase.

Two-dimensional diffusion of guest molecules is required by the channel patterns in heulandite, clinoptilolite and levynite. The pattern for levynite is shown in Fig. 26c; the guest molecule must move periodically up and down in following this 2-D channel system. Wide parallel channels occur in mordenite, dachiardite, zeolite *L*, offretite, mazzite, cancrinite hydrate and gmelinite. In some of these there may be narrow channels cross-linking the main ones. For example in gmelinite and in offretite the wide parallel channels are connected through gmelinite cages (14-hedra of Type II). Access into the main channels is through 12-rings; access between channels, via the gmelinite cages, is governed by 8-rings. Thus to large molecules the diffusion would be 1-D; for suitably small molecules in these two zeolites it is 3-D. In other parallel channel zeolites such as zeolite *L*, cancrinite hydrate or mazzite there are no openings between channels wide enough for molecule diffusion so that for all guest species 1-D diffusion only is possible.

At attempt has been made in Table 12 to summarize information regarding the nature and free dimensions of channel systems for various members of the zeolite groups. Channels to which access is controlled by 6-rings are not included since 6-rings allow only the smallest polar molecules such as water to enter, and then only very slowly at normal temperatures. In considering molecule diffusion and sieving, proper attention must be paid to a number of features. Firstly the free dimensions given in Table 12 refer to the hydrated zeolites; dehydration produces small but definite changes in unit cell dimensions and channel dimensions. These changes can be significant in such zeolites as those of the natrolite group, where chains are cross-linked with single bonds to other chains to yield relatively flexible networks. They are least where structure-forming blocks are cross-linked by sharing 4-, 6- or 8- ring faces; or by way of 4-, 6- or 8-ring prisms. This multiple bonding between blocks helps to give the relatively rigid anionic frameworks most desirable in zeolite sorbents and catalysts.

The free dimensions also refer to structurally ideal zeolites. It is relatively easy for stacking faults to occur in gmelinite for example, so that the *AB* . . . sequence of hexagonal prism layers may be interrupted by *ABC* . . . sequences as in chabazite. This, repeated several times, effectively breaks the wide channels along [001] and reduces access to them to 8-ring apertures. A natural sample of gmelinite was indeed found to sorb like chabazite rather than behaving as a wide pore zeolite.[115] Cancrinite hydrate can similarly be blocked by stacking faults involving sodalite sequences of layers of single 6-rings. It has been found to be a very poor sorbent, rather than behaving as an open pore zeolite.[116] On the other hand, offretite can be synthesized free of erionite-type stacking faults[117, 118] and behaves as an open pore

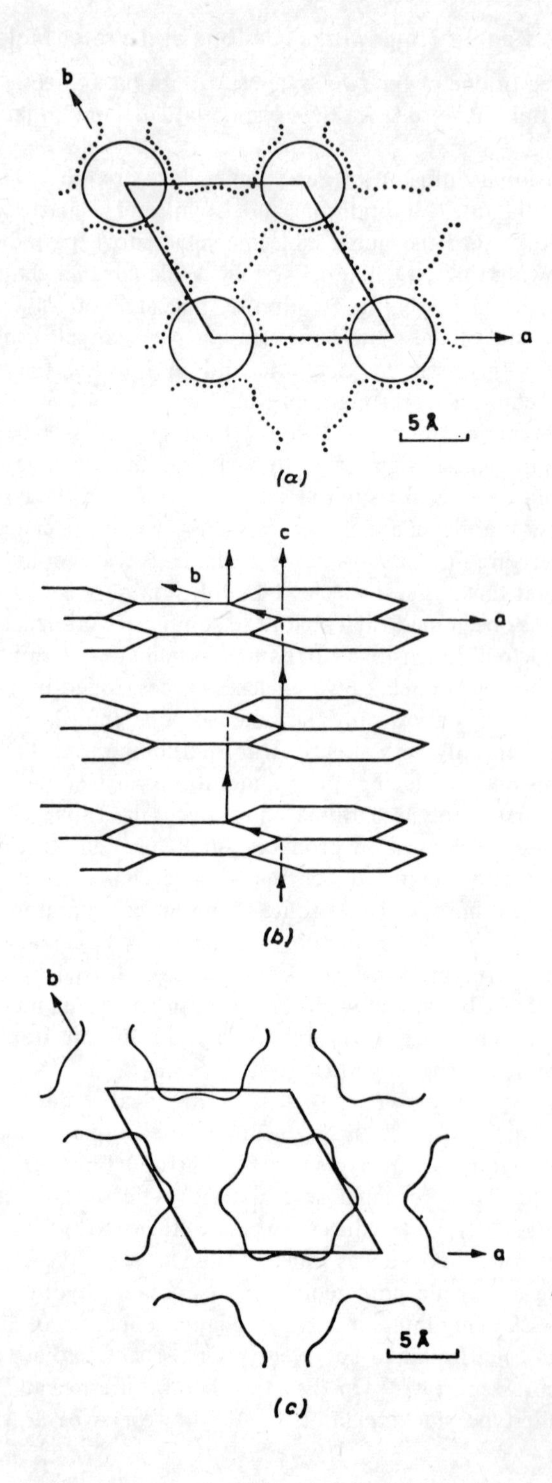
b
a
5 Å
(a)
c
b
a
(b)
b
a
5 Å
(c)

sorbent just as predicted from the free dimensions of the channels given in Table 12. The same is true of zeolite *L*.

Inclusion of impurities within the channel systems can be a further reason why a given zeolite sorbent fails to behave in the manner expected from the free dimensions of its channels. Discrete molecules of $NaAlO_2$, Na_2SiO_3, $Al(OH)_3$, $AlO(OH)$, salts or alkalis may be present in zeolite channels under certain circumstances, and, being relatively immobile, act as high energy barriers impeding or blocking altogether the passage of guest molecules (Chapter 6, § 7). This is a possible explanation of reported differences in openness[119] of some synthetic mordenites. There is also analytical evidence of the presence of discrete silicate anions in the cancrinite hydrate channels.[56, 120] Because of the 1-D character of molecule diffusion in these zeolites, sorption can be very sensitive to the presence of entrained impurities.

Finally, the positions, sizes and numbers of the exchangeable cations must be considered. In Na-mordenite there are Na^+ ions in the distorted 8-ring windows[36] which nominally could provide access between the parallel wide channels (Table 12). These ions block the 8-rings, thus effectively transforming the behaviour even for small molecules from that of a 2-D to that of a 1-D sieve. The Ca-forms of zeolite *A*,[121] chabazite[122] and zeolite *ZK-5*[123] behave as expected from the free dimensions of the framework; the Na- and K- forms do not: the cations evidently obstruct the 8-ring openings through which diffusing molecules must pass. The *H*-form of a synthetic mordenite admitted molecules,[124] including neopentane, expected from the channel free dimensions of Table 12, but Na-, K-, Ca- and Ba- forms of mordenite behaved as much less open sorbents.[125,126] H-clinoptilolite and dealuminated H-clinoptilolite admitted molecules of the expected dimensions but a natural (Na,Ca)-clinoptilolite did not.[79, 127] In all these examples one sees the important role played by cations in modifying molecule sieving and sorption by zeolites. On the other hand faujasite-type zeolites, zeolite *L*[128,129,130] and offretite[118] remain very open whether in H-, Na-, Ca-, K- or other forms involving simple inorganic ions.

Where cations, stacking faults and intercalated impurities do not obstruct the channels the free dimensions of these channels provide only a rough limit to the sizes of the molecules which can permeate the crystals. The guest molecules and framework oxygens are not hard spheres and the lattice

---

FIG. 26. Formal representation of channel patterns in several zeolites of the chabazite group: (a) Gmelinite—the wide channels run in the *c*-direction, and are shown in cross-section by the full lines surrounded by dashed lines. The gmelinite cages (outlined as dashed lines) serve via 8-ring windows to allow passage from one wide channel to another; (b) Erionite—each 23-hedron provides six 8-ring windows allowing molecules to pass from a central 23-hedron to any of six other 23-hedra (Table 5); (c) Levynite—a 17-hedron is linked via 8-ring windows to each of three other 17-hedra. Two-dimensional diffusion occurs.[48]

TABLE 12

Channel systems and free diameters of apertures governing access to channels, assuming radius of oxygen = 1·35Å[4]

Zeolite	Channels	Access apertures and free dimensions (Å)		
Natrolite	⊥ [001]	8-rings	2·6 × 3·9	
Thomsonite	⊥ [001]	8-rings	2·6 × 3·9	
Edingtonite	⊥ [001]	8-rings	3·5 × 3·9	
Heulandite	[100]	8-rings	4·0 × 5·5	interconnected
	[001]	10-rings	4·4 × 7·2	
Stilbite	[100]	10-rings	4·1 × 6·2	interconnected
	[001]	8-rings	2·7 × 5·7	
Brewsterite	[100]	8-rings	2·3 × 5·0	interconnected
	[001]	8-rings	2·7 × 4·1	
Phillipsite	[100]	8-rings	4·2 × 4·4	interconnected
	[010]	8-rings	2·8 × 4·8	
	[001]	8-rings	3·3	
Yugawaralite	[100]	8-rings	$3{\cdot}1_5$ × 3·5	interconnected
	[001]	8-rings	2·8 × 3·6[a]	
Gismondine	[100]	8-rings	3·1 × 4·4	interconnected
	[010]	8-rings	2·8 × 4·9	
Mordenite	[001]	12-rings	6·7 × 7·0	interconnected
	[010]	8-rings	2·9 × 5·7	
Ferrierite	[001]	10-rings	4·3 × 5·5	interconnected
	[010]	8-rings	3·4 × 4·8	
Dachiardite	[010]	10-rings	3·7 × 6·7	interconnected
	[001]	8-rings	3·6 × 4·8	
Epistilbite	[100]	10-rings	3·2 × 5·3	interconnected
	[001]	8-rings	3·7 × 4·4	
Chabazite	⊥ [001]	8-rings	3·6 × 3·7	
Gmelinite	[001]	12-rings	7·0	interconnected
	⊥ [001]	8-rings	3·6 × 3·9	
Erionite	⊥ [001]	8-rings	3·6 × 5·2	
Offretite	[001]	12-rings	6·4	interconnected
	⊥ [001]	8-rings	3·6 × 5·2	
Mazzite	[001]	12-rings	7·1	
Zeolite *L*	[001]	12-rings	7·1	
Cancrinite hydrate	[001]	12-rings	6·2	
Faujasite	⟨111⟩	12-rings	7·4	
Zeolite *A*	⟨100⟩	8-rings	4·1	
Zeolite *ZK-5*	⟨100⟩	8-rings	3·9	not interconnected
	⟨100⟩	8-rings	3·9	
Paulingite	⟨100⟩	8-rings	3·9	not interconnected
	⟨100⟩	8-rings	3·9	
Laumontite	[100]	10-rings	4·0 × 5·6	

[a] A small correction of the values given by Meier and Olsen.[4]

oxygens are also in a state of vibration which includes breathing frequencies of the windows. The most open and the most porous zeolites include faujasite and its variants zeolites X and Y, and its probable hexagonal polymorphs, zeolites ZSM-2[131] and ZSM-3.[60] The openness and estimated free distances in the channel system of faujasite is illustrated in projection in Fig. 27. Less porous zeolites with almost equally accessible channel systems to those in faujasite include mordenite, zeolite *L* and zeolite Ω.

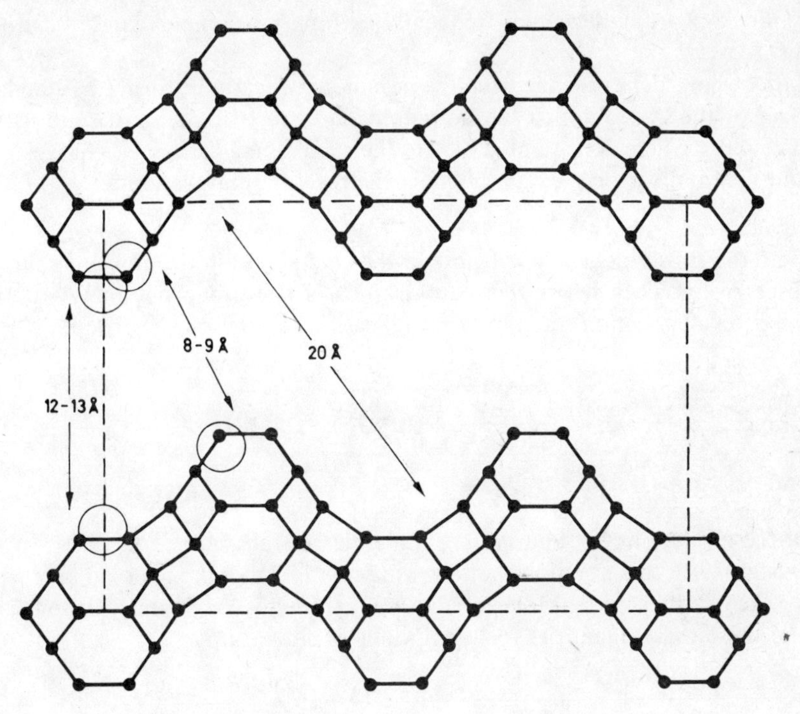

FIG. 27. The projection of Al + Si atoms in the sodalite cages of faujasite centred on the (011) plane. The free dimensions and effective continuity of the channel are shown.[103]

9. CATION POSITIONS

Zeolite frameworks usually provide more than one kind of site for the exchangeable cations. The number of electrochemical equivalents of cations needed to balance the anionic framework charge may also be considerably less than the total number of available cation sites of all kinds. Thus when the cations distribute themselves among the sites so as to minimize the free energy of the system there may be partial occupancy of some or all of the kinds of site available. The distribution equilibrium is expected to be a function of temperature, of the cationic species present and of the state of hydration or dehydration of the zeolite. If all water is removed and other guest molecules are introduced these guest species may also cause a re-distribution of some cations, especially if they are polar like NH_3 or SO_2. Another variable is the Si/Al ratio which determines the anionic charge per unit cell and so fixes the number of cations. As the Si/Al ratio changes so may the distribution of the cations among the kinds of site present in the zeolite. Accordingly zeolites frequently exemplify a complex kind of cationic disorder.

If we consider exchange from the homoionic A cationic form (the standard state of the A-zeolite) to the homoionic B-form (the standard state of the B-zeolite), where A^{ν_A+} and B^{ν_B+} are the exchanging cations of charges $\nu_A{}^+$ and $\nu_B{}^+$ and valences ν_A and ν_B then the reaction may be written

$$\nu_B A_c^{\nu_A+} + \nu_A B_s^{\nu_B+} \rightarrow \nu_B A_s^{\nu_A+} + \nu_A B_c^{\nu_B+}.$$

Here the subscripts c and s denote in the crystal and in the ambient solution respectively. Then where there are n kinds of site group one may write for the equivalent cation fraction, X_i of all cations on the ith site group

$$X_i = \frac{\nu_i^A m_i^A + \nu_B m_i^B}{\sum_{i=1}^{n} (\nu_A m_i^A + \nu_B m_i^B)}$$

and $\sum_{i=1}^{n} X_i = 1$.

In these relations m_i^A and m_i^B are molalities of ions A^{ν_A+} and B^{ν_B+} in the crystal. With this notation the following relationships can be established[132] for the overall standard free energy, $\Delta G^\ominus$, enthalpy, $\Delta H^\ominus$ and entropy $\Delta S^\ominus$, and these three quantities for the n kinds of site group:

$$\Delta G^\ominus = \sum_{i=1}^{n} X_i \Delta G_i^\ominus,$$

$$\Delta H^\ominus = \sum_{i=1}^{n} X_i \Delta H_i^\ominus,$$

$$\Delta S^\ominus = \sum_{i=1}^{n} X_i \Delta S_i^\ominus.$$

From the first of these it follows that the overall thermodynamic equilibrium constant, K, and those for the n kinds of site group are related by

$$K = \prod_{i=1}^{n} K_i^{X_i}.$$

Formally therefore, even in this complex situation, relationships exist which describe the overall cation exchange equilibria in terms of the equilibrium distributions over all kinds of site. Unfortunately the reverse process of finding the K_i from K is not possible without auxiliary information, for example, from X-ray structural studies which may establish the cation distributions.

On the X-ray crystallographic side, however, the fractional filling of sites and the relatively small numbers of cations compared with other atoms in the structure make the degree of filling of a given kind of site and even the kinds of site occupied difficult to determine. Despite this the study of cation siting has made considerable progress. Table 13 illustrates for some zeolites

TABLE 13

Some cation sites and occupancies in typical zeolites

Zeolite	Site	Coordinates x/a	y/b	z/c	Fraction occupied (%)
(a) Hydrated zeolites at room temperature					
Li-*ABW*[35]	Li1	0·3125	0·1970	−0·2600	64
	Li2	0·1850	0·2714	−0·2190	27
Stilbite[25]	Ca	0·2805	0	0·0949	100
	Na	0·5055	0·0659	0·0392	22
Stellerite[27]	Ca	0·500	0	0·2910	100
Barrerite[28]	C1	0·250	0	0·0417	72
	C1P	0·250	0	0·4558	61
	C2	0·0482	0·0624	0·0446	14
	C2P	0·0369	0·0634	0·4792	25
	C3	0·1611	0	0·2386	25
Mazzite[50]	(K, Ca, Na)	0·500	0	0	46
	Mg	0·333	0·667	0·25	100
	Ca	0	0	0·68	21
Levynite[49]	C1	0	0	0·1389	100
	C2	0	0	0·2782	28
	C3	0	0	0·4092	27
	C4	0	0	0·4498	15
	C5	0	0	0·500	25
Zeolite *A*[65]	Na	0·1064	0·1064	0·1064	100
(b) Dehydrated zeolites at room temperature					
Zeolite *A*[133]	Na1	0·200	0·200	0·200	100
	Na2	0	0·429	0·429	25
	Na3	0·204	0·204	0·500	8
Chabazite[134]	Ca1	0	0	0	60
	Ca2	0·169	0·169	0·169	35
Mazzite[52]	K	−0·773	0·1473	0·25	
	Mg	−0·333	0·333	0·1681	
	Ca	0	0	0·68	

both cation positions and the estimated degrees of occupancy. The sites given in Table 13, at least in the instances of hydrated zeolites *A*, stilbite and stellerite, are not the only kinds present, and the multiplicity of site groups sometimes involved is illustrated in barrerite and in levynite.

In the dehydrated zeolites of the table more than one kind of occupied site also appears. From the viewpoint of sorption the cation locations and degrees of occupancy of the dehydrated zeolite are of greater interest than those of the hydrated form. At least for rather weakly sorbed non-polar species the cations in the water-free sorbents are not likely to change position. Cation

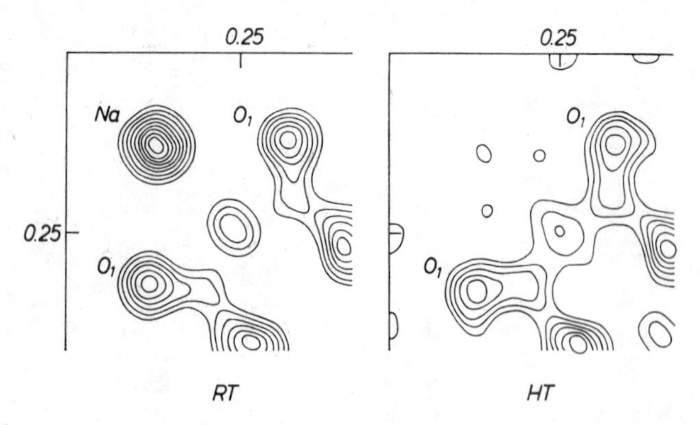

FIG. 28. Fourier sections in zeolite *ZK-5* at $Z = 0 \cdot 13$ at room temperature (*RT*) and at 150°C (*HT*). Contrours start at around 4e/Å^3 and are at equal but arbitrary intervals.[67]

resiting due to dehydration is accompanied by small shifts in the positions of the framework atoms, as exemplified for hydrated[43] and anhydrous chabazites.[134] The displacement of a Na^+ ion in zeolite *ZK-5* when the crystals were partially dehydrated by heating at 150°C is clearly seen in Fig. 28,[3, 67] where electron density contours are shown at room temperature and at 150°C.

Coordination numbers of cations and bond lengths of cations to framework oxygens and water molecules are illustrated in Table 14 for several zeolites. Because bond lengths vary in the coordination shell the concept of a coordination number is not precise. Also the number may be lower than that given if, when some of the water positions are occupied, others, for lack of space, are necessarily unoccupied. For example it is not possible that Mg^{2+} in hydrated mazzite can have 11 water molecules around it all at the same time and at the close distances given in column 5. Mg^{2+} and Ca^{2+} in zeolites with wide channels or voids may be solvated by water only (ferrierite and mazzite); but mostly the solvation is shared by water and framework oxygens. In dehydrated zeolites solvation, except for residual zeolitic water, is necessarily through anionic oxygens of the framework.

TABLE 14

Coordination numbers (CN) of ions, and ion-oxygen, ion-water distances in some zeolites

(a) Hydrated zeolites

Ion	Zeolite	CN	Distances (Å) To framework oxygen	Distances (Å) To water
Li	Li-*ABW*[35]	$8(6 \times O, 2 \times H_2O)$	1·81, 1·93, 2·03, 2·05, 2·20, 2·38	2·06, 2·18
Na	Zeolite *A*[65]	$4(3 \times O, 1 \times H_2O)$	2·36	2·28
	Stilbite[25]	$5(2 \times O, 3 \times H_2O)$	2·46, 2·63	2·53, 2·25
	Natrolite[16]	$6(4 \times O, 2 \times H_2O)$	2·36, 2·39, 2·51, 2·61	2·36, 2·40
K	Phillipsite[29]	$12(8 \times O, 4 \times H_2O)$	2·941, 2·985, 3·353, 3·482, 3·505	3·088, 3·110, 3·287
Mg	Ferrierite[37]	$6(6 \times H_2O)$	–	1·96, 2·00, 2·09
	Mazzite[50]	$11(11 \times H_2O)$	–	1·97, 2·03, 2·06
Ca	Mazzite[50]	$6(6 \times H_2O)$	–	2·50, 2·64
	Yugawaralite[34]	$8(4 \times O, 4 \times H_2O)$	2·449, 2·486, 2·546, 2·564	2·414, 2·435, 2·474, 2·530
Sr	Brewsterite[24]	$9(4 \times O, 5 \times H_2O)$	2·85, 2·87	2·66, 2·68, 2·71, 2·81
Ba	Harmotome[29]	$10(6 \times O, 4 \times H_2O)$	2·899, 3·007, 3·089, 3·286, 3·678	2·826, 2·861, 2·890
(Ca, Na)	Heulandite[21]	$5(2 \times O, 3 \times H_2O)$	2·65	2·45, 2·47, 2·49
		$6(1 \times O, 5 \times H_2O)$	2·59	2·40, 2·56, 2·62
	Phillipsite[29]	$7(3 \times O, 4 \times H_2O)$	2·510, 2·584, 2·607	2·203, 2·298, 2·468, 2·632
(K, Na, Ca)	Mazzite[50]	$8(6 \times O, 2 \times H_2O)$	2·89, 3·07	2·82

(b) Dehydrated zeolites

Ion	Zeolite	CN	Distances (Å) To framework oxygen	Distances (Å) To water
Na	Zeolite *A*[133]	$4(4 \times O)$	2·47, 2·51	–
		$6(6 \times O)$	2·32, 2·90	–
		$8(8 \times O)$	2·40, 2·64	–
K	Mazzite[52]	$4(4 \times O)$	2·603, 2·776	–
Mg	Mazzite[52]	$4(3 \times O, 1 \times H_2O)$	2·286	2·385
Ca	Chabazite[134]	$6(6 \times O)$	2·38	–
		$6(6 \times O)$	2·37, 2·84	–

10. CATIONS IN ZEOLITE *A* AND IN FAUJASITE-TYPE ZEOLITES

By far the largest number of investigations has been devoted to zeolite *A* and to faujasite and its variants, zeolites X and Y. These studies give further information about the role of intrazeolitic cations. Table 13 gives only one position for Na^+ in zeolite *A*, although other positions which may be occupied by either water or Na^+ were found by Gramlich and Meier.[65] Possible sites in this zeolite (comprising cubic or 4–4 units, sodalite cages, and 26-hedra, Type I) may be codified as follows.

In the 8-ring	S1	3 per pseudo unit cell
In the 6-ring	S2	8 per pseudo unit cell
In the sodalite cage but adjacent to the 6-ring	S2′	8 per pseudo unit cell
In the 26-hedra, adjacent to the 6-ring	S2*	8 per pseudo unit cell
Against the 4-rings	S3	12 per pseudo unit cell
Centre of sodalite cages	SU	1 per pseudo unit cell
Centre of the 26-hedra	S4	1 per pseudo unit cell

In faujasite (composed of hexagonal prisms, sodalite cages, and 26-hedra of Type II) the cation sites are codified as below.

Centre of prisms	I	16 per unit cell
In 14-hedra, adjacent to hexagonal prisms	I′	32 per unit cell
Centre of sodalite cages	U	8 per unit cell
In 6-rings linking sodalite cages and 26-hedra	II	32 per unit cell
Near 6-rings of sites II but inside sodalite cage	II′	32 per unit cell
Near 6-rings of sites II but in 26-hedron	II*	32 per unit cell
Against 4-rings of "ribs" of 26-hedron	III	48 per unit cell
Centre of 26-hedron	IV	8 per unit cell
In 12-rings of 26-hedron	V	16 per unit cell

Studies made of cation sites in zeolite *A* have been concerned, among others, with the K-form,[135] an eleven twelfths exchanged Rb-*A*,[136] a seven twelfths exchanged Cs-*A*[137], the Tl-form[138] and mixed (Ag, Ca)-forms.[139] In dehydrated K-*A* three K^+ ions per pseudo cell were reported near the centres of 8-rings (i.e. on Sl sites) and six equivalent K^+ lay on three of the four 3-fold axes opposite 6-rings in the large cavity (S2* sites). The remaining three K^+ were non-equivalent, located along the fourth 3-fold axis, two inside the sodalite cage (S2′ sites) and one well out in the large cavity (S4). In hydrated K-*A* it was reported that each sodalite cage contained eight water molecules. However this would seem physically impossible because eight waters, even if heavily compressed, occupy more free volume than a sodalite cage provides. In sodalite hydrate this cage is filled by four H_2O. It could be that of eight possible sites only alternate ones are occupied.

The S4 site was also occupied by a Rb^+ ion in dehydrated $Rb_{11}Na$-*A*, confirming the zero coordination of the twelfth K^+ in K_{12}-*A*. Three more Rb^+ were at Sl, five on S2* displaced 1·25Å from the plane of the O_{III} oxygens of the 6-ring, and two were on S2′, 1·69 and 0·99Å respectively from the (111) plane of the O_{III} oxygens. Both the K^+ and the Rb^+ at S4 are ≈4·3Å distant from the nearest framework oxygens, indeed the nearest neighbours are other cations. Both ions at S4 have large thermal ellipsoids indicative of a rather flat potential energy minimum.

The Cs_7Na_5-*A* in both hydrated and anhydrous forms was reported to have three Cs^+ per pseudo cell at Sl, three on S2* and one on S2′. In each structure four Na^+ were assigned to S2. The remaining Na^+ and the H_2O molecules were not located. In Tl_{12}-*A* Tl^+ ions were located on Sl, S2′ and S2* in both the hydrated and anhydrous states. In Rb-, Cs- and Tl-forms the ions are too large to be present in S2 positions.

In faujasite type zeolites, positions and degrees of occupancy of sites have been summarized, up to 1970, by Smith(59) This information, supplemented by some later results, is given in Tables 15 and 16 for a number of dehydrated, partially dehydrated and hydrated exchange forms. The results for a given ion are not always consistent as between different samples and authors. In part this may reflect different histories of samples and different Si/Al ratios, but in part it reflects difficulties in locating non-framework atoms. Not all the cations present are always accounted for on the sites referred to in the table as is particularly the case for K^+ in the forms of K-Y and K-X with 69·8 and 86·5 Al per unit cell. Table 15 shows a strong tendency of the exchange cations to congregate in or against the more condensed parts of the framework. In the Cu-faujasite in addition to the locations given Cu^{2+} was reported at sites III.

In the hydrated faujasite-type zeolites of Table 16 a first difference from Table 15 arises from the presence of multivalent ions in the 12-ring sites, V. These sites are vacated by the ions when the water is removed. In forms containing a considerable number of monovalent ions many of these have not been located. Among the K-forms, as the Al and therefore the K contents increase the sites I and II tend to fill and I′ to empty. When the K^+ locations in the hydrated zeolites of Table 16 are compared with those of the corresponding water-free K-faujasites of Table 15 it is seen that sites I, I′ and II are always more heavily populated in the dehydrated crystals, so that in hydrated faujasites more K^+ ions are present in unspecified positions in the 26-hedra.

Numerous studies have been made of the environment and nature of transition metal ions in zeolites. In hydrated $(Co^{II}_{0\cdot33}\ Na_{0\cdot33})$-*A*, Co^{2+} ions were reported on SU at the centre of the sodalite cages, octahedrally co-ordinated to water molecules.[157] The $Co^{2+}—H_2O$ centre to centre distance was given as 2·11Å, as in $CoSO_4$, $6H_2O$. The other three Co^{2+} ions per unit cell were on S2* sites, on the threefold axes. Two Na^+ were on S2 and one on S2*, and the fourth Na^+ was considered to be well out in the 26-hedral

TABLE 15

Reported occupancies of some sites in faujasite-type zeolites

(a) Strongly dehydrated

Zeolite	Al per unit cell	Sites I	I′	II′	II
Na-Y[140]	57	7·8 Na	20·2 Na		31·2 Na
Ag-Y[140]	57	16 Ag	11 Ag		29·2 Ag
K-Y[140]	57	12·0 K	14·6 K		31·0 K
K-Y[141]	48·2	6·4 K	14·1 K		26·1 K
K-Y[141]	54·7	5·4 K	18·1 K		26·8 K
K-Y[141]	69·8	9·4 K	16·6 K		13·6 K
K-X[141]	86·5	9·2 K	13·6 K		25·6 K
K-faujasite[59]		8·6 K	12·9 K		31·7 K
Ba-faujasite[59]		7·3 Ba	5·0 Ba		11·3 Ba
Ca-faujasite[142]	57	14·2 Ca	2·6 Ca		11·4 Ca
Cu-faujasite[143]	56	1·5 Cu	14·2 Cu	0·8 Cu	5·3 Cu
Ni-faujasite[144]		10·6 Ni	{3·2 Ni, 5·8 H_2O}	{1·9 Ni, 1·9 H_2O}	6·4 Ni
La-faujasite[145]		11·8 La	2·5 La		1·5 La
La-faujasite (420°C)[146]		11·7 La	2·5 La		1·4 La

(b) Partially dehydrated

Zeolite	Al per unit cell	I	I′	II′	II
(Na, Ce^{III})-faujasite[147]	59	3·4 Na	11·5 Ce	16 H_2O	10·7 Na
La-Y (725°C)[148]	55	5·2 La	8·9 La		5·5 La
La-X[147]	86·8		30 La	32 H_2O	
La-X (425°C)[149]		5·0 La	15·2 La		4·9 La
La-X (735°C)[149]		5·2 La	14·1 La		6·3 La
Ca-X[150]		7·5 Ca	17·3 Ca	{9 Ca, 10·5 H_2O}	17·3 Ca
Sr-X (16 hr at 400°C)[150]		11·2 Sr	7·0 Sr	{4·2 Sr, 5·4 H_2O}	19·5 Sr
Sr-X (16 hr at 680°C)[150]		6·1 Sr	12·0 Sr	6·4 Sr, 7·7 H_2O	20·3 Sr

TABLE 16

Reported occupancies of some cation sites in hydrated faujasite-type zeolites

Zeolite	Al per unit cell	Sites I	I′	II′	II	II*	V
(Na, Ca)-faujasite[58]	57·6		16(Na, Ca)	32 H_2O		11 H_2O	
Ca-faujasite[151]			9·7 Ca	11·5 Ca		23 H_2O	2·2 Ca
La-faujasite[152]			3·3 La	28 H_2O		14 H_2O	10·3 La
(Ce^{III}, Na, Ca)-faujasite[153]	59		18 Na	32 H_2O		26 H_2O	5·8 Ce
La-X[147]	86·8		12 La	32 H_2O	17 La		4 La
Na-X[154]		9 Na	{8 Na, 11 H_2O}	26 H_2O	{24 Na, 8 H_2O}		
$Sr_{42}Na$-X[155]	85	2·1 Sr	11·1 Sr	32 H_2O	15 Sr		
$Sr_{30}Na_{24}$-X[155]	85	12 Na	7·3 Sr	26 H_2O	11·5 Sr		
K-Y[156]	48·2		13·6 K		17·8 K		
K-Y[156]	54·7	1·3 K	13·3 K		20·0 K		
K-Y[156]	69·8	7·0 K	12·0 K		24·3 K		
K-X[156]	86·5	8·9 K	7·2 K		23·2 K		

cavity. The Co^{2+} ions on S2* had four nearest oxygen neighbours at 2·36Å, but were 2·86 and 3·31Å distant from framework oxygens. It was suggested that they satisfy their coordination by breaking a framework bond and generating Bronsted acidity

$$\rangle Al - O - Si \langle + [Co(H_2O)_n]^{2+} \rightarrow \rangle Al \quad \overset{HO}{\diagdown} Si \langle + [Co(OH)H_2O_{(n-1)}]^{+}.$$

In hydrated zeolite *A* containing Mn^{2+} and Zn^{2+} these ions are also found on S2* sites,[158, 159] but are much closer to the 6-rings. The $Mn^{2+} - O$ distance for the three nearest framework oxygens is 2·28Å and the $Zn^{2+} - O$ distance is 2·24Å. Mn^{2+} favours a trigonal-bipyramidal coordination number of five (three framework oxygens of the 6-ring and two water molecules, one in the 26-hedron and one in the sodalite cage) and Zn^{2+} a tetrahedral one (three framework oxygens and one water molecule).

In fully hydrated Co^{II}-*A* the electronic spectrum matches that of $Co(H_2O)_6^{2+}$ except for a hypsochromic shift of $\approx$650 cm^{-1}.[160] Six coordinated cobalt ions may be in the 26-hedra as well as in the previously assigned[157] location in the sodalite cages. As the crystals are dehydrated, Co^{2+} moves nearer to the walls. Partially dehydrated $Na_{0\cdot83}[Co^{II}-H_2O]_{0\cdot0\,83}$–*A* has the intense blue and the electronic spectrum and magnetic susceptibility typical of tetrahedral Co^{II}.[161]

The Ni^{2+} ion in a fully hydrated $Ni^{II}_{1\cdot7}Na_{8\cdot6}$–*A* had, like the Co^{II}-enriched form of zeolite *A*, an electronic spectrum which, except for a hypsochromic shift of 500 cm^{-1}, matched that of octahedrally coordinated Ni^{2+}.[162] Complete dehydration under vacuum at 350°C gave a yellow product the electronic spectrum of which was appropriate for Ni^{2+} stabilized, like Cr^{2+},[163] in planar or near-planar trigonal coordination by the three nearest oxygens of the 6-ring (sites S2). In anhydrous zeolite Y, Ni^{2+} showed a strong but not total preference for site I, with the six nearest oxygens of the hexagonal prism in approximately octahedral coordination.[164, 165] The same preference was shown in natural faujasite, but the degree of occupancy of I depended critically upon the extent of hydration.[144] Residual water, or else NH_3 and to a lesser extent NO, could draw the Ni^{II} to sites I′ or II′.[164, 166] Ni^{II}_{15}-X dehydrated at 350°C was reported to be beige yellow, but became rose-mauve after sorbing pyridine, with an infra-red band characteristic of pyridine coordinated to Ni^{II}.[167] Since pyridine is too large to enter sodalite cages and does not cause appreciable Ni^{II} migration in anhydrous zeolite Y,[164, 166] at least some Ni^{2+} ions in zeolite X may occupy sites II′ or II* in trigonal coordination with the three nearest oxygens of the adjacent 6-ring. Support for this view comes from the observation that anhydrous Ni^{II}-*A* and Ni^{II}-X have the same colour[168] although in anhydrous Ni^{II}-*A* octahedral coordination is impossible. The rose-mauve colour with pyridine is consistent with 4-coordination of Ni^{II} (three framework oxygens and one pyridine).

In anhydrous zeolites X and Y Cu^{II} showed a strong preference for trigonal coordination in I′ (zeolite Y) or in I′ and II′ (zeolite X). On the other hand fully hydrated $Cu^{II}_{(1\text{-}2)}$-Y gave the electronic spectrum characteristic of $Cu(H_2O)_6^{2+}$.[169] Allowing for the usual hypsochromic shift ($\approx$1500 cm^{-1}) of a fully hydrated Cr^{II}-*A*, the electronic spectrum was also that characteristic of $Cr(H_2O)_6^{2+}$.[170] Thus a part of the pattern of behaviour for divalent Co, Cu, Ni and Cr is very similar. The presence of Cu^{2+} ions in the large cages of hydrated (Na, Cu)-Y was confirmed by X-ray study of three samples containing respectively 7, 12 and 15 Cu^{2+} ions per unit cell.[171] Sites II and III were favoured and at high copper content part of the Cu^{2+} was considered to be unlocalized in the 26-hedra of Type II.

Two chemical states of iron in $Fe^{II}_{6\text{-}13}$-Y have been identified by Mossbauer spectroscopy and are attributed to Fe^{II} in sites I and to Fe^{II} in a tetrahedral environment.[172] The tetrahedral environment could arise for Fe^{II} bonded to the nearest three oxygens of a 6-ring and to one H_2O or OH ion, i.e. $Ox_3Fe(H_2O)^{2+}$ or $Ox_3Fe(OH)^{+}$. It is suggested that this latter ion either remains stable at high temperatures or undergoes further reaction with like ions:

$$2Ox_3Fe(OH)^+ \rightarrow Ox_3Fe\text{-}O\text{-}FeOx_3^{2+} + H_2O.$$

Such bridging ions are similar to bridging ions also postulated for oxidized Cr^{II}-X and Cr^{II}-Y.(173) They would for example be expected to form between ions on I′ and II′ in sodalite cages.

Among trivalent ions in zeolites there have been investigations of La, Ce, Cr and Ti. The siting of La^{3+} and Ce^{3+} has been indicated in Tables 15 and 16. It is interesting that ions notionally of high charge appear to prefer sites I′ and V to sites I where octahedral coordination to the six nearest framework oxygens could occur, with some charge neutralization also by the remaining six oxygens of the hexagonal prism. It has however been proposed that rare earth ions in dehydrated or partially dehydrated faujasites form bridged structures within sodalite cages in the manner shown in the stereo pair of Fig. 29.(147) In this way each rare earth ion in sites I′ achieves

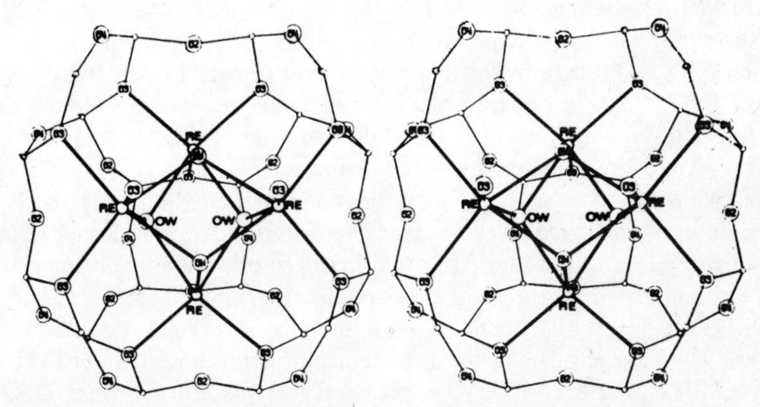

FIG. 29. Steoscopic pair showing the rare earth (RE) + sodalite cage + water oxygen (Ow) complex in partially dehydrated faujasite-type zeolite.(147)

coordination with the three nearest oxygens (Ox) of the adjacent 6-ring and with the oxygens (Ow) of several water molecules. In the case of Ce^{III} in dehydrated Ce^{III}-faujasite the Ce^{III}-Ox and Ce^{III}-Ow distances are respectively 2·52Å and 2·44Å. In dehydrated La-X each La in I′ has the La-Ox distance of 2·5Å as it is also attached to three Ow at the same distance. In the La-X sample there were ≈3·8 La^{3+} ions in each sodalite cage. The arrangement indicated in Fig. 29, satisfactorily utilizes the geometry of the 17-hedral cavity and the space within this cavity. The Ce^{3+} ions are about 4Å apart and are screened from each other by the bridging Ow; a like situation exists with La^{3+} ions which are similar in size to Ce^{3+}.

In fully hydrated Cr_6^{III}-X and Cr_{10}^{III}-Y the electronic spectra were characteristic of $Cr(H_2O)_6^{3+}$. After vacuum dehydration at 300°C the spectrum was unchanged apart from a hypsochromic shift of ≈2000 cm^{-1}.(174) The outgassing should have removed most of the zeolitic water so that one possibility is that Cr^{III} on dehydration migrated into sites I and so retained octahedral coordination. However migration to sites I′ with bridging Ow could achieve a similar result (cf. Fig. 29). Zeolite Y has also been exchanged with Ti^{III} under oxygen-free conditions to give $Ti^{III}_{4\cdot2}$-Y and Ti^{III}_{10}-Y.(175) At room temperature these forms then sorbed oxygen which it was suggested gave Ti^{IV}-O_2^-. It was proposed that oxygen does not enter sodalite cages so that Ti^{IV} was coordinated to the three nearest oxygens of a 3-ring at II′, II or II* and an O^{2-} ion. However it is by no means certain that oxygen at room temperature could not slowly gain access to sodalite cages.

When two kinds of cation compete for the available kinds of site group it is of interest to know the relative site group populations as functions of cationic composition; and for a given cationic composition, including homoionic forms, to know how cation distributions change with temperature.

Information relating to the first of these questions has been obtained by measuring uptakes, under standard conditions, of C_2H_6 in dehydrated zeolite *A*(176) as a function of extent of exchange of Na^+ by K^+ and by Ag^+; and of C_3H_8(177) as a function of extent of exchange of Ca^{2+} by the same two ions. In (K, Na)-*A* ethane uptake declined rapidly and was almost zero for less than 30% exchange by K^+. On the other hand (Ag, Na)-*A* sorbed ethane with little change in capacity for more than 80% exchange by Ag^+. These observations were taken to mean that entering K^+ first displaced Na^+ from 8-ring S1 sites; but that Ag^+ went firstly to other sites. This interpretation was supported by the similar behaviour of propane sorbed in (K, Ca)- and (Ag, Ca)-forms of zeolite *A*, as well as by X-ray powder diffraction studies. The change of intensity in certain diffraction lines chosen to correspond with cation-bearing crystal planes ((111) plane for 6-ring S2 sites; (110) plane for S3 sites against 4-rings; and (100) plane for 8-ring S1 sites). Thus in (Ag, Na)-*A* for instance as exchange proceeded, the intensity for (100) changed little up to 80% exchange of Na^+ by Ag^+, but the intensities for (111) and (110) decreased rapidly from the beginning of exchange.

Some examples of the effects of heating on cation distributions have already been referred to in §9 and in this section. Heat treatments cause some or all of the zeolitic water to be lost and this as well as the temperature can change the cation distributions. An alternative experiment would consist in holding an outgassed zeolite of known cationic composition at each of a series of temperatures and determining the cation distributions and relative populations in each site group at each temperature. However the way in which investigations have normally been conducted has been to give the zeolite the requisite heat treatment, cool it to room temperature and then investigate the cation distribution.

In one such study(165) samples of $Ni_{14}Na_{23}H_5$-Y were heated to 140, 200, 280, 300 and 600°C; samples of $Ni_{10}Na_{31}H_5$-Y to 200 and 600°C; and of

TABLE 17

Ion distributions in $Ni_{14}Na_{23}H_5$–Y as a function of temperature of heat treatment[a]

Treatment (°C)	Unit cell edge in Å	I	I′	I′	II′	II′	II
140	24·625	3·2 Ni^{2+} (0·0)	2·5 Ni^{2+} (0·058)	3·5 Ni^{2+} (0·078)	12·5 Ow (0·167)	1·9 Ni^{2+} (0·196)	23·5 Na^{+} (0·236)
200	24·615	4·1 Ni^{2+} (0 0)	3·6 Ni^{2+} (0·055)	2·9 Ni^{2+} (0·076)	9·5 Ow (0·166)	2·1 Ni^{2+} (0·209)	20·0 Na^{+} (0·237)
280	24·565	6·9 Ni^{2+} (0·0)	3·3 Ni^{2+} (0·053)	2·5 Ni^{2+} (0·078)	5·0 Ow (0·176)	1·5 Ni^{2+} (0·208)	20·0 Na^{+} (0·236)
300	24·51	10·0 Ni^{2+} (0·0)	1·5 Ni^{2+} (0·055)	1·3 Ni^{2+} (0·078)			23·2 Na^{+} (0·234)
600	24·47	11·7 Ni^{2+} (0·0)	1·1 Ni^{2+} (0·065)				25·9 Na^{+} (0·236)

[a] Figures in brackets are the $x = y = z$ coordinate of the ions and water oxygens (Ow).

$Ni_{19}Na_{15}H_3$-Y also to 200 and 600°C. The Ni^{2+} distribution was investigated by X-ray crystallography in each heat-treated sample. The progressive loss of water was accompanied by migration of Ni^{2+} into hexagonal prism sites, I, up to the number 12 Ni^{2+} ions per unit cell. The Ni^{2+} population in sites I increased very rapidly over the interval 200° to 300°C, in which most of the water loss took place. The migration of Ni^{2+} to I therefore is probably more a result of water loss than of temperature change. Nickel was also reported on other sites as the data in Table 17 show for the $Ni_{14}Na_{23}H_5$-Y samples, but the strong drift to sites I accompanied the progressive loss of water and increase of heating temperature. The unit cell edge was a linear function of the Ni^{2+} content of the sites I, in which the Ni^{2+} can achieve octahedral coordination with its six nearest oxygen neighbours. Throughout these changes in the Ni^{2+} populations the Na^+ ions in sites II remained relatively constant and accounted for nearly all the available Na^+. Thus the Ni^{2+} ions have almost undisputed occupancy of sites I, I′ and II′ while the Na^+ ions are selectively present in the 6-rings between sodalite cages and 26-hedra.

Barry and Lay[178,179] monitored Mn^{2+} locations in zeolites X and Y using electron spin resonance spectra. The original Na-forms were first enriched by exchange in various cations and then a little Mn^{2+} was introduced, also by exchange. The products were heated in dry nitrogen at selected temperatures. From the e.s.r. spectra five different centres for Mn^{2+} were detected which were designated as follows:

A Mn^{2+} in hydrated sites
I Mn^{2+} in sites I
II Mn^{2+} in II, I′ or possibly II′
(i) $MnOH^+$, probably in I′
(ii) MnO, probably in I′.

In the various parent forms of zeolite Y and in the heat-treated products the e.s.r. spectra then indicated the Mn^{2+} sitings of Table 18. The centres (i) and (ii) are considered to arise by hydrolysis:

$$Mn^{2+} + H_2O \rightarrow MnOH^+ + H^+ \rightarrow MnO + 2H^+.$$

The protons can then become attached to the framework by breaking Al–O–Si bonds. Oxygen-bearing Mn is expected to form in sodalite cages because these should retain residual zeolitic water longer than the 26-hedra. The hydrated state, A, of Mn^{2+} is normal in all parent cationic forms save that containing La^{3+}. The small amounts of Mn^{2+} present in the zeolites are not likely to perturb significantly the distributions of the main cations, about which, however, the e.s.r. method gave no information.

TABLE 18

Classification of Mn^{2+} spectra

Cations in zeolite Y	Parent hydrated	Treatment 200°C	300°C	400°C	600°C
Li, Mn	*A*	I	I	I	
Na, Mn	*A*	(i)	(i)	II	I
K, Mn	*A*	(i)	(i)	II, (i)	
Cs, Mn	*A*	I, (ii)	I, (ii)	I	
Mg, Mn	I, *A*	I, II	II	II	II
Ca, Mn	*A*	(ii)	II (ii)	II	II, I
Zn, Mn	*A*	I	I	I	
La, Mn	I	I	I	I	I, II

Pin Pah Lai and Rees[180] used radio-tracers and Szilard-Chalmers recoil to investigate cation siting in a series of ion exchanged and heat-treated forms of zeolites X and Y. They described sites in 26-hedra as "open" and those in 14-hedra and hexagonal prisms as "locked in". The elution behaviour

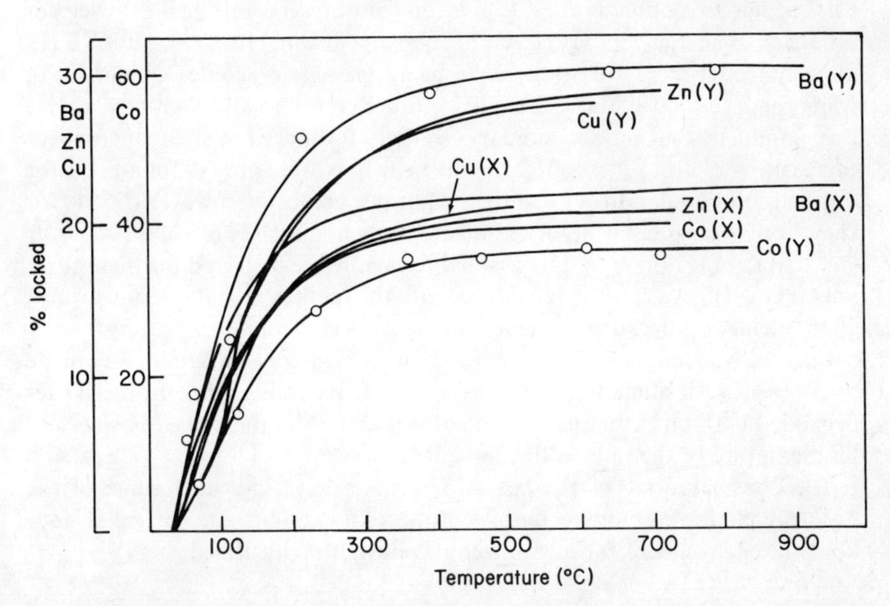

FIG. 30. The fraction (%) of divalent cation locked in sodalite cage and hexagonal prism sites of zeolites X and Y as functions of temperature of calcination.[180] Experimental points are shown on two of the curves to indicate the quality of the data.

was first studied with the radio-tracers. When the ions to be introduced by exchange were too large to enter the sodalite cages and hexagonal prisms the Na-form of the zeolite was the starting point. When the entering ions had access to these sites (e.g. Ag^+ and K^+) such sites were first blocked by replacing Na^+ by Ba^+ which became locked in the 14-hedra and prisms. The Ba-form then served to prepare the other forms. More than 98·5% of open site cations could then be eluted from the zeolite for Cs^+, Rb^+, K^+, Ba^{2+}, Co^{2+}, Zn^{2+} and Cu^{2+}. Elution of La^{3+} and Ag^+ was less complete. The fraction of cations in locked in sites increased as the temperature of calcination increased up to about 220°C and then levelled off, as shown in Fig. 30 for the divalent cations. The final levels tend to be established when most of the zeolitic water has been lost.

Following irradiation with thermal neutrons the Szilard-Chalmers recoil of Cs^+, Rb^+, Na^+, Ba^{2+}, Co^{2+}, Zn^{2+}, Cu^{2+} and La^{3+} from locked in sites in sodalite cages to open sites in 26-hedra had a probability of ≈90%; from hexagonal prism sites to 26-hedra this probability was between 40 and 50%. It was thus possible to determine the preferences shown as between prisms and 14-hedra as functions of the temperature of calcination, T_c, and of the concentrations and kinds of other cations also in these sites. The cations present in the 26-hedra before irradiation were usually NH_4^+, although Ba^{2+}, Ca^{2+} and Na^+ were also used in these sites. For T_c above 400°C Rb^+ and Cs^+ began to populate sites I, the temperature for this being lower for zeolite X than for zeolite Y. As T_c increased from 110°C to 220°C Ba^{2+} moved strongly to I, the preference being greater in zeolite X than in Y; while for $T_c > 400$°C Ba^{2+} occupied I in preference to all other sites. Co^{2+}, Zn^{2+} and Cu^{2+} all showed similar selectivity for locked in sites. In competition with Na^+ for $T_c > 500$°C the divalent ions were always found in I for divalent ion populations of less than eight per unit cell. However if NH_4^+ or Ca^{2+} ions were present in locked in sites only about 50% of Co^{2+}, Zn^{2+} or Cu^{2+} were in I. For $T_c = 110$°C sodalite cages were preferred for these three ions. The NH_4^+ was always introduced into the zeolites at room temperatures after calcination because of the instability of NH_4-zeolites to heating.

Finally, Na^+ either in trace amounts or in high concentrations was made to compete with numerous cations for sites I. Its ability to be in these sites depended both on the nature of the competing ions in the locked in sites and on the nature of the ions in the 26-hedra.

The considerations of the last two sections have indicated some of the progress made in studying mobile intrazeolitic cations. In particular four kinds of environment for ions in zeolites have been identified:

(i) Ions are complexed to or stabilized by frameworks oxygens only.
(ii) Ions are coordinated only by ligands such as water and are without direct contact with the walls of the cages and channels.

(iii) The coordination of ions is shared by framework oxygens and ligand molecules.
(iv) The ion is uncoordinated by any near neighbour of anionic character (e.g. one K^+ or Rb^+ in dehydrated K- and Rb-*A*).

It is also seen that the three nearest oxygens of 6-rings in anhydrous zeolites *A*, X and Y often provide trigonal coordination. This, especially for 2- and 3-valent ions, is in a coordinative sense highly unsaturated and offers sites to which residual water, or other ligands intentionally added, can attach themselves and so increase the coordination number.

11. BINDING ENERGY OF CATIONS IN ZEOLITES

It is possible, from heats of wetting of zeolites, from heats of exchange, and known values of hydration energies of gaseous ions, to evaluate[181] the heats, ΔH, of the reaction

$$BZ + (1/\nu)A^{\nu+}\text{ (gas)} \xleftarrow{\Delta H} (1/\nu)AZ_\nu + B^+\text{ (gas)}$$

where Z denotes the anhydrous zeolite framework, ν and ν^+ are valency and charge of ion $A^{\nu+}$, and the ion B^+ is the ion initially in the zeolite, which was chosen to be Na^+. ΔH follows from the thermochemical cycle

$$\begin{array}{cccc} BZ + (1/\nu)A^{\nu+}_{(gas)} & \xrightarrow{\Delta H} & (1/\nu)AZ_\nu + B^+_{(gas)} & \\ \Delta H^w_{BZ}\downarrow \quad (1/\nu)\Delta H^h_{A^{\nu+}}\downarrow & & (1/\nu)\Delta H^w_{AZ}\downarrow \quad \Delta H^h_{B^+}\downarrow & \\ BZ, n_B H_2O + (1/\nu)A^{\nu+}_{(aq)} & \xrightarrow{\Delta H^\ominus} & (1/\nu)[AZ_\nu], n_A H_2O + B^+_{aq} & \end{array}$$

Here ΔH^w_{BZ} and ΔH^w_{AZ} are respectively heats of wetting for the amount of zeolite containing one gram ion of B^+ and $A^{\nu+}$; $\Delta H^A_{B^+}$ and $\Delta H^h_{A^{\nu+}}$ are the heats of hydration of gaseous ions and $\Delta H^\ominus$ is the standard heat of complete exchange. Heats of wetting[182,183] and heats of exchange[184, 185, 186, 187] are known for a number of ion-zeolite systems and so are heats of hydration of gaseous ions.[188, 189] From the above thermochemical cycle one obtains

$$\Delta H = \Delta H^w_{BZ} - (1/\nu)\,\Delta H^w_{AZ} + (1/\nu)\,\Delta H^h_{A^{\nu+}} - \Delta H^h_{B^+} + \Delta H^\ominus$$

To obtain from ΔH the heats of solvation of individual gaseous cations by the zeolite one needs to know the heat of solvation of gaseous Na^+ in the zeolite. To estimate this heat the assumption was made that

$$\frac{\Delta H^s_{Na^+}}{\Delta H^h_{Na^+}} = \frac{(1/\nu)\Delta H^s_{A^{\nu+}}}{(1/\nu)\Delta H^h_{A^{\nu+}}}$$

This relation leads to

$$\frac{\Delta H^s_{Na^+}}{\Delta H^h_{Na^+}} = \frac{-\Delta H}{\Delta H^h_{Na^+} - (1/\nu)\Delta H^h_{A^{\nu+}}}$$

where all quantities save $\Delta H^s_{Na^+}$ are known, and so to $\Delta H^s_{Na^+}$. The reasonable consistency of the assumption involved was checked by choosing each of a series of ions $A^{\nu+}$, such as Li^+, Na^+, Rb^+ and Cs^+, and showing that for a given zeolite the resultant values of $\Delta H^s_{Na^+}$ did not vary very much *inter se.*[181]

For the heats $\Delta' H$ of the process

$$BZ, n_B H_2O + (1/\nu)A^{\nu+}_{(gas)} \xrightarrow{\Delta' H} (1/\nu)AZ_\nu, n_A H_2O + B^+_{(gas)}$$

one has

$$\Delta' H = (1/\nu)\Delta H^h_{A^{\nu+}} - \Delta H^h_{B^+} + \Delta H^\ominus = (1/\nu)\Delta' H^s_{A^{\nu+}} - \Delta' H^s_{B^+}$$

where as before the reference ion B^+ was taken to be Na^+. To obtain $\Delta' H^s_{Na+}$ the assumption was made that

$$\frac{\Delta' H^s_{Na^+}}{\Delta H^h_{Na^+}} = \frac{(1/\nu)\Delta' H^s_{A^{\nu+}}}{(1/\nu)\Delta H^h_{A^{\nu+}}}.$$

This, with the expression for $\Delta' H$, leads to

$$\frac{\Delta' H^s_{Na^+}}{\Delta H^h_{Na^+}} = 1 - \frac{\Delta H^\ominus}{\Delta H^h_{Na^+} - (1/\nu)\Delta H^h_{A^{\nu+}}}.$$

The consistency in values of $\Delta' H^s_{Na+}$, choosing $A^{\nu+}$ as Li^+, K^+, Rb^+ and Cs^+, was again reasonable in each of several zeolites.

Values of $\Delta H^s_{A^{\nu+}}$ and $\Delta' H^s_{A^{\nu+}}$ are given for a number of ions and zeolites in Tables 19 and 20, using mean values for $\Delta H^s_{Na^+}$ and $\Delta' H^s_{Na^+}$ for evaluating solvation heats for $A^{\nu+}$ ions. For the alkali metal ions the binding energies are in the sequence, in each zeolite

$$Li^+ > Na^+ > K^+ > Rb^+ > Cs^+$$

and the values of $\Delta H^s_{A^+}$ are less negative than $\Delta' H^s_{A^+}$ for each ion A^+. Thus, irrespective of the resiting of cations often associated with removal of water from a zeolite, the ions are more exothermally solvated in the hydrated forms of the zeolites. It is also seen however that solvation is never so exothermal in the zeolites as it is in pure water. For the four zeolites in the table the sequences of exothermicities are:

Hydrated: water > zeolite Y > zeolite X > zeolite *A* > chabazite.
Dehydrated: water > zeolite *A* > zeolite Y > zeolite X > chabazite.

TABLE 19

Heats of solvation of gaseous ions by zeolites[182] (kJ $(equiv)^{-1}$)

Cation	Water	Heat in							
		Chabazite		Zeolite X		Zeolite Y		Zeolite *A*	
		Hydrated	Dehydrated	Hydrated	Dehydrated	Hydrated	Dehydrated	Hydrated	Dehydrated
Li^+	−552	−493	–	−512	−466	−516	−489	−496	−491
Na^+	−442	−396	−286	−412	−374	−415	−393	−398	−396
K^+	−358	−321	−232	−333	−303	−335	−318	−322	−319
Rb^+	−334	−298	−216	−310	−282	−312	−296	−300	–
Cs^+	−301	−269	−194	−279	−255	−282	−267	−270	−268

TABLE 20

Heats of solvation of gaseous ions by anhydrous zeolites[83] (kJ $(equiv)^{-1}$)

Cation	Heat[a] in				
	Water[100]	Zeolite X	Zeolite Y	Zeolite *A*	Zeolite *L*
Tl^+	−342·7	–	−322 (50·6)	–	–
Ag^+	−489·0	–	−460 (99·7)	−480 (100)	–
Mg^{2+}	−976·0	−968 (61·7)	−918 (68·8)	−959 (48.1)	–
Mn^{2+}	−938·4	−930 (79·1)	−883 (72·1)	–	–
Co^{2+}	−1042·9	−1034 (74·2)	−981 (71·1)	−1025 (81·0)	−1010 (31·4)
Ni^{2+}	−1067·9	−1059 (70·0)	−1005 (71·8)	−1050 (68·8)	−1035 (27·3)
Cu^{2+}	−1063·8	−1055 (26·1)	−1001 (79·3)	−1046 (30·8)	−1031 (41·3)
Zn^{2+}	−1036·6	−1028 (80·8)	−975 (78·2)	−1019 (88·5)	–
Cd^{2+}	−917·5	−910 (88·7)	−863 (83·0)	−902 (96·2)	−889 (31·1)
Cr^{3+}	−2053·2	−2036 (28·4)	−1933 (24·7)	−2019 (22·5)	–
Fe^{3+}	−1471·3	−1459 (33·7)	−1385 (34·9)	–	–
Y^{3+}	−1221·6	−1211 (27·7)	−1150 (20·6)	−1201 (27·5)	–
Li^+	−530·8	–	–	–	−514 (23·9)
Na^+	−422·1	−418 (100)	−403 (100)	−415 (100)	−409 (38·8)
K^+	−338·5	–	–	–	−328 (88·9)
Rb^+	−313·5	–	–	–	−303 (43·0)
Cs^+	−280·0	–	–	–	−271 (57·5)

[a] Figures in brackets denote the extent of exchange in % for the ion in question.

Coughlan and Carroll[183] made similar calculations to those in Table 19 for dehydrated zeolites, except that in most instances they neglected the heat of exchange, since this was not known. These heats do not usually exceed about 12 kJ (g ion)$^{-1}$ and can be positive or negative according to the direction of the exchange and the ions involved. Their estimates of the binding energy associated with solvation, given in Table 20, are therefore in this respect less certain than those in Table 19, but are of particular interest in that they include a range of transition metal ions. Because the zeolites were not fully exchanged the extent of exchange of each is included in the table. The solvation heats in Table 20 for Na^+ ions in zeolites *A*, X and Y differ from those in Table 19 largely because different values were used for the hydration energy of gaseous Na^+ by pure water (columns 2 of the tables). Because of the different value taken for the hydration heat of each alkali metal ion, given in column 2 of Tables 19 and 20, the solvation heats for the alkali metal forms of zeolite *L* in Table 20 are not directly comparable with those given for the alkali metal forms in Table 19. However, for the alkali metal ions the sequence of exothermicity for zeolite *L* remains the same as in Table 19. The extent of exchange of K^+ in zeolite *L* is limited because much of the K^+ is locked into difficultly accessible sites, for example in cancrinite cages.[53] The exchangeable part is largely that in the main channels, so that even though the exchange is limited, most of the entering ions are exposed in the wide channels.

From Table 21 it is seen that exothermicity per equivalent follows the sequence $M^{III} > M^{II} > M^{I}$ for ions M of valence three, two and one; that for a given ion all solvation heats are less exothermic than that in water; and that among the zeolites the least exothermic heats are found with zeolite Y. The nature of the divalent and trivalent metal ions in the zeolites is a complicating factor, since the evidence of the previous section indicates that in dehydrated zeolites they are not always present as bare M^{3+} and M^{2+} ions. However the results of the present section indicate clearly that the binding energy of cations in zeolites is always large and is a strong function of ion valence. The estimated heats of solvation in the tables are all integral heats per equivalent, and so are weighted averages over all the kinds of site occupied by the ion in question in the given zeolite. It has been seen in previous sections that the ion distributions can leave one or more kinds of site virtually unpopulated, and that different ions are often differently distributed among the kinds of site. All these factors may influence the heats of solvation per equivalent of cations.

12. ZEOLITIC WATER AND FRAMEWORK OXYGEN DISTANCES

From the viewpoint of zeolitic sorbents, the positions of water molecules, unlike those of the cations, are not important, because to activate the zeolites the first objective is to remove all or as much as possible of this water. In the most open zeolites the problem of locating all water is at least as difficult as is that of locating all the cations. The water is very mobile at or near room temperatures so that the positions occupied are time averages, and considerable disorder in the water clusters must be expected. The water molecules are frequently in association with cations, as indicated by the electronic spectra referred to in §11, and the nearest neighbour atoms given in Table 14. Water molecules are also often hydrogen-bonded to anionic framework atoms, and can form bridges between pairs of cations or between cations and other waters or framework oxygens. Some positions to which water molecules have been assigned are indicated in Tables 15 and 16 for faujasite and its variants, zeolites X and Y.

The refinement by Gramlich and Meier[65] of the structure of zeolite *A* revealed five occupied sites which had not previously been reported, and which were considered to be occupied primarily by water. Of these, the first was in the sodalite cages and provided equivalent positions for four waters, in a distorted tetrahedron with four edges of 2·9Å and the remaining two of 3·3Å. The second and third sites were in the 26-hedron of Type I and ideally accommodate 20 water molecules in a cluster of about 3·85Å radius. This cluster represents the pentagonal dodecahedron which is one of the

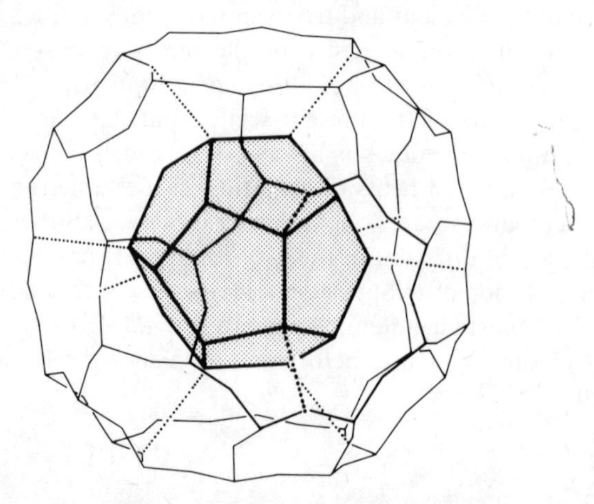

FIG. 31. The proposed dodecahedral water cluster (slightly idealized) in the 26-hedron of zeolite *A*. The shortest approach distances of the cluster and framework oxygens are indicated by the dotted lines.[65]

kinds of water cage found in clathrate hydrates.[186] The configuration of the cluster, within the 26-hedron, is shown in Fig. 31.[65] Of the water molecules in the cluster those in sites two form part of the Na^+ coordination. As previously noted, there is Na^+ near the 6-rings and a water in position two is coordinated to this Na^+. The ion thereby achieves tetrahedral coordination with the three nearest framework oxygens and water ($Ox_3Na^+H_2O$). The water molecules in positions three are coordinated to framework oxygens. The water on positions four is in the 8-ring windows and provides linkages between the dodecahedral water clusters in adjacent cages. These sites probably contain Na^+ also, and ideally may hold one Na^+ and one H_2O per window. The fifth position appeared as a fairly broad peak at the centre of the dodecahedron. The distance of 3·85Å to the nearest neighbours on sites two and three and unreal thermal parameters and site occupancy values suggested that this site involved a less symmetrical arrangement of more than one atom, such as one H_2O and one Na^+.

Water to water and water to framework oxygen distances in zeolites are much more variable than oxygen–oxygen distances in the framework, as seen from the examples in Table 21. The subscripts in the table denote that the water molecules or oxygen atoms are in different crystallographic sites. If one considers natrolite, where the Al and Si atoms are ordered on tetrahedral sites, the oxygen–oxygen distances are substantially larger in the AlO_4 than in the SiO_4 tetrahedra. Thus, the more aluminous the zeolite, such as zeolite *A* or sodalite hydrate ($Al/Si = 1$) the larger should be the average O–O distance. If one averages the O–O distances tabulated (or calculated from Si–O and Al–O bond lengths) one obtains for several zeolites the values in Table 22. The results indicate, with some irregularities, the expected trend. As noted previously stilbite, barrerite and stellerite have the same framework topology, as do heulandite and clinoptilolite. There are no irregularities in the rule that decreasing average O–O distance goes with increasing silica content. In faujasite-type zeolites in a given homoionic form the unit cell edge may be used to estimate the Si/Al ratio.[59]

In the hydrogen bond, O–H...O, the centre to centre distance of the oxygens is usually considered as a measure of the strength of the bond. Several bond lengths are:[192]

$(COOH)_2.2H_2O$	$H_2O...OH$	2·49Å
	$H_2O...O=$	2·88Å
HCOOH	OH...O=	2·58Å
Ice	OH...O	2·78Å.

The shortest distances, 2·49 and 2·58Å, denote strong hydrogen bonds; that in ice is of moderate strength; 2·88Å indicates a weak bond. Table 21 then shows many very weak interactions between H_2O and H_2O and between

TABLE 21

Some bond distances in Å for H_2O–H_2O, H_2O–O and O–O in zeolites. Only values $\leqslant 3{\cdot}30$Å are given

Oxygen–Oxygen				Water–Oxygen		Water–Water	
(a) *Chabazite*[43]							
O_I–O_{II}	2·67			H_2O_I–O_I	2·90	H_2O_I–H_2O_{IV}	2·78
O_I–O_{III}	2·73			H_2O_I–O_{III}	3·07	H_2O_{III}–H_2O_{IV}	3·23
O_I–O_{IV}	2·72			H_2O_{II}–O_I	3·30	H_2O_{IV}–H_2O_{IV}	2·55
O_{II}–O_{III}	2·64			H_2O_{III}–O_{IV}	2·65		
O_{II}–O_{IV}	2·79			H_2O_{III}–O_{III}	3·01		
O_{III}–O_{IV}	2·65						
(b) *Levynite*[49]							
O_I–O_I	2·74			H_2O_I–O_{III}	2·96,	H_2O_I–H_2O_I	3·07
O_I–O_{II}	2·74				3·08	H_2O_I–H_2O_{II}	3·03
O_I–O_{III}	2·68			H_2O_{II}–O_I	2·98	H_2O_I–H_2O_{III}	3·29
O_I–O_{IV}	2·67			H_2O_{IV}–O_I	2·74	H_2O_{II}–H_2O_{II}	3·02
O_{II}–O_{III}	2·64			H_2O_{IV}–O_V	2·54	H_2O_{II}–H_2O_{III}	2·92
O_{II}–O_{IV}	2·75					H_2O_{II}–H_2O_{IV}	2·13,
O_{III}–O_{IV}	2·74						2·02
O_I–O_V	2·63, 2·72					H_2O_{III}–H_2O_{III}	3·15
O_V–O_V	2·64					H_2O_{III}–H_2O_{IV}	2·57
						H_2O_{IV}–H_2O_{IV}	2·85
(c) *Natrolite*[16] (ordered Al and Si distribution on tetrahedral sites)							
O_I–O_I	2·67	O_I–O_{II}	2·86	H_2O–O_I	2·86		
O_I–O_V	2·64	O_I–O_{III}	2·82	H_2O–O_{II}	3·26		
O_{II}–O_{III}	2·60	O_I–O_{IV}	2·91	H_2O–O_V	2·99		
O_{II}–O_{IV}	2·66	O_{II}–O_{III}	2·88				
O_{II}–O_V	2·60	O_{II}–O_{IV}	2·78				
O_{III}–O_{IV}	2·67	O_{III}–O_{IV}	2·86				
O_{III}–O_V	2·68						
O_{IV}–O_V	2·63						
(d) *Clinoptilolite*[23] (from Agoura)							
O_I–O_{II}	2·67	O_V–O_{VII}	2·66	H_2O_I–O_{II}	3·26	H_2O_I–H_2O_{II}	2·86
O_I–O_{IV}	2·72	O_V–O_{VIII}	2·61	H_2O_I–O_{III}	3·05	H_2O_I–H_2O_{III}	2·93
O_I–O_X	2·63	O_V–O_X	2·58	H_2O_I–O_{IV}	3·07	H_2O_{II}–H_2O_V	2·81,
O_{II}–O_{III}	2·58	O_{VI}–O_{IX}	2·65	H_2O_{II}–O_{III}	3·19		2·65
O_{II}–O_{IV}	2·69	O_{VII}–O_{VIII}	2·61	H_2O_{III}–O_{VIII}	3·06	H_2O_{II}–H_2O_{VI}	2·74
O_{II}–O_X	2·70	O_{VII}–O_{IX}	2·65	H_2O_{III}–O_X	3·20	H_2O_{III}–H_2O_{III}	2·95,
O_{II}–O_{VII}	2·63	O_{VII}–O_X	2·62	H_2O_{IV}–O_V	3·19		3·08
O_{II}–O_{IX}	2·64	O_{VIII}–O_X	2·62	H_2O_V–O_{VII}	3·06	H_2O_V–H_2O_V	1·12,
O_{III}–O_{IV}	2·61			H_2O_{VI}–O_{VI}	3·24		3·2
O_{III}–O_{VI}	2·67					H_2O_V–H_2O_{VI}	2·71
O_{III}–O_{VII}	2·64						
O_{III}–O_{IX}	2·69						
O_{IV}–O_{VI}	2·64						
O_{IV}–O_{IX}	2·66						
O_{IV}–O_X	2·71						

TABLE 22

Average distances between framework oxygens in some zeolites

Zeolite	Si/Al ratio	Mean O–O distance in Å
Natrolite[16]	1·5	2·73
Edingtonite[20]	1·5	2·73
Levynite[49]	1·86	2·70
Chabazite[43]	2·08	2·70
Stilbite[25]	2·48	2·67
Mazzite[50]	2·71	2·68
Heulandite[21]	2·85	2·67
Barrerite[28]	3·37	2·66
Stellerite[27]	3·5	2·65
Clinoptilolite[23]	4·57	2·65

H_2O and framework oxygen. Some instances of strong bonds are those between H_2O_{IV} and H_2O_{IV} and H_2O_{III} and O_{IV} in chabazite; and also H_2O_{IV} and O_V and H_2O_{III} and H_2O_{IV} in levynite. In some other instances the bond lengths compare with those in ice (e.g. H_2O_{IV}–O_I in levynite and several H_2O–H_2O distances in clinoptilolite). In a few instances the H_2O–H_2O distances in the table are unrealistically short (H_2O_{II}–H_2O_{IV} in levynite and one of the H_2O_V–H_2O_V distances in clinoptilolite). This implies that when one of such water positions is occupied the other is necessarily empty. In clinoptilolite all H_2O-framework oxygen distances are very long so that the corresponding interactions are particularly weak and probably reflect the small net negative charge of this silica-rich zeolite. The molar binding energy of water in the zeolites can be evaluated by calorimetry, for example from heats of wetting, and has been measured[182,183] for zeolites *A*, X, Y, *L* and chabazite, enriched in various alkali, alkaline earth and transition metal ions. Taking the reference state as that of water vapour the binding energies, referred to in Chapter 4, normally exceed the latent heat of condensation of water vapour to liquid water.

13. ANGLES (Al, Si)–O–Si AND O–(Al, Si)–O

The ways in which (Al, Si)O_4 can be linked to other such tetrahedra to yield different tectosilicate frameworks depend very much upon the flexibility of the (Al, Si)–O–Si bond. These angles are illustrated for a number of zeolites in Table 23. There is a maximum range, in the examples selected, from 129·1°

TABLE 23

Reported bond angles in several zeolites[a]

Zeolite	Bond and angle		Zeolite	Bond and angle	
Na-*A*[65]	Si–O_I–Al	145·5°	Dehydrated Na-*A*[133]	Si–O_I–Al	145·1°
	Si–O_{II}–Al	159·5°		Si–O_{II}–Al	165·6°
	Si–O_{III}–Al	144·1°		Si–O_{III}–Al	145·5°
Faujasite[58]	(Si, Al)–O_I–Si	140·6°	Natrolite[16]	Si_I–O_V–Si_{II}	143·4°
	(Si, Al)–O_{II}–Si	140·3°		Si_I–O_I–Al	162·3°
	(Si, Al)–O_{III}–Si	145·1°		Si_{II}–O_{II}–Al	129·1°
	(Si, Al)–O_{IV}–Si	140·6°		Si_{II}–O_{III}–Al	138·7°
Ferrierite[37]	T_{II}–O_I–T_{II}'	169·2°	Mazzite[50]	T_I–O_I–T_I	149·2°
	T_{III}–O_{II}–T_{III}'	152·7°		T_I–O_{II}–T_I	171·2°
	T_I–O_{III}–T_{II}	153·6°		T_I–O_{IV}–T_{II}	144·8°
	T_I–O_{IV}–T_{III}	157·9°		T_{II}–O_{III}–T_{II}	146·5°
	T_{IV}–O_V–T_{IV}'	180·0°		T_{II}–O_V–T_{II}	137·4°
	T_{IV}–O_{VI}–T_{IV}'	153·3°		T_{II}–O_{VI}–T_{II}	136·6°
	T_{II}–O_{VII}–T_{IV}	152·9°	Levynite[49]	T_I–O_I–T_{II}	148·5°
	T_{IV}–O_{VIII}–T_{III}	147·4°		T_I–O_{II}–T_I	135·6°
Edingtonite[20]	Si_I–O_{IV}'–Al'	138·3°		T_I–O_{III}–T_I	140·4°
	Si_I–O_V–Si_{II}	142·0°		T_I–O_{IV}–T_{II}	150·1°
	Si_{II}–O_V–Si_{II}	132·5°		T_{II}–O_V–T_{II}	155·3°
	Si_{II}–O_{III}–Al	134·8°			

[a] In Tables 23 and 24 T denotes Al or Si on a tetrahedral site. The subscripts indicate crystallographically distinguishable atoms.

to 180°, although in any one zeolite the range is less than this. The average *T*–O–*T* angles for various zeolites are.

Zeolite Na-*A*	149·7°	Edingtonite	136·9°
Dehydrated Na-*A*	152·1°	Levynite	146·0°
Faujasite	141·7°	Zeolite Li-*ABW*	134·8°
Natrolite	143·4°	Laumontite	135·9°
Stilbite	144·4°	Yugawaralite	146·0°
Stellerite	145·5°	Mazzite	147·6°
Barrerite	147·7°	Ferrierite	158·1°

There are significant differences in average bond angle even for the isotypes stilbite, stellerite and barrerite.

The O–*T*–O angles in the individual tetrahedra, where *T* denotes a Si or Al atom also show small variations (Table 24) and in this respect contribute to

TABLE 24

Representative angles O–*T*–O in zeolites

Zeolite	Range in angles	Mean value of angle
Zeolite *A*[65]	SiO_4 : 108·0–110·9°	109·47°
	AlO_4 : 107·0–112·0°	109·42°
Edingtonite[20]	Si_IO_4 : 108·2–113·2°	109·45°
	$Si_{II}O_4$: 106·3–113·2°	109·43°
	AlO_4 : 107·0–112·0°	109·43°
Ferrierite[37]	T_IO_4 : 102·5–111·2°	109·47°
	$T_{II}O_4$: 107·7–112·5°	109·43°
	$T_{III}O_4$: 105·8–111·4°	109·47°
	$T_{IV}O_4$: 108·0–110·8°	109·26°
Mazzite[50]	T_IO_4 : 108·4–110·4°	109·47°
	$T_{II}O_4$: 106·4–110·7°	109·50°
Stilbite[25]	T_IO_4 : 106·09–111·66°	109·44°
	$T_{II}O_4$: 106·47–113·51°	109·42°
	$T_{III}O_4$: 106·39–112·24°	109·44°
	$T_{IV}O_4$: 105·35–111·44°	109·44°
	T_VO_4 : 104·55–111·69°	109·48°

the diversity of zeolite tectosilicate frameworks. However the variations are small and the average O–*T*–O angle in each TO_4 is nearly the value expected for an ideal tetrahedron.

Progress in the quantitative understanding of sorption equilibria, energetics and kinetics depends in part upon good structural information including detailed knowledge of charge distributions. Although this knowledge is still inadequate it is hoped to show in subsequent chapters that satisfactory progress is being made in interpreting the behaviour of molecular sieve sorbents.

REFERENCES

1. L. Bragg and G. F. Claringbull, "Crystal Structures of Minerals", Bell and Sons, London (1965).
2. W. A. Deer, R. A. Howie and J. Zussman, "Rock Forming Minerals", Vol. IV Longmans (1963).
3. W. M. Meier *in* "Molecular Sieves", Soc. Chem. Ind., London (1968) p. 10.
4. W. M. Meier and D. H. Olson, *in* "Molecular Sieve Zeolites—I" Advances in Chemistry Series No. 101, Amer. Chem. Soc. (1971 p. 155.
5. D. W. Breck, "Zeolite Molecular Sieves", Wiley-Interscience (1974).
6. J. V. Smith, Miner. Soc. of America, Special Paper 1 (1963) 281.
7. K. F. Fischer and W. M. Meier, *Fortschr. Miner.* (1965) **42**, 50.
8. S. Merlino, *Soc. Ital. di Mineralog. e Petrolog., Rendicorti* (1975) **31**, 513.
8a. S. Merlino, *Izvj. Jugoslav. centr. krist.* (*Zagreb*) (1976) **11**, 19.
9. W. H. Taylor, *Zeit. Krist.* (1930) **74**, 1.
10. W. H. Taylor, *Proc. Roy. Soc.* (1934) A, **145**, 80.
11. A. Steiner, *Mineralog. Mag* (1955) **30**, 691.
12. D. S. Coombs, *Mineralog. Mag.* (1955) **30**, 699.
13. D. McConnell, *Amer. Mineralog.* (1952) **37**, 609.
14. D. McConnell, *Mineralog. Mag.* (1964) **33**, 799.
15. W. H. Taylor, C. A. Meek and W. W. Jackson, *Zeit. Krist.* (1933) **84**, 373.
16. W. M. Meier, *Zeit. Krist.* (1960) **113**, 430.
17. M. H. Hey and F. A. Bannister, *Mineralog. Mag.* (1936) **24**, 227.
18. M. H. Hey and F. A. Bannister, *Mineralog. Mag.* (1933) **23**, 421.
19. M. H. Hey and F. A. Bannister, *Mineralog. Mag.* (1932) **23**, 51.
20. E. Galli, *Acta Cryst.* (1976) B **32**, 1623.
21. A. Alberti, *Tschermaks Mineralog. Pet. Mitt.* (1972) **18**, 129.
22. A. B. Merkle and M. Slaughter, *Amer. Mineralog.* (1968) **53**, 1120.
23. A. Alberti, *Tschermaks Mineralog. Pet. Mitt.* (1975) **22**, 25.
24. A. J. Perotta and J. V. Smith, *Acta Cryst.* (1964) **17**, 857.
25. E. Galli, *Acta Cryst.* (1971) B **27**, 833.
26. E. Galli and G. Gottardi, *Mineralog. Pet. Acta* (1966) **12**, 1.
27. E. Galli and G. Gottardi, *Bull. Soc. fr. Mineral. Crist.* (1975) **98**, 11.
28. E. Galli and G. Gottardi, *Bull. Soc. fr. Mineral. Crist.* (1975) **98**, 331.
29. R. Rinaldi, J. J. Pluth and J. V. Smith, *Acta Cryst.* (1974) B **30**, 2426.
30. K. Fischer and V. Schramm, *in* "Molecular Sieve Zeolites—I", Advances in Chemistry Series, No. 101, Amer. Chem Soc. (1971) p. 250.
31. C. Baerlocher and W. M. Meier, *Zeit. Krist.* (1972) **135**, 339.
32. G. P. L. Walker, *Mineralog. Mag.* (1962) **33**, 173.
33. R. M. Barrer, F. W. Bultitude and I. S. Kerr, *J. Chem. Soc.* (1959) 1521.
34. I. S. Kerr and D. J. Williams, *Acta Cryst.* (1969) B **25**, 1183.
35. I. S. Kerr, *Zeit. Krist.* (1974) **139**, 186.
36. W. M. Meier, *Zeit. Krist.* (1961) **115**, 439.
37. P. A. Vaughan, *Acta Cryst.* (1966) **21**, 983.
38. G. Gottardi and W. M. Meier, *Zeit. Krist.* (1963) **119**, 53.
39. I. S. Kerr, *Nature* (1964) **202**, 589.
40. A. J. Perotta, *Mineralog. Mag.* (1967) **36,** 480.
41. V. Kozman, R. J. Gait and J. Rucklidge, *Amer. Mineralog.* (1974) **59**, 71.
42. W. M. Meier, *in* "Molecular Sieves", Soc. Chem. Ind., London (1968) pp. 12–16.
43. J. V. Smith, F. Rinaldi and L. S. Dent Glasser, *Acta Cryst.* (1963) **16**, 45.

44. H. S. Dent and J. V. Smith, *Nature* (1958) **181**, 1794.
45. L. W. Staples and J. A. Gard, *Mineralog. Mag.* (1959) **32**, 261.
46. J. M. Bennett and J. A. Gard, *Nature* (1967) **214**, 1005.
47. J. A. Gard and J. M. Tait, *Acta Cryst.* (1972) B **28**, 825.
48. R. M. Barrer and I. S. Kerr, *Trans. Faraday Soc.* (1959), **55**, 1915.
49. S. Merlino, E. Galli and A. Alberti, *Tschermaks Mineralog. Pet. Mitt.* (1975) **22**, 117.
50. E. Galli, *Cryst. Struct. Comm.* (1974) **3**, 339.
51. E. Galli, E. Passaglia and D. Pongiluppi, *Contr. Mineralog. and Pet.* (1974) **45**, 99.
52. R. Rinaldi, J. J. Pluth and J. V. Smith, *Acta Cryst.* (1975) B **31**, 1603.
53. R. M. Barrer and H. Villiger, *Zeit. Krist.* (1969) **128**, 352.
54. D. W. Breck, "Zeolite Molecular Sieves", Wiley-Interscience (1974) p. 155.
55. R. M. Barrer and J. F. Cole, *J. Chem. Soc.*, A (1970) 1516.
56. R. M. Barrer, J. F. Cole and H. Villiger, *J. Chem. Soc.*, A (1970) 1523.
57. W. Sieber and W. M. Meier, *Helv. chim. Acta* (1974) **57**, 1533.
58. W. H. Bauer, *Amer. Mineralog.* (1964) **49**, 697.
59. J. V. Smith, *in* "Molecular Sieve Zeolites—I", Advances in Chemistry Series, No. 101, Amer. Chem. Soc. Ed. R. F. Gould (1971) p. 171
60. G. T. Kokotailo and J. Ciric, *in* "Molecular Sieve Zeolites—I", Advances in Chemistry Series, No. 101, Amer. Chem. Soc. Ed. R. F. Gould (1971) p. 109.
61. W. B. Kamb and W. C. Oke, *Amer. Mineralog* (1960) **45**, 79.
62. E. K. Gordon, S. Samson and W. B. Kamb, *Science* (1966) **154**, 1004.
63. T. B. Reed and D. W. Breck, *J. Amer. Chem. Soc.* (1956) **78**, 5972.
64. L. Broussard and D. P. Shoemaker, *J. Amer. Chem. Soc.* (1960) **82**, 1041.
65. V. Gramlich and W. M. Meier, *Zeit. Krist.* (1971) **133**, 134.
66. H. E. Robson, D. P. Shoemaker, R. A. Ogilvie and P. C. Manor, *in* "Molecular Sieves", Advances in Chemistry Series, No. 121, Amer. Chem. Soc. Eds. W. M. Meier and J. B. Uytterhoeven (1973) p. 106.
67. W. M. Meier and G. T. Kokotailo, *Zeit. Krist.* (1965) **121**, 211.
68. H. Bartle and K. F. Fischer, *Neues Jahrb. Mineralog. Mh.* (1967) 2/3, 33.
69. V. Schramm and K. F. Fisher, *in* "Molecular Sieve Zeolites—I", Advances in Chemistry Series, No. 101, Amer. Chem. Soc. Ed. R. F. Gould (1971) p. 259.
70. R. M. Barrer and R. Beaumont, *J. Chem. Soc., Dalton* (1974) 405.
70a. Mobil Co., British Patent, 1,117,768 (1968).
70b. R. J. Argauer and G. R. Landolt, US Patent, 3,702,886 (1972).
70c. Mobil Co., Netherlands Patent, 7,014,807 (1971).
70d. J. Ciric, US Patent, 3,692,470 (1972).
71. S. P. Zhdanov, *in* "Molecular Sieves", Soc. Chem. Ind., London (1968) p. 62.
72. R. M. Barrer and D. E. Mainwaring, *J. Chem. Soc., Dalton* (1972) 1254.
73. C. Baerlocher and W. M. Meier, *Helv. chem. Acta* (1969) **52**, 1853.
74. R. M. Barrer and J. W. Baynham, *J. Chem. Soc.* (1956) 2882.
75. A. J. Gude, III, and R. A. Sheppard, *Amer. Mineralog.* (1966) **51**, 909.
76. H. Kacirek and H. Lechert, *J. Phys. Chem.* (1976) **80**, 1291.
77. D. W. Breck, "Zeolite Molecular Sieves", Wiley-Interscience (1974) p. 134.
78. W. Loewenstein, *Amer. Mineralog.* (1954) **39**, 92.
79. R. M. Barrer and M. B. Makki, *Can. J. Chem.* (1966) **42**, 1481.

80. R. M. Barrer and J. A. Lee, *J. Colloid and Interface Sci.* (1969) **30**, 111.
81. D. K. Thakur and S. W. Weller, *in* "Molecular Sieves", Advances in Chemistry Series, No. 121, Amer. Chem. Soc. Eds W. M. Meier and J. B. Utterhoeven (1973) p. 596.
82. J. R. Goldsmith, *J. Geol.* (1950) **58**, 518.
83. J. R. Goldsmith, *Mineralog. Mag.* (1952) **29**, 952.
84. R. M. Barrer, J. W. Baynham, F. W. Bultitude and W. M. Meier, *J. Chem. Soc.* (1959) 195.
85. R. M. Barrer and E. F. Freund, *J. Chem. Soc., Dalton* (1974) 1049, 2055, 2060 and 2123.
86. R. M. Barrer and D. J. Marshall, *J. Chem. Soc.* (1965) 6616 and 6621.
87. G. H. Kuhl *in* "Molecular Sieves", Soc. Chem. Ind., London (1968) p. 85.
88. G. H. Kuhl *in* "Molecular Sieve Zeolites—I", Advances in Chemistry Series, No. 101, Amer. Chem. Soc., Ed. R. F. Gould (1971) p. 63.
89. E. M. Flanigen and R. W. Grose *in* "Molecular Sieve Zeolites—I", Advances in Chemistry Series, No. 101, Amer. Chem. Soc., Ed, R. F. Gould (1971) p. 76.
90. R. M. Barrer and M. Liquornik, *J. Chem. Soc., Dalton* (1974) 2126.
91. R. M. Barrer, *Chem. Brit.* (1966) p. 380.
92. R. M. Barrer and N. McCallum, *J. Chem. Soc.* (1953) 4035.
93. R. M. Barrer and E. A. D. White, *J. Chem. Soc.* (1952) 1561.
94. R. M. Barrer, *J. Chem. Soc.* (1948) 127.
95. R. M. Barrer and C. Marcilly, *J. Chem. Soc.*, A (1970) 2735.
96. J. V. Smith and F. Rinaldi, *Mineralog. Mag.* (1962) **33**, 202.
97. C. L. Frye and W. T. Collins, *J. Amer. Chem. Soc.* (1970) **92**, 5586.
98. K. Olsson, *Arkiv Kemi* (1958) **13**, 367.
99. J. F. Brown, Jr., L. H. Vogt., Jr., and P. I. Prescott, *J. Amer. Chem. Soc.* (1964) **86**, 1120.
100. A. J. Barry, W. H. Daut, D. J. Domicone and J. W. Gilkey, *J. Amer. Chem. Soc..* (1955) **77**, 4248.
101. L. H. Vogt, Jr., and J. F. Brown, *J., Inorg. Chem.* (1963) **2**, 189.
102. A. F. Wells "The Third Dimension in Chemistry", Oxford University Press (1968) p. 57.
103. R. M. Barrer, Chem. and Ind. (1968) 1203.
104. R. M. Barrer and H. Villiger, Chem. Comm. (1969) 659.
105. P. B. Moore and J. V. Smith, *Mineralog. Mag.* (1964) **33**, 1008.
106. R. C. Evans "An Introduction to Crystal Chemistry", Cambridge University Press (1964) p. 145.
107. G. T. Kokotailo and S. L. Lawton, *Nature* (1964) **203**, 621.
108. W. M. Meier, Symposium on Natural Zeolites, Tucson, Arizona, June 6th to 13th (1976).
109. L. P. Solov'eva, S. V. Borisov and V. V. Bakakin, *Sov. Phys-Cryst.* (1972) **16**. 1035.
110. J, V. Smith, *Mineralog Mag.* (1968) **36**, 640.
111. N. Haga, *Mineralog. Jour.* (1973) **7**, 262.
112. I. S. Kerr, *Nature* (1963) **197**, 1194.
113. A. Alberti and G. Gottardi, *Neues Jahrb. Mineralog. Mh.* (1975) 396.
114. D. P. Shoemaker, H. E. Robson and L. Broussard, *in* "Proc. of 3rd Internat. Conference on Molecular Sieves", Ed. J. B. Uytterhoeven, Zurich, September 3rd–7th (1973) p. 138.
115. R. M. Barrer, *Trans. Faraday Soc.* (1944) **40**, 555.

116. R. M. Barrer and D. E. W. Vaughan, *J. Phys. and Chem. of Solids* (1971) **32**, 731.
117. R. Aiello and R. M. Barrer, *J. Chem. Soc.*, A (1970) 1470.
118. R. M. Barrer and D. A. Harding, *Separation Science* (1974) **9**, 195.
119. L. B. Sand, *in* "Molecular Sieves", Soc. Chem. Ind., London (1968) p. 71.
120. J. L. Guth, *Rev. Chim. Minerale* (1965) **2**, 127.
121. D. W. Breck, "Zeolite Molecular Sieves", Wiley Interscience (1974) p. 672.
122. R. M. Barrer and D. A. Ibbitson, *Trans. Faraday Soc.* (1944) **40**, 206.
123. R. M. Barrer and T. Burstein, unpublished results.
124. R. M. Barrer and D. L. Peterson, *Proc. Roy. Soc.* (1964) A **280**, 466.
125. R. M. Barrer, *J. Soc. Chem. Ind.* (1945) **44**, 130.
126. R. M. Barrer, *Trans. Faraday Soc.* (1949) **45**, 358.
127. R. M. Barrer and B. R. Wheeler, unpublished results.
128. R. M. Barrer and J. A. Davies, *Proc. Roy. Soc.* (1971) A **322**, 1.
129. R. M. Barrer and I. M. Galabova, *in* "Molecular Sieves", Advances in Chemistry Series No. 121, Amer. Chem. Soc., Eds W. M. Meier and J. B. Uytterhoeven (1973) p. 356.
130. R. M. Barrer and J. A. Lee, *Surface Sci.* (1968) **12**, 354.
131. R. M. Barrer and W. Sieber, *J. Chem. Soc., Dalton*, (1977) 1020.
132. R. M. Barrer, J. Klinowski and H. S. Sherry, *J. Chem. Soc., Faraday II*, (1973) **69**, 1669.
133. R. Y. Yanagida, A. A. Amaro and K. Seff, *J. Phys. Chem.* (1973) **77**, 805.
134. J. V. Smith, *Acta Cryst.* (1962) **15**, 835.
135. P. C. W. Leung, K. B. Kunz and K. Seff, *J. Phys. Chem.* (1975) **79**, 2157.
136. R. L. Firor and K. Seff, *J. Amer. Chem. Soc.* (1976) **98**, 5031.
137. T. B. Vance Jr., and K. Seff, *J. Phys. Chem.* (1975) **79**, 2163.
138. P. E. Riley, K. Seff and D. P. Shoemaker, *J. Phys. Chem.* (1972) **76**, 2593.
139. M. Nitta, K. Ogawa and K. Aomura, *Bull. Chem. Soc., Jap.* (1975) **48**, 1939.
140. G. R. Eulenberger, J. G. Keil and D. P. Shoemaker, *J. Phys. Chem.* (1967) **71**, 1812.
141. W. J. Mortier, H. J. Bosmans and J. B. Uytterhoeven, *J. Phys. Chem.* (1972) **76**, 650.
142. J. M. Bennett and J. V. Smith, *Mater. Res. Bull.* (1968) **3**, 633.
143. I. E. Maxwell and J. J. de Boer, *J. Phys. Chem.* (1975) **79**, 1874.
144. D. H. Olson, *J. Phys. Chem.* (1968) **72**, 4366.
145. J. M. Bennett and J. V. Smith, *Mater. Res. Bull.* (1968) **3**, 865.
146. J. M. Bennett and J. V. Smith, *Mater. Res. Bull.* (1969) **4**, 7.
147. D. H. Olson, G. T. Kokotailo and J. F. Charnell, *J. Coll. Interface Sci.*, (1968) **28**, 305.
148. J. V. Smith, J. M. Bennett and E. M. Flanigen, *Nature* (1967) **215**, 241.
149. J. M. Bennett, J. V. Smith and C. L. Angell, *Mat. Res. Bull.* (1969) **4**, 77.
150. D. H. Olson and E. Dempsey, *J. Catal.* (1969) **13**, 221.
151. J. M. Bennett and J. V. Smith, *Mater. Res. Bull.* (1968) **3**, 933.
152. J. M. Bennett and J. V. Smith, *Mater. Res. Bull.* (1969) **4**, 343.
153. D. H. Olson, G. T. Kokotailo and J. F. Charnell, *Nature* (1967) **215**, 271.
154. D H Olson, *J. Phys. Chem.* (1970) **74**, 2758.
155. D. H. Olson and H. S. Sherry, *J. Phys. Chem.* (1968) **72**, 4095.
156. W. J. Mortier and H. J. Bosmans, *J. Phys. Chem.* (1971) **75**, 3327.
157. P. E. Riley and K. Seff, *J. Phys. Chem.* (1975) **79**, 1594.
158. R. Y. Yanagida, T. B. Vance, Jr., and K. Seff, *Inorg. Chem.* (1974) **13**, 723.
159. A. A. Amaro, C. L. Kovaciny, K. B. Kunz, P. E. Riley, T. B. Vance, Jr.,

R. Y. Yanagida and K. Seff, *in* "Proceedings of 3rd International Conference on Molecular Sieves", Ed. J. B. Uytterhoeven, Zurich, Sept. 3rd–7th (1973) p. 113.

160. K. Klier, R. Kellerman and P. J. Hutta, *J. Chem. Phys.* (1974) **61**, 4225.
161. R. Kellerman and K. Klier, Surface and Defect Properties of Solids, Chem. Soc. (London) (1975) **5**, 1.
162. K. Klier and M. Ralek, *J. Phys. Chem. Solids* (1968) **29**, 951.
163. R. Kellerman, P. J. Hutta and K. Klier, *J. Amer. Chem. Soc.* (1974) **96**, 5946.
164. P. Gallezot, Y. Ben Taarit and B. Imelik, *J. Catal.* (1972) **26**, 481.
165. P. Gallezot and B. Imelik, *J. Phys. Chem.* (1973) **77**, 652.
166. P. Gallezot, Y. Ben Taarit and B. Imelik, *J. Phys. Chem.* (1973) **77**, 2556.
167. M. F. Guilleux, J. F. Tempere and D. Delafosse, *Chim. Phys.* (1974) **71**, 42.
168. R. Polak and K. Klier, *J. Phys. Chem. Solids* (1969) **30**, 2231.
169. R. Kellerman and K. Klier, Surface and Defect Properties of Solids, Chem. Soc. (London) (1975) **5**, 31.
170. R. Kellerman and K. Klein, unpublished results.
171. J. Marti, J. Soria and F. H. Cano, *J. Phys. Chem.* (1976) **80**, 1776.
172. R. L. Garten, W. N. Delgass and M. Boudart, *J. Catal.* (1970) **18**, 90.
173. T. Kubo, H. Tominaga and K. Kunugi, *Bull. Chem. Soc. Jap.* (1973) **46**, 3549.
174. Yu. S. Khodakov, I. D. Mikheikin, V. S. Nakhshunov, V. A. Shvets, V. B. Kazanskii and Kh. M. Minachev, *Izvest. Akad. Nauk S.S.S.R., Ser. Khim.* (1969) 523.
175. Y. Ono, K. Suzuki and T. Keii, *J. Phys. Chem.* (1974) **78**, 218.
176. M. Nitta, S. Matsumoto and K. Aomura, *Chem. Comm.* (1974) 552.
177. M. Nitta, K. Ogawa and K. Aomura, *Bull. Chem. Soc. Jap.* (1975) **48**, 1939.
178. T. I. Barry and L. A. Lay, *J. Phys. Chem. Solids* (1966) **27**, 1821.
179. T. I. Barry and L. A. Lay, *J. Phys. Solids* (1968) **29**, 1395.
180. P. P. Lai and L. V. C. Rees, *J. Chem. Soc., Faraday I* (1976) **72**, 1809, 1818 and 1827.
181. R. M. Barrer and J. A. Davies, *J. Phys. Chem. Solids* (1969) **30**, 1921.
182. R. M. Barrer and P. J. Cram, *in* "Molecular Sieve Zeolites—II", Advances in Chemistry Series No. 102, Amer. Chem. Soc., Ed. R. F. Gould (1971) p. 105.
183. B. Coughlan and W. M. Carroll, *J. Chem. Soc., Faraday I* (1976) **72**, 2016.
184. R. M. Barrer, L. V. C. Rees and D. J. Ward, *Proc. Roy. Soc.* (1963) A **273**, 180.
185. R. M. Barrer, L. V. C. Rees and M. Shamsuzzoha, *J. inorg. nucl. Chem.* (1966) **28**, 629.
186. R. M. Barrer, J. A. Davies and L. V. C. Rees, *J. inorg. nucl. Chem.* (1968) **30**, 3333.
187. R. M. Barrer, J. A. Davies and L. V. C. Rees, *J. inorg. nucl. Chem.* (1969) **31**, 219.
188. L. Benjamin and V. Gold, *Trans. Faraday Soc.* (1954) **50**, 797.
189. D. R. Rosseinsky, *Chem. Rev.* (1965) **65**, 467.
190. V. P. Vasilev, E. K. Zolotarev, A. F. Kapustinski, K. P. Mischinko, E. A. Podgornaya and E. A. Yatsimirskii, *Russ. J. Phys. Chem.* (1960) **34**, 840.
191. R. M. Barrer *in* "Molecular Sieves", Advances in Chemistry Series No. 121, Amer. Chem. Soc. Eds W. M. Meier and J. B. Uytterhoeven (1973) p. 1.
192. R. C. Evans, "An Introduction to Crystal Chemistry," Cambridge University Press, (1964) Chapters 12 and 14.

3

Equilibrium

As a result of systematic studies of equilibria and energetics a considerable insight has been gained concerning the state of guest molecules in host crystals and the nature of the bonds involved. This insight can be used to correlate the behaviour with the structures of zeolites and other porous crystals, with the nature of the cations present, if any, with lattice stability and modification, with intracrystalline electrostatic fields and field gradients, and with size, electric moments, polarizability and other properties of the guest molecules. To some extent porous crystals may approach ideal behaviour in that the pore structures are almost wholly regular and are known, so that *a priori* calculations of bonding and of equilibrium constants are at least potentially possible. When in addition the industrial and laboratory uses of porous crystals are considered, it is not surprising that a very large number of papers have appeared dealing with sorption kinetics, equilibria, energetics, selectivity and modification of porous crystalline sorbents. In accordance with the plan of this account no overall coverage of the vast literature will be attempted, but areas of general interest are considered from viewpoints which have been found especially interesting. Where equilibria are concerned the interpretation must rest firmly upon thermodynamics, on statistical thermodynamics, and on kinetic theory.

1. EQUILIBRIUM DISTRIBUTION OF GUEST BETWEEN HOST CRYSTAL AND GAS PHASE

The system comprises a gas phase and a porous zeolite phase between which the guest molecules distribute themselves. The host zeolite plus the sorbed guest mixture may be regarded as a solution, or the zeolite may be considered as an inert medium containing regular intracrystalline pores which are occupied by the guest. This latter view regards sorption as a volume filling process. Whichever view is taken, the equilibrium condition for the guest molecules is

$$\mu_s = \mu_g \tag{1}$$

and also at equilibrium

$$(\bar{S}_s - \bar{S}_g) = (\bar{H}_s - \bar{H}_g)/T. \tag{2}$$

In these expressions the subscripts s and g denote "sorbed" and "in the gas phase" respectively. The μ are chemical potentials and the $\bar{S}$ and $\bar{H}$ are differential molar entropies and enthalpies respectively. For a pure gaseous component $\bar{S}_g = \tilde{S}_g$ and $\bar{H}_g = \tilde{H}_g$ where $\tilde{S}_g$ and $\tilde{H}_g$ are integral molar entropy and enthalpy.

We now introduce the standard states through the general relation

$$\mu = \mu^\ominus + RT \ln a/a^\ominus \tag{3}$$

a is the activity and where $\mu^\ominus$ and $a^\ominus$ are the chemical potential and the activity in the standard state. When eqns 3 and 1 are combined one obtains

$$\left.\begin{aligned} \Delta\mu^\ominus &= (\mu_s^\ominus - \mu_g^\ominus) = -RT \ln K \\ K &= \frac{a_s a_g^\ominus}{a_s^\ominus a_g} \end{aligned}\right\} \tag{4}$$

where K is the equilibrium constant for the distribution of guest molecules between gas phase and host crystal. By definition, in the standard states $a_g^\ominus = 1$ and $a_s^\ominus = 1$. If the gas is considered to be ideal $a_g/a_g^\ominus = p/p^\ominus$ where p denotes pressure. $p^\ominus$ is taken to be one atmosphere and so K can be written as

$$K_p = (a_s/p)_{\mathrm{eq}} \tag{5}$$

where the subscript "eq" denotes at equilibrium.

Like K in eqn 4, K_p in eqn 5 is dimensionless because a_s is divided by unit activity ($a_s^\ominus = 1$) and p by unit pressure ($p^\ominus = 1$). For the ideal gas we may also write $p = C_gRT$ and $p^\ominus = 1 = C'_gRT$ where C_g and C'_g are the gas phase concentrations corresponding with p, and with $p^\ominus = 1$ respectively. We thus have

$$K_p = \left(\frac{a_s}{C_g}\right)_{\text{eq}} C'_g = K_c C'_g$$
$$K_c = \left(\frac{a_s}{C_g}\right)_{\text{eq}}. \tag{6}$$

In K_c so derived C_g is no longer divided by C'_g so that K_c will have the dimensions of C^{-1}. Whether the equilibrium ratio is dimensionless or not depends on how it is obtained. As a further example, the equilibrium constant, $\theta/p(1-\theta)$, for Langmuir's isotherm, derived kinetically, has the dimensions of p^{-1}. If the gas phase is not ideal p and $p^\ominus$ in K_p may be replaced by the corresponding fugacities, f and $f^\ominus$. In terms of the virial equation of state with coefficients $B, C, D \ldots$

$$pV/RT = 1 + Bp + Cp^2 + Dp^3 + \ldots \tag{7}$$

one obtains the fugacity from

$$\ln(f/p) = Bp + Cp^2/2 + Dp^3/3 + \ldots \tag{8}$$

Additional relationships may be obtained from K_p or K_c. For example one may define a standard energy of sorption, $\Delta E^\ominus$, and a corresponding entropy of sorption, $\Delta S^\ominus_{K_c}$, by the relations

$$\left.\begin{aligned} \frac{\mathrm{d}\ln K_c}{\mathrm{d}T} &= \frac{\Delta E^\ominus}{RT^2} \\ \Delta S^\ominus_{K_c} &= (\Delta E^\ominus + RT\ln K_c)/T \end{aligned}\right\} \tag{9}$$

$\Delta E^\ominus$ can be identified with the constant energy of sorption in the Henry's law range in the following circumstances. One has $\mathrm{d}\ln K_c/\mathrm{d}T = \left(\frac{\partial \ln K_c}{\partial T}\right)_{C_s}$ since K_c does not depend upon C_s, the concentration of sorbed molecules. If in addition the sorption volume is constant (volume filling of intracrystalline pores) or the volume of the mixture host crystal plus sorbed molecules (solution) is constant, then fixing C_s also fixes n_s, the number of mols of guest within the crystals. We now consider the Henry's law range of sorption where $a_s = C_s$ and so $K_c = C_s/C_g$. Then, with $p = C_gRT$,

$$\frac{\mathrm{d}\ln K_c}{\mathrm{d}T} = -\left(\frac{\partial \ln C_g}{\partial T}\right)_{n_s} = -\left(\frac{\partial \ln p}{\partial T}\right)_{n_s} + \frac{1}{T} = \frac{\Delta \bar{H} + RT}{RT^2} = \frac{\Delta \bar{E}}{RT^2} \tag{10}$$

Here $\Delta\bar{H} = (\bar{H}_s - \tilde{H}_g)$ and $\Delta\bar{E} = (\bar{E}_s - \tilde{E}_g)$ are the constant heat and energy of sorption appropriate to the Henry's law range. Thus $\Delta E^\ominus = \Delta\bar{E}$ as stated above. In equating $(\Delta\bar{H} + RT)$ to $\Delta\bar{E}$ it is assumed that the volume change per mol sorbed is $-V$, the molar volume of the gaseous molecules, so that in $\Delta H = \Delta E + p\Delta V$, $p\Delta V = -pV = -RT$. With this interpretation of $\Delta E^\ominus$, both this quantity and $\Delta S^\ominus_{K_c}$ are available for interpretation in terms of molecular interactions and statistical thermodynamics.

The activity, a_s, and activity coefficient, γ_s, defined by $a_s = C_s\gamma_s$, may be found from isotherms in which C_s is plotted against C_g. It is only in porous crystals like zeolites, however, that the sorption volume V_s is known with any accuracy and hence it is only for such sorbents that C_s may be obtained. Of the five isotherm types given by Brunauer[1] three have been observed in zeolites for strictly intracrystalline sorption: Type I, Type IV and Type V. In Fig. 1 a Type IV isotherm is drawn. The dashed line gives the limiting

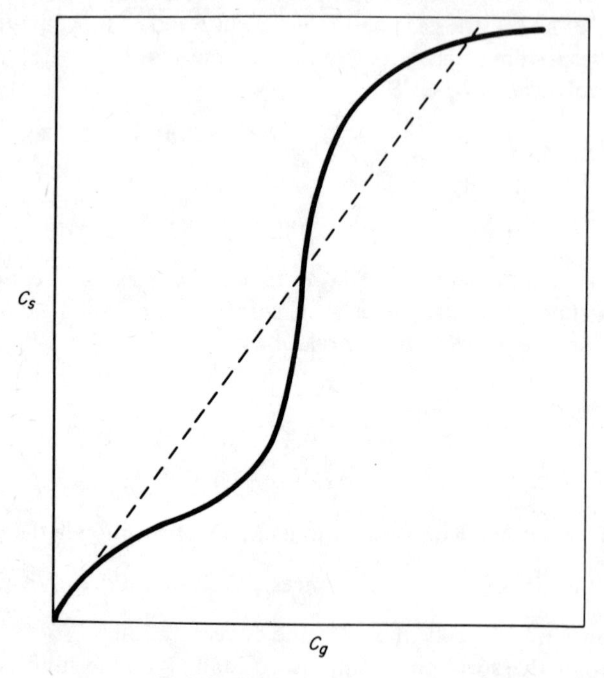

FIG. 1. The type IV isotherm.

slope as C_s (or C_g) $\rightarrow 0$. The slope of this line gives the thermodynamic equilibrium constant, K_c. At any point outside this Henry's law range we have $\gamma_s = K_c C_g / C_s$. In general and without considering the problem of finding V_s and C_s for amorphous sorbents one has from considerations like the above the following correlations between isotherm type and γ_s.

Type I $\gamma_s \geqslant 1$
Type II $\gamma_s \geqslant 1$ (usually)
Type III $\gamma_s \leqslant 1$
Type IV $\gamma_s \geqslant 1$ initially but possibly in some ranges of uptake $\gamma_s < 1$ (cf. Fig. 1)
Type V $\gamma_s \leqslant 1$, but eventually $\gamma_s > 1$

Figure 2 gives examples of activity coefficients for Ar, Kr, O_2 and CO_2 at

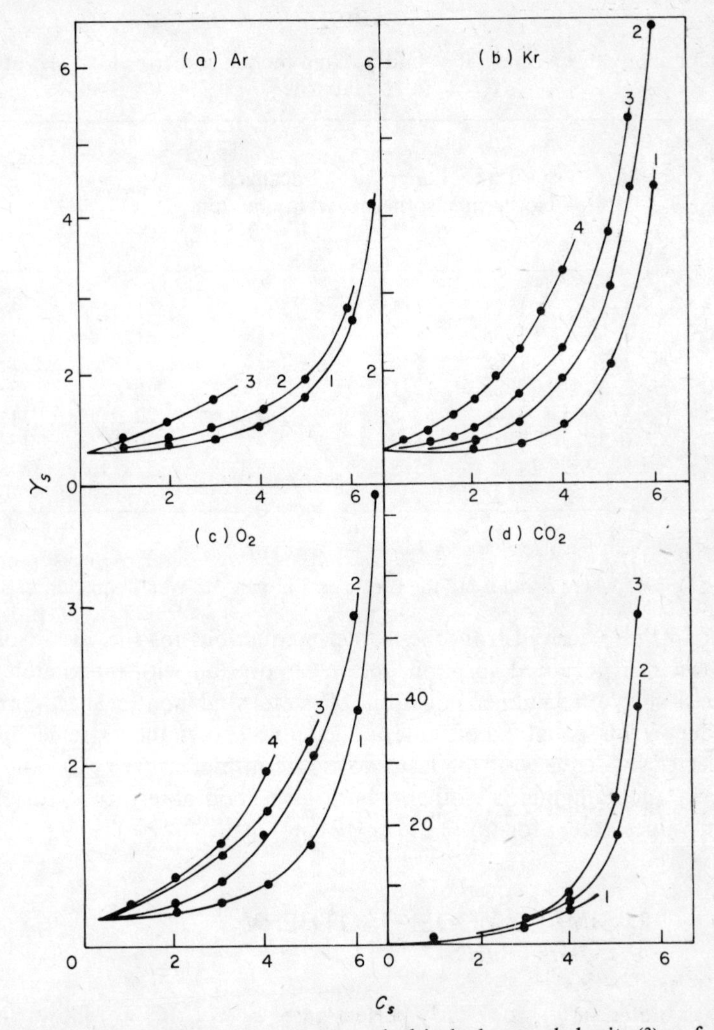

FIG. 2. Activity coefficients for some gases sorbed in hydrogen chabazite,[2] as functions of C_s (mol dm^{-3}). (a): (1) 137·8K, (2) 151·7K, (3) 183·2K; (b): (1)144·9K, (2) 163·1K, (3) 183·8K, (4) 209·2K; (c): (1) 135·2K, (2) 156·4K, (3) 165·4K, (4) 175·2K; (d): (1) 229·6K, (2) 252·6K, (3) 298·2K.

various temperatures in a decationated chabazite.[2] All these sorbates gave isotherms of Type I and γ_s begins at 1 and increases very rapidly as C_s approaches the saturation values, C_{sat}. The activity coefficients can be used to examine the validity of isotherm models. For example for that which gives Langmuir's isotherm equation $\gamma_s = (C_{sat} - C_s)^{-1}$. In anticipation of the several model isotherm equations referred to in the next and later sections Table 1[3] compares for C_2H_6 at 0°C in decationated chabazite actual values

TABLE 1

Comparison of experimental γ_s and γ_s from model isotherms for C_2H_6 at 0°C in H-chabazite[3]

$\frac{10^{-3}\,C_s}{\text{mol m}^{-3}}$	Exptl	Virial Isotherm	Langmuir Isotherm	γ_s Localized[a] w. interaction $b = 0{\cdot}5$	Volmer	van der Waals $\alpha = 1$
0·5	1·16	1·18	1·15	1·08	1·34	1·18
1·0	1·38	1·38	1·36	1·19	1·97	1·50
1·5	1·70	1·68	1·67	1·36	3·27	2·19
2·0	2·22	2·23	2·15	1·54	6·8	$3{\cdot}9_9$
2·5	3·16	$3{\cdot}4_1$	3·03	2·17	22·9	11·7
3·0	6·35	6·3	$5{\cdot}9_8$	$3{\cdot}4_0$	30·1	13·5
3·5	22·9	14·8	$15{\cdot}8_7$	$9{\cdot}9_4$	$4{\cdot}5 \times 10^7$	$1{\cdot}8 \times 10^7$

[a] The coefficient b is its value in $K_2' = \dfrac{\theta}{p(1-\theta)} \exp(b\theta)$

[b] $\alpha = 2a/bRT$ where a and b are the coefficients in van der Waals equation of state.

of γ_s and those derived from the isotherm equations for the virial isotherm (§ 6), for ideal localized sorption, localized sorption with interaction, non-localized sorption (Volmer equation of state) and non-localized sorption (van der Waals equation of state). The table shows the extreme but not unexpected inadequacy of the last two isotherm models (see § 2). The virial isotherm and Langmuir's isotherm both give good agreement with experimental values except for $C_s = 3{\cdot}5 \times 10^3$ mol m^{-3}.

2. SIMPLE MODEL ISOTHERM EQUATIONS

Since, as noted in § 1, for nearly perfect gases $a_g \approx C_g/C_g'$, it follows that to interpret K_c one may consider a_s in terms of statistical thermodynamic models. Various idealized isotherm models have been proposed which lead to values of a_s as follows:

Model	Activity
1. Ideal localized sorption (Langmuir;[4] Fowler[5])	$\frac{\theta}{1-\theta} = \frac{C_s}{C_{sat} - C_s}$
2. Localized sorption with interaction (Lacher;[6,7] Fowler and Guggenheim[8])	$\frac{\theta}{1-\theta}\left\{\frac{2-2\theta}{\beta+1-2\theta}\right\}^z$
3. Mobile Sorbate[9,10] obeying Volmer equation ($p(V-b) = C_gRT$)	$\frac{\theta}{1-\theta}\exp\{\theta/(1-\theta)\}$
4. Mobile Sorbate[11,12] obeying van der Waals equation ($(p + a/V^2)(V-b) = C_gRT$)	$\frac{\theta}{1-\theta}\exp\{\theta/(1-\theta) - \alpha\theta\}$

In the above relations, θ for models 1 and 2 is the fraction of sites occupied by guest molecules; for models 3 and 4 it is the degree of saturation. In model 2, z is the coordination number of a sorption site and

$$\beta = \left\{1 - 4\theta(1-\theta)\left[1 - \exp - \frac{2\omega}{zRT}\right]\right\}^{1/2} \tag{11}$$

where $2\omega/z$ is the extra energy when two sorbate molecules occupy adjacent sites. An approximate alternative[13] for a_s for model 2 is

$$a_s = \frac{\theta}{1-\theta}\exp 2\omega\theta/RT. \tag{12}$$

In models 1 and 2 all sites are assumed to be identical and in models 3 and 4 the pore space is assumed to provide a uniform sorption potential. The coefficient α in model 4 is given by

$$\alpha = 2a/bRT, \tag{13}$$

where a and b are the interaction and covolume coefficients in an appropriate analogue of van der Waals equation of state.

If the equilibrium constants for models 1 to 4 are denoted by K_1 to K_4, and that for eqn 12 by K_2', then

1. Langmuir isotherm: $\ln\frac{\theta}{p(1-\theta)} = \ln K_1$

2. Lacher isotherm: $\ln\frac{\theta}{p(1-\theta)} = \ln K_2 - z\ln\left(\frac{2-2\theta}{\beta+1-2\theta}\right)$

2a. Equation 12: $\ln\frac{\theta}{p(1-\theta)} = \ln K_2' - 2\omega\theta/RT$

3. Volmer isotherm: $$\ln\frac{\theta}{p(1-\theta)} = \ln K_3 - \frac{\theta}{1-\theta}$$

4. van der Waals isotherm: $$\ln\frac{\theta}{p(1-\theta)} = \ln K_4 - \frac{\theta}{1-\theta} + \alpha\theta.$$

The best way to examine the validity of Langmuir's isotherm or that of eqn 12 involves plotting $\frac{\theta}{p(1-\theta)}$ or the logarithm of this quotient against θ. Either a horizontal line or a straight line of slope $-2\omega/RT$ are found for the two cases. Volmer's and van der Waals' isotherms are examined by plotting $\ln\left(\frac{\theta}{1-\theta}\right) + \frac{\theta}{1-\theta}$ against θ, giving respectively a line parallel to the axis of θ or a line of slope α.

If $p_{1/2}$ denotes the equilibrium pressure when $\theta = \frac{1}{2}$ the following relations give the equilibrium constants:

Langmuir:	$\ln K_1 = -\ln p_{1/2}$
Lacher:	$\ln K_2 = -\ln p_{1/2} - z \ln \beta_{1/2}$
Volmer:	$\ln K_3 = -\ln p_{1/2} + 1$
van der Waals:	$\ln K_4 = -\ln p_{1/2} + 1 - \alpha/2.$

In the case of Lacher's and of van der Waals isotherm equations for suitably exothermal values of $2\omega/z$ or of α there may be two coexisting sorbed phases[8] and corresponding metastable regions in the contours of the theoretical isotherms. The actual isotherms would then show vertical steps such as those noted when phosphorus was sorbed in Na-faujasite (Na-X)[14] and mercury in Ag-faujasite (Ag-X).[15]

The validity of some or all the isotherm equations for the models 1 to 4 has been tested for example for hydrocarbons, rare gases and CO_2 in H-chabazite;[2] argon in chabazite;[16] hydrocarbons in Ca-*A*,[17] H-*L*[3], Na-X[18, 19] and (Na, K)-*L*;[20] SF_6, fluorocarbons C_1 to C_4, and chlorofluorocarbons in Na-X;[21] NH_3 in ion-exchanged forms of Na-X;[22] I_2 in Na-X, Ca-*A* and chabazite;[23] CO_2 and Kr in erionite and H-clinoptilolite;[24] N_2, O_2 and Ar in mordenite,[25] and hydrocarbons in mordenite.[26] Figure 3 shows the equilibrium quotients K_1 to K_4 for Ar in chabazite at 90·2K plotted against θ.[16] The most nearly constant function is $\frac{\theta}{p(1-\theta)} \exp\left(\frac{\theta}{1-\theta}\right)$; usually this quotient and that of the van der Waals isotherm represent sorption equilibria in zeolites very poorly, as shown in Table 2 which gives K_1 and K_3 for C_2H_6 in Na-X at 297·9K.[19] The unsatisfactory nature of isotherm equations based on such equations of state as those of Volmer or van der Waals does not necessarily arise because the guest

TABLE 2

Langmuir and Volmer isotherm quotients[19] for C_2H_6 at 297·9K in Na-faujasite (Na-X), p in cm Hg

θ	$1/K_1 = \dfrac{p(1-\theta)}{\theta}$	$1/K_3 = \dfrac{p(1-\theta)}{\theta} \exp - \dfrac{\theta}{(1-\theta)}$
0·187	23	19
0·250	19	14
0·312	18	12
0·375	18	9·8
0·438	17	7·8
0·500	17	6·1
0·562	16	4·5
0·625	17	3·2
0·688	18	2·0
Average	18·1	

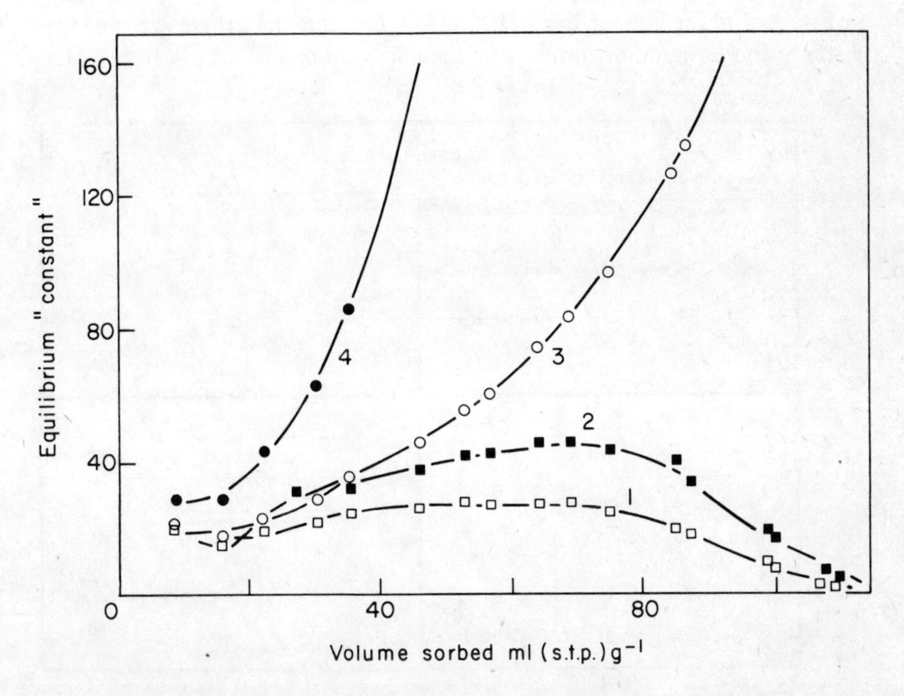

FIG. 3. Argon sorbed in natural chabazite.[16] The equilibrium quotients K_1 to K_4 are plotted against the amount sorbed. Curve 1: Volmer isotherm quotient (K_3). Curve 2: Van der Waals isotherm quotient (K_4). Curve 3: Langmuir isotherm quotient (K_1). Curve 4: Lacher isotherm quotient (K_2).

molecules are not mobile, but because neither Volmer nor van der Waals equations of state are valid even for bulk fluids at the densities of guest molecules which, for larger values of θ, exist inside the cavities and channels of porous crystals. If these equations do describe actual isotherms the agreement is best considered as due largely to chance compensation among contributing factors. Isotherms of O_2, N_2, Ar and CO_2 in mordenite[25, 27] are additional examples of systems which have been interpreted as one-dimensional gases, using the Volmer isotherm equation. In this zeolite wide channels run parallel to the *c*-axis.

As regards localized sorption models, Table 2 shows an example where the Langmuir isotherm equation is reasonably valid over a considerable range in θ. Linear plots of $\log \frac{p(1-\theta)}{\theta}$ and of $\frac{p(1-\theta)}{\theta}$ (or of the inverse functions $\log \frac{\theta}{p(1-\theta)}$ or $\frac{\theta}{p(1-\theta)}$) against θ are not uncommon. Instances where, for a range in θ, these plots run nearly parallel with the axis of θ (Langmuir isotherm equation) are illustrated in Fig. 4 for Kr in H-clinoptilolites dealuminated to various degrees;[24] for CO_2 in erionite[24] in Fig. 5; and for I_2 in chabazite[23] in Fig. 6. Sometimes there are ranges in θ where the Langmuir quotients or their logarithms are nearly linear func-

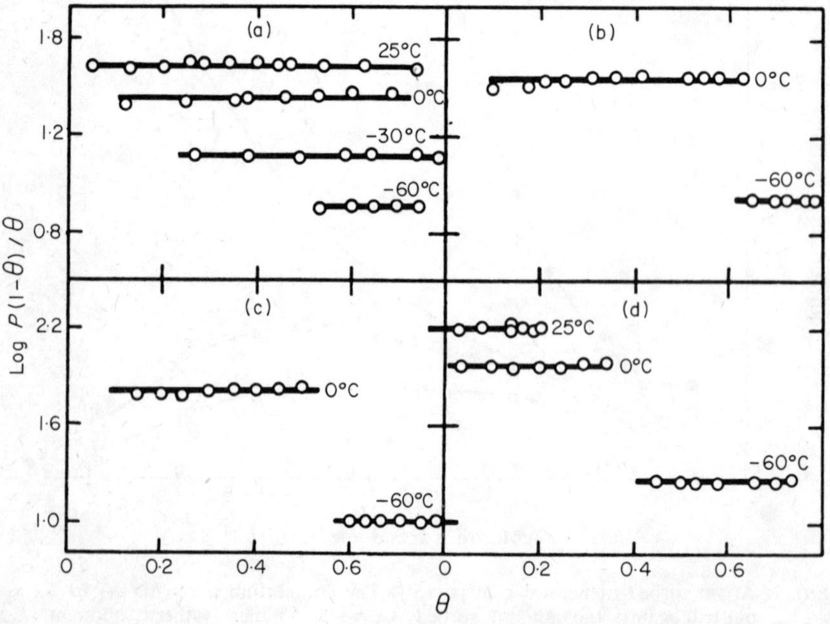

FIG. 4. The quotient, $p(1-\theta)/\theta$, as a function of θ for Kr at different temperatures in clinoptilolites treated with (a) 0·25N, (b) 0·5N, (c) 1N and (d) 2N hydrochloric acid.[24]

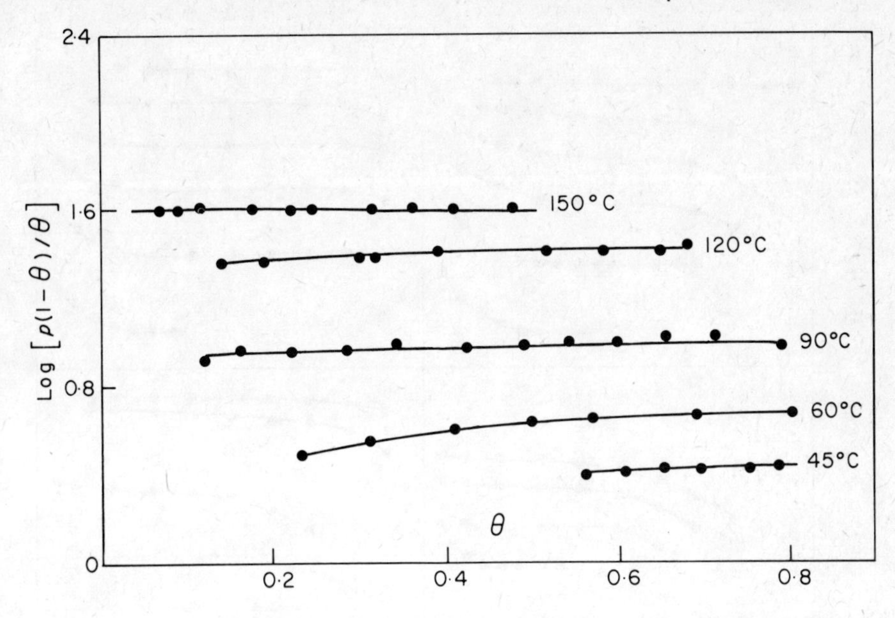

FIG. 5. Log $[p(1 - \theta)/\theta]$ against θ for CO_2 in erionite at different temperatures.[24]

tions of θ, as seen for NH_3 in (Cs, Na)-X[22] at several temperatures in Fig. 7. This shows the reasonable validity of the equilibrium quotients K_2' (the first approximation to the isotherm equation for localized sorption with interaction). In other systems the situation is more complex, as shown in Fig. 8[20] and 9[3] for hydrocarbons in (Na, K)-*L* and in H-chabazite respectively.

3. REASONS FOR DEVIATIONS FROM IDEAL LOCALIZED ISOTHERM EQUATIONS

More complex behaviour than that of localized sorption models is by no means unexpected. The differential heats of sorption are usually functions of θ, in part because the Langmuir–Lacher condition of uniformity of sorption "sites" is not satisfied. Interaction energies of the guest molecules with each other contribute exothermally to these differential heats and tend to offset, and in some instances to reverse, the decline in heat associated with site heterogeneity alone. Examples of both effects for I_2 in Na-X, Ca-*A* and chabazite[28] are shown in Fig. 10; the I_2–I_2 interactions cause q_{st}, the isosteric heat, to increase for larger θ. The horizontal dashed line gives the heat of evaporation of molecular iodine. Situations where heterogeneity is

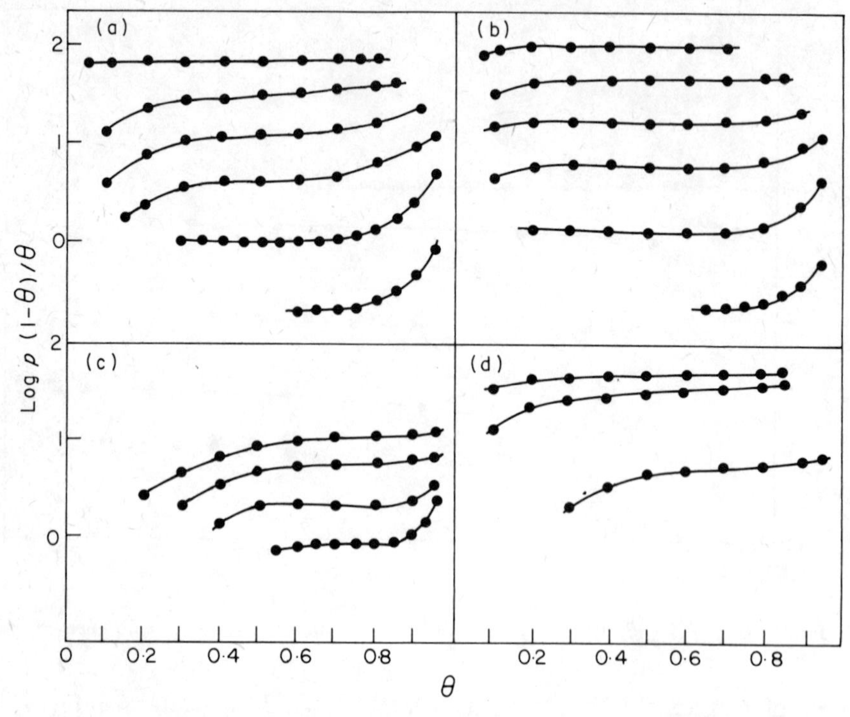

FIG. 6. Log $[p(1-\theta)/\theta]$ against θ for I_2 in (a) zeolite Na-X, (b) zeolite Ca-*A* and (c) chabazite.[23] For (a) and (b) the curves reading from the top downwards are at 573, 538, 503, 468, 428 and 393K respectively. For (c) the temperatures are 573, 538, 501 and 468K respectively. (d) compares these quotients for Na-X, Ca-*A* and chabazite at 538K (top, middle and bottom curves respectively).

marginally less important or marginally more important than sorbate–sorbate interactions are illustrated in Fig. 11 for some hydrocarbons in H-*L*.[3] For CH_4 interaction energies with other CH_4 molecules are understandably small, so site heterogeneity prevails. For n-C_4H_{10} molecules these energies are considerable, and the isosteric heats increase almost from the beginning. They decline only when the crystals are approaching saturation so that it becomes difficult for spatial reasons to insert more n-C_4H_{10} into the zeolite. All such heats must necessarily approach the latent heat of vaporization when capillary condensation between crystallites takes over from intracrystalline sorption.

An additional property which may play an important part in determining the degree of validity of the isotherm equations of Langmuir and of Lacher is the thermal entropy of the sorbate. The variation of $q_{st} = -\Delta\bar{H}$ with amount sorbed, shown in Fig. 10 for I_2 in several zeolites, would be expected to lead to deviations from the Langmuir isotherm equation even for localized

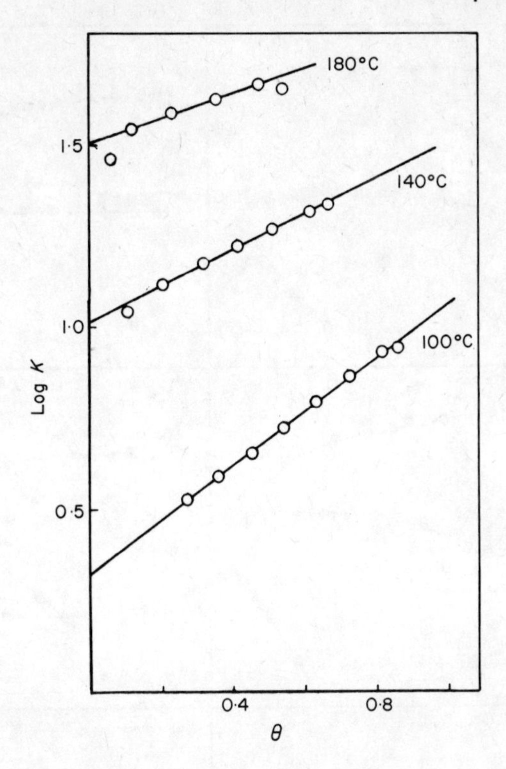

FIG. 7. Log K against θ for NH_3 in (Cs, Na)-X,[22] where $K = 1/K_1 = p(1 - \theta)/\theta$.

sorption. In fact Fig. 6 shows that this equation can describe the uptake of I_2 reasonably over considerable ranges in θ. At equilibrium $\Delta\bar{G} = 0$ and $\Delta\bar{H}/T = \Delta\bar{S} = \bar{S}_s - \tilde{S}_g$ where $\bar{S}_s$ is the differential entropy of sorbed I_2 and $\tilde{S}_g$ is its molar entropy in the gas phase. Thus, if $\tilde{S}_g^{\ominus}$ is the standard entropy of gaseous I_2 at temperature T, one has

$$\Delta\bar{H}/T = \bar{S}_s - \tilde{S}_g^{\ominus} + R \ln p/p^{\ominus} \tag{14}$$

and so

$$-\bar{S}_c - R \ln p/p^{\ominus} = -\Delta\bar{H}/T + (\bar{S}_s - \bar{S}_c) - \tilde{S}_g^{\ominus} = -\Delta\bar{H}/T + \bar{S}_{Th} - \tilde{S}_g^{\ominus} \tag{15}$$

where $\bar{S}_c$ is the configurational entropy and $\bar{S}_{Th}$ the thermal entropy of I_2 in the crystals and $p^{\ominus}$ is standard pressure. If

$$\Delta\bar{H} - T\Delta\bar{S}_{Th} = \text{constant} \tag{16}$$

where $\Delta S_{Th} = \bar{S}_{Th} - \tilde{S}_g^{\ominus}$, then the probability of occupation by I_2 will be the same for all sites even though these are *energetically* heterogeneous,

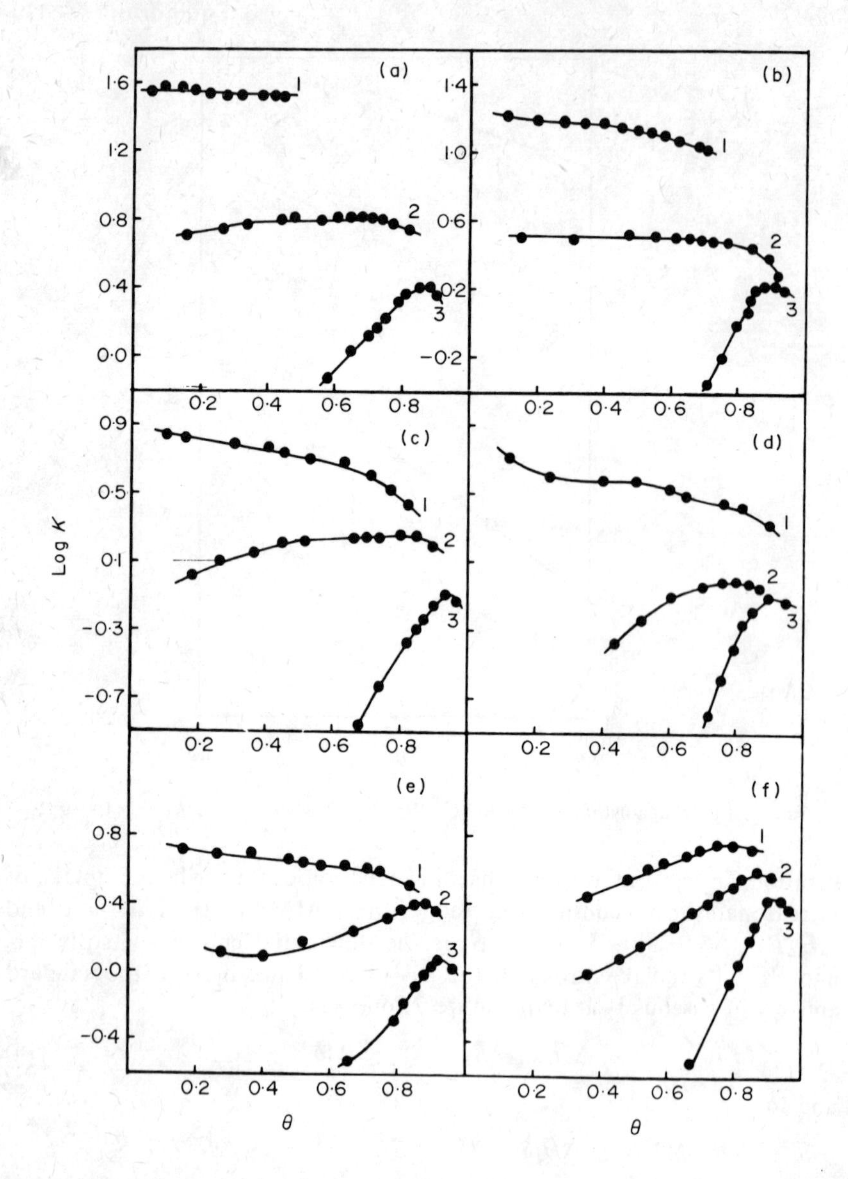

FIG. 8. Plots of log K vs θ for isotherms of hydrocarbons in zeolite L, where $K = 1/K_1 = p(1-\theta)/\theta$.[20] (a) CH_4: (1) $-30 \cdot 0$°C, (2) $-80 \cdot 6$°C, (3) $-117 \cdot 2$°C; (b) C_2H_6: (1) $46 \cdot 2$°C, (2) $0 \cdot 0$°C, (3) -62°C; (c) C_3H_8 (1) $87 \cdot 1$°C, (2) $42 \cdot 2$°C, (3) -20°C; (d) $n-C_4H_{10}$: (1) 130°C, (2) $71 \cdot 0$°C, (3) $0 \cdot 0$°C; (e) iso-C_4H_{10}: (1) 132°C, (2) $81 \cdot 0$°C, (3) $11 \cdot 0$°C; (f) neo-C_5H_{12}: (1) $143 \cdot 0$°C, (2) $98 \cdot 0$°C, (3) $30 \cdot 0$°C.

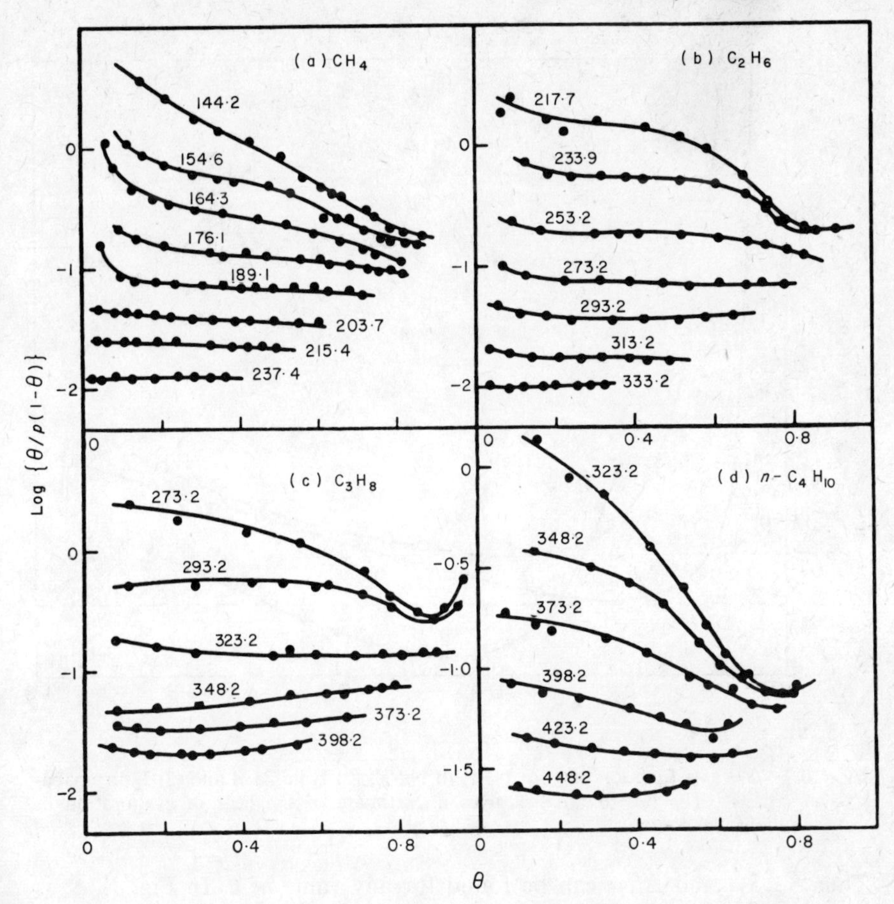

FIG. 9. Log $[\theta/p(1-\theta)]$ vs θ for some hydrocarbons in H-chabazite.[3] Figures on curves are absolute temperatures.

and so random distribution will be found with

$$\bar{S}_c = R \ln (1-\theta)/\theta \tag{17}$$

as for ideal localized sorption. Thus from eqns 15 and 16 and with $p^{\ominus} = 1$ atm

$$\frac{\theta}{p(1-\theta)} = \exp\left[-\Delta\bar{H}/RT + \Delta\bar{S}_{Th}/R\right] = \text{constant} \tag{18}$$

over all the range in θ for which eqn 16 holds.

If C_p is the constant pressure heat capacity of gaseous molecules the differential entropy of the sorbed molecules can be obtained from the relation

$$\bar{S}_s = (\tilde{S}_g^{\ominus})_{298} + C_p \ln T/298 + R \ln p^{\ominus}/p + \Delta\bar{H}/T. \tag{19}$$

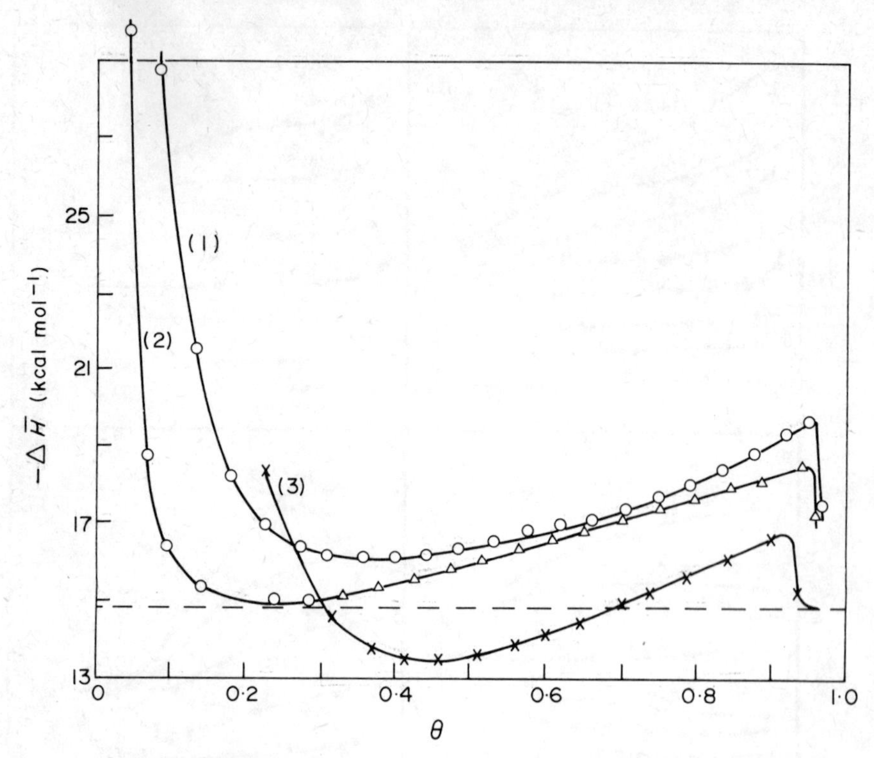

FIG. 10. $-\Delta\bar{H}$ as a function of θ for (1) I_2 in Na-X, (2) I_2 in Ca-*A* and (3) I_2 in chabazite.[28] The horizontal line gives an estimate of the heat of evaporation of molecular iodine.

Thus $\bar{S}_s$, $\bar{S}_{Th}$ and $\Delta\bar{S}_{Th}$ can be found for any value of θ. In Fig. 12 $\bar{S}_s$ is plotted for I_2 as a function of θ at each of a series of temperatures in Na-X, Ca-*A* and chabazite.[23] $\bar{S}_s$ has a positive temperature coefficient and in each system there is a maximum qualitatively corresponding with the minimum in q_{st} (Fig. 10). Eqn 16 can be rewritten as

$$\bar{S}_{Th} = \Delta\bar{H}/T + \tilde{S}_g^{\Theta} + \text{constant} \tag{20}$$

and plots of $\bar{S}_{Th}$ against $\Delta\bar{H}$, at 503K were indeed linear for I_2 in each of the three zeolites.

Other systems in which there is at least partial compensation between $\Delta\bar{S}_{Th}$ and $\Delta\bar{H}/T$ have been observed for gas–zeolite systems by Barrer and Rees,[29] for hydrocarbons in Ca-*A* by Schirmer *et al.*[17] and for other sorbents by Everett.[10] The tendency to compensation is a general one, since $\Delta\bar{H}$ and $\Delta\bar{S}_{Th}$ are each negative, and may explain why some isotherms obey Langmuir's isotherm equation over considerable ranges in θ. The correlation between $\Delta\bar{H}/T$ and $\Delta\bar{S}_{Th}$ extends not only to a given gas on a

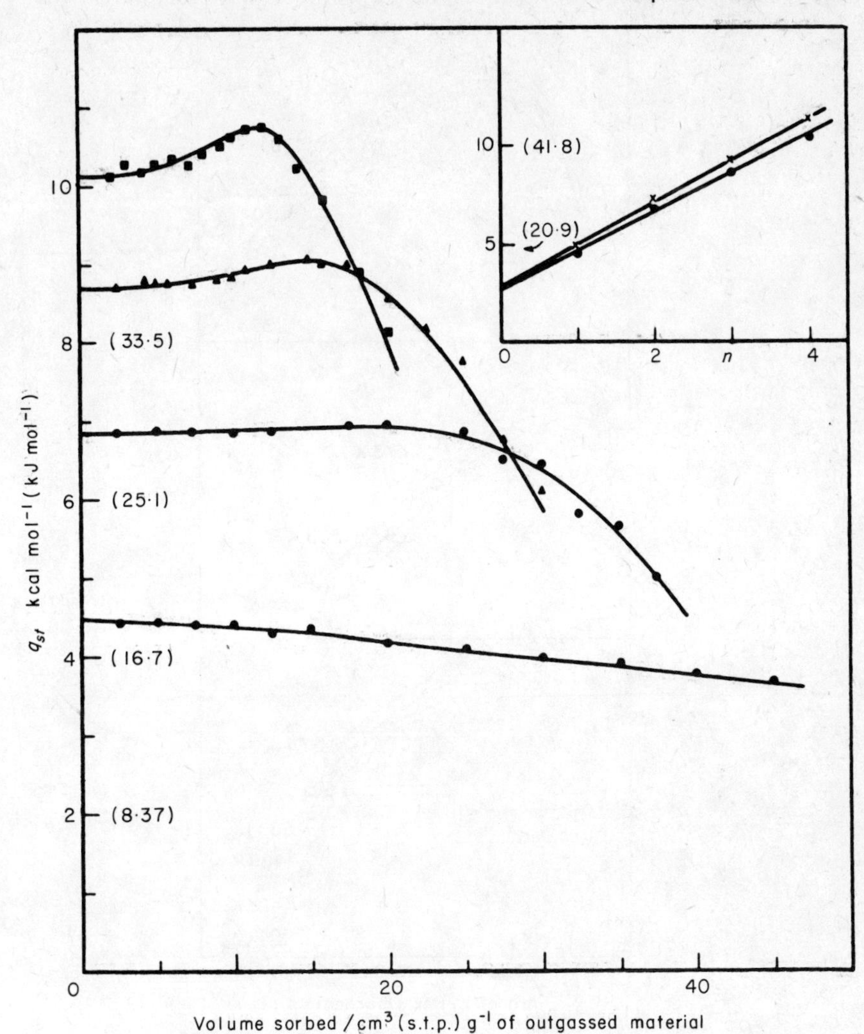

FIG. 11. $q_{st} = -\Delta\bar{H}$ in H-*L* as a function of $\theta^{(3)}$ for CH_4 (bottom) to *n*-C_4H_{10} (top). Inset: plots of initial value of q_{st} vs carbon number for K-*L* (top curve) and for H-*L* (bottom curve). Figures in brackets are heats in kJmol^{-1}.

given sorbent over a range in θ, but to different sorbates in a given sorbent. The more negative $\Delta\bar{H}$ is, the less thermal entropy the sorbate possesses and so the more negative $\Delta\bar{S}_{Th}$ becomes.

Langmuir's and Lacher's isotherm equations each depend upon the concept of a fixed number of sites with one molecule per site. However, in zeolites the number of such sites must vary with the size of the guest molecules: fewer large than small balls fill a bucket. This is illustrated in

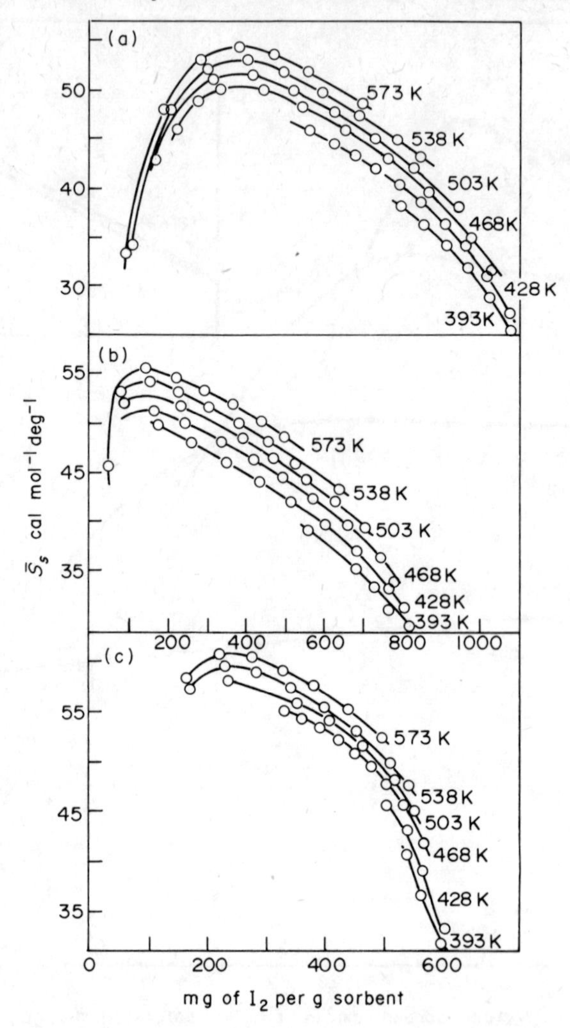

FIG. 12. $\bar{S}_s$ as a function of amount sorbed for (a) zeolite Na-X, (b) zeolite Ca-*A* and (c) chabazite.[23] Scale of the abscissa of (a) and of (b) is the same.

Table 3 by the numbers of molecules needed to fill the large 26-hedral cavity in faujasite.

4. HETEROGENEITY OF SORBENTS

It is of interest to consider heterogeneity in greater detail. One may have energetic heterogeneity in which the isosteric heat of sorption, $q_{st} = -\Delta\bar{H}$, decreases as the amount sorbed increases, or one may have heterogeneity in the quantity $\Delta\bar{H} - T\Delta\bar{S}_{Th}$, or both. If there is energetic heterogeneity but if $\Delta\bar{H} - T\Delta\bar{S}_{Th}$ is constant ideal isotherm behaviour can be preserved, as noted in the previous section. However one may consider the situation which arises when there is heterogeneity in both $\Delta\bar{H}$ and $\Delta\bar{H} - T\Delta\bar{S}_{Th}$.

As an approximation one may take a sorbent which is composed of n

TABLE 3

Numbers of molecules filling a large 26-hedral cavity in faujasite[30]

Guest molecule	Approximate saturation number
H_2O	32 (28 + 4[a])
Ar, N_2, O_2	17 to 19
I_2	7·5
CF_4	7·8
SF_6	6·5
C_2F_6	5·8
cyclopentane	5·6
benzene	5·4
toluene	4·6
n-pentane	4·5
cyclohexane	4·1
perfluorocyclobutane	4·1
$C_2F_4Cl_2$	4·1
n-heptane	3·5
perfluoropropane	3·4
n-perfluorobutane	2·8
isooctane	2·8
perfluoromethylcyclohexane	2·3
perfluorodimethylcyclohexane	2·1

[a] The four molecules are in a sodalite cage; the 28 are in the 26-hedral cage.

homogeneous parts, each part being the sum of volume elements distributed within each unit cell and so throughout the pore volume of the crystal. Thus the ith part, having volume V_i, is composed of elements $\delta V_i(\Sigma \delta V_i = V_i)$. Then if V_s is the total accessible intracrystalline sorption volume $\sum_{i=1}^{n} V_i = V_s$. For the guest molecules present in each part at equilibrium with the gas phase

$$\mu_g = \mu_1 = \mu_2 = \ldots = \mu_i = \ldots \mu_n \tag{21}$$

while the corresponding equilibrium constants $K_1, K_2, \ldots K_i, \ldots K_n$ are related by

$$\frac{1}{a_g} = \frac{K}{a_s} = \frac{K_1}{a_1} = \frac{K_2}{a_2} = \ldots = \frac{K_i}{a_i} = \ldots = \frac{K_n}{a_n}. \tag{22}$$

In eqn 22 the overall equilibrium constant, K, is included, and the subscript "eq" has been dropped.

If n_s is the total number of mols sorbed and n_i is this number in V_i then $C_s = n_s/V_s$, $C_i = n_i/V_i$ and $\sum_{i=1}^{n} n_i = n_s$. In the Henry's law range

$$K = K_c = \frac{C_s}{C_g} = \sum_{i=1}^{n} C_i V_i / C_g V_s, \tag{23}$$

while outside the Henry's law range, but assuming a perfect gas in the gas phase

$$K_c = C_s \gamma_s / C_g = \sum_{i=1}^{n} (C_i V_i) \gamma_s / C_g V_s \tag{24}$$

which defines the overall activity coefficient, γ_s, and hence activity a_s. The relations between the K's of eqn 22 may be variously rearranged, added etc. to give, among other possibilities,

$$\left.\begin{aligned} K &= a_s \sum_{i=1}^{n} K_i \Big/ \sum_{i=1}^{n} a_i \\ K &= a_s \left(\frac{K_1 K_2 \ldots K_n}{a_1 a_2 \ldots a_n}\right)^{1/n} \\ \Delta\mu^\ominus &= \frac{1}{n} \sum_{i=1}^{n} \Delta\mu_i^\ominus + \frac{RT}{n} \ln\left\{\prod_{i=1}^{n} (a_i)\right\} - RT \ln a_s. \end{aligned}\right\} \tag{25}$$

Such relations are of course all embodied in the original eqn 22.

Hobson and Armstrong[31] measured the adsorption of N_2 on pyrex glass between 90·2 and 63·3 K and down to 10^{-8} cm Hg and found that the slope $s = \left(\frac{\partial \ln\theta}{\partial \ln p}\right)_T$ had not even then reached the Henry's law limit of unity. This behaviour was considered by Sparnaay[32] to arise as a result of heterogeneity among the sorption sites, so that when Henry's law was approached for adsorption on some, on others sorption still fell outside this range. Although for the most part measured at higher temperatures and pressures the family of isotherms for N_2 in H-mordenite obtained by Barrer and Peterson[33] and shown in Fig. 13 reproduces the behaviour reported by Hobson and Armstrong. Sparnaay's analysis evidently has relevance for zeolite sorbents.

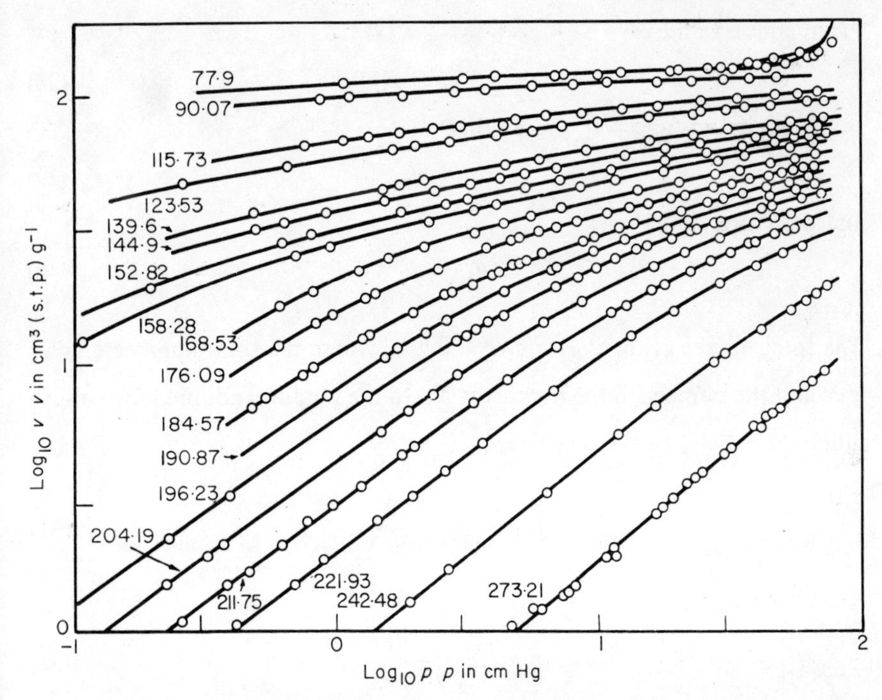

FIG. 13. Isotherms for N_2 in H-mordenite.[33] Numbers on curves are absolute temperatures.

Sparnaay considered a surface composed of *n* parts on each of which ideal localized sorption occurred with a different affinity. Thus eqn 21 represents the equilibrium condition. He then generalized the treatment for continuous distributions of sorption affinities and, in a second paper,[34] for localized sorption with interaction according to eqn 12. For ideal localized sorption on the *i*th part of the homotattic sorbent

$$\Delta\mu_i^\ominus = \mu_g^\ominus - \mu_i^\ominus = RT\left[\ln\frac{\theta_i}{1-\theta_i} - \ln p\right] \tag{26}$$

where $\Delta\mu_i^\ominus$ is the standard free energy for desorption. The heterogeneities were so arranged that $\Delta\mu_1^\ominus$ has the smallest and $\Delta\mu_n^\ominus$ the largest value ($\Delta\mu_i^\ominus < \Delta\mu_{i+1}^\ominus$ and so at the same pressure and temperature $\theta_i < \theta_{i+1}$). Further parameters were defined as follows:

$$\left.\begin{aligned} A_i = (\mu_1^\ominus - \mu_i^\ominus)/RT &= \ln\frac{\theta_i}{1-\theta_i} - \ln\frac{\theta_1}{1-\theta_1} \\ \chi_i &= \exp - A_i\,(1 \geq \chi_i \geq 0) \\ u &= \exp - \Delta\mu_1^\ominus/RT \\ \chi_i u &= \exp - \Delta\mu_i^\ominus/RT. \end{aligned}\right\} \tag{27}$$

From eqns 26 and 27

$$\theta_i = p/(p + \chi_i u). \tag{28}$$

In many circumstances $\theta_1 \ll 1$ so that

$$\theta_1 = p/u \tag{29}$$

and therefore

$$\theta_i = \theta_1/(\theta_1 + \chi_i). \tag{30}$$

The total uptake is then $m = \theta N = \sum_{i=1}^{n} \theta_i N_i$ where the total number of sites is N and the number in the ith part is N_i. In the notation adopted for volume filling $m = C_s V_s = \sum_{i=1}^{n} C_i V_i$. Thus

$$m = p \sum_{i=1}^{n} \frac{N_i}{p + \chi_i u} \tag{31}$$

and for $\theta_1 \ll 1$

$$m = \sum_{i=1}^{n} \frac{N_i \theta_1}{\theta_1 + \chi_i}. \tag{32}$$

Differentiation of eqn 26 at constant temperature gives $\partial \ln p = \dfrac{1}{1 - \theta_i} \partial \ln \theta_i$ and differentiation of $m = \sum_{i=1}^{n} \theta_i N_i$ gives $\mathrm{d}m = \sum_{i=1}^{n} N_i \theta_i \, \partial \ln \theta_i$. Combination of these two relations gives

$$s = \left(\frac{\partial \ln m}{\partial \ln p}\right)_T = 1 - \frac{\sum_{i=1}^{n} N_i \theta_i^2}{\sum_{i=1}^{n} N_i \theta_i}. \tag{33}$$

For a homogeneous sorbent the slope, s, reduces to $(1 - \theta)$; and so to unity as $\theta \to 0$. When the sorbent is heterogeneous, since $0 < \theta_i < 1$, eqn 33 also shows that the slope cannot exceed unity and again approaches unity in the Henry's law limit.

Equations 31 and 33 may be expressed in forms suitable for integration by writing

$$N_i = Nbf(\chi_i)\mathrm{d}\chi_i \tag{34}$$

where N_i is now the number of sites with χ_i-values between χ_i and $\chi_i + \mathrm{d}\chi_i$

and where $X_i = \exp - A_i$ is related to N_i through the function $f(X_i)$. The normalization factor b is given by

$$1/b = \int_0^1 f(X_i)\mathrm{d}X_i \tag{35}$$

Dropping the subscript "i" one obtains

$$\theta = pb\int_0^1 \frac{f(X)\mathrm{d}X}{p + Xu} \tag{36}$$

$$s = 1 - p\frac{\displaystyle\int_0^1 \frac{f(X)}{(p + Xu)^2}\mathrm{d}X}{\displaystyle\int_0^1 \frac{f(X)}{(p + Xu)}\mathrm{d}X} \tag{37}$$

The integration limits arise because for $A_i = 0$, $X = 1$ and for $A_i = \infty$, $X = 0$. Isotherms are given in Table 4 for six values of $f(X)$, and also the slopes when $\theta_1 \ll 1$. The first four of these cases as well as the sixth give the Henrys law slope of unity in the limit $p \to 0$. The fifth gives the slope of a Freundlich isotherm, while the sixth is essentially the slope of the Tempkin isotherm, which was given a theoretical basis by Brunauer, Love and Keenan.[35] The energetic heterogeneities for the Tempkin isotherm were accounted for by writing, in the present notation,

$$\Delta\mu_i^\ominus = \Delta\mu_1^\ominus + \omega\left(1 - \sum_{i=1}^{n} \frac{N_i}{N}\right). \tag{38}$$

Then $\Delta\mu_n^\ominus = \Delta\mu_1^\ominus + \omega$, and $\ln a$ in Table 4 is given by

$$\ln a = -\omega/RT. \tag{39}$$

The site distribution function is $N_i = (1/\ln a)\,N\mathrm{d}A_i$. For the first of the relations in Table 4 ($f(X) = X^2$) the distribution is, $N_i = -3N\exp(-3A_i)\mathrm{d}A_i$ so that sites with large A_i are almost absent. In the limit as $m \to 0$ and in a real system one must always expect Henry's law because even for θ_n as $m \to 0$, $\theta_n \ll 1$. This suggests that Freundlich and Tempkin isotherm equations and associated heterogeneity distributions do not apply at the lowest uptakes.

5. INTRACRYSTALLINE FLUIDS

In the foregoing discussion it was seen that, while equations derived from simple localized models of intracrystalline sorption can sometimes, but not

TABLE 4

Isotherm equations for various values of $f(\chi)$[32]

Case	$f(X)$	b	Isotherm $\theta = \theta(p)$	Slope[a] for $p/u = \theta_1 \ll 1$
1	X^2	3	$\theta = \frac{3p}{2u}\left[1 - \frac{2p}{u} + 2\left(\frac{p}{u}\right)^2 \ln\left(\frac{p+u}{p}\right)\right]$	$1 - \frac{2p}{u}$
2	X	2	$\theta = \frac{2p}{u}\left[1 - \frac{p}{u}\ln\left(\frac{p+u}{p}\right)\right]$	$1 - \frac{p}{u}\ln\frac{u}{p}$
3	$X^{1/2}$	3/2	$\theta = \frac{3p}{u}\left[1 - \sqrt{\frac{p}{u}}\tan^{-1}\sqrt{\frac{u}{p}}\right]$	$1 - \frac{\pi}{2}\sqrt{\frac{p}{u}}$
4	1	1	$\theta = \frac{p}{u}\left[\ln\left(\frac{p+u}{p}\right)\right]$	$1 - 1/\ln(u+p)$
5	$X^{-1/2}$	$\frac{1}{2}$	$\theta = \sqrt{\frac{p}{u}}\tan^{-1}\sqrt{\frac{u}{p}}$	1/2
6	X^{-1}	$\frac{-1}{\ln a}$	$\theta = -\frac{1}{\ln a}\left[\ln\left(\frac{u+p/a}{u+p}\right)\right]$	$p\left[\frac{\frac{1}{au+p} - \frac{1}{u+p}}{\ln\left(\frac{u+p/a}{u+p}\right)}\right]$

[a] The slope is defined as $\partial \ln\theta / \partial \ln p$. Where possible the slopes are given as the Henry's law limiting value minus the first correction term.

always, describe the experimental isotherms, those derived from Volmer's or van der Waals equations of state are less adequate. However their limitations arise essentially because these simple equations of state are unsatisfactory for dense fluids. Accordingly more comprehensive equations of state must be used to check the quantitative usefulness and accuracy of the model of sorption as a volume filling process.

Guest molecules in the sorption volume of a zeolite are usually mobile in that they diffuse rapidly and are hence fluid-like. This fluid varies continuously in density with the equilibrium pressure of sorbate up to the approximate saturation capacity of the sorption volume. One property of a dense fluid is a covolume per molecule which increases with rising temperature. Therefore the saturation uptakes as judged from the flat tops of rectangular isotherms decline slowly as the temperature at which the isotherm is measured increases. This behaviour is illustrated in Fig. 14 for n-C_6H_{14} in zeolite (Na, Ca)-A at 323, 348, and 373 K. The parallel changes in saturation capacity and bulk density of liquid sorbate are shown in

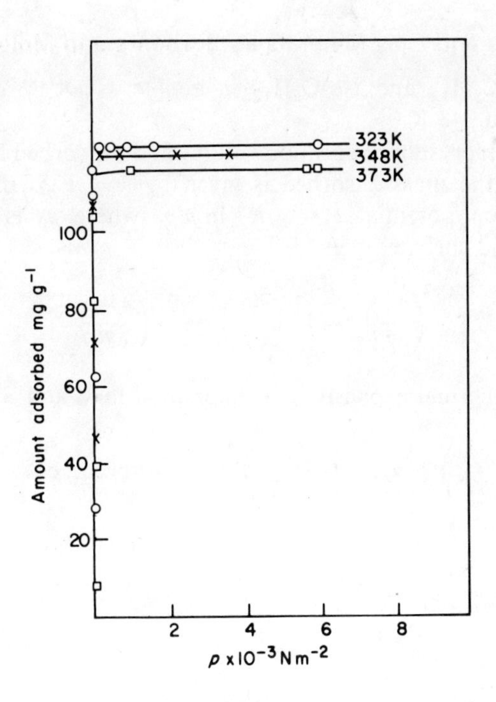

FIG. 14. Isotherms for n-C_6H_{14} in 30·55% Ca-exchanged zeolite A at three temperatures.[36]

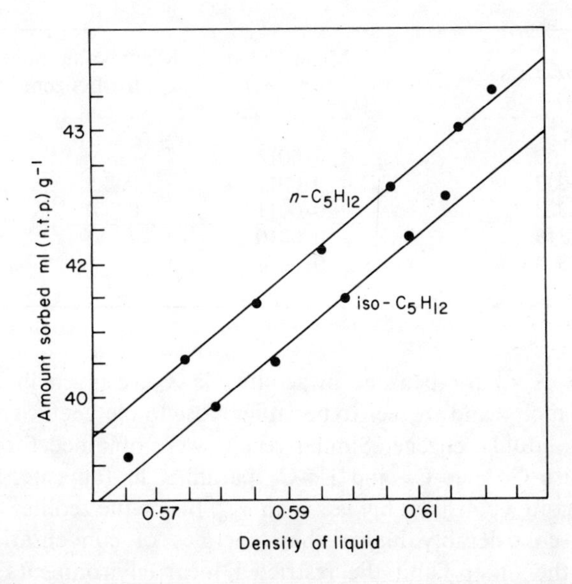

FIG. 15. The saturation capacities at each of a series of temperatures between 298 and 343K plotted against density of bulk liquids for n- and iso-C_5H_{12} in zeolite Na-X.[18]

Fig. 15 for n-C_5H_{12} and iso-C_5H_{12} in zeolite Na-X[18] at temperatures between 298 and 343 K.

If V_s is the intracrystalline sorption volume and the sorbed fluid has a mean density δ then the mass x sorbed is given by $x = V_s\delta$. If we consider a constant sorption potential $\epsilon = -RT \ln p/p_0$ where p_0 is the saturation pressure of liquid sorbate, then

$$\left(\frac{\partial x}{\partial T}\right)_\varepsilon = V_s\left(\frac{\partial \delta}{\partial T}\right)_\varepsilon + \delta\left(\frac{\partial V_s}{\partial T}\right)_\varepsilon \tag{40}$$

and if α is the thermal expansivity of interstitial fluid and α_s that of

$V_s \left(\alpha = -\frac{1}{\delta}\left(\frac{\partial \delta}{\partial T}\right)_\varepsilon ; \alpha_s = \frac{1}{V_s}\left(\frac{\partial V_s}{\partial T}\right)_\varepsilon\right)$ eqn 40 becomes

$$-\frac{1}{x}\left(\frac{\partial x}{\partial T}\right)_\varepsilon = (\alpha - \alpha_s). \tag{41}$$

TABLE 5

Comparison of $(\alpha - \alpha_s)$ at $\epsilon = 1500$ cal mol^{-1} and the thermal expansivity, α_{liq}, of liquid benzene.[19] Sorbent is Na-faujasite (Na-X)

T/K	Mean Value $(\alpha - \alpha_s)$	Mean Value of α_{liq} for Benzene
343	0·0012	
333	0·0011	
323	0·0011	0·0013
313	0·0010	
303	0·0010	

Values of $(\alpha - \alpha_s)$ for benzene in zeolite Na-X are given in Table 5 for $\epsilon = 1500$ cal mol^{-1} and are seen to be rather close to the coefficient of expansion of bulk liquid benzene. Similar results were obtained for n-C_5, n-C_6, n-C_7, n-C_8, iso-C_5, neo-C_5 and iso-C_8 paraffins in faujasite, except that $(\alpha - \alpha_s)$ was often considerably less than α_{liq}. Inside the zeolites the sorption energies are considerably higher than energies of condensation of bulk liquids. On the other hand the restricted local environments may make spatially economical packing of the molecules more difficult than in bulk fluids.

6. ISOTHERM FOR INTRACRYSTALLINE FLUIDS

One may proceed from this view of an intracrystalline fluid and interpret sorption isotherms using more accurate equations of state than hitherto. Barrer and Rees[29] used the equation of Hirschfelder *et al.*[37] for this purpose:

$$\left(p + \frac{a}{V^2} + \frac{a'}{V^3}\right)\left(V - b + \frac{b'}{V}\right) = RT, \tag{42}$$

where a, a', b and b' are coefficients and V is the molar volume at pressure p. It is claimed that this equation can represent $p - V - T$ relations of gases up to the critical density ($\theta \approx 0 \cdot 3$) and to pressures of several thousand atmospheres. The coefficients a, a', b and b' are known for gases such as Ar and N_2, as functions of T.

If the gas is sorbed into a uniform field of sorption potential ϕ the isotherm relating the molar volume in the gas phase to the molar volume V_i of sorbed fluid is

$$V = (V_i^2 - bV_i + b')^{1/2} \exp\left[\frac{2a}{RTV_i} + \frac{3a'}{2RTV_i^2} - \frac{\phi}{RT} - \frac{(bV_i - b')}{(V_i^2 - bV_i + b')} - b\left\{\frac{1}{2V_i - b} - \frac{1}{3}\frac{(4b' - b^2)}{(2V_i - b)^3} + \frac{1}{5}\frac{(4b' - b^2)^2}{(2V_i - b)^5} - \ldots\right\}\right]. \tag{43}$$

This, at sufficiently low pressures, reduces to Henry's law in the form

$$V = V_i \exp - \phi/RT \tag{44}$$

where

$$\phi = \Delta H^0 + RT. \tag{45}$$

and ΔH^0 is the limiting value of the heat of sorption in the Henry's law range.

Two adjustments were made to allow for the restricted intracrystalline environment of the fluid. Firstly when molecules were transferred from the gas into the restricted pores and channels in the crystal the thermal entropy is changed more than by compression of bulk fluid to the same V_s. Thus eqn 44 was rewritten as

$$V/V_i = K_c = \exp(\Delta S^0/R) \exp - \phi/RT. \tag{46}$$

With these adjustments to eqn 44, one obtained in the Henry's law range, for example for Ar in Ba-faujasite

$$\Delta S^0/R = -2 \cdot 23_6;\ \phi = -11 \cdot 79 \text{ kJ mol}^{-1}.$$

The second adjustment, to eqn 42 and hence to eqn 43, allowed for a decreased average coordination number, z, of a sorbed molecule relative to this number, 12, in bulk liquid, because of the restricted space inside the crystal. Accordingly the term $(a/V_i^2 + a'/V_i^3)$ in eqn 42 for sorbed fluid was multiplied by $z/12$ where z was treated as an adjustable parameter. To obtain its value, the extreme experimental point was chosen, representing the largest amount sorbed at the lowest experimental temperature above the critical temperature and z adjusted until the modified isotherm equation (eqn 43 with $\frac{za}{12}$ and $\frac{za'}{12}$ replacing a and a' and containing the term $\exp(\Delta S^0/R)$ in the right hand side) gave the correct value of p ($pV = RT$). Again for the example of Ar in Ba-faujasite one then obtained $z/12 = 0{\cdot}310_3$.

This procedure was applied to isotherms for Ar in Na- and Ba-faujasite and for N_2 in K-faujasite. The observed and calculated isotherms are illustrated in Fig. 16 for Ar in the two ion-exchanged faujasites.[29] The

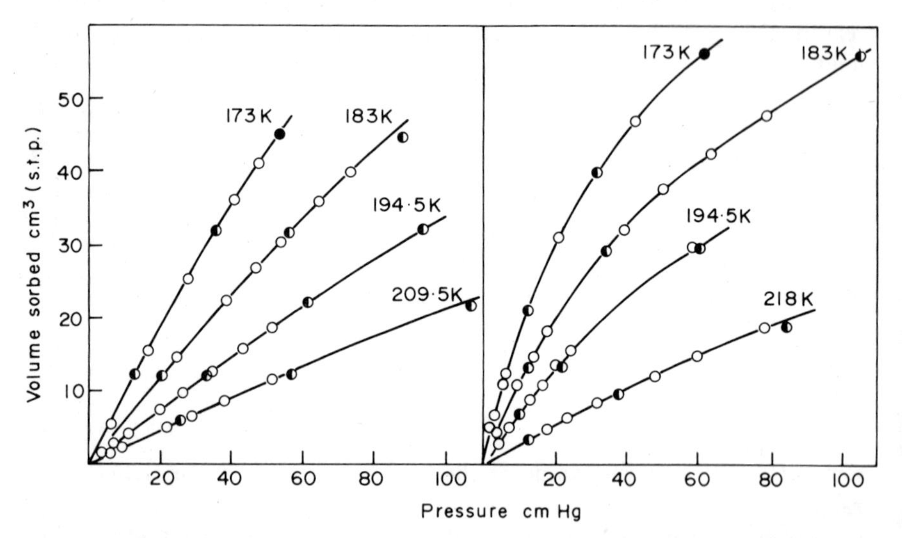

FIG. 16. Observed isotherms and isotherms calculated using an equation of state for the sorbed fluid[29]: (a) Ar in Na-X; (b) Ar in Ba-X. ● = point from which argon-argon coordination number was obtained; ◐ = calculated isotherm point; ○ = experimental isotherm point. The scale of the ordinate is the same in both halves of the diagram.

agreement is certainly good, using the bulk values of a, a', b and b' and the above two adjustments for a reduced coordination number and thermal entropy of sorbed as compared with bulk-phase fluids of the same density (same V_i). The mean pressures p_s developed in the intracrystalline pore volumes, calculated from the modified eqn 42 are given in Table 6.

TABLE 6

T/K	p_s (atm)	p (cm Hg)	Amount sorbed (cm^3 at s.t.p./g)
(a) Ar in Ba-faujasite			
173	$6{\cdot}9_7$	1·36	2·51
	63·6	12·9	21·0
	136	32·4	39·6
	222	62·6	56·2
218	12·3	12·8	3·48
	34·4	36·7	9·40
	72·2	79·1	18·7
(b) N_2 in K-faujasite			
173	11·6	0·46	4·2
	172	$9{\cdot}0_2$	45·5
	236	15·5	55·9
273	18·6	32·0	4·2
	46·3	$83{\cdot}_8$	10·0

7. THE VIRIAL SORPTION ISOTHERM

Equation 42 upon which the foregoing treatment is based is inadequate for densities beyond the critical ($\theta \approx 0{\cdot}3$). For fluids denser than this a virial equation of state with sufficient terms can be used to relate surface pressure and surface concentration of adsorbed films on external non-porous surfaces; or mean pressure, p_s, in micropores (volume filling of micropores), with mean concentration C_s in the pores; or osmotic pressure* and concentration of intracrystalline guest species in mol per unit volume of crystal (zeolitic solid solution). All three views of sorption lead to the same form of isotherm. For volume filling for example we start with

$$p_s/C_sRT = 1 + A_1C_s + A_2C_s^2 + A_3C_s^3 + \ldots \tag{47}$$

where the A are coefficients independent of C_s but dependent as a rule upon temperature. A virial equation can in general be used to represent pressure-density relations even of solids. In this sense therefore, no view of the guest species as localized, non-localized, liquid-like or solid-like need be implied in eqn 47. The isotherm, by a thermodynamic argument,[22] is then, with a perfect gas in the gas phase ($p = C_gRT$),

* The osmotic pressure is equated to the hydrostatic pressure which would raise the chemical potential of the lattice-forming units of host crystal containing the guest molecules to the value it would have in absence of guest molecules.

$$K_p = \frac{C_s}{p} \exp \{2A_1C_s + (3/2)A_2C_s^2 + (4/3)A_3C_s^3 + \ldots\} \tag{48}$$

or

$$K_c = \frac{C_s}{C_g} \exp \{2A_1C_s + (3/2)A_2C_s^2 + (4/3)A_3C_s^3 + \ldots\} \tag{49}$$

where

$$K_c = K_pRT \tag{50}$$

p (in atm) and C_g (e.g. mol dm^{-3} of gas phase) are the gas pressure and concentration respectively in equilibrium with C_s (in the same units as C_g, e.g. mol dm^{-3} of intracrystalline pore volume).

To evaluate K_p or K_c and the coefficients A one may plot $\ln(C_s/p)$ or $\ln(C_s/C_g)$ respectively against C_s. Intercepts at $C_s = 0$ give K_p or K_c, and values of $\ln(C_s/p)$ or $\ln(C_s/C_g)$ are read from the curves for as many appropriately spaced values of C_s as the numbers of coefficients A which are required.[38] Three or four coefficients normally prove sufficient for representing even strongly curved isotherms. Their values may be found by solving for the A the same number of simultaneous equations $\ln(K_cC_g/C_s) = 2A_1C_s + (3/2)A_2C_s^2 + (4/3)A_3C_s^3 + \ldots$, one for each value of C_s. A variant of this procedure has been used by Kiselev.[39]

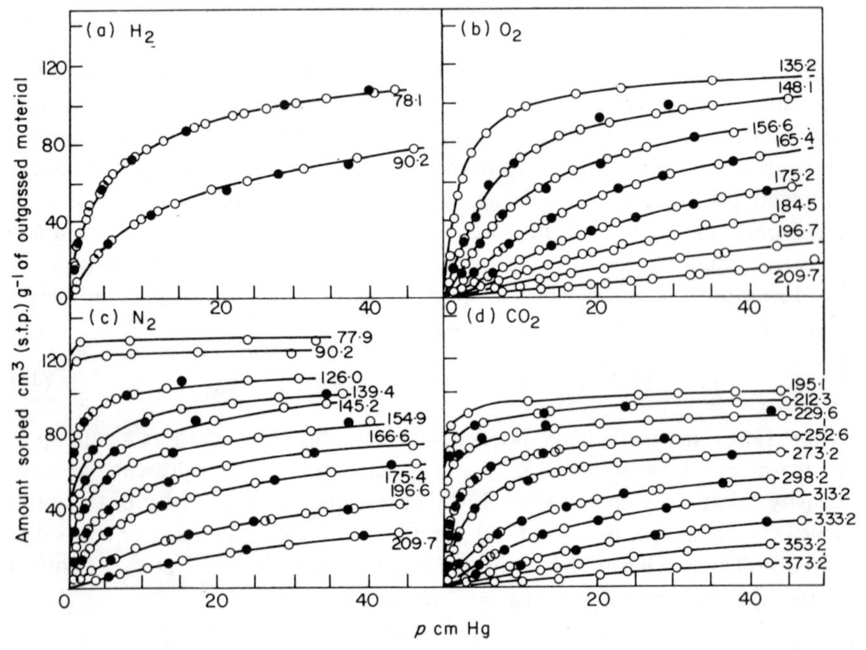

FIG. 17. Isotherms for H_2, O_2, N_2 and CO_2 in H-chabazite.[2] ○ = experimental point; ● = points calculated using eqn 50, with appropriate values of the A and of K_p. Not more than three coefficients A were required.

TABLE 7

The constants A_i^0 and a_i for gases in H-chabazite[2]

Sorbate	$10^5A_1^0$	10^2a_1	$10^8A_2^0$	10^2a_2	$10^{12}A_3^0$	10^3a_3
O_2	10	−0·60	−10·6	−0·61	16·1	−5·3
N_2	24·3	−0·21	7·0	−4·11	—	—
Ar	22·6	−0·51	−17·2	−1·05	44·4	−5·9
Kr	10·3	0·00	−26·6	−0·49	44·4	−4·4
Xe	72·0	−0·29	−25·7	−0·35	48·4	−3·9

Figure 17 illustrates for some gases in H-chabazite[2] the degree of success obtained using eqn 48 with three coefficients, A_1, A_2 and A_3, in representing the isotherms. Table 7 gives some values of the A expressed as linear functions of T, i.e.

$$A_i = A_i^0(1 + a_iT) \tag{51}$$

The numerical values of A depend on whether C_s is expressed as mol per unit volume of crystal (solution) or as mol per unit intracrystalline pore volume (volume filling). The values of A in Table 7 refer to unit volume of crystal; in Table 9, C_s refers to unit intracrystalline pore volume. The ratio of values of C_s expressed in these two ways is the density of hydrated zeolite times the pore volume per hydrated gramme.

8. THERMODYNAMIC ANALYSIS USING THE VIRIAL ISOTHERM

Equation 5 can be written as

$$K_p = C_s\gamma_s/p \tag{52}$$

where γ_s is the activity coefficient of the intracrystalline fluid. Some values of γ_s have been shown in Fig. 2 as functions of C_s for gases in H-chabazite. The standard state for sorbed gas was taken as that when $C_s\gamma_s = 1$ and that in the gas phase when $p = 1$ atm ($1{\cdot}03 \times 10^5$ N m^{-2}). Comparison with eqn. 48 shows that

$$\gamma_s = \exp\{2A_1C_s + (3/2)A_2C_s^2 + (4/3)A_3C_s^3 + \ldots\} \tag{53}$$

Also (cf eqns 9 and 10)

$$\left.\begin{aligned} \Delta H^{\ominus} &= RT^2 \,\mathrm{dln}\, K_p/dT \\ \Delta S^{\ominus}_{K_p} &= R\ln K_p + RT\,\mathrm{dln}\, K_p/\mathrm{d}T. \end{aligned}\right\} \tag{54}$$

Values of these standard thermodynamic quantities and of $C_s^\ominus$, the value of C_s for which $a_s^\ominus = 1$, are illustrated for K_p and K_c in Table 8 and 9. Some plots of ln K_p against $1/T$ from which $\Delta H^\ominus$ was derived in Table 8 are shown in Fig. 18.[2] By re-arranging* eqn 48 and differentiating ln$p$ with respect to

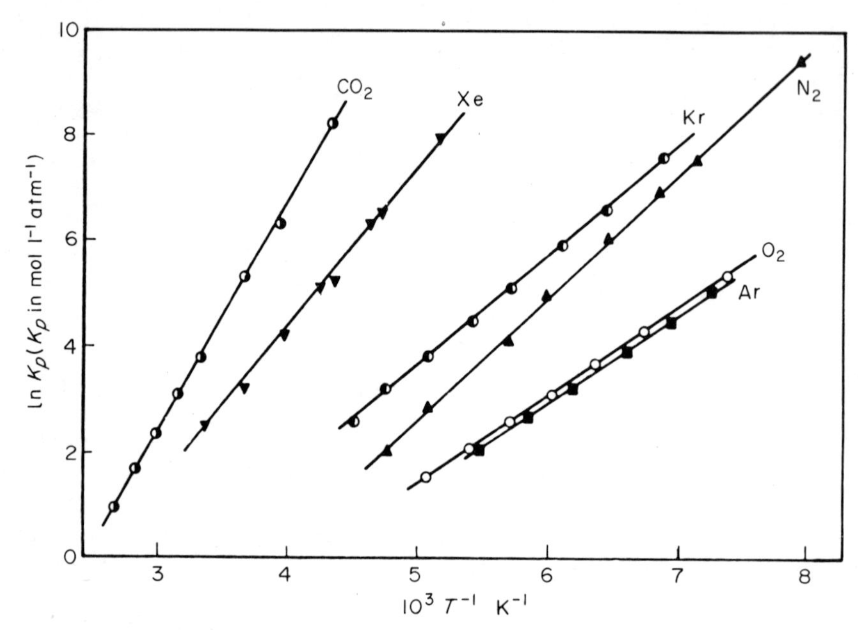

FIG. 18. Plots of ln K_p against reciprocal absolute temperature for each of a series of gases sorbed in H-chabazite.[2]

T keeping C_s constant one obtains for the variation in the heat of sorption with amount sorbed

$$\Delta \bar{H} = \Delta H^\ominus - RT^2 \left[2C_s \left(\frac{\partial A_1}{\partial T}\right)_{C_s} + (3/2)C_s^2 \left(\frac{\partial A_2}{\partial T}\right)_{C_s} + (4/3)C_s^3 \left(\frac{\partial A_3}{\partial T}\right)_{C_s} + \ldots \right] \quad (55)$$

A similar relation was derived by Kiselev *et al.*[39, 41] For those A_i given in Table 7, $(\partial A_i/\partial T)_{Cs} = A_i^\circ a_i$. Substitution of these values in eqn 55 and of the relations of eqn 51 into eqn 48 gives

$$K_p = \frac{C_s}{p} \exp \frac{(\Delta H^\ominus - \Delta \bar{H})}{RT} \exp\left(2A_1^\circ C_s + (3/2)A_2^\circ C_s^2 + (4/3)A_3^\circ C_s^3 + \ldots\right) \quad (56)$$

* It is assumed as in §1 that neither sorption volume nor the volume of the sorbent depends upon amount sorbed or on temperature. Thus constant C_s represents a constant amount sorbed for either sorption thermodynamics (volume filling) or solution thermodynamics.

In this relation the coefficients A_i° are independent both of temperature and of C_s.

Reference to Tables 8 and 9 also shows that $\Delta S^{\ominus}_{K_p}$ does not depend significantly upon temperature. This behaviour can be expected from the relation

$$\frac{\partial \Delta S^{\ominus}}{\partial T} = \frac{1}{T}\frac{\partial \Delta H^{\ominus}}{\partial T} \tag{57}$$

TABLE 8

Standard states and thermodynamic functions for gases in H-chabazite[2]

Sorbate	T/K	$10^{-3}C^{\ominus}$ mol m^{-3} of crystal	$-10^{-3}RT\ln K_p$ J mol^{-1}	$10^{-3}\Delta H^{\ominus}$ J mol^{-1}	$\Delta S^{\ominus}_{K_p}$ J mol^{-1}K^{-1}
H_2	78·1	0·75	−3·60	−8·12	−57·7
	90·2	0·75	−3·05		−56·2
O_2	135·2	≈0·99	−5·95	$-14{\cdot}3_5$	−62·1
	148·1	≈0·98	−5·32		−61·2
	156·4	0·97	−4·73		−61·2
	165·4	0·95	−4·23		−61·2
	175·2	0·94	−3·77		−60·4
	184·5	0·92	−3·23		−60·4
	196·7	—	−2·51		−60·4
N_2	126·0	0·75	−9·84	$-20{\cdot}0_5$	−81·3
	139·4	0·75	−8·77		−81·0
	145·2	0·76	−8·37		−80·4
	154·9	0·76	−7·83		−79·3
	166·6	0·76	−6·95		−78·9
	175·4	0·75	−6·00		−80·4
	196·6	0·77	−4·69		−78·5
	209·7	0·77	−3·48		−79·3
Ar	137·8	≈0·98	−5·74	$-14{\cdot}2_0$	−62·0
	144·3	0·97	−5·40		−62·4
	151·7	0·96	−4·93		−61·0
	162·0	0·93	−4·36		−61·0
	171·4	0·92	−3·85		−60·6
	183·2	0·89	−3·26		−59·8
Xe	211·4	—	−11·41	$-24{\cdot}7_0$	−63·3
	216·8	0·78	−11·31		−62·0
	228·5	0·82	−9·96		−64·5
	234·0	0·82	−9·96		−62·8
	251·2	0·76	−9·21		−62·0
	273·2	0·78	−7·28		−63·7
	296·0	—	−6·08		−63·3

because, as seen in Fig. 18, there was no observable curvature in plots of $\ln K_p$ against $1/T$. Thus, within the error of measurement $\partial \Delta H^{\ominus}/\partial T$ is zero, and therefore so is $\partial \Delta S^{\ominus}/\partial T$. Nevertheless $\Delta H^{\ominus}$ should have a temperature coefficient, normally small and corresponding with the change in heat capacity upon sorption and upon special factors which will be considered later (Chapter 5).

The above results may be taken as typical of others which have been obtained for hydrocarbons,[3,20,42] ammonia,[22] carbon dioxide[24] and nitrogen[43] in various porous crystals, using the virial isotherm to evaluate

TABLE 9

K_c, $\Delta E^{\ominus}$ and $\Delta S^{\ominus}_{K_c}$ for Kr in some zeolites[a(40)]

Zeolites	T(°K)	K_c	K_0	$-\Delta E^{\ominus}$ kJ/mol	$-\Delta S^{\ominus}_{K_c}$ kJ/mol-deg
H-offretite	144·4	$4{\cdot}7_4 \times 10^5$	$2{\cdot}8_8 \times 10^{-2}$	20·0	29·5
(channel, 1-D)	163·8	$6{\cdot}6_1 \times 10^4$			29·7
	190·7	$8{\cdot}6_1 \times 10^3$			29·4
	213·8	$2{\cdot}19 \times 10^3$			29·5
	231·8	$8{\cdot}8_1 \times 10^2$			29·8
				Average	29·6
H-erionite	141·9	$4{\cdot}6_8 \times 10^4$	$3{\cdot}7_6 \times 10^{-2}$	17·0	30·6
(cage, 3-D)	171·3	$7{\cdot}0_2 \times 10^3$			25·2
	183·4	$2{\cdot}4_5 \times 10^3$			27·4
	211·5	$7{\cdot}2_4 \times 10^2$			25·4
	229·3	$1{\cdot}80 \times 10^2$			30·9
				Average	27·9
H-*L*	142·5	$1{\cdot}78 \times 10^4$	$3{\cdot}2_3 \times 10^{-2}$	15·4	26·6
(channel, 1-D)	161·7	$4{\cdot}0_5 \times 10^3$			27·1
	189·9	$6{\cdot}9_2 \times 10^2$			26·7
	210·1	$2{\cdot}6_8 \times 10^2$			26·8
	230·6	$1{\cdot}21 \times 10^2$			26·8
				Average	26·8
Li-*L*	143·5	$2{\cdot}6_7 \times 10^4$	$8{\cdot}6_5 \times 10^{-2}$	15·1	20·4
(channel, 1-D)	162·1	$6{\cdot}0_8 \times 10^3$			20·5
	176·6	$2{\cdot}4_0 \times 10^3$			20·7
	195·3	$9{\cdot}1_4 \times 10^2$			20·5
	221·0	$3{\cdot}0_9 \times 10^2$			20·5
				Average	20·5

[a] In calculating K_c, C_s was expressed in mol dm^{-3} of intracrystalline pore volume and C_g in mol dm^{-3} of gas phase. Also 1-D and 3-D mean respectively that guest molecules can migrate in one or in all three directions in the crystals.

equilibrium constants and thermodynamic quantities. It can be hoped in this way that sorption in zeolites may be progressively systematized. The virial isotherm does not involve the concept of a strict saturation capacity as do site models.

In a study of various gases and simple hydrocarbons sorbed in zeolite Na-A, Eagan and Anderson[43] considered a somewhat different form of two-dimensional virial equation which was an extension of Volmer's equation of state. For volume filling of intracrystalline pores the surface concept is inappropriate, but the same isotherm equation is obtained if their virial equation is rewritten as

$$p_s(V_i - b) = RT\left(1 + \frac{\beta}{V_i - b} + \frac{\gamma}{(V_i - b)^2} + \ldots\right) \tag{58}$$

where p_s is the mean pressure of the intracrystalline fluid when the molecular volume is V_i, and the molecular volume when the pore space is saturated is b. With $\theta = b/V_i$ the isotherm equation is then

$$-\ln K = \ln \frac{\theta}{p(1-\theta)} + \frac{\theta}{1-\theta} + \frac{\beta}{b}\frac{\theta(2-\theta)}{(1-\theta)^2} + \frac{\gamma}{2b^2}\frac{\theta^2(3-\theta)}{(1-\theta)^3} + \ldots \tag{59}$$

which is seen as an extension of Volmer's isotherm. It is again emphasized that eqn 58 need not imply a localized or a non-localized sorption process. Like eqn 47, eqn 58 is a formal representation of the pressure-density relations of the intracrystalline sorbate whatever its state. As with eqn 48 or 49 coefficients β, γ, . . . can be chosen in sufficient numbers to represent isotherms adequately, as the authors have demonstrated. Equation 59 also reduces to Henry's law in the limit and so lends itself to thermodynamic analysis in the same way as eqn 48 or 49. The isotherm equation was successfully fitted to the experimental isotherms of various gases by a linear least squares procedure. For C_2H_6 the constants K, β/b and γ/b^2 were required, but for the other sorbates only K and β/b were needed for adequate representation of the experimental isotherms. Values of ln K and the constants β/b and γ/b^2 are given in Table 10. For molecules without quadrupole moments β/b is negative, but positive for N_2, CO and CO_2 which possess such moments. The equilibrium constant K as defined by Eagan and Anderson refers to desorption rather than sorption as considered in defining K_c and K_p in § 1, and K_1 to K_4 in § 2. It thus increases with rising temperature and is numerically greater for weakly sorbed than for strongly sorbed gases.

9. ALTERNATIVE ISOTHERMS FOR VOLUME FILLING

A general non-mechanistic approach was developed by Polanyi.[45] Sorbed films form in the field of the surface. The sorption potential ϵ_i at a point where the molar volume of sorbed fluid is V_i is defined as

TABLE 10

K, β/b and γ/b^2 (eqn 61) for gases in zeolite Na-A[43]

Gas	Saturation uptake (cm^3 at s.t.p. g^{-1} anhydrous	T(°C)	ln K (K in mm)	β/b	γ/b^2
Ar	$219{\cdot}_3$	−100	$7{\cdot}4_0$	$-0{\cdot}42_4$	
		−90	$7{\cdot}8_1$	$-0{\cdot}42_0$	
		−80	$8{\cdot}2_2$	$-0{\cdot}34_1$	
		−78	$8{\cdot}3_4$	$-0{\cdot}28_4$	
CH_4	$162{\cdot}_9$	−20	$7{\cdot}1_7$	$-0{\cdot}31_8$	
		−10	$7{\cdot}5_3$	$-0{\cdot}39_1$	
		0	$7{\cdot}9_0$	$-0{\cdot}32_2$	
C_2H_6	$119{\cdot}_7$	0	$3{\cdot}6_2$	−0·121	0·088
		10	$4{\cdot}3_9$	$-0{\cdot}35_2$	$0{\cdot}23_1$
		20	$4{\cdot}8_7$	$-0{\cdot}37_2$	$0{\cdot}24_1$
O_2	$226{\cdot}_0$	−150	$2{\cdot}0_2$	−0·017	
		−130	$4{\cdot}5_1$	−0·014	
		−120	$5{\cdot}4_7$	−0·047	
		−110	$6{\cdot}0_2$	−0·053	
		−100	$6{\cdot}9_3$	−0·254	
N_2	$181{\cdot}_6$	−78	$4{\cdot}5_4$	$0{\cdot}54_0$	
		−70	$5{\cdot}3_7$	0·178	
		−60	$6{\cdot}1_4$	$0{\cdot}40_1$	
		−50	$6{\cdot}7_6$	$0{\cdot}29_8$	
CO	$183{\cdot}_0$	−80	$2{\cdot}9_2$	$0{\cdot}27_0$	
		−70	$2{\cdot}9_6$	$0{\cdot}42_2$	
		−60	$3{\cdot}3_0$	$0{\cdot}55_1$	
CO_2	$223{\cdot}_8$	−70	$0{\cdot}073_5$	$0{\cdot}026_2$	
		−60	$1{\cdot}10_2$	$0{\cdot}037_6$	
		−30	$2{\cdot}2_6$	$0{\cdot}20_5$	
		−20	$2{\cdot}0_2$	$0{\cdot}38_3$	
		−10	$2{\cdot}2_4$	$0{\cdot}49_9$	

$$\epsilon_i = \int_{V_0}^{V_i} \left(V \frac{\mathrm{d}p}{\mathrm{d}V} \right) \mathrm{d}V \tag{60}$$

where V_0 is the molar volume of the gas. Thus ϵ_i is the work of compression from V_0 to V_i by the surface forces. Below the critical temperature the film is assumed to be liquid-like and incompressible, to grow in thickness as amount sorbed increases, and to have a sorption potential $RT \ln p_0/p$ where p_0 is the saturation vapour pressure of the liquid film. This is the negative of the free energy of transfer of sorbate from bulk liquid to the zeolite over which the equilibrium pressure is p.

For a suitable isotherm below the critical point one may plot the sorption potential against the quantity $v=a/\rho$ where a is the weight sorbed when the equilibrium pressure is p, ρ is assumed to be the liquid density and v is the volume sorbed estimated as liquid. This curve is the characteristic curve assumed independent of temperature. From it, one may therefore evaluate the isotherm at any other temperature. The Polanyi method is thus a possible way of interpolating or extrapolating isotherms provided its premises are correct. It sometimes appears that the characteristic curve for vapours in non-polar sorbents such as charcoal is indeed the same for isotherms covering a considerable range in temperature.[46, 47, 48] For polar sorbates in polar sorbents such as zeolites the situation is less clear. Dubinin, Kadlec and Zukal[49] in a study of water uptake by zeolite Na-X chose molar volumes of liquid water between the boiling point and the critical temperature in such a way as to obtain a single characteristic curve. While the molar volumes were reasonable there remains an element of adjustment in fitting the results to the single curve of ϵ against a/ρ.

Dubinin and his coworkers[50, 51, 52] extended the Polanyi method of describing sorption data to give an empirical isotherm equation with two adjustable parameters. In micropore systems such as zeolites or certain charcoals the equation was

$$\theta = a/a_0 = \exp - (A/E)^2 \tag{61}$$

where a is the amount sorbed at pressure p, a_0 (the first adjustable parameter) is the uptake at saturation of the micropore volume and E (the second adjustable parameter) is a coefficient having the dimensions of energy. Also $A = RT \ln (p_0/p)$. According to eqn 61, plots of $\ln a$ against A^2 should be straight lines of slope $-E^2$ and intercept $\ln a_0$ when $p = p_0$. Straight lines plots are often found over a range in A^2 but deviations can appear as p approaches p_0 and also for large values of A.[53,54, 55, 40, 56, 57] The linear parts of the curves have been used successfully to estimate saturation capacities of micropores in zeolites.[53, 55, 40, 56]

For larger molecules sorbed in the voids in zeolites it is reported[58] that isotherms are better represented, as might be expected, by an equation with three adjustable parameters:

$$a/a_0 = \exp - (A/E)^n \tag{62}$$

where n is an integer greater than 2, usually 3, 4, 5 or 6. Equations 61 and 62, like the Freundlich isotherm ($C_s = kC_g^m$, $m < 1$) fail to reduce to Henry's law as $C_s \to 0$ and so they do not give the equilibrium constants K_c and the derived quantities $\Delta A^\ominus$, $\Delta E^\ominus$ and $\Delta S^\ominus$ which can serve as the basis for subsequent statistical thermodynamic interpretations.

Peterson and Redlich[59] reported that in zeolite Ca-A the sorption of n-paraffins approached Langmuir's isotherm equation in a limited range of low pressures while Freundlich's isotherm equation was obeyed at higher pressures. They therefore proposed for the quantitity q sorbed the expression

$$q = \frac{Ap}{1 + Bp^g} \tag{63}$$

where A, B and g are coefficients which are functions of temperature and of the carbon number of the n-paraffin. The coefficient g lies in the range $0 < g < 1$. The isotherm reduces to Henry's law when the pressure is small enough and at high pressures becomes

$$q = \frac{A}{B} p^{(1-g)} \tag{64}$$

which is the Freundlich equation. Values of the coefficients were given as functions of carbon number and of reduced temperature for n-paraffins from pentane to hexadecane.

Schirmer *et al.*[17] also sought to correlate sorption data for n-alkanes up to octadecane in zeolite Ca-A by means of another empirical relation due to Grossman and Fiedler[56]:

$$\theta = \frac{p}{b + p}\left(\frac{p}{p_0}\right)^m, \tag{65}$$

where b and m are coefficients and p_0 is the saturation vapour pressure. For light hydrocarbons, m lay between 0·3 and 0·6 but for heavy hydrocarbons m became very small and the term $(p/p_0)^m$ approached unity. This equation does not reduce to Henry's law as θ and p become small. The heavy hydrocarbons, which tended to follow Langmuir's isotherm equation ($m \to 0$) were considered to do this because of compensation between $\Delta\bar{H}$ and $T\Delta\bar{S}_{Th}$ as discussed in § 3.

Aristov, Bosaček and Kiselev[60] used the equation

$$K_\theta = \frac{\theta}{p(1 - \theta)(1 + K_n\theta)} \tag{66}$$

TABLE 11

K_θ and K_n (computer values) for Xe and Kr in Li-X and Na-X[60]

Zeolite	Sorbate	T°C	$K_\theta \times 10^3$(mm^{-1})	K_n
Li-X	Xe	−30	1·89	4·53
		−45	2·91	6·07
		−60	4·11	7·75
		−80	11·23	13·41
	Kr	−78	0·17	4·16
Na-X	Xe	−30	3·90	3·06
		−45	7·78	3·66
		−60	16·6	4·08
		−90	91·91	6·88

in an analysis of the sorption of Xe and Kr by zeolites Li-X and Na-X. The term $(1 + K_n\theta)^{-1}$ allows for sorbate-sorbate interaction in a semi-empirical way. The coefficient K_n depends on temperature according to the relation $K_n = A_n \exp Q_n/RT$, where Q_n is the heat of association of sorbate molecules with each other. The isotherm reduces to Henry's law and so lends itself to the establishment of equilibrium constants K_θ and the derivation therefrom of standard thermodynamic quantities. Some values of K_θ and K_n are illustrated in Table 11. Heats $\Delta H^\ominus$ for K_θ were $-16{\cdot}2$ and $-18{\cdot}8$ kJ mol^{-1} for Xe in Li-X and Na-X respectively.

10. A STATISTICAL THERMODYNAMIC FORMULATION OF ISOTHERMS FOR INTRACRYSTALLINE SORPTION

The regular pore structure of zeolites makes them convenient for application of the statistical method. A general formulation by Brauer *et al.*[61] will be used to illustrate this method. The zeolite contains B cavities each of which may contain up to m guest molecules. It is assumed that the interaction energy between guest molecules in different cavities is negligible. If n_s is the number of cavities containing s molecules (where s can be 1, 2, . . . m) the canonical ensemble partition function Q is given by[62, 63]

$$Q(N, B, T) = \Sigma \frac{B! Q_1^{n_1} \ldots Q_s^{n_s} \ldots Q_m^{n_m}}{n_0! n_1! \ldots n_s! \ldots n_m!}, \tag{67}$$

where N is the total number of guest molecules in the zeolite cavities and Q_s is the partition function for s molecules in one cavity. The summation is carried out over all distributions $(n_0, \ldots n_s, \ldots n_m)$ compatible with the condition

$$n_1 + 2n_2 + \ldots + sn_s + \ldots + mn_m = N \tag{68}$$

The grand partition function, Ξ, for this system is

$$\Xi = [1 + Q_1\lambda + \ldots + Q_s\lambda^s + \ldots + Q_m\lambda^m]^B \tag{69}$$

or

$$\Xi = [1 + Z_1\xi + \ldots + Z_s\xi^s + \ldots + Z_m\xi^m]^B \tag{70}$$

where $\lambda = \exp \mu/kT$ is the absolute activity, $\xi \equiv \lambda(2\pi mkT/h^2)^{3/2}$ is the activity which for an ideal gas is p/kT and Z_s is the configuration integral for the system comprising s molecules in one cavity. In the quasi-classical approximation

$$Z_s = \frac{1}{s!} \int_v \exp\left\{-\frac{U_s(r_1, \ldots r_s)}{kT}\right\} dr_1 \ldots dr_s \tag{71}$$

where v is the volume of the cavity and U_s is the potential energy of s molecules adsorbed in the cavity.

If the average number of molecules in the crystal is $\langle N \rangle$ then the degree of saturation, θ, of the crystal is given by

$$m\theta = \phi = \frac{\langle N \rangle}{B} = \frac{Z_1\xi + 2Z_2\xi^2 + \ldots + mZ_m\xi^m}{1 + Z_1\xi + Z_2\xi^2 + \ldots + Z_m\xi^m}. \quad (72)$$

The quantity Z_1/kT is the limiting Henry law constant. This equation can be rearranged to give

$$-\phi + (1-\phi)Z_1\xi + \ldots + (s-\phi)Z_s\xi^s + \ldots + (m-\phi)Z_m\xi^m = 0 \quad (73)$$

After differentiating this expression with respect to T keeping $\langle N \rangle$, and therefore ϕ, constant, the isosteric heat can be obtained as

$$q_{st} = RT - \left[\frac{(1-\phi)\langle U_1 \rangle Z_1\xi + (2-\phi)\langle U_2 \rangle Z_2\xi^2 + \ldots + (m-\phi)\langle U_m \rangle Z_m\xi^m}{(1-\phi)Z_1\xi + 2(2-\phi)Z_2\xi^2 + \ldots + m(m-\phi)Z_m\xi^m}\right] \quad (74)$$

where $\langle U_s \rangle$ is the average potential energy of the system which consists of s molecules in one cavity.

At sufficiently low values of ξ the expression for ϕ may be written in virial form:

$$\phi = Z_1\xi + (2Z_2 - Z_1^2)\xi^2 + (3Z_3 - 3Z_1Z_2 + Z_1^3)\xi^3 + \ldots \quad (75)$$

By reversion of series, this equation can be rearranged to give ξ as a function of ϕ, and the result can be inserted in the equation for q_{st} to give

$$q_{st} = RT - \langle U_1 \rangle - 2\frac{Z_2}{Z_1^2}(\langle U_2 \rangle - 2\langle U_1 \rangle)\phi - 3\frac{Z_3}{Z_1^3}\left[(\langle U_3 \rangle - 3\langle U_1 \rangle) + \frac{1}{Z_3}\left(Z_1Z_2 - 4\frac{Z_2^2}{Z_1}\right)(\langle U_2 \rangle - 2\langle U_1 \rangle\right]\phi^2 + \ldots \quad (76)$$

If ρ_s denotes the average fraction of cavities containing s guest molecules then ρ_s is given by

$$\rho_s = \frac{Z_s t_+^s}{1 + Z_1t_+ + Z_2t_+^2 + \ldots + Z_mt_+^m} \quad (77)$$

where t_+ is the single positive root of the equation

$$-\phi + (1-\phi)Z_1t + (2-\phi)Z_2t^2 + \ldots + (m-\phi)Z_mt^m = 0. \quad (78)$$

By comparing eqns 78 and 73 one sees that $t_+ = \xi$. If one does not consider the interaction of the guest molecules with each other even though they are in the same cavity the expression for ρ_s becomes

$$\rho_s = \frac{\frac{1}{s!}x_+^s}{1 + x_+ + \frac{1}{2!}x_+^2 \ldots + \frac{1}{m!}x_+^m} \tag{79}$$

where x_+ is the single positive root of

$$-\phi + (1-\phi)x + (2-\phi)\frac{x^2}{2!} + \ldots + (m-\phi)\frac{x^m}{m!} = 0 \tag{80}$$

with $x = Z_1 t$. When ϕ is small and m is large the distribution of eqn 79 becomes the Poisson distribution. Figure 19 shows $\rho_s(\phi)$ against ϕ when $m = 6$, the values of s being indicated on the curves.(61)

11. APPLICATION OF THE STATISTICAL RELATIONSHIPS

If the degree of filling of cavities is very low, eqn 72 may be used, omitting all save Z_1 and Z_2 ($\rho_s = 0$ for all $s > 2$). Equation 72 with Z_1 and Z_2 only is also valid for any θ if $m = 2$. This could be true for larger guest molecules when the cavities are not too big. Z_1, Z_2, $\langle U_1 \rangle$ and $\langle U_2 \rangle$ were estimated for Ar in zeolite Na-A and for CH_4 in Ca-A. Z_1 was evaluated from eqn 71. The energy $U_1(r)$ was calculated for many positions of the guest molecule in a given 26-hedral cavity, so that it was possible to replace the integral of eqn 71 by the corresponding integral sum.(39) The energy calculations required the evaluation of the sum of dispersion, close-range repulsion and polarization energies for each position so that energy contours could be constructed within the cavity for a given probe molecule. The detailed procedures for the evaluation of these energies are given in the original papers.(64, 65, 39) Calculations of dispersion and repulsion energies, will be referred to again later in this chapter. In the Russian work for zeolite A coordinates of cations and oxygen atoms as given by Howell(66) were employed. The negative framework charge was assumed to be distributed evenly among all oxygen atoms. The average $\langle U_1 \rangle$ was next determined from U_1 by means of the equation

$$\langle U_1 \rangle = \int_v U_1(r) \exp\left\{-\frac{U_1(r)}{kT}\right\} \mathrm{d}r \Big/ \int_v \exp\left\{-\frac{U_1(r)}{kT}\right\} \mathrm{d}r. \tag{81}$$

To calculate Z_2 in eqn 72 or 75 the actual potential energy contour in a 26-hedral cavity was replaced by an energy well of radius r_0 with vertical sides and flat base of a depth $-U$ such that Z_1 as already calculated would be equal to

$$Z_1 = \frac{4\pi}{3} r_0^3 \exp - U/kT. \tag{82}$$

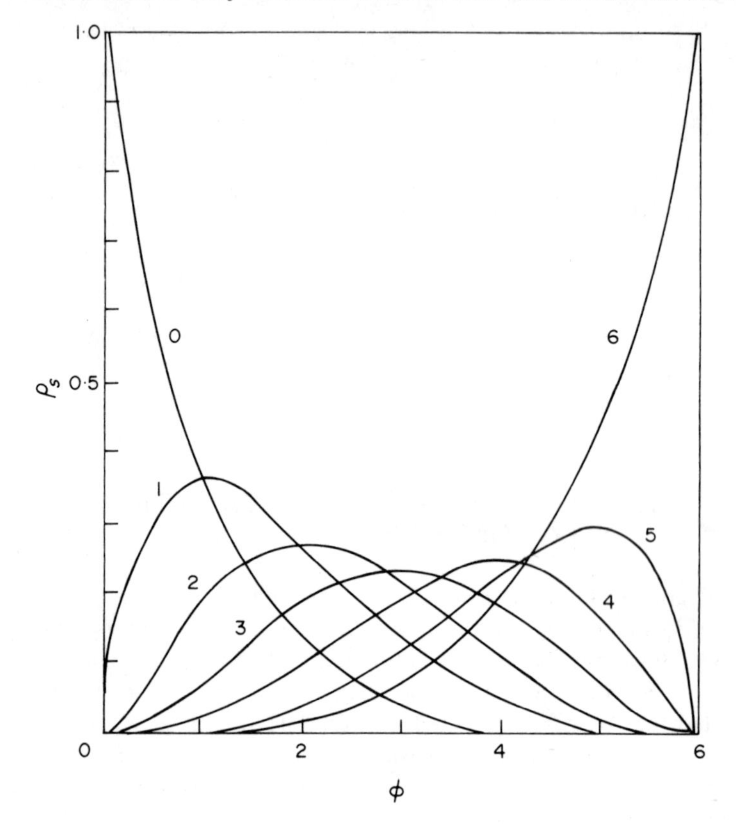

FIG. 19. Dependence of ρ_s on ϕ for a system in which $m = 6$.[62] The values of s are indicated on the corresponding curves.

For two guest molecules in the same cavity their interaction with each other was assumed to be

$$\left.\begin{aligned} F(r_{12}) &= +\infty \text{ for } r_{12} \leq l \\ &= -\omega \text{ for } t < r_{12} < R_0 l \\ &= 0 \text{ for } R_0 l \leq r_{12} \end{aligned}\right\} \tag{83}$$

where r_{12} is the distance between the centres of the interacting pair and $\omega = 0$ corresponds with the hard sphere approximation. The values assigned to ω, l and R_0 were

Ar $\omega = 138$ cal mol^{-1}; $R_0 = 1{\cdot}85$ Å; $l = 3{\cdot}16$ Å

CH_4 $\omega = 285$ cal mol^{-1}; $R_0 = 1{\cdot}60$ Å; $l = 3{\cdot}35$ Å.

The values of r_0 were taken as $4{\cdot}30$ Å for Ar in Na-A and $3{\cdot}81$ Å for CH_4 in Ca-A. With the above potential $F(r_{12})Z_2$ may be obtained explicitly from eqn 71, and thence the ratio $2Z_2/Z_1^2$ required in the expression 76 for q_{st}. If ln (Z_2/Z_1^2) is differentiated with respect to temperature and the differential is multiplied by RT^2 one also obtains the quantity $[\langle U_2 \rangle - 2\langle U_1 \rangle]$ needed in equation 76. Some results of the calculations are given in Table 12.

Calculated isosteric heats and the Henry's law parts of the isotherm agree with experimental values for Ar in Na-*A*, although, as pointed out later, these calculations can be much influenced by the choice of London, Slater–Kirkwood or Kirkwood–Muller approximations for the dispersion energy contribution to $\langle U_1 \rangle$.[67] The latter approximation was used.[65]

TABLE 12

Values of $\langle U_1 \rangle$ and of $q_1 = -2Z_2/Z_1^2\{\langle U_2 \rangle - 2\langle U_1 \rangle\}$ in kcal mol^{-1}[62]

System	T/K	50	100	150	200	250	300
Ar in Na-*A*	$\langle U_1 \rangle$	2·81	2·64	2·51	2·43	2·40	2·38
	q_1	0·30	0·15	0·12	0·11	0·10	0·09
CH_4 in Ca-*A*	$\langle U_1 \rangle$	5·30	5·15	5·02	4·85	4·70	4·55
	q_1	2·32	0·55	0·34	0·27	0·23	0·21

12. SIMPLIFIED FORMULATIONS OF THE STATISTICAL THERMODYNAMIC ISOTHERM

Although the statistical thermodynamic method is elegant in its formulation, the complexity of the physical situation is such that it cannot be applied without various simplifications,[68] some of which can be seen from the foregoing illustration of its use. These limiting cases nevertheless can be of interest and significance. Langmuir's isotherm follows from eqn 72 if $m = 1$ (1 guest molecule fills a cavity). This situation can be encountered in some porous crystals, and is considered in the next section.

One may consider in simple kinetic terms the situation which may arise if one again has a total of B identical sorption centres (these could be cations, or individual cavities) each capable of binding a maximum of m molecules. At any given pressure let there be N_0 centres unoccupied, N_1 singly occupied, N_2 doubly occupied, and so on. It is also assumed that guest molecules clustering at or in a sorption centre (ion or cavity) do not interact with one another. At equilibrium, with invariant rate constants k_1 for sorption and k_2 for evaporation of a molecule, one has, by detailed balancing,[69]

$$\left.\begin{aligned} mk_1pN_0 &= k_2N_1 \\ (m-1)k_1pN_1 &= 2k_2N_2 \\ &\cdots\cdots \\ (m-\overline{m-1})k_1pN_{m-1} &= mk_2N_m. \end{aligned}\right\} \quad (84)$$

From eqn 84

$$\left.\begin{aligned} N_1 &= mrN_0 \\ N_2 &= \frac{m(m-1)}{1\cdot 2} r^2N_0 \\ &\cdots\cdots \\ N_m &= \frac{m(m-1)\ldots 2\cdot 1}{1\cdot 2\ldots m} r^mN_0 \end{aligned}\right\} \quad (85)$$

where $r = (k_1/k_2)p$. The total number, N_s, of molecules sorbed, divided by the total number of centres (or cavities) is then

$$\left.\begin{aligned} \frac{N_s}{B} &= \frac{N_1 + 2N_2 + \ldots + mN_m}{N_0 + N_1 + \ldots + N_m} \\ &= \frac{mr(1-r)^{(m-1)}}{(1+r)^m} = \frac{mr}{(1+r)} \end{aligned}\right\} \quad (86)$$

on substituting for $N_1, N_2 \ldots N_m$ from eqn 85. Thus

$$\theta = \frac{N_s}{mB} = \frac{r}{1+r}$$

which is Langmuir's isotherm.

On the other hand, if the ratios r depend on the size of the cluster, as could happen if guest molecules interact with each other, eqn 85 become

$$\left.\begin{aligned} N_1 &= mr_1N_0 \\ N_2 &= \frac{m(m-1)}{1\cdot 2} r_1r_2N_0 \\ &\cdots\cdots \\ N_m &= \frac{m(m-1)\ldots 2\cdot 1}{1\cdot 2\ldots m} (r_1r_2\ldots r_m)N_0 \end{aligned}\right\} \quad (87)$$

where $r_1, r_2, \ldots r_m$ are no longer equal to each other as in eqn 85, and so

$$\theta = \frac{r_1\left\{1 + (m-1)r_2 + \frac{(m-1)(m-2)}{1\cdot 2} r_2r_3 + \ldots + (r_2r_3\ldots r_m)\right\}}{1 + mr_1 + \frac{m(m-1)}{1\cdot 2} r_1r_2 + \ldots + (r_1r_2\ldots r_m)} \quad (88)$$

This equation has the same form as eqn 72 and has term by term identity if

$$\left.\begin{aligned}
Z_1\xi &= mr_1 \\
Z_2\xi^2 &= \frac{m(m-1)}{1\cdot 2} r_1 r_2 \\
Z_3\xi^3 &= \frac{m(m-1)(m-2)}{1\cdot 2\cdot 3} r_1 r_2 r_3 \\
&\cdots\cdots\cdots\cdots \\
Z_m\xi^m &= r_1 r_2 \ldots r_m.
\end{aligned}\right\} \quad (89)$$

Thus a kinetic theory treatment analogous to eqn 72 can readily be developed. Equation 88 could in principle describe not only the filling of cavities in zeolites, but also the solvation of ions in zeolites, or solvation of gaseous ions.

Ruthven[70] proposed replacing the configuration integral of eqn 71 (for s molecules in a single zeolite cavity of volume V) by the approximation

$$Z_s = \frac{Z_1^s}{s!}(1 - sv_m/V)^s \exp\left(\frac{sv_m\epsilon}{VkT}\right). \quad (90)$$

He then obtained the isotherm equation

$$m\theta = \phi = \frac{Kp + (K/p)^2(1-2v_m/V)^2 \exp\left(\frac{2v_m\epsilon}{VkT}\right) + \ldots + \frac{(Kp)^m}{(m-1)!}(1-mv_m/V)^m \exp\left(\frac{mv_m\epsilon}{VkT}\right)}{1 + Kp + \frac{1}{2!}(Kp)^2(1-2v_m/V)^2 \exp\left(\frac{2v_m\epsilon}{VkT}\right) + \ldots + \frac{(Kp)^m}{m!}(1-mv_m/V)^m \exp\left(\frac{mv_m\epsilon}{VkT}\right)} \quad (91)$$

In this expression v_m is the effective covolume of a sorbed molecule, ϵ is the depth of the potential energy well when two isolated sorbate molecules interact and K is a Henry law constant. That is, for small enough ϕ, $\phi \to Kp$ and m, θ and ϕ have the same significance as in eqn 72. Ruthven and Loughlin[71] applied this equation successfully in describing experimental isotherms of light paraffins in zeolite Ca-A, and Derrah, Loughlin and Ruthven[55] showed that isotherms of simple alkenes in the same zeolite could also be described by eqn 91. Since v_m, ϵ and V can be obtained independently, and also m, provided the saturation capacity for the sorbate can be determined, this leaves only K to be found. If sorption can be measured down to the Henry law range K is also obtainable from the experimental data. However, eqn 90 is only a rough approximation for the configuration integral of eqn 71. Among other approximations involved in eqn 91 is the assumption that m is an integer and is constant, independent of temperature. This assumption is not in accord with experiment (§ 5). Nevertheless eqn 91 is relatively simple and, as already seen, independent evaluation of most of the coefficients involved is possible. As previously noted, the larger the guest molecule, the smaller m becomes and the fewer are the terms in the isotherm equation. Equation 91 (and eqn 88 and 72) become very involved for rounded values of m taken from Table 3 for small molecules in the large cavities of faujasite.

13. FURTHER CONSIDERATION OF EQUILIBRIUM CONSTANTS

It was pointed out in § 7 that the virial isotherm does not involve any choice between localized and non-localized sorption. While some properties of guest species suggest it can be fluid-like (§ 5), as also do certain *nmr* studies of intracrystalline mobilities of guest molecules[72, 73, 74, 75] and direct studies of diffusion[76, 77, 78, 36, 79] and tracer diffusion,[77, 80] this does not mean that the equilibrium properties may not correspond reasonably with those of oscillators. To have a high diffusion rate in a zeolite, only a small fraction of the molecules need be mobile at any moment. The equilibrium properties of the fluid would then be dominated by the great majority of the immobile but oscillating molecules. One may therefore consider how far K_p or K_c can be interpreted in terms of an oscillator model.

The situation most ideally close to that envisaged for surfaces by Langmuir arises if the intracrystalline pore volume consists of cells each able to accommodate one guest molecule only and in which all such cells are identical (Z_2, Z_3 . . . in eqn 72 are then all zero). The porous crystalline silicas, tridymite and cristobalite, have been shown to sorb He and Ne at high pressure and temperature[81] and should closely approach Langmuir's model. Likewise outgassed sodalite hydrate can sorb Ar and Kr at high pressures and temperatures and only one rare gas atom is present per sodalite cage.[82] The van der Waals diameters of Ar and Kr are respectively $3{\cdot}8_3$ and $3{\cdot}9_4$ Å and the free diameter of the sodalite cage is $\approx 6{\cdot}6$ Å. Again the conditions are appropriate for the validity of Langmuir's isotherm equation

$$K_1 = \theta/f(1-\theta) \tag{92}$$

with f denoting fugacity. The fugacity was determined from eqn 8 of § 1 with appropriate virial data. Some values of K_1 are given in Table 13. The samples S1 and S2 may differ a little in purity and in the amounts of NaOH adventitiously incorporated during crystal growth.

If the inert gas atoms are regarded as a set of simple harmonic oscillators within the sodalite vibrating with mean frequency ν then K_1 is given by

$$K_1 = \frac{1}{k_0 T}\frac{h^3}{(2\pi mkT)^{3/2}}\left(2\sinh\frac{h\nu}{kT}\right)^{-3}\exp - \phi_m/RT. \tag{93}$$

In this equation k_0 is the value of Boltzmann's constant in the relation

TABLE 13

Equilibrium Constants K_1 for Ar and Kr in Sodalite Hydrates.[82] Fugacity in cm Hg

Sodalite hydrate sample	$K_1 = \theta/f(1-\theta)$			
	T/K	Ar	T/K	Kr
S_1	534	$6 \cdot 0 \times 10^{-5}$	723	$3 \cdot 3 \times 10^{-5}$
	603	$3 \cdot 5 \times 10^{-5}$	816	$2 \cdot 5 \times 10^{-5}$
	673	$2 \cdot 7 \times 10^{-5}$		
	713	$1 \cdot 7 \times 10^{-5}$		
S_2	590	$4 \cdot 1 \times 10^{-5}$	599	$8 \cdot 5 \times 10^{-5}$
	717	$3 \cdot 0 \times 10^{-5}$	723	$5 \cdot 4 \times 10^{-5}$
			816	$3 \cdot 1 \times 10^{-5}$

$f = C_g \gamma_g k_0 T$ appropriate for the chosen units of f and C_g (C_g in molecules per unit volume); m is the mass of a guest molecule; k is the ordinary value of Boltzmann's constant; and ϕ_m is the binding energy at the minimum of the potential energy well for the guest molecule. The standard heat of sorption is related to ϕ_m by

$$\Delta H^\ominus = \phi_m - 5RT/2 + (3R/2)(h\nu/k) \coth h\nu/2kT. \tag{94}$$

Since the inert gases have no permanent electric moments the most important contributions to ϕ_m arise from dispersion and close range repulsion energies, respectively ϕ_D and ϕ_R. Polarization energy ϕ_P may contribute a little, but in the symmetrical intracrystalline environment this should be slight and was neglected. $\phi = \phi_D + \phi_R$ was calculated using the Lennard–Jones 12: 6 potential and pair-wise summation of inert gas–oxygen and inert gas–Na^+ ion interactions:

$$\phi = -A_1 \Sigma \left(\frac{1}{r_i^6} - \frac{r_{0i}^6}{2r_i^{12}} \right) - A_2 \Sigma \left(\frac{1}{r_j^6} - \frac{r_{0j}^6}{2r_j^{12}} \right) \tag{95}$$

A_1 and A_2 are the dispersion energy constants for inert gas–oxygen and inert gas–Na^+ interactions respectively and r_i and r_j are the centre-to-centre distances for these two kinds of interaction. Finally r_{0i} and r_{0j} are the equilibrium separations approximated by

$$\left. \begin{aligned} r_{0i} &= \tfrac{1}{2}(r_{\text{oxygen}} + r_{\text{inert gas}}) \\ r_{0j} &= \tfrac{1}{2}(r_{\text{Na}} + r_{\text{inert gas}}) \end{aligned} \right\} \tag{96}$$

where $r_{Na^+} = 0 \cdot 95$ Å; $r_{\text{oxygen}} = 1 \cdot 35$ Å; $r_{Ar} = 1 \cdot 91_5$ Å; and $r_{Kr} = 1 \cdot 97$ Å. Dispersion energy constants A_1 and A_2 have been obtained using London, Slater–Kirkwood and Kirkwood–Muller approximations. The first of these for inert gas–oxygen leads to too small and the last to too large energies when compared with experiment.[67] The Slater–Kirkwood value is usually intermediate between the other two. In the Ar- and Kr-sodalite systems London and Slater–Kirkwood values were each used, with previously given polarizabilities and characteristic energies.[83]

From space group and unit cell data the coordinates of framework oxygens and Na^+ ions were found. By moving the probe molecule in small increments along chosen directions across the sodalite cages and by pairwise summation of interactions, energy profiles along [111], [110] and [001] axes were found. The calculations involved a group of $3 \times 3 \times 3$ unit cells for each traverse point. Interactions with Al and Si were not included because their numbers total half those of the oxygens, their polarizabilities are low, and being buried in tetrahedra of oxygens they never make direct contact with the guest molecules. Figure 20 shows for London's approximation the energy profiles for the above three directions.[82] These calculations may be taken as typical of those made for guest molecules in zeolite *A*,[39, 61]

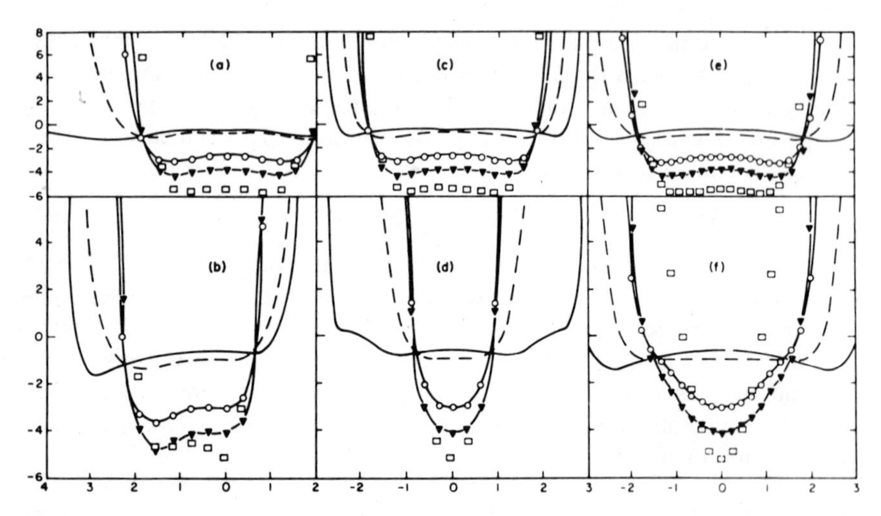

FIG. 20. Sodalite hydrate-inert gas interaction energies[85] (London dispersion energy constant).

He, ——— Kr ▼ Ne, - - - - - Xe □ Ar ○

(a) [111] traverse, inert gas–oxygen interaction. (b) [111] traverse, inert gas–(oxygen + Na^+) interaction (c) [110] traverse, inert gas–oxygen. (d) [110] traverse, inert gas–(oxygen + Na^+) (e) [001] traverse, inert gas–oxygen (f) [001] traverse, inert gas–(oxygen + Na^+). The ordinate is the energy in kcal mol^{-1} and the abscissa gives the distance from the centre of the sodalite cage in Å.

gas hydrates,[67] quinol clathrates,[83] tridymite and cristobalite,[81] cancrinite,[82] faujasite-type host crystals[28, 84, 85, 86] and mordenite.[33]

The next step involved estimating mean vibration frequencies. These frequencies may be found from the energy contours of traverses[67] such as those in Fig. 20. A classical oscillator at temperature T has average energy RT per mode ($\frac{1}{2}RT$ each of potential and kinetic energy). At the extreme of its vibration when its centre is R_0 distant from the minimum of the energy well the RT is all potential; at the minimum of the well it is all kinetic. At some intermediate point the kinetic energy is $\Delta\phi = \frac{1}{2}mv^2$. If r is the corresponding distance from the centre of the well, since $dr/dt = v(r)$ one must have for the time $t_{1/4}$ of a quarter of a vibration

$$t_{1/4} = \int_0^{R_0} \frac{dr}{v(r)} = \int_0^{R_0} \frac{dr}{(2\Delta\phi/m)^{1/2}} \tag{97}$$

$\Delta\phi$ was read off from the energy contours and $(2\Delta\phi/m)^{-1/2}$ was plotted against r. Graphical integration then gave $t_{1/4}$ and hence ν. Table 14 gives these frequencies for Ar and Kr and the traverses in the [111], [110] and [001] directions across the sodalite cage. The frequencies in Table 13 are reasonable, but they are only some of those possible for other traverses, so that some uncertainty remains as to the value to employ in eqn 93. Alternatively the experimental values of K_1 (Table 13) and the calculated ϕ_m may be used to find ν. ϕ_m may also be used with eqn 94 to evaluate $\Delta H^\ominus$, the standard

TABLE 14

Calculated Frequencies and Values of ϕ_m for Ar and Kr in Sodalite Cavities

Guest molecule		[111]	[110]	[001]
(a) London approximation for obtaining ϕ				
Ar	$-\phi_m$(cal mol^{-1})	3660	3060	3060
	$\nu \times 10^{-11}$(s^{-1})	7·0	8·3	9·7
Kr	$-\phi_m$(cal mol^{-1})	4820	4110	4110
	$\nu \times 10^{-11}$(s^{-1})	5·8	6·0	6·9
(b) Slater–Kirkwood approximation for obtaining ϕ				
Ar	$-\phi_m$(cal mol^{-1})	5750	4860	4860
	$\nu \times 10^{-11}$(s^{-1})	6·4	9·0	10·0
Kr	$-\phi_m$(cal mol^{-1})	7240	6160	6160
	$\nu \times 10^{-11}$(s^{-1})	4·7	7·2	7·5

TABLE 15

Frequencies, ν, calculated from K_1 and ϕ_m for sodalite hydrate, and heats of encapsulation, $\Delta H^\ominus$ calculated from ϕ_m

System \ Dispersion Constant	T/K	London ν (10^{-11}s^{-1})	London $-\Delta H^\ominus$ (kJ mol^{-1})	Slater–Kirkwood ν (10^{-11}s^{-1})	Slater–Kirkwood $-\Delta H^\ominus$ (kJ mol^{-1})
Sl-Ar	638	6·6	12·5	11·6	21·3
Sl-Ar	770	5·8	17·1	9·5	27·2
S2-Ar	654	5·9	12·5	10·0	21·3
S2-Kr	713	5·0	17·1	8·9	27·2

heat of encapsulation of the inert gas, which may then be compared with $\Delta H^{\ominus}$ obtained from the temperature coefficient of ln K_1. The results are summarized in Table 15. The values of ν are similar to those derived from eqn 97, and the experimental and calculated values of $\Delta \bar{H}$ are also comparable. The experimental heats for Sl-Ar and Sl-Kr were respectively $18 \cdot 0 \pm 2 \cdot 5$ and $24 \cdot 7 \pm 2 \cdot 5$ kJ mol^{-1}; and for S2-Kr this heat was $19 \cdot 6 \pm 2 \cdot 5$ kJ mol^{-1}. The success achieved is encouraging, and other studies[81] of this kind have also had a measure of success. Since $kT \gg h\nu$ the equations 93 and 94 reduce to

$$K_1 = \frac{T^{1/2}}{k_0(2\pi m)^{3/2}} \frac{k^{3/2}}{\nu^3} \exp -\phi_m/RT \qquad (98)$$

and

$$\Delta H^{\ominus} = \phi_m + \tfrac{1}{2}RT \qquad (99)$$

in which forms they were employed.

14. ABSOLUTE AND GIBBS EXCESS SORPTION

In zeolite molecular sieves guest molecules in all positions in the intracrystalline pores are within the range of forces from the cavity walls. Thus the whole of the intracrystalline pore volume is sorption volume. Also over most of the range of the relative pressure, p/p_0, adsorption upon external surfaces of the crystallites is minimal compared with the intracrystalline sorption, and can usually be neglected. Accordingly, zeolites are almost unique in providing known sorption volumes and as a result both absolute and Gibbs excess sorptions can be measured. Helium is normally used as calibrating gas to determine dead spaces. These spaces can be found with the intracrystalline pore volume full of zeolitic water, so that helium is excluded. The known intracrystalline pore volume can be added after removal of zeolitic water. This minimizes complications which could arise from sorption of helium during calibration of the apparatus and so allows even the sorption of this gas to be studied. The absolute sorption of a gas (n_s mols) may be found from the Gibbs excess uptake (n_s' mols) by adding the number n_g of mols which would be present at the equilibrium pressure in a volume of the gas phase equal to the sorption volume:

$$n_s = n_s' + n_g. \qquad (100)$$

Whereas n_s for intracrystalline sorption increases asymptotically towards an approximate saturation limit as the pressure increases, the Gibbs excess uptake increases to a maximum and then declines asymptotically to zero, as the density in the gas phase becomes more and more nearly equal to the density of the fluid in the intracrystalline pores.

Absolute and Gibbs excess sorption isotherms have been determined at high pressures and temperatures in a number of zeolites for krypton and xenon.[38,87] Examples of isotherms are shown for krypton and xenon in Ca-X in Fig. 21. The differences between absolute and Gibbs excess uptakes

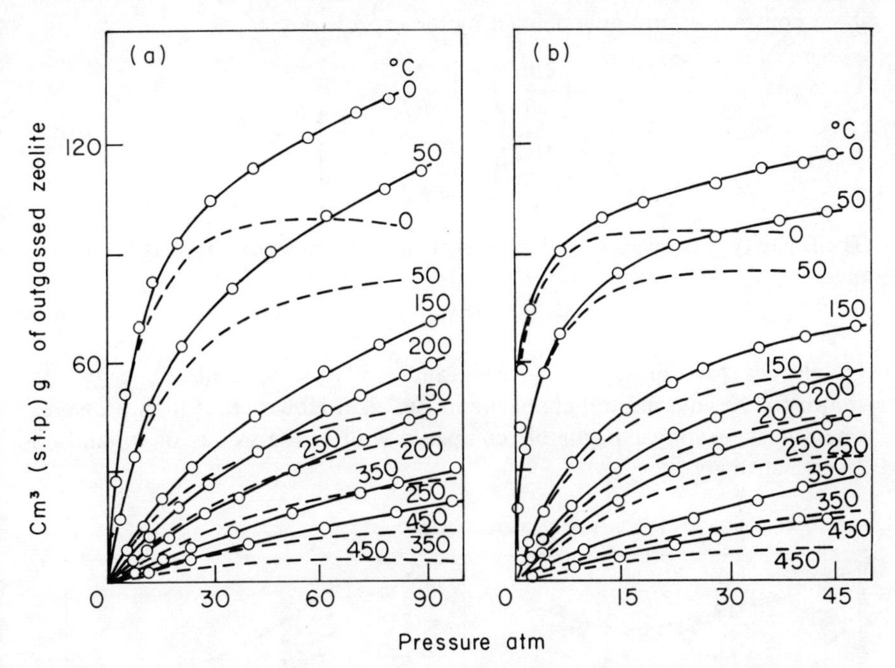

FIG. 21. Absolute (full lines) and Gibbs excess isotherms (dashed lines) of (a) Kr in Ca-X and (b) Xe in Ca-X.[87] The temperatures in °C are the numbers on the curves.

are slight at low pressures but become very significant as the pressures increase. At the lowest temperature the Gibbs excess sorptions show the expected maxima.

Isosteric heats of sorption are in general evaluated for constant Gibbs excess, n_s', in a fixed amount of sorbent. However for sorption considered either as solution (host crystal + guest molecules) or as a volume filling process the heat should be found for constant absolute sorption, n_s. Thus

$$\left(\frac{\partial p}{\partial T}\right)_{n_s} = -\frac{\Delta \bar{H}}{T(V - \bar{V}_s)} \tag{101}$$

where $\Delta\bar{H}$ is the differential heat of sorption per mol, rather than

$$\left(\frac{\partial p}{\partial T}\right)_{n_s'} = -\frac{\Delta\bar{H}'}{T(V-\bar{V}_s)} \tag{102}$$

where $\Delta\bar{H}'$ is an apparent differential heat. In these two relations $\bar{V}_s$ is the partial molar volume of sorbed guest molecules. Since neither the sorption volume V_s nor the volume of the crystal changes appreciably when guest molecules are imbibed, $\bar{V}_s \approx 0$. Two isosteric heats, q_{st} at constant n_s and q_{st}' at constant n_s', are now defined by the expressions

$$\left.\begin{aligned}\left(\frac{\partial \ln f}{\partial T}\right)_{n_s} &\equiv \frac{q_{st}}{RT^2}\\ \left(\frac{\partial \ln f}{\partial T}\right)_{n_s'} &\equiv \frac{q_{st}'}{RT^2}\end{aligned}\right\} \tag{103}$$

The fugacity f, is given by eqn 8 so that, with $\bar{V}_s \approx 0$ eqns 103, 101 and 102 give

$$q_{st} = -\Delta\bar{H} \text{ and } q_{st}' = -\Delta\bar{H}'. \tag{104}$$

Usually at low pressures $n_g \ll n_s'$ and so $n_s \approx n_s'$ and $q_{st} \approx q_{st}'$, as shown for Xe in a natural chabazite in Fig. 22a. However, at high pressures where n_g is no longer negligible compared with n_s' the values of q_{st} and q_{st}' diverge strongly.

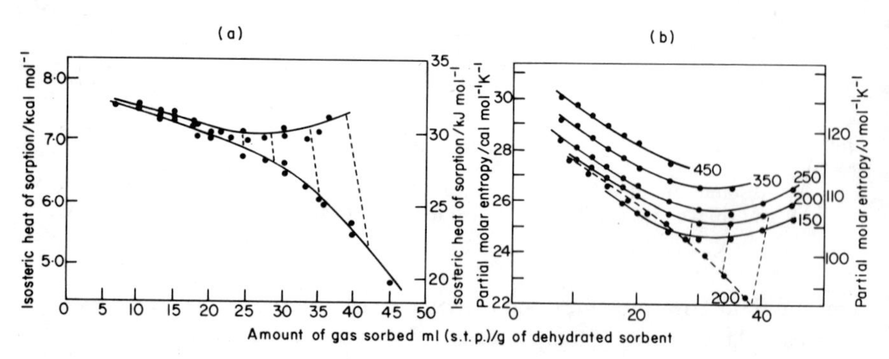

FIG. 22. (a) Isosteric heats q_{st}' (upper curve) and q_{st} (lower curve) for Xe in a natural chabazite[38], at a mean temperature of 200°C. (b) $\bar{S}_s$ (full lines) and $\bar{S}_s'$ (dashed line) for the sorption of Xe in the chabazite. Temperatures are given in °C by the numbers on the curves. Typical corresponding points of absolute and excess sorptions in both (a) and (b) are joined by the dashed lines.

For a constant amount of sorbent and for the three variables n_s, p and T fixing any two of these determines the third. The same is true of n_s', p and T. Accordingly

$$\left.\begin{aligned}\left(\frac{\partial p}{\partial T}\right)_{n_s} &= -\left(\frac{\partial n_s}{\partial T}\right)_p \Big/ \left(\frac{\partial n_s}{\partial p}\right)_T \\ \left(\frac{\partial p}{\partial T}\right)_{n_s'} &= -\left(\frac{\partial n_s'}{\partial T}\right)_p \Big/ \left(\frac{\partial n_s'}{\partial p}\right)_T\end{aligned}\right\} \quad (105)$$

and therefore

$$\frac{q_{st}'}{q_{st}} = \frac{(\partial n_s'/\partial T)_p(\partial n_s/\partial p)_T}{(\partial n_s/\partial T)_p(\partial n_s'/\partial p)_T}. \quad (106)$$

This expression gives the ratio q_{st}'/q_{st} in terms of the slopes of isobars and isotherms for n_s' and n_s. At the maximum in the isotherm of n_s' against p (Fig. 22), the slope $(\partial n_s'/\partial p)_T$ is zero and so at that value of T $q_{st}' = +\infty$ (eqns 102, 104 and 106). Beyond the maximum this slope becomes negative, so that q_{st}' returns from $-\infty$ to finite negative values. This behaviour demonstrates the artificial nature of the apparent isosteric heat $q_{st}' = -\Delta\bar{H}$. Equation 105 shows that q_{st} by contrast is always finite and positive, since isotherms for absolute sorption always have positive slopes decreasing asymptotically towards zero as $p \rightarrow \infty$.

From eqn 100 one has $dn_s = dn_s' + dn_g$. It is recalled that n_g is the number of molecules which would in the gas phase occupy a constant volume equal to the sorption volume V_s at pressure p and temperature T. It is then possible to relate the partial differentials in n_s and n_s' in eqn 106. The virial eqn 7 applies to n_g with $V = V_s/n_g$. Then[38]

$$\left(\frac{\partial n_s'}{\partial T}\right)_p = \left(\frac{\partial n_s}{\partial T}\right)_p + \frac{n_g}{T}(1 + T\phi_p) \quad (107)$$

where

$$\phi_p = \left\{\frac{\partial \ln (1 + Bp + Cp^2 + \ldots)}{\partial T}\right\}_p \quad (108)$$

and

$$\left(\frac{\partial n_s'}{\partial p}\right)_T = \left(\frac{\partial n_s}{\partial p}\right)_T + \frac{n_g^2 RT}{V_s}\left(-\frac{1}{p^2} + C + 2Dp + \ldots\right). \quad (109)$$

At a given temperature the maximum in the isotherm of n_s' against p thus occurs at a pressure p such that

$$-\left(\frac{\partial n_s}{\partial p}\right)_T = \frac{n_g^2 RT}{V_s}\left(-\frac{1}{p^2} + C + 2Dp + \ldots\right) \quad (110)$$

Equations 109 and 107 may be substituted into eqn 106 to give

$$\frac{q'_{st}}{q_{st}} = \frac{1 + (n_g/T)[1 + T\phi_p](\partial T/\partial n_s)_p}{1 + (n_g^2/RTV_s)[-1/p^2 + C + 2Dp + \ldots](\partial p/\partial n_s)_T}. \tag{111}$$

As the pressure falls and the ideal gas law is approached eqn 111 reduces to

$$\frac{q'_{st}}{q_{st}} = \frac{1 + (pV_s/RT^2)(\partial T/\partial n_s)_p}{1 - (V_s/RT)(\partial p/\partial n_s)_T} \tag{112}$$

and finally, in the Henry's law range, where $n_s = Kn_g$ and $K = K_0 \exp -\Delta E^\ominus/RT$, one obtains, with $\Delta H^\ominus = \Delta E^\ominus - RT$,

$$\frac{q'_{st}}{q_{st}} = \frac{K + RT/\Delta H^\ominus}{K - 1}. \tag{113}$$

$\Delta H^\ominus$ is negative but is expected to be greater than RT and so the right hand side of eqn 113 exceeds unity. Usually $K \gg 1$ and so in the Henry's law range q'_{st} and q_{st} are nearly equal. Thus, for Xe in chabazite at 150°C, q'_{st}/q_{st} was estimated as 1·004, although for very weakly sorbed gases such as helium q'_{st}/q_{st} in the Henry's law range might differ considerably from unity. These considerations place upon a quantitative basis the differences between q'_{st} and q_{st} shown in Fig. 22a.

It is sometimes considered that Gibbs excess sorptions at the gas solid interface are always positive. In this context some measurements of the absolute sorption of helium in zeolite *A* in the temperature range 0°C to 75°C are of interest.[88] The very small uptakes corresponded with values of $K < 1$ and so with negative Gibbs excess. In the relation $-RT\ln K = \Delta E^\ominus - T\Delta S^\ominus$ the values of $\Delta E^\ominus$ were -780 and -1300 cal mol^{-1} in Na-*A* and Ca-*A* respectively and $\Delta S^\ominus$ was $-3{\cdot}5_2$ and $-5{\cdot}4_9$ cal mol^{-1} deg^{-1}. The term $T\Delta S^\ominus$ is thus numerically larger than $\Delta E^\ominus$. For helium at still higher temperatures the corresponding K would increasingly diminish below unity, and such high temperature behaviour where $T\Delta S^\ominus$ exceeds $\Delta E^\ominus$ should be true for this gas on all sorbents. At low temperatures K will become greater than unity and the Gibbs excess will again be positive.

Differential molar entropies, $\bar{S}_s$ for $q_{st} = -\Delta\bar{H}$ and $\bar{S}'_s$ for $q'_{st} = -\Delta\bar{H}'$ may be evaluated according to eqn 14. Then there will be differences between $\bar{S}_s$ and the apparent entropy $\bar{S}'_s$ given by

$$\bar{S}_s - \bar{S}'_s = -\frac{q_{st}}{T}\left\{1 - \frac{(\partial n'_s/\partial T)_p(\partial n_s/\partial p)_T}{(\partial n_s/\partial T)_p(\partial n'_s/\partial p)_T}\right\} \tag{114}$$

where the quotient of partial differentials is given in the most general case by the right hand side of eqn 111. In the Henry's law range and for a perfect gas the counterpart of eqn 113 is

$$\bar{S}_s - \bar{S}'_s = \frac{q_{st}}{T}\left\{\frac{RT/\Delta H^\ominus + 1}{K - 1}\right\}. \tag{115}$$

Since $RT/\Delta H^{\ominus}$ is negative and usually less than unity and K is usually much greater than unity, the RHS of eqn 115 is in these circumstances expected to be positive, and $\bar{S}_s$ to exceed $\bar{S}'_s$. However, wherever the ratio $q_{st}/T(K-1)$ is much less than unity the difference between $\bar{S}_s$ and $\bar{S}'_s$ becomes small in the Henry law limit. Some values of $\bar{S}_s$ at each of a series of different temperatures, and of $\bar{S}'_s$ at 200°C are shown in Fig. 22b. As with the isosteric heats, and in accordance with the above analysis, $\bar{S}_s$ and $\bar{S}'_s$ converge as the sorption decreases. They show considerable differences for larger uptakes.

From the previous illustrations it is seen that zeolite sorbents, with known intracrystalline porosities equal to the available sorption volumes V_s, provide opportunities for studies and analyses of absolute sorption. Such studies are not possible with carbon blacks, silica gels, oxides and other standard sorbents, in which only Gibbs excess uptakes can be measured. In particular, the thermodynamic analysis to give isosteric heats and entropies and the interpretation of these in statistical thermodynamic terms relies upon identity between q_{st} and q'_{st} and between $\bar{S}_s$ and $\bar{S}'_s$. This identity is most nearly approached in the Henry's law range, but for uptakes outside this range the heats and the entropies can diverge significantly. Statistical thermodynamic formulations of isotherms like those considered in §§ 2 and 10 are made in terms of the absolute rather than the Gibbs excess populations on sites or in cavities.

15. "HYDRAULIC RADII" AND "SURFACE AREAS" OF MICROPORES

For porous sorbents the hydraulic radius $r_h = \epsilon/A$ (ϵ = porosity in cm^3 per g of sorbent and A = surface area in cm^2 per g of sorbent). The hydraulic radius is often considered to be a length related to the mean pore dimensions of the porous medium. For macroporous materials this is reasonable but for microporous sorbents the idea requires examination.[89] If the covolume of a molecule is b, the number of molecules a zeolite can accommodate is $N_{sat} = \epsilon/b$. This number multiplied by the co-area, $b^{2/3}$, is the monolayer equivalent area, A. Accordingly $r_h = \epsilon/A = b^{1/3}$ and is a characteristic not of the pore dimensions of the sorbent but of the molecular dimensions of the guest molecules. For example, with N_2 as guest species one obtained for several zeolites the values of r_h given below.

faujasite	$r_h = 3{\cdot}6$Å
zeolite A	$3{\cdot}5$Å
chabazite	$3{\cdot}6$Å
mordenite	$4{\cdot}0$Å

Nitrogen is an ellipsoidal molecule with major and minor axes 3·0Å and 4·1Å, based on van der Waals dimensions. Thus the values of r_h are in accord with the view that r_h is a mean molecular dimension and not a mean cavity dimension. As r_h increases, its value will reflect both these dimensions, while for macroporous materials only the pore dimensions will be involved.

For strongly curved cavity walls of molecular dimensions and for volume filling of these cavities the significance of the term "surface area" also needs consideration. Such surface areas are widely referred to in the literature. However what is involved in these estimates is not the true surface area of the cavity wall, but the area which would be covered by the number of guest molecules which completely fills the intracrystalline pores if these molecules were removed and placed in a close-packed monolayer upon an ideally smooth plane surface. This area has been termed the monolayer equivalent area.[90] Similar considerations apply to microporous sorbents, other than zeolites wherever the sorption process is partially or largely one of volume filling of micropores.

16. CONCLUDING REMARKS

Many of the examples discussed in this chapter, although drawn from the investigation of zeolite sorbents, are of interest for both the gas–solid interface and for microporous sorbents in general. In the distribution of a gas between the gas phase and the surface or micropores one seeks, just as in the distribution coefficient for a gas dissolving in a liquid, a quantitative and *a priori* theoretical derivation of the equilibrium constant. Two main obstacles to this are, firstly, the inadequacy of the approximations which have to be made to evaluate dispersion energies (those of London, Slater–Kirkwood and Kirkwood–Muller) and the difficulty of using rigorous quantum mechanical procedures; and secondly the heterogeneity of sorbents and the complexity of quantitative treatments of equilibria involving such sorbents. At best one must assume empirical distributions of sorption affinities in different volume elements of the interior of micropores, or in different elements of the adsorbing surfaces. Porous crystals offer real advantages over amorphous micropore sorbents in that the coordinates of the lattice atoms are known, so that interactions between these atoms and guest molecules can be evaluated. Also, the micropore volume is known, and so is its geometry. Even so, there may be more than one kind of channel and/or cavity in a given crystal; and if more than one cation is present there may be variable proportions between the cations in different cavities, even where these cavities are all of the same kind. Moreover, the sorption energy varies from point to point within a given cavity. For these reasons and also on account of structural defects the zeolites can also pose problems, even though they may come nearer to being model systems than some sorbents. Further consideration of the guest-zeolite bond is deferred to Chapter 4.

REFERENCES

1. S. Brunauer, "The Adsorption of Gases and Vapours", Oxford University Press (1944) p. 150.
2. R. M. Barrer and J. A. Davies, *Proc. Roy. soc.* (1970) A, **320**, 289.
3. R. M. Barrer and J. A. Davies, *Proc. Roy. Soc.* (1971) A, **322**, 1.
4. I. Langmuir, *J. Amer. Chem. Soc.* (1918) **40**, 1361.
5. R. H. Fowler, *Proc. Camb. Phil. Soc.* (1935) **31**, 260.
6. J. R. Lacher, *Proc. Camb. Phil. Soc.* (1937) **33**, 518.
7. J. R. Lacher, *Proc. Roy. Soc.* (1937) A, **161**, 525.
8. R. H. Fowler and E. A. Guggenheim, "Statistical Thermodynamics", Cambridge University Press (1939) p. 441.
9. M. Volumer, *Zeit. Phys. Chem.* (1925) **115**, 253.
10. D. H. Everett, *Trans. Faraday Soc.* (1950) **46**, 942 and 952.
11. T. L. Hill, *J. Chem. Phys.* (1946) **14**, 441.
12. T. L. Hill, *J. Chem. Phys.* (1946) **15**, 767.
13. R. H. Fowler and E. A. Guggenheim, "Statistical Thermodynamics" Cambridge University Press (1939) p. 430.
14. R. M. Barrer and J. L. Whiteman, *J. Chem. Soc.* (1967) A 14.
15. R. M. Barrer and J. L. Whiteman, *J. Chem. Soc.* (1967) A 19.
16. L. A. Garden, G. L. Kington and W. Laing, *Proc. Roy. Soc.* (1956) A **234**, 35.
17. W. Schirmer, G. Fiedrich, A. Grossmann and H. Stach, *in* "Molecular Sieves", Soc. Chem. Ind. (1968) p. 276.
18. R. M. Barrer and J. W. Sutherland, *Proc. Roy. Soc.* (1956) A **237**, 439.
19. R. M. Barrer, F. W. Bultitude and J. W. Sutherland, *Trans. Faraday Soc.* (1957) **53**, 1111.
20. R. M. Barrer and J. A. Lee. *Surface Sci.* (1968) **12**, 354.
21. R. M. Barrer and P. J. Reucroft, *Proc. Roy. Soc.* (1960) A **258**, 431.
22. R. M. Barrer and R. M. Gibbons, *Trans. Faraday Soc.* (1963) **59**, 2875.
23. R. M. Barrer and S. Wasilewski, *Trans. Faraday Soc.* (1961) **57**, 1153.
24. R. M. Barrer and B. Coughlan, *in* "Molecular Sieves", Soc. Chem. Ind. (1968) p. 141 and p. 233.
25. T. Takaishi, A. Yusa and F. Amakasu, *Trans. Faraday Soc.* (1971) **67**, 3565.
26. P. E. Eberly Jr., *J. Phys. Chem.* (1963) **67**, 2404.
27. T. Takaishi, A. Yusa, Y. Ogino and S. Ozawa, *Jap. J. App. Phys.* (1974) Suppl. 2, Pt. 2, 279.
28. R. M. Barrer and S. Wasilewski, *Trans. Faraday Soc.* (1961) **57**, 1140.
29. R. M. Barrer and L. V. C. Rees, *Trans. Faraday Soc.* (1959) **55**, 992.
30. See "Non-Stoichiometric Compounds", Ed. L. Mandelcorn, (Academic Press) (1963) p. 393.
31. J. P. Hobson and R. A. Armstrong, *J. Phys. Chem.* (1963) **67**, 2000.
32. M. J. Sparnaay, *Surface Sci.* (1968) **9**, 100.
33. R. M. Barrer and D. L. Peterson, *Proc. Roy. Soc.* (1964) A **280**, 466.
34. M. J. Sparnaay, *in* "Clean Surfaces", Ed. G. Goldfinger, Marcel Dekker (1970) p. 153.

35. S. Brunauer, K. S. Love and R. G. Keenan, *J. Amer. Chem. Soc.* (1942) **64**, 751.
36. R. M. Barrer and D. J. Clarke, *J. Chem. Soc., Faraday I* (1974) **70**, 535.
37. J. A. Hirschfelder, R. J. Buehler, H. A. McGee Jr. and J. R. Sutton, *Ind. Eng. Chem.* (1958) **50**, 375.
38. R. M. Barrer and R. Papadopoulos, *Proc. Roy. Soc.* (1972) A **326**, 315.
39. A. V. Kiselev *in* "Molecular Sieve Zeolites", Advances in Chemistry Series, Amer. Chem. Soc., No. 102 (1971) p. 37.
40. R. M. Barrer and I. M. Galabova, *in* "Molecular Sieves", Advances in Chemistry Series, Amer. Chem. Soc., No. 121 (1973) p. 356.
41. A. G. Bezus, A. V. Kiselev, Z. Sedlacek and Pham Quang Du. *Trans. Faraday Soc.* (1971) **67**, 468.
42. D. Barthomeuf and Baik-Hyon Ha, *J. Chem. Soc., Faraday I* (1973) **69**, 2147.
43. T. Takaishi, A. Yusa, Y. Ojina and S. Ozawa, *J. Chem. Soc., Faraday I*, (1974) **70**, 671.
44. J. D. Eagan and R. B. Anderson, *J. Coll. and Interface Sci.* (1975) **50**, 419.
45. M. Polanyi, *Verh. deut. physik. Ges.* (1914) **16**, 1012; and (1916) **18**, 55.
46. L. Berenyi, *Zeit. Phys. Chem.* (1923) **105**, 55.
47. L. Berenyi, *Zeit. Phys. Chem.* (1920) **94**, 628.
48. H. H. Lowry and P. S. Olmstead, *J. Phys. Chem.* (1927) **31**, 1601.
49. M. M. Dubinin, O. Kadlec and A. Zukal, *Coll. Czech. Chem. Comm.* (1966) **31**, 406.
50. M. M. Dubinin, Chem. Rev., (1960) **60**, 235 and Progress in Surface and Membrane Science, Academic Press (1975) Vol. 9, p. 1.
51. M. M. Dubinin and L. V. Raduschkewitsch, *Dokl. Akad. Nauk SSSR* (1947) **55**, 327.
52. B. P. Bering, M. M. Dubinin and V. V. Serpinski, *J. Coll. and Interface Sci.* (1966) **21**, 368.
53. W. Schirmer, G. Meinert and A. Grossmann, *Monatsberichte* (1969) **11**, 886.
54. P. Collin and R. Wey, *C.R. Acad. Sci., Paris* (1970) **270**, 457.
55. R. I. Derrah, K. F. Loughlin and D. M. Ruthven, *J. Chem. Soc., Faraday I.* (1972) **68**, 1947.
56. A. Grossmann and K. Fiedler, *Zeit. Phys. Chem.* (1967) **236**, 38.
57. N. N. Avgul, A. G. Bezus, E. S. Dobrova and A. V. Kiselev, *J. Coll. Interface Sci.* (1973) **42**, 486.
58. M. M. Dubinin and V. A. Astakov, *in* "Molecular Sieve Zeolites", Advances in Chemistry Series, Amer. Chem. Soc., No. 102, 1971, p. 69.
59. D. L. Peterson and O. Redlich, *J. Chem. Eng. Data* (1962) **7**, 570.
60. B. G. Aristov, V. Bosacek and A. V. Kiselev, *Trans. Faraday Soc.* (1967) **63**, 2057.
61. P. Braüer, A. A. Lopatkin and G. Ph. Stepanez, *in* "Molecular Sieve Zeolites", Advances in Chemistry Series, Amer. Chem. Soc., No. 102 (1971) p. 97.
62. V. A. Bakajev, *Dokl. Akad. Nauk USSR* (1966) 169, 369.
63. T. L. Hill, "Statistical Mechanics", McGraw Hill (1956) p. 413.
64. P. Brauer, A. V. Kiselev, E. A. Lesnik and A. A. Lopatkin, *Zh. Fiz. Khim.* (1968) **42**, 2556.
65. A. V. Kiselev and A. A. Lopatkin, *in* "Molecular Sieves", Soc. Chem. Ind. (1968) p. 252.
66. P. A. Howell, *Acta Cryst.* (1960) **13**, 737.
67. R. M. Barrer and D. J. Ruzicka, *Trans. Faraday Soc.* (1962) **58**, 2253.

68. M. M. Dubinin, *in* "Molecular Sieve Zeolites", Advances in Chemistry Series, Amer. Chem. Soc., No. 102 (1971) p. 62.
69. R. M. Barrer and L. V. C. Rees, *Trans. Faraday Soc.* (1954) **50**, 852.
70. D. M. Ruthven, *Nature, Phys. Sci.* (1971) **232**, 70.
71. D. M. Ruthven and K. F. Loughlin, *J. Chem. Soc., Faraday I* (1972) **68**, 696.
72. P. Ducros, *Bull. Soc. Franc. Mineralog. Crist.* (1960) **83**, 85.
73. D. Deininger, H. Pfeifer, F. Przyborowski, W. Schirmer and H. Stach, *Zeit. Phys. Chem.* (1970) **245**, 68.
74. P. A. Egelstaff, J. S. Downes and J. W. White, *in* "Molecular Sieves", Soc. Chem. Ind. (1968) p. 306.
75. H. Resing and J. K. Thomson, *in* "Molecular Sieve Zeolites", Advances in Chemistry Series, Amer. Chem. Soc. No. 101 (1971) p. 473.
76. R. M. Barrer and B. E. F. Fender, *J. Phys. Chem. Solids* (1961) **21**, 12.
77. R. M. Barrer *in* "Molecular Sieve Zeolites", Advances in Chemistry Series, Amer. Chem. Soc., No. 102 (1971) p. 1.
78. C. N. Satterfield and J. R. Katzer *in* "Molecular Sieve Zeolites", Advances in Chemistry Series, Amer. Chem. Soc., No. 102 (1971) p. 193.
79. R. L. Gorring, *J. Cat.* (1973) **31**, 13.
80. A. Quig and L. V. C. Rees, *in* Proc. of Third International Zeolite Symposium, Zurich, Sept. (1973) p. 277.
81. R. M. Barrer and D. E. W. Vaughan, *Trans. Faraday Soc.* (1967) **63**, 2275.
82. R. M. Barrer and D. E. W. Vaughan, *J. Phys. Chem. Solids* (1971) **32**, 731.
83. J. H. van der Waals and J. C. Platteeuw, *Adv. Chem. Physics* (1959) **2**, 2.
84. R. M. Barrer and R. M. Gibbons, *Trans. Faraday Soc.* (1963) **59**, 2569.
85. R. M. Barrer and R. M. Gibbons, *Trans. Faraday Soc.* (1965) **61**, 948.
86. E. Dempsey, *in* "Molecular Sieves", *Soc. Chem. Ind.* (1968) p. 293.
87. R. M. Barrer, R. Papadopulos and J. D. F. Ramsay, *Proc. Roy. Soc.* (1972) A **326**, 331.
88. R. M. Barrer and J. H. Petropoulos, *Surface Sci.* (1965) **3**, 126.
89. R. M. Barrer *in* "The Structure and Properties of Porous Materials", Eds D. H. Everett and F. S. Stone, Butterworth (1958) p. 239.
90. R. M. Barrer *in* "The Structure and Properties of Porous Materials", Eds D. H. Everett and F. S. Stone, Butterworth (1958) p. 49.

4

Energetics of Sorption

One of the most informative ways of investigating the guest molecule–host crystal bond is through the measurement and interpretation of heats of sorption. These heats are evaluated by direct calorimetry, from the shift of reversible equilibrium with temperature, or, under appropriate conditions, from retention times in adsorption chromatography. Heats may be differential or integral, and their magnitude and variation with amount sorbed can be considered in terms of different contributory interactions and of the energetic heterogeneity of the sorbent. Investigations involving many sorbates have been made on the faujasite-type zeolites X and Y, on zeolite *A*, mordenite, chabazite and zeolite *L*. More limited studies have been made on other zeolites including clinoptilolite, offretite, erionite, gmelinite, synthetic ferrierite and zeolite *ZK-5* (Species P and Q).

1. HEAT OF SORPTION AS A FUNCTION OF AMOUNT SORBED

The isosteric heats are defined by

$$q'_{st} \equiv RT^2 \left(\frac{\partial \ln f}{\partial T}\right)_{n_s'} \tag{1}$$

or

$$q_{st} = RT^2\left(\frac{\partial \ln f}{\partial T}\right)_{n_s} \tag{2}$$

where the sorption occurs on a fixed amount of sorbent, f denotes the fugacity and the subscripts n_s' and n_s indicate respectively constant Gibbs excess or constant absolute sorption in mols.[1,2] In most systems the difference between q_{st}' and q_{st} is negligible and the fugacity may be replaced by the equilibrium pressure, p. At higher pressures the relation is more complex:

$$\ln(f/p) = Bp + (C/2)p^2 + (D/3)p^3 + \ldots \tag{3}$$

where B, C and D are the virial coefficients of the equation of state.

The isosteric heat is positive in sign and so corresponds with the differential heat of desorption. In what follows it is assumed that $q_{st} \approx q_{st}'$. Then

$$q_{st} = -\Delta\bar{H} = -(\bar{H}_s - \tilde{H}_g) \tag{4}$$

where $\Delta\bar{H}$ is the differential heat of sorption, $\bar{H}_s$ is the partial molar enthalpy of sorbed gas and $\tilde{H}_g$ is the integral molar enthalpy of the sorbate in the gas phase. The integral heat of sorption, $\Delta\tilde{H}$, is related to $\Delta\bar{H}$ by any of the relations

$$\Delta\tilde{H} = \frac{1}{n_s}\int_0^{n_s} \Delta\bar{H}\,\mathrm{d}n_s \tag{5}$$

$$\Delta\bar{H} = \Delta\tilde{H} + n_s\frac{\mathrm{d}(\Delta\tilde{H})}{\mathrm{d}n_s} \tag{6}$$

$$\frac{\mathrm{d}(\Delta\bar{H})}{\mathrm{d}n_s} = 2\frac{\mathrm{d}(\Delta\tilde{H})}{\mathrm{d}n_s} + n_s\frac{\mathrm{d}^2(\Delta\tilde{H})}{\mathrm{d}n_s^2}. \tag{7}$$

Equation 6 shows that when $n_s = 0$ or when the curve of $\Delta\tilde{H}$ against n_s passes through a maximum or minimum $\Delta\tilde{H} = \Delta\bar{H}$. Equation 7 indicates that when $n_s = 0$ or when there is an inflexion in the curve of $\Delta\tilde{H}$ against n_s then the gradient of $\Delta\bar{H}$ is twice that of $\Delta\tilde{H}$. These relations establish the form of the curve of $\Delta\bar{H}$ vs uptake from that of $\Delta\tilde{H}$, or vice versa.

The form of plots of q_{st} against amount sorbed is illustrated for I_2 in chabazite, zeolite Ca-*A* and zeolite Na-X in Fig. 10, Chapter 3; and for C_1 to *n*-C_4 paraffins in the hydrogen form of zeolite *L* in Fig. 11 of the same chapter. Other examples are shown in Figs 1 to 5 below. Figure 1 illustrates differential heats for CO_2, CO, N_2, O_2 and Ar in a natural (Ca, Na)-chabazite, determined calorimetrically.[3] In Fig. 2, isosteric heats are plotted for C_2H_6 and C_2H_4 in various ion-exchanged forms of zeolite X.[4] Figure 3a gives these heats in zeolite (K, Na)-*L* for CH_4, C_2H_6, C_3H_8, *n*-C_4H_{10} iso-C_4H_{10} and neo-C_5H_{12}, as functions of $\phi = v/v_{sat}$ where v is the amount sorbed at the experimental temperature and v_{sat} is the saturation capacity measured at $-183 \cdot 1°C$, $-84 \cdot 5°C$, $-4 \cdot 0°C$, $0°C$, $-5 \cdot 8°C$ and $30°C$ for the hydrocarbons in the order given above.[5] Figure 3b gives integral heats for these same hydrocarbons as functions of v. Differential heats are plotted in Fig. 4 against degree of saturation of zeolite Na-X for CF_4, SF_6, C_2F_6 and C_3F_8.[6] Finally,

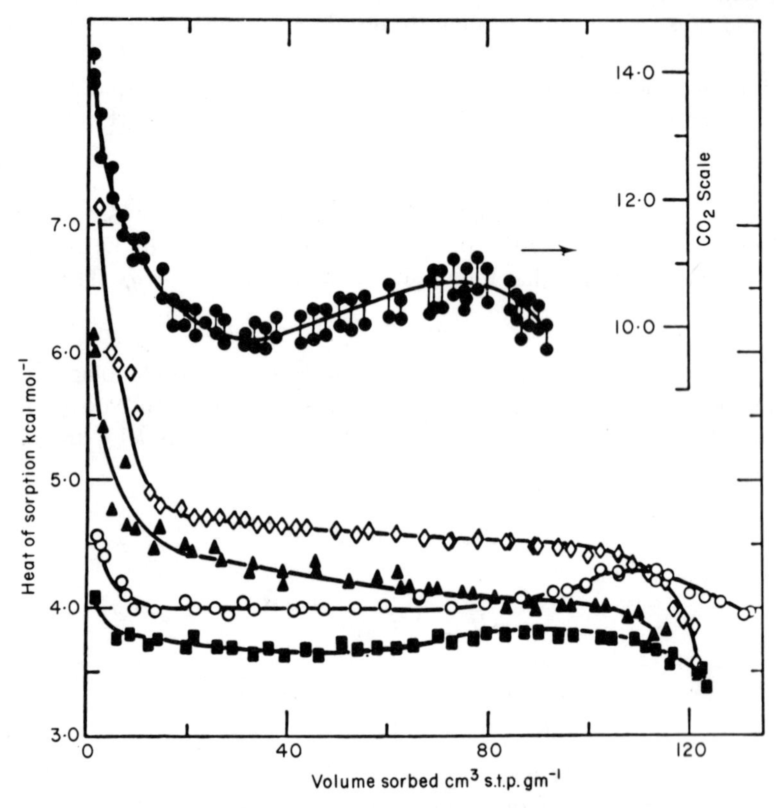

FIG. 1. Differential heats of sorption in natural chabazite.[3]
▲ = N_2; ■ = Ar; ○ = O_2; ◇ = CO; ● = CO_2

Fig. 5 gives curves of differential heats for N_2, O_2, Ar and H_2 in Na- and H-mordenites.[7] All these results, typical of many which could serve as illustrations, demonstrate a number of features:

(i) For small uptakes, q_{st} sometimes decreases rather strongly with amount sorbed (Fig. 10 in Chapter 3 and Figs 1 and 5). This indicates that in such systems there are some local intracrystalline positions where the guest molecules are preferentially sorbed more exothermally than they are in the rest of the intracrystalline volume; and that these energetically sorbing positions may be proportionately few in number. The heats do not however

always show this behaviour, particularly for larger guest molecules, as seen for hydrocarbons and fluorocarbons (Fig. 11, Chapter 3, and Figs 3a and 4).
(ii) Where q_{st} initially decreases with uptake the negative slope often becomes smaller as the amount sorbed increases (Fig. 10 in Chapter 3 and Fig. 1). At intermediate uptakes the curves sometimes become almost

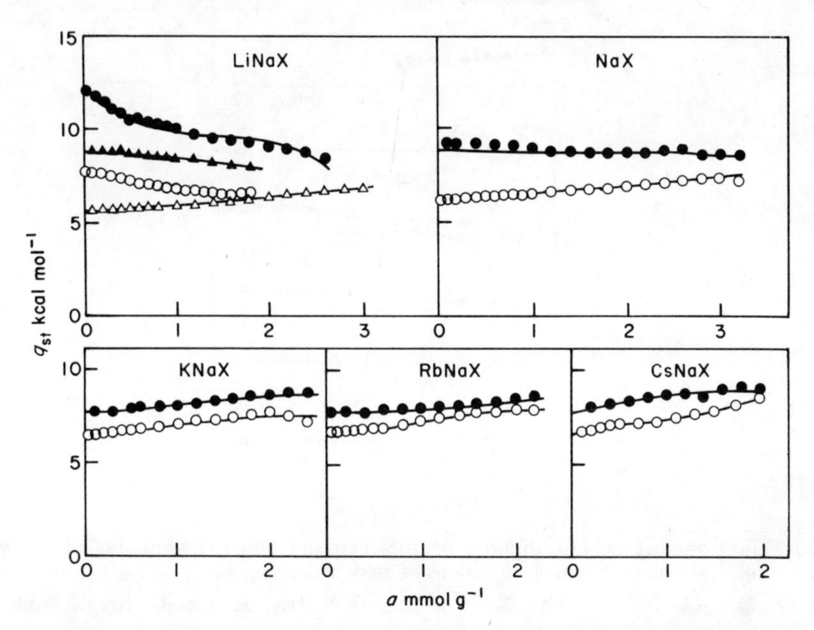

FIG. 2. Differential heats of sorption of C_2H_6 (open points) and of C_2H_4 (filled points) on various ion-exchanged forms of zeolite X.(4) Two different specimens of (Li, Na)-X were used.

horizontal and thereafter q_{st} may actually increase, an effect ascribed to exothermal sorbate-sorbate interactions as the intracrystalline fluids become denser (Fig. 10 in Chapter 3 and Figs 1 and 3). If the initial energetic heterogeneity is small, this increase is perceptible even at rather low uptakes (some curves in Fig. 11 in Chapter 3 and in Figs 2 and 3), but where the heterogeneity is very marked, the continued fall in the sorbate-lattice interaction masks the sorbate-sorbate interaction (Fig. 5, and C_2H_4 in (Li, Na)-X in Fig. 2).
(iii) For uptakes approaching saturation of the intracrystalline volume the isosteric heats all decline once more and eventually, as multilayer sorption and capillary condensation on external surfaces of the zeolites take over, these heats all approach $-\Delta H$ for liquefaction (some curves in Figs 1, 3 and 5).

Observations (i), (ii) and (iii) are also sometimes valid for non-crystalline

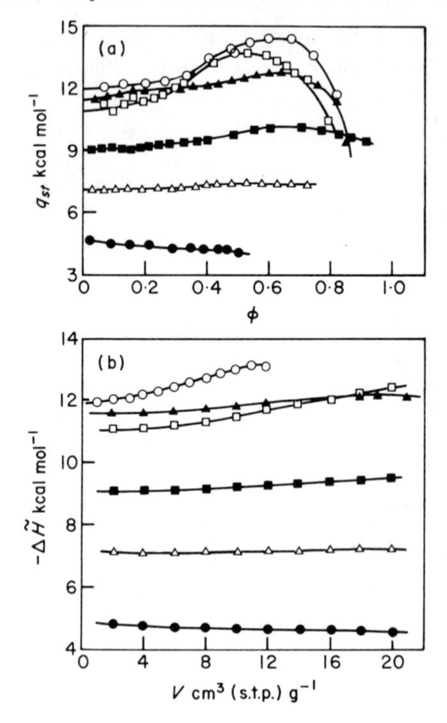

FIG. 3. (a) Isosteric heats of sorption of hydrocarbons in zeolite K-*L* plotted as a function of ϕ. (*b*) Integral heats as functions of amount sorbed.[5]

● = CH_4; △ = C_2H_6; ■ = C_3H_8; □ = nC_4H_{10}; ▲ = iso-C_4H_{10}; ○ = neo-C_5H_{12}

sorbents such as silica gels and carbons. In addition, in zeolites energetic heterogeneity is usually more pronounced when the exchange ions are small (cf Fig. 2); but heats were smaller in H-mordenite than in the Na-form (Fig. 5). In outgassed H-mordenite the hydrogen is not present as H_3O^+ but as silanol hydroxyl groups. Thus the intracrystalline environment is less polar in H- than in Na-mordenite. These aspects will be considered in more detail in § 8.

2. ENERGETIC HETEROGENEITY

The model of a homotattic surface on which there are patches of sites, each patch being different, has been referred to in Chapter 3, in which Sparnaay's[8] analysis of the resultant heterogeneity was outlined. This treatment did not consider sorbate-sorbate interaction which, we have seen in § 1, may influence the course of curves of q_{st} plotted against uptake. However Sparnaay[9]

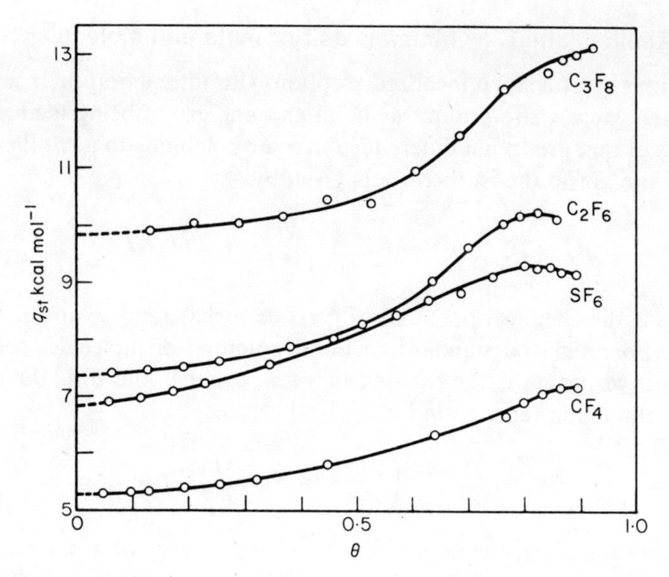

FIG. 4. Isosteric heats of sorption of CF_4, SF_6, C_2F_6 and C_3F_8 in zeolite Na-X as functions of θ.(6)

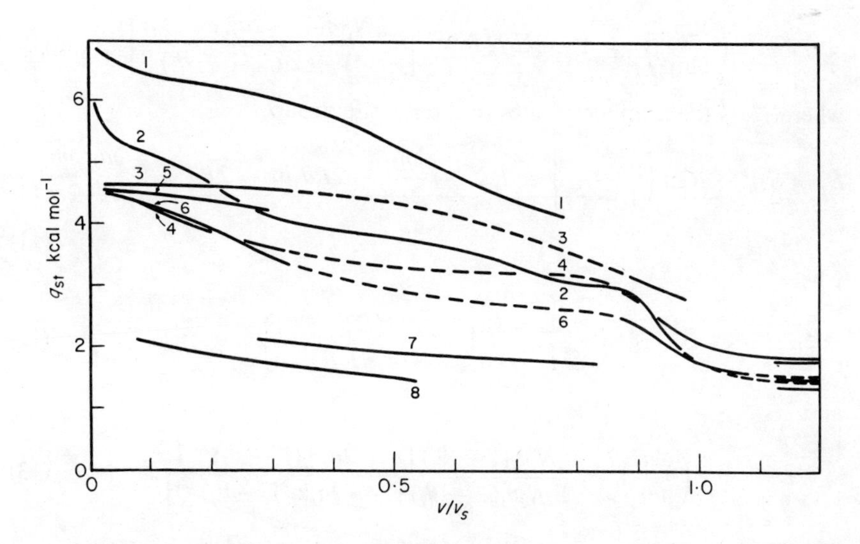

FIG. 5. Differential heats of sorption of N_2, O_2, Ar and H_2 in Na- and H-mordenites.(7) 1 = N_2 in Na-mordenite; 2 = N_2 in H-mordenite; 3 = O_2 in Na-mordenite; 4 = O_2 in H-mordenite; 5 = Ar in Na-mordenite; 6 = Ar in H-modenite; 7 = H_2 in Na-mordenite; 8 = H_2 in H-mordenite.

extended his treatment to localized sorption with interaction on a surface in which there were n site groups, a site in any one group being identical with all others of that group but different from a site belonging to any other group. On each site group the isotherm was given by

$$\mu_g = \mu_i = \mu_i^\ominus + RT \ln \frac{\theta_i}{1-\theta_i} + 2a_i\theta_i RT. \tag{8}$$

where μ_g is the chemical potential of gaseous sorbate and μ_i and $\mu_i^\ominus$ are the chemical potential and standard chemical potential of molecules sorbed on the ith site group. θ_i is the fraction of sites occupied and a_i is the sorbate-sorbate interaction term given by

$$a_i = \frac{N}{2A}\int_{d(i)}^{\infty}\left(1 - \exp - \frac{E(r)}{kT}\right) 2\pi r dr \tag{9}$$

A/N is here the surface area per site; $E(r)$ is the energy of a pair of isolated sorbate molecules as a function of their distance r apart and $d(i)$ is the distance of separation of nearest neighbour sites of type i. The site groups 1 to n were considered in order of increasing affinity of sorption.

If the total amount sorbed is m and the amount on the ith site group is m_i Sparnaay established the following relations:

$$\left(\frac{\partial \ln m}{\partial \ln p}\right)_T = 1 - [\sum_i N_i\theta_i]^{-1}\left[\sum_i \frac{N_i\theta_i^2(1 + 2a_i(1 - \theta_i))}{1 + 2a_i\theta_i(1 - \theta_i)}\right] \tag{10}$$

where N_i is the number of sites in the ith site group,

$$\bar{S}_s = \bar{S}_{Th} - R\ln\left[\frac{\theta_1}{1-\theta_1}\right] - R\sum_i A_i \frac{\partial m_i}{\partial m} - 2Ra_1\theta_1 - 2RT\sum_i \theta_i \frac{\partial a_i}{\partial T}\frac{\partial m_i}{\partial m} \tag{11}$$

where

$$\left.\begin{aligned} A_i &= (\mu_1^\ominus - \mu_i^\ominus)/RT \\ \Delta A_i &= (A_{i+1} - A_i)/RT \end{aligned}\right\} \tag{12}$$

and

$$\left(\frac{\partial m_i}{\partial m}\right)_T = \frac{N_i\theta_i(1 - \theta_i)\,[1 + 2a_i\theta_i(1 - \theta_i)^{-1}]}{\sum_i N_i\theta_i(1 - \theta_i)\,[1 + 2a_i\theta_i(1 - \theta_i)^{-1}]}. \tag{13}$$

Also since at equilibrium for the heat of *desorption*, $\Delta\bar{H}_d$,

$$q_{st} = \Delta\bar{H}_d = T\Delta\bar{S}_d = T(\tilde{S}_g - \bar{S}_s) \tag{14}$$

one may use eqn 11 with eqn 14 to evaluate $q_{st} = \Delta\bar{H}_d$. Non-zero values of the a_i affect $\bar{S}_s$ and hence $\Delta\bar{H}_d$ in three ways: by altering the value of θ_i

(eqn 8); through their explicit appearance in $\partial m_i/\partial m$; and through their temperature dependence. For the dispersion energy, by expanding the exponential in eqn 9 as a series and taking only the first term one finds $\partial a_i/\partial T = -a_i/T$, while for dipole-dipole energy $\partial a_i/\partial T = -2a_i/T$.[9]

Model calculations were made taking $n=5$; $[\mu_i^\ominus - \mu_{i+1}^\ominus]/RT = \Delta A_i = 3$; $N_{i-1} = 3N_i$; and $\partial a_i/\partial T = -a_i/T$. Two sets of values of the a_i were considered. Firstly, $a_1 = a_3 = a_2 = -1$ (exothermal sorbate-sorbate interaction) and $a_4 = a_5 = 0$. Secondly $a_1 = a_2 = -1$ and $a_3 = a_4 = a_5 = 0$. The basis for these assignments was that with the above assumptions site groups 1 and 2 together comprise about 90% of all sites and are reasonably assumed to form coherent parts of the surface. Thus $d(1)$ and $d(2)$ in eqn 9 are small and a_1 and a_2 differ considerably from zero. Site groups 4 and 5 represent about 3% of the total surface and were assumed to be scattered over the surface so that $d(4)$ and $d(5)$ are large and a_4 and a_5 close to zero. Site group 4 was alternatively assumed to form a coherent part with $a_3 = -1$; and then to be scattered with $a_3 = 0$. Figure 6[9] shows the resultant curves for $\bar{S}_s/R$ and

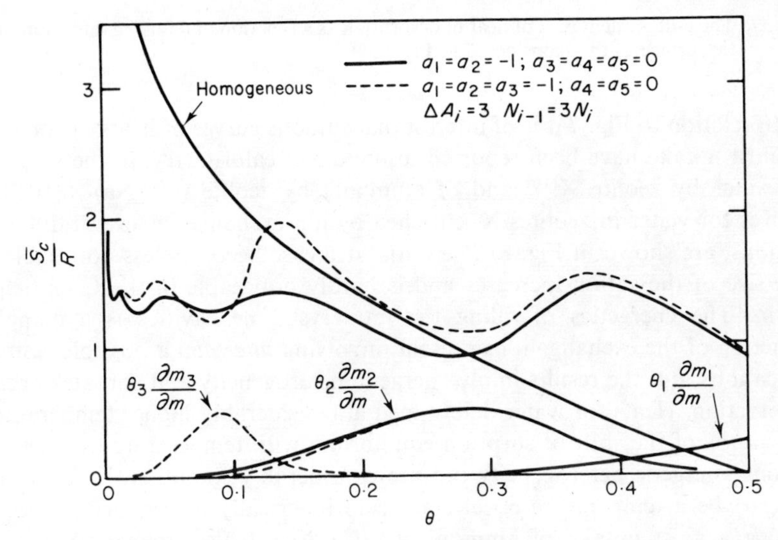

FIG. 6. A model homotattic surface and localized sorption with interaction. Differential entropy and $\theta_i \mathrm{d}m_i/\mathrm{d}m$ as functions of θ. The full curve shows the entropy for a homogeneous surface.[9]

for $\theta_i \partial m_i/\partial m$ against θ. It also shows $\bar{S}_s/R$ for ideal localized sorption on a homogeneous surface. Figure 7[9] gives the θ-dependent part of $\Delta\bar{H}_d/RT$ plotted against θ. The curves clearly reflect the heterogeneities through their sinous forms. Isotherms are however much smoother whether the a_i are zero or not.[8, 9] Although derived for surfaces Sparnaay's treatment has relevance to site distributions in three dimensions, such as arise in porous crystals.

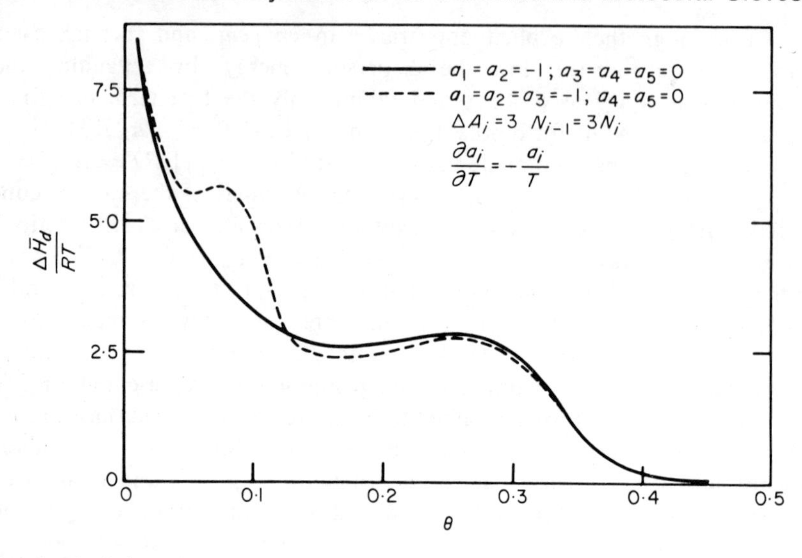

FIG. 7. The θ-dependent part of heat of desorption as a function of θ for the same homotattic surface as that considered in Fig. 6.[9]

In relation to Fig. 7 it is of interest that sinuous curves of heats of sorption against uptake have been reported, using direct calorimetry, in the sorption of water by zeolite X[10] and of ammonia by zeolite (Ca, Na)-*A*.[11] The curves for water in zeolites X, enriched by ion exchange in several different cations, are shown in Fig. 8. The initial decrease becomes less noticeable as the size of the cation increases and is hardly noticeable in the Cs-enriched form. The energetics of filling the intracrystalline cavities is a complex function of the exchange ions present, involving *inter alia* a possible resiting of cations, but the results imply energetic heterogeneity and sorbate-sorbate interaction. Heats for water determined at considerably higher temperatures by means of the shift of sorption equilibrium with temperature have shown strong energetic heterogeneity but are considerably smoother.[12] This may in part be a temperature effect since calorimetrically determined curves of $-\Delta\bar{H}$ against uptake of ammonia in (Ca, Na)-*A* also appear to become smoother the higher the temperature.[11] There are however considerable differences in the magnitudes of reported heats of sorption of water in a given zeolite[13] which reflect not only variations among the zeolite samples in composition, pretreatment and outgassing procedures but also the considerable experimental difficulties when water is the sorbate. In calorimetry at room temperature it is difficult to ensure a truly uniform water distribution throughout the mass of powder. Because of the strong affinity between water and zeolites the water tends to remain where it is first sorbed in the bed. On the other hand at the high temperatures of some isotherms the redistribution

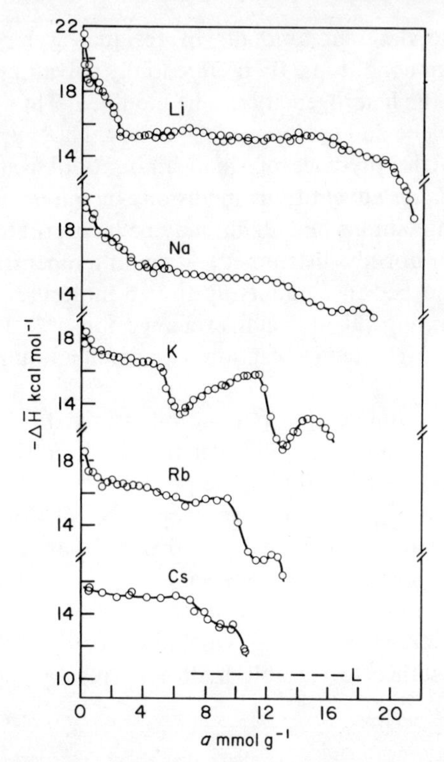

FIG. 8. Differential heats of sorption of water in (Li, Na)-, Na-, (K, Na)-, (Rb, Na)- and (Cs, Na)-X as functions of amount sorbed.(10)

problem is avoided but side reactions involving the water and the crystal may occur with resultant irreversibility of isotherms.(14)

One may summarize some general causes of hetergeneity in zeolite sorbents as follows.

1. In some zeolites there is more than one kind of cavity to which small guest molecules such as water may have access. Examples are mordenite (main channels and side pockets); zeolite *A* (26-hedra of Type I and 14-hedral sodalite cages); faujasite including zeolites X and Y (26-hedra of Type II and sodalite cages), and gmelinite and offretite (wide channels and 14-hedral Type II cages of gmelinite type).
2. In larger cavities, such as the 26-hedra in zeolite *A* or in faujasites, each channel or cavity can provide a variety of environments for a guest molecule, and it requires considerable numbers of such molecules to saturate the cavity (cf Table 3, Chapter 3). For large molecules the smaller cavities may be inaccessible and the large cavity may have sufficient volume to provide only one or two possible environments. Thus for large molecules the heterogeneity may be much reduced.

3. There may be different amounts of residual water consequent upon different outgassing conditions. By preferentially solvating the most energetically sorbing sites the heterogeneity can be modified. Thus the initial sorption of propane, propylene and cyclopropane in Li-X, Na-X and K-X has proved to be sensitive to the presence of small amounts of water.[15] In addition there may be variable adventitious inclusions of impurities. Those, such as aluminate or silicate anions and alkali, may be incorporated during synthesis and may be non-uniformly distributed. At room temperature carbon dioxide was slowly chemisorbed in amounts up to 0·5 molecules per large cavity in a zeolite X containing alkali metal exchange ions.[16] The carbonate-type products could be attributed to reaction with small amounts of alkali present in the cavities.

4. If the intracrystalline cation population is mixed (e.g. Na^+ and Ca^{2+}) sites in the vicinity of a cation will differ for Na^+ and Ca^{2+} whether or not they occupy equivalent crystallographic positions.

5. In a mixed cation population the proportion of, for example, Na^+ to Ca^{2+} may vary from one cavity to another so that the behaviour of the cavities as multiple sorption sites may vary throughout the crystal.[17]

6. The local environments within the crystal may be modified by lattice defects such as stacking faults (e.g. gmelinite, zeolite T[18]; or by chemical defects due for instance to variable hydrolysis resulting from washing and other treatments:

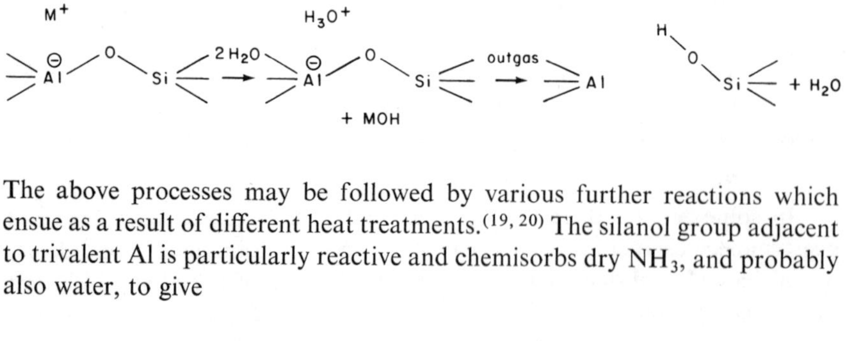

$$\overset{M^+}{\geqslant\overset{\ominus}{Al}}\!-\!O\!-\!Si\!\leqslant \xrightarrow{2H_2O} \overset{H_3O^+}{\geqslant\overset{\ominus}{Al}}\!-\!O\!-\!Si\!\leqslant + MOH \xrightarrow{\text{outgas}} \geqslant Al \quad H\!-\!O\!-\!Si\!\leqslant + H_2O$$

The above processes may be followed by various further reactions which ensue as a result of different heat treatments.[19, 20] The silanol group adjacent to trivalent Al is particularly reactive and chemisorbs dry NH_3, and probably also water, to give

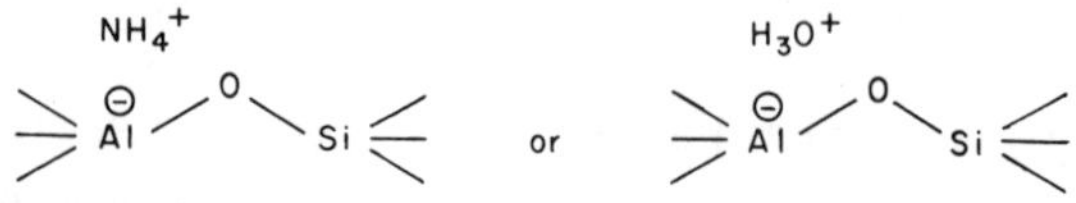

$$\overset{NH_4^+}{\geqslant\overset{\ominus}{Al}}\!-\!O\!-\!Si\!\leqslant \quad \text{or} \quad \overset{H_3O^+}{\geqslant\overset{\ominus}{Al}}\!-\!O\!-\!Si\!\leqslant$$

in amounts depending upon the variable concentrations of silanols generated. Chemisorption processes have then preceded physical sorption and will contribute to heterogeneity.

7. In zeolites where, for example, the Al : Si ratio is less than unity there

may be disorder in the distribution of Al and Si on the tetrahedral framework sites. Thus some cavities or parts of cavities may have a higher framework charge than others, necessitating more equivalents of cations locally to neutralize this charge. Accordingly for a homoionic zeolite the distribution of, for example, Na^+ ions in Na-X among the sites of types I, II and III may vary from one intracrystalline location to another. Moreover Habgood[15] has pointed out that for a small proportion of exchange ions it may in such circumstances not be possible to neutralize fully the local cationic charge. For polar molecules especially the initial sorption seems to be associated with the cations because, as seen in Fig. 8, it is the initial heat which is most sensitive to the exchange ion. In particular for the small proportion of these cations with locally unbalanced charge the heat of binding of polar molecules will be abnormally high. This is a consequence of the very high electrostatic fields and field gradients near such exposed cations and their interaction respectively with dipoles and quadrupoles, and of higher polarization energy contributions. One instance of limited strong sorption was reported in the uptake of N_2 by zeolite Ag-X[21] where, initially, the room temperature separation factor over O_2 approached 50, with $q_{st} \approx 33$ kJ mol^{-1}. This strong preference for N_2 was limited to less than 2 molecules per cavity because between 1 and 2 molecules the heat decreased to ≈ 21 kJ mol^{-1} and the separation factor to a value between 2 and 3.

Egerton and Stone[22] have also found high initial selectivities for CO sorbed in zeolite (Ca, Na)-Y after 50 to 55% of the Na^+ had been exchanged. The initial isosteric heat then rose from ≈ 25 in the parent Na-form to 40 to 45 kJ mol^{-1}. The results were interpreted as due to an association of CO with Ca^{2+} when these ions occupied sites exposed to the large 26-hedral cages (i.e. Ca^{2+} ions on sites of type II). The first 16 Ca^{2+} ions per unit cell ($\approx 55\%$ exchange) were considered to occupy sites of types I or I′ which were in hexagonal prisms or in the sodalite cages and were not accessible to CO. The limiting amount of specific adsorption on the zeolite outgassed at 350°C corresponded with one CO molecule per exposed Ca^{2+}. When Na-Y was partially exchanged with Zn^{2+}, Mn^{2+}, Co^{2+}, Cu^{2+}, Ba^{2+}, UO_2^{2+} and Ce^{3+} ions all the divalent ions produced sites specific for CO adsorption even at low % exchange.[23] This was interpreted as meaning that, unlike Ca^{2+}, the above divalent ions did not initially have a total preference for "buried" type I or type I sites, but that positions accessible to CO such as type II sites were also occupied to some extent. With the Ce-exchanged zeolite however very little specific CO uptake was found, suggesting that Ce was preferentially located in the "buried" sites.

3. INTERACTIONS CONTRIBUTING TO HEATS OF SORPTION

Contributions to heats or energies of physical sorption arise from dispersion energy (ϕ_D), close range repulsion (ϕ_R), polarization energy (ϕ_P), field-dipole interaction ($\phi_{F\mu}$), field gradient-quadrupole interaction ($\phi_{\dot{F}Q}$) and sorbate-sorbate interaction (ϕ_{SP} where "SP" denotes self-potential). Thus for the total, ϕ, of these terms

$$\phi = \phi_D + \phi_R + \phi_P + \phi_{F\mu} + \phi_{\dot{F}Q} + \phi_{SP} \,. \tag{15}$$

The first three terms on the right side of eqn 15 are always present and comprise the "non-specific" part of ϕ. The next two, $\phi_{F\mu}$ and $\phi_{\dot{F}Q}$, depend upon the presence or absence of permanent dipoles or quadrupoles respectively in the guest molecule. The term ϕ_{SP}, which allows for molecule-molecule interaction as discussed in § 4, goes to zero for small uptakes. It includes dispersion and close range repulsion energies between pairs of sorbate molecules and, for polar molecules, dipole-dipole, dipole-induced dipole, quadrupole-quadrupole and similar energy contributions.

The isosteric heat, q_{st}, is given by*

$$q_{st} = \tilde{E}_g - \bar{E}_s + RT = -(\bar{\phi} - \bar{\phi}_0) + RT - F(T) \tag{16}$$

where $\tilde{E}_g$ is the molar energy of gaseous sorbate and $\bar{E}_s$ is the differential molar energy of sorbed material. $\bar{\phi}$ is the differential molar value of ϕ in eqn 15 and $\bar{\phi}_0$ the differential molar zero point energy of sorbed molecules. $F(T)$, a function of temperature, occurs because as the temperature rises the sorbed molecules acquire vibrational energy in excess of the zero-point energy while the gaseous molecules also gain in translational energy. Thus for monatomic classical oscillators $F(T) = 3RT/2$ since a vibrational mode involves two "square terms" and a translation only one. Accordingly, at absolute zero

$$q^0_{st} = -(\bar{\phi} - \bar{\phi}_0) \tag{17}$$

so that

$$q_{st} - q^0_{st} = RT - F(T) = \int_0^T \Delta\bar{C}.\mathrm{d}T \tag{18}$$

where

$$\Delta\bar{C} = (\tilde{C}_g - \bar{C}_s) \tag{19}$$

with $\tilde{C}_g$ the molar heat capacity of gaseous sorbate and $\bar{C}_s$ the differential

* Negative values for the ϕ are here taken to correspond with the convention of exothermic interactions or processes.

molar heat capacity of sorbed molecules. In terms of the integral values of ϕ (eqn 15) and ϕ_0 for n_s mols sorbed in a fixed amount of sorbent

$$q_{st} = -(\phi - \phi_0) - n_s \frac{\partial(\phi - \phi_0)}{\partial n_s} + RT - F(T)\,. \tag{20}$$

Thus $q_{st} - RT + F(T) = -(\bar{\phi} - \bar{\phi}_0) = -(\phi - \phi_0)$ either for $n_s = 0$ or when $\partial(\phi - \phi_0)/\partial n_s = 0$. Since as $n_s \to 0$ also $\phi_{SP} \to 0$, the simplest situation for comparing calculated and observed values of q_{st} is the initial heat. However the initial isosteric heat may be atypical because it may refer to a limited number of unusual very energetically sorbing centres whereas the calculations normally involve averaged situations within an ideal crystal. Nevertheless calculated heats can also take into account particularly exposed intracrystalline cations.

It is important to be able to separate the non-specific contributions to the heat of sorption (ϕ_D, ϕ_R and ϕ_P) from the specific contributions ($\phi_{F\mu}$, $\phi_{\dot{F}Q}$ and sometimes ϕ_{SP}). The following procedure has been developed for this purpose.[24] For non-polar molecules (the inert gases, paraffins) only the non-specific terms can contribute. Accordingly if the initial heats for such mole-

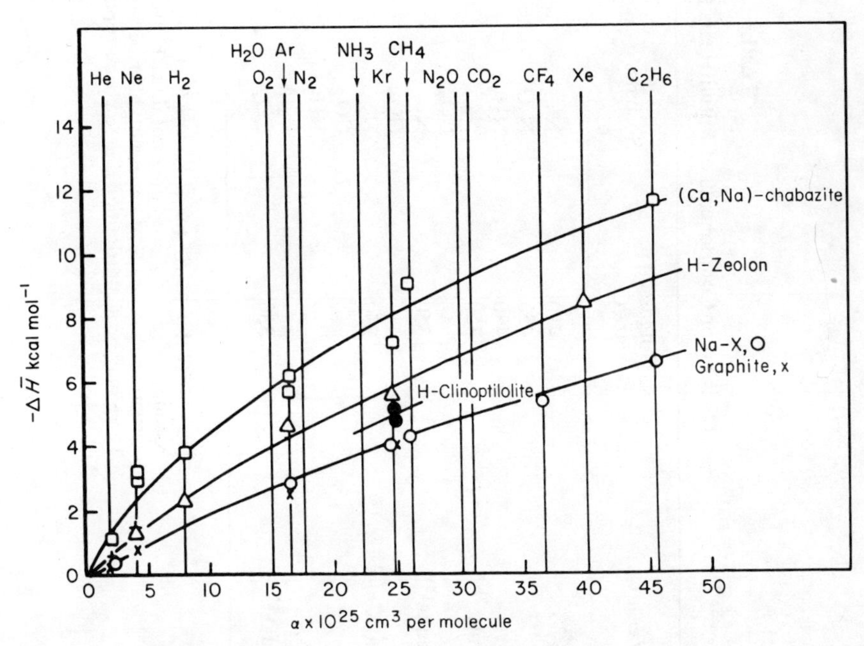

FIG. 9. Characteristic curves of initial heats against polarizability of sorbate for small relatively symmetrical adsorbed species.[24]

□ = (Ca, Na)-chabazite; △ = H-mordenite (H-Zeolon); + = N-amordenite (Na-Zeolon); ○ = zeolite Na-X; X = graphitized carbon; ● = H-clinoptilolite.

TABLE 1

Division of Components of Initial Heats (kJ mol^{-1}) in Zeolite Sorbents[24]

Zeolite	Outgassed at °C	Sorbate	Contributions to $q_{st}(n_s \to 0)$ from[a]		
			$q_{st}(n_s \to 0)$	$-(\phi_D + \phi_R + \phi_P)$	$-(\phi_{F\mu} + \phi_{FQ})$
Chabazite[25, 26]	480	N_2	37·6	27·0	10·7
	450	N_2O	64·$_0$	38·0	25·9
	480	NH_3	132	31·4	100[d]
H-mordenite[7]	350	N_2	25·9	18·8	7·1
		CO_2	46·$_4$	28·2	18·2
Na-mordenite[7]	350	N_2	29·3	18·8[b]	10·5
		CO_2	65·$_6$	28·2[b]	37·4
Zeolite Na-X[27, 28, 29, 12]	350	N_2	27·2	13·0	14·2
		CO_2	51·$_0$	17·6	33·4
		NH_3	75·$_2$	15·7	59·$_6$[d]
		H_2O	≈142	11·1	≈131[d]
Zeolite Na-Y[28]	350	CO_2	34·3	20·3[c]	14·0

[a] $[RT - F(T)]$ is $-\frac{1}{2}RT$ for oscillators (eqn. 16), this small term has been neglected.
[b] Assuming $\phi_D + \phi_R + \phi_P$ does not differ between Na- and H-mordenites.
[c] Assuming $\phi_D + \phi_R + \phi_P$ does not differ between Na-X and Na-Y.
[d] There may be a contribution due to chemisorption in these initial values.

cules in a given porous crystal or other sorbent are plotted against a parameter determining sorbability, such as the polarizability, α, a reference curve is obtained for each sorbent. The heats for polar molecules of known polarizabilities lie above the relevant point on the reference curve and the differences can be attributed to specific contributions. Figure 9[24] shows reference plots of initial heats against α for (Ca, Na)-chabazite, H-mordenite (H-Zeolon), zeolite Na-X, and graphite. A point for H-clinoptilolite is also

TABLE 2

Differential Heats of sorption for $\theta \to 0$ in (Na, Ca) - *A*,[30] and the Components $\phi_D + \phi_R + \phi_P$ and $\phi_{F\mu} + \phi_{\dot{F}Q}$

Sorbate	μ (Debye)	$-\Delta\bar{H}$	$-(\phi_D+\phi_R+\phi_P)$	$-(\phi_{F\mu}+\phi_{\dot{F}Q})$
			(in kJ mol^{-1})	
C_2H_6	0	25	25	0
C_3H_8	0	35	35	0
n-C_4H_{10}	0	44	44	0
Propylene	0·35	47	35	12
CH_3NH_2	1·33	80	30	50
n-$C_3H_7NH_2$	1·39	109	50	59
NH_3	1·46	105	12·5	92·5
CH_3OH	1·68	82	18	64

indicated. On the basis of Fig. 9 the components of some initial heats can be divided as indicated in Table 1. For the polar molecules CO_2, NH_3 and H_2O the terms $\phi_{F\mu} + \phi_{\dot{F}Q}$ are together more important than the sum of the non-specific terms. The initial value for water in Na-X is not very accurate because of necessary extrapolation, but the big initial contribution to the heat of sorption by $\phi_{F\mu}$ is not in doubt. Schirmer *et al.*[30] made a similar division of the initial heats for a number of sorbates in (Na, Ca)-*A*. Their evaluation of the energy components is given in Table 2, and shows again the large electrostatic energy contribution as the dipole moment increases.

One may compare the "non-specific" and "specific" contributions to $\Delta\bar{H}$ for graphitic carbon (a non-polar but semi-metallic sorbent) with those in the zeolites (very polar but non-metallic). This comparison can be made using the results of Table 3, where for graphitic carbon the "specific" components of $\Delta\bar{H}$ partially arise because electric moments in polar sorbates induce mirror-image charge distributions in the graphite and interact with these mirror image distributions. For nitrogen, with a relatively small quadrupole moment, "specific" contributions are much larger for the zeolites than for graphite, and the same appears true for the polar species NH_3 and H_2O.

However the "specific" terms for NH_3 and H_2O are by no means small for graphite.

In zeolite Na-X the isosteric heat for argon was nearly constant up to at least 40 cm^3 at s.t.p. g^{-1} of sorbed gas. For argon, the only terms involved are ϕ_D, ϕ_R and ϕ_P. Accordingly if for other guest molecules in the same zeolite (or in its silica rich variant Na-Y) $(\phi_D + \phi_R + \phi_P)$ is also nearly independent of amount sorbed one may use the reference curve for Na-X in Fig. 9 to evaluate the contributions of $(\phi_D + \phi_R + \phi_P)$ to $-\Delta\bar{H}$ and so to find that part which is due to the electrostatic terms $\phi_{F\mu}$ and $\phi_{\dot{F}Q}$ and to the

TABLE 3

Components of initial Heats, $\Delta\bar{H}$ on Graphic Carbons[24]

Sorbate	$-\Delta\bar{H}$ (kJ mol^{-1})	$-(\phi_D + \phi_R)$ (kJ mol^{-1})	−Mirror Image and other Energy (kJ mol^{-1})
N_2[31]	$12{\cdot}9_6$	$11{\cdot}5 - 10{\cdot}5^b$	$1{\cdot}46 - 2{\cdot}51$
	$10{\cdot}8_7{}^a$	$11{\cdot}5 - 10{\cdot}5^b$	
H_2O[32]	$\approx 43{\cdot}9$	$11{\cdot}1$	$\approx 32{\cdot}8$
NH_3[33]	$30{\cdot}1$	$14{\cdot}6$	$15{\cdot}5$
	$25{\cdot}9^a$		
CH_3OH[34]	$50{\cdot}_2$	$20{\cdot}3$	$29{\cdot}9$
	$43{\cdot}9^a$		$23{\cdot}6$
SO_2[35]	$25{\cdot}1$	$22{\cdot}4 - 17{\cdot}6^b$	$2{\cdot}72 - 7{\cdot}5_2$

[a] Curves of $-\Delta\bar{H}$ vs amount sorbed show relatively flat portions independent of uptake, preceded by a small rise as amounts sorbed decrease toward zero. Values of $-\Delta\bar{H}$ refer to the flat portions.

[b] Anisotropy of these molecules means direction dependent values of polarizability. α. The smaller figure results if α normal to length (N_2) or to the plane of OSO (SO_2) is used with Fig. 9 to determine $\phi_D + \phi_R$. The larger figure corresponds with the mean value of α for each molecule.

self-potential ϕ_{SP}. Values for the electrostatic and self-potential parts of $-\Delta\bar{H}$ are given in Table 4. The self-potential component increases with amount sorbed; the other components decrease. The comparison of CO_2 in the faujasites Na-X and Na-Y shows that $\phi_{\dot{F}Q}$ is much less in the silica-rich faujasite Na-Y, than in the aluminous variant, Na-X. The density of charges in Na-Y is lower than in Na-X, a distinction which strongly influences the electrostatic energy $\phi_{\dot{F}Q}$ for carbon dioxide. Further consideration of the role of cations and of electrostatic components of the heat of sorption is given in § 8.3.

TABLE 4

Electrostatic Energy plus Self-Potential (kJ mol^{-1}) in Relation to Uptake for some polar Sorbates in Zeolite Na-X[24]

Sorption (cm^3 at s.t.p. g^{-1} of outgassed Na-X)	N_2[27]	NH_3[29]	H_2O[12]	CO_2[28]	CO_2[28] in Na-Y
≈0	14·6	$60\cdot_2$	≈131	33·4	16·7
10	12·1	$56\cdot_4$	105	30·1	14·2
20	10·7	$52\cdot_3$	$85\cdot_7$	27·6	13·0
30	9·4	$49\cdot_3$	$71\cdot_9$	26·3	13·0
40	8·2	$46\cdot_8$	$61\cdot_4$	24·2	13·0
50	$7\cdot3_2$	$44\cdot_3$	$53\cdot_5$	24·2	13·4
60	$6\cdot6_9$	$42\cdot_2$	$46\cdot_0$	23·4	13·4
80	$6\cdot2_7$	$39\cdot_3$	$42\cdot_2$	21·7	13·8
100	—	36·4	38·9	—	—

4. SORBATE-SORBATE INTERACTION ENERGY

Isosteric heats sometimes show a rise for larger uptakes which, as noted in § 1, has been attributed to sorbate-sorbate interaction. When n_s mols have been sorbed in a fixed amount of sorbent the heat, q_{st}, is related to the integral energy of sorption, $\Delta\tilde{E}$, by the relation

$$q_{st} - RT = -\frac{\partial}{\partial n_s}(n_s\Delta\tilde{E}) = -\frac{\partial}{\partial n_s}\{n_s(\Delta\tilde{E}_L + f(n_s))\} \tag{21}$$

where $\Delta\tilde{E}_L$ is the integral energy of interaction between guest molecules and zeolite lattice and $f(n_s)$ is the integral sorbate-sorbate interaction energy. This, for the initial heat $(q_{st})_0$ when $n_s \to 0$, becomes

$$(q_{st})_0 - RT = (\Delta\tilde{E}_L)_{n_s=0}. \tag{22}$$

From eqns 21 and 22

$$\Delta q_{st} = q_{st} - (q_{st})_0 = (\Delta\tilde{E}_L)_0 - \Delta\tilde{E}_L - n_s\frac{\partial\Delta\tilde{E}_L}{\partial n_s} - f(n_s) - n_s\frac{\partial f(n_s)}{\partial n_s} \tag{23}$$

and for an energetically homogeneous sorbent where $\Delta\tilde{E}_L = (\Delta\tilde{E}_L)_0$ one has

$$\Delta q_{st} = -f(n_s) - n_s\frac{\partial f(n_s)}{\partial n_s}. \tag{24}$$

In such a sorbent Δq_{st} rises to a maximum at a value near to $\theta = 1$ and then declines when further additions involve packings which call increasingly into play the close range repulsion energies, or which result in multilayer sorption or condensation upon external surfaces of the zeolite crystals. In the latter situation q_{st} will decline towards a value corresponding with the latent heat of condensation of bulk liquid.

Several treatments have been given which, in the case of homogeneous sorbents, show how Δq_{st} may change with uptake. Two are based upon site models; one is not. The site models will first be outlined. For a lattice of identical fixed sites the energy ϕ_{SP} will be determined by the number of nearest neighbour pairs of sorbed molecules and the energy $u = 2w/z$ per pair, where z is the coordination number of a site. All interactions save nearest neighbour ones are neglected and the pair interaction energy is additive. The first approximation to the sorption isotherm is (Chapter 3).

$$K_2' = \frac{\theta}{p(1-\theta)} \exp\left[2\theta w/RT\right] \tag{25}$$

where K_2' is the equilibrium constant. This relation gives

$$\Delta q_{st} = -zu\theta \tag{26}$$

which follows from eqn 24 when $f(n_s) = zu\theta/2$. The factor $\frac{1}{2}$ appears in order to avoid counting each molecule twice when evaluating $f(n_s)$.

The linear increase in q_{st} with θ which eqn 26 predicts is sometimes approximated in experimental studies, for example with gases in H-chabazite[36] and in faujasite,[37] at least over a limited range in θ. For I_2 in zeolites Na-X and Ca-A, and in chabazite there is a linear range in q_{st} after the minimum value of q_{st}[38] (Fig. 10, Chapter 3). If the I_2 molecules interact with one another so that the long axes are parallel the Lennard–Jones 6-12 potential gives $u = -7{\cdot}4$ kJ mol^{-1} at the minimum of the interaction energy curve for a pair of molecules. From the latent heat of sublimation and the crystal structure of I_2 a value of $-7{\cdot}5$ kJ mol^{-1} was found. With the coordination number taken as 4 and with $u = -7{\cdot}4$ kJ mol^{-1} the calculated and observed values of $(q_{st})_{max} - (q_{st})_{min}$ are compared in Table 5. The subscripts "max" and "min" refer to the maximum in q_{st} near $\theta = 1$, and the minimum in q_{st} following the high initial q_{st}. For example with I_2 in zeolite Na-X the minimum occurred at $\theta \approx 0{\cdot}35$. The observed and calculated values show reasonable correspondence. In the table are also given values of $(q_{st})_{max} - (q_{st})_0 = \Delta q_{st}$ for some other systems which did not show the initial high values of q_{st} found with I_2-zeolite systems. The simple treatment leading to eqn 26 assumes a random distribution of molecules on sites, which cannot be true if molecules interact with one another and can arrange themselves so as to give the most stable assembly. When there are N_s sites and N_A sorbed molecules ($\theta = N_A/N_s$) the sorbate-sorbate interaction energy is

$$N_s w \left\{ \theta - \frac{2\theta(1-\theta)}{\beta+1} \right\} \text{ where } \beta = \{1 - 4\theta(1-\theta)(1 - \exp - 2w/zRT)\}^{1/2}.$$

The plot of q_{st} against θ is then no longer a linear one in the quasi-chemical approximation.

A different interpretation of the increase in q_{st} with θ was given by Parsonage[39, 40] who again considered a site model. The sorbent was considered to provide two kinds of site (α, of energy 0; β, of energy U) such that simultaneous occupation of an α and its nearest neighbour β site led to a

TABLE 5

Sorbate-sorbate Interactions

Zeolite	Sorbate	$[(q_{st})_{max} - (q_{st})_{min}]$ in kJ mol^{-1} Observed	Calculated
Na-X[37]	I_2	15·5	19·2
Ca-*A*[37]	I_2	15·0	20·5
Chabazite[37]	I_2	13·8	17·6
		$[(q_{st})_{max} - (q_{st})_0]$ in kJ mol^{-1} Observed	
Na-X[6]	CF_4	7·9	—
	SF_6	10·5	—
	C_2F_6	12·1	—
	C_3F_8	18·4	—
(K, Na)-*L*[5]	C_3H_8	5·0	—
	n-C_4H_{10}	12·1	—
	iso-C_4H_{10}	5·8	—
	neo-C_5H_{12}	10·0	—

repulsion contribution E. In each cavity a model was considered in which there were two sites of each kind in a square configuration, with α and β sites alternating at the corners of the square.[39] At very low coverages a number of β sites, determined by the Boltzmann factor, $\exp - U/kT$, are occupied. However as the average occupancy of a cell approaches 2 (saturation of α-sites) this number is reduced so as to avoid repulsions with molecules in the now almost fully occupied α-sites, and there is a consequent reduction in energy of the system and so an increase in q_{st}. Beyond two molecules per cavity ($\theta > 0{\cdot}5$) the next molecules cannot avoid β sites and so the energy of the system rises or q_{st} decreases. The model was treated quantitatively and was subsequently extended to monolayer adsorption on a plane surface.[40]

Although it appears to be somewhat artificial, it provides an interesting alternative to the usual explanation.

In Chapter 3 evidence was presented that the zeolitic guest molecules can have properties analogous to those of fluids. Peters and Tappe[41] considered volume filling of an energetically uniform intracrystalline free volume, within which the molecules interacted with each other according to an inverse power law, such as the 6-12 potential; for each pair of guest molecules. The sorption potential, $\phi(v)$, was thus taken to have the form

$$\phi(v) = \phi_L + \Lambda\left[\left(\frac{v_0}{v}\right)^a - c\left(\frac{v_0}{v}\right)^b\right] \tag{27}$$

where ϕ_L arises from the interaction of guest with sorbent, and is assumed to be independent of the uptake, at which the molecular volume is v. The remaining term arises from the sorbate-sorbate interaction, and v_0 is the molar volume at saturation of the intracrystalline pore space, the total volume of which is V. Thus $v = V/n$ where n denotes the number of mols sorbed. Λ, a, b and c in eqn 27 are constants. The difference in the Helmholtz free energy, A, of a mol of sorbed gas and the corresponding free energy of the guest molecules as ideal non-sorbed gas, A_{id}, was taken to be

$$A - A_{id} = -RT\ln(v_f/v) + \phi(v) \tag{28}$$

where v_f, the molar free volume, was defined by

$$v_f^{1/3} = (v^{1/3} - v_0^{1/3}) \tag{29}$$

and

$$A_{id} = A^{\ominus} + RT\ln RT - RT\ln v \tag{30}$$

with $A^{\ominus}$ the molar free energy of gas at temperature T and pressure $p = 1$ atm. From the above relations expressions were obtained for the chemical potential of sorbed molecules, the sorption isotherm, and the isosteric heat. For the latter

$$q_{st} = (q_{st})_0 + \Lambda[(a+1)\theta^a - (b+1)c\,\theta^b]\,. \tag{31}$$

The experimental observations have shown that $\Delta q_{st} = q_{st} - (q_{st})_0$ increases to a maximum at a value of $\theta = \theta_m$ and then declines through zero at $\theta = \theta_0$, and finally becomes negative. At $\theta = \theta_m$, $d(\Delta q_{st})/d\theta = 0$ and when $\theta = \theta_0$, the second term on the right-hand side of eqn 31 is zero. In numerical applications the authors took c, a geometrical factor, to be unity and $b = 2a$. With these simplifications and the relations at θ_m and θ_0 one obtains

$$\left.\begin{aligned} a &= b/2 = \ln 2/\ln(\theta_0/\theta_m) \\ 1/\theta_0 &= \left(\frac{2a+1}{a+1}\right)^{1/a} \\ \Lambda &= (\Delta q_{st})_{\max}\,4(2a+1)/(a+1)^2 \end{aligned}\right\} \tag{32}$$

where $(\Delta q_{st})_{max}$ is the value of Δq_{st} at θ_m. The value of Δq_{st} as a function of θ is shown in Fig. 10 for values of a between 1 and 10. The variation of q_{st} with θ is characteristically non-linear, and qualitatively comparable in form with some of the observed variations of q_{st} with θ (e.g. Fig. 3a and Fig. 4).

The treatments of sorbate-sorbate interaction based on localized sorption, or on volume filling start with the idea of energetically homogeneous sorbents.

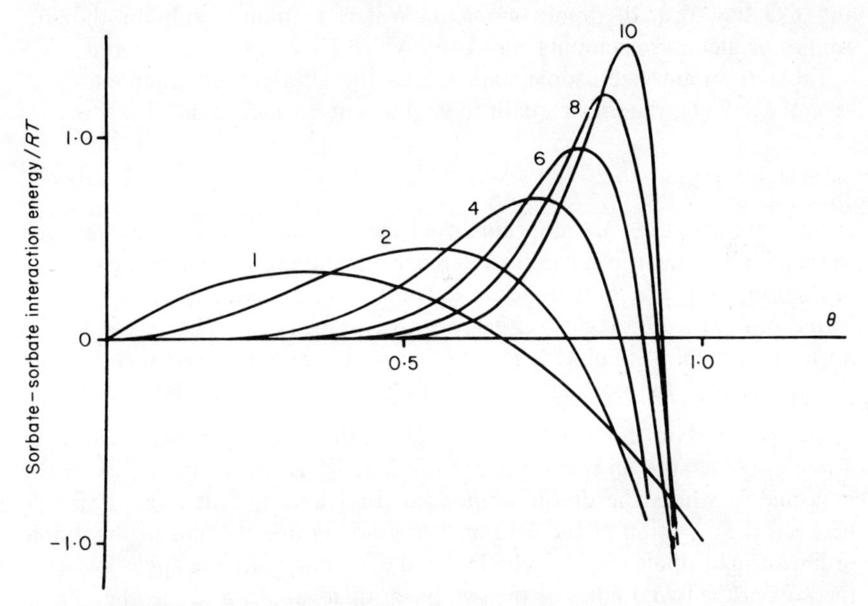

FIG. 10. Sorbate-sorbate interaction contributions to the heat of sorption for various values of the parameter a according to the treatment of Peters and Tappe.[41] The values of a are given by the numbers on the curves.

In the previous section, however, some results of Sparnaay[9] were referred to, which considered localized sorption with interaction in a sorbent comprizing a number of uniform sets of sites, each set differing from any other set. When all the treatments are considered, reasonable understanding is achieved, although quantitative rigour for real sorbents is not attained.

5. THE BASIS OF ELECTROSTATIC ENERGY

The effect of any distribution of charges around a chosen reference point within it, at a chosen point external to it, can always be reduced to the effect, at the external point, of net charge, a point dipole, a point quadrupole, and a point octupole, etc., centred at the internal reference point. Thus the interaction of any two charge distributions (i.e. molecules or ions) can be evaluated as the sum of the interactions of net charges and point electric moments, one set being located at a reference point within the first distribution and the other set at a reference point within the second distribution. For neutral molecules there is of course no net charge to consider. For a spherically symmetrical molecule such as an inert gas, there is also no permanent electric moment. Molecules such as N_2, CO and CO_2 have permanent quadrupole moments and CO has a small dipole moment. Water, ammonia, sulphur dioxide, amines or alcohols exemplify molecules which have large dipole moments.

Table 6 summarizes some expressions for electrostatic interactions of several kinds. Q, the linear quadrupole moment, is defined as[42]

$$Q = \tfrac{1}{2} \int \rho(r, \psi)(3 \cos^2 \psi - 1)r^2 \mathrm{d}v \tag{33}$$

r and ψ are here the polar coordinates of the element of volume, dv, referred to the centre of the molecule, the reference direction for $\psi = 0$ being the axis of the molecule. $\rho(r, \psi)$ is the charge density and the integration is over the whole molecule. Q may be either positive or negative. The restriction to linear quadrupoles simplifies the approach and does not detract greatly from its usefulness. An important example of a molecule with a linear quadrupole moment is carbon dioxide, $\overset{\delta-}{O} = \overset{\delta+\delta+}{C} = \overset{\delta-}{O}$. In the interactions 1, 2 and 3 of Table 6, the charge on the ion is ze. α in 2 and 5 is the polarizability of the molecule in which the dipole is induced. In 1 and 3, θ denotes the angle between the direction of the field or field gradient and the axis of the dipole or linear quadrupole respectively. In 4 and 6, θ_1 and θ_2 are the angles between the axes of the two dipoles or the two linear quadrupoles and the line joining their centres; and ω is the angle between the planes containing this line and the axes of the two dipoles or linear quadrupoles. In 1 to 6 r is the distance between the centres of the interacting pairs.

The energies in Table 6 refer to particular orientations (except for 2). Where these energies are functions of orientation the average energy over all orientations, $\langle\phi\rangle$, is a weighted mean. For example, keeping r constant, and when θ is the only angle to be considered, this mean is

$$\langle\phi\rangle = \int E \exp(-E/RT) \sin\theta \, \mathrm{d}\theta \Big/ \int \exp(-E/RT) \sin\theta \, \mathrm{d}\theta \tag{34}$$

where the integrations are over all possible orientations and E denotes an appropriate one of the ϕ of Table 6. Several examples of averages obtained from eqn 34 are:

$$\langle\phi\rangle_{F\mu} = -C \coth(C/kT) + kT \tag{35}$$

$$\langle\phi\rangle_{FQ} = -\frac{(3BkT/\pi)^{1/2}}{v\{(3B/kT)^{1/2}\}} + B + kT/2 \tag{36}$$

TABLE 6

Some electrostatic interaction energies

Interaction	Energy
1. Ion with point dipole, μ	$\phi_{F,\mu} = -F\mu\cos\theta = \frac{-ze\mu\cos\theta}{r^2(4\pi\epsilon_0)}$
2. Ion with induced point dipole μ_i	$\phi_P = -\alpha F^2/2 = -\mu_i^2/2\alpha = -\frac{\alpha(ze)^2}{2r^4(4\pi\epsilon_0)^2}$
3. Ion with linear point quadrupole, Q	$\phi_{\dot{F}Q} = Q\dot{F}/2 = -\frac{Qze(3\cos^2\theta - 1)}{4r^3(4\pi\epsilon_0)}$
4. Point dipole, μ_1, with point dipole, μ_2	$\phi_{\mu_1\mu_2} = -\frac{\mu_1\mu_2}{r^3(4\pi\epsilon_0)}\{2\cos\theta_1\cos\theta_2 - \sin\theta_1\sin\theta_2\cos\omega\}$
5. Point dipole, μ, with point induced dipole μ_i	$\phi_{\mu\mu_i} = -\frac{\mu^2\alpha(3\cos^2\theta + 1)}{2r^6(4\pi\epsilon_0)^2}$
6. Linear point quadrupole, Q_1, with linear point quadrupole, Q_2	$\phi_{Q_1Q_2} = \frac{3}{4}[Q_1Q_2/r^5(4\pi\epsilon_0)][1 - 5\cos^2\theta_1 - 5\cos^2\theta_2 - 15\cos^2\theta_1\cos^2\theta_2 + 2(4\cos\theta_1\cos\theta_2 + \sin\theta_1\sin\theta_2\cos\omega)^2]$

$$\langle\phi\rangle_P + \langle\phi\rangle_{\mu\mu_i} = -A - \frac{(3DkT/\pi)^{1/2}}{v\{(3D/kT)^{1/2}\}} - D + kT/2 \qquad (37)$$

where

$$A = \frac{\alpha(ze)^2}{2r^4(4\pi\epsilon_0)^2}; \qquad B = \frac{Q(ze)}{4r^3(4\pi\epsilon_0)}$$

$$C = -\frac{ze\mu}{r^2(4\pi\epsilon_0)}; \qquad D = \frac{\mu^2\alpha}{2r^6(4\pi\epsilon_0)^2}$$

and where $v\{x\}$ is the error function of complex argument.[43] When such components contribute significantly to the overall energy part of the temperature coefficient of the energy will arise from them. Thus from eqn 35

$$\frac{d\langle\phi\rangle_{F\mu}}{dT} = k\left[1 - \left(\frac{C}{kT}\right)^2 \operatorname{cosech}^2(C/kT)\right] \qquad (38)$$

and from eqn 36 and 37

$$\frac{d\langle\phi\rangle}{dT} = k/2 + \frac{k[(x-\frac{1}{2})(x/\pi)^{1/2}v(\sqrt{x}) - x/\pi]}{[v(\sqrt{x})]^2} \qquad (39)$$

where for eqn 36 $$x = \frac{3Q(ze)}{4r^3(4\pi\epsilon_0)kT}$$

and for eqn 37 $$x = \frac{3\mu^2\alpha}{2r^6(4\pi\epsilon_0)^2kT}.$$

The field inside the channels of a zeolite is more complex than that arising from an isolated ion or dipole and must be calculated from the distribution of charges within the zeolite. For the quadrupole energy, ϕ_{FQ}, the magnitude of the quadrupole energy decreases when the molecule rotates as compared with its value when the molecule is not rotating.[44] This was considered to be the main reason for several appreciable temperature coefficients in q_{st} for nitrogen in zeolite X (faujasite). Nitrogen has a significant quadrupole moment, and when 15 cm^3 at s.t.p. g^{-1} of this gas were sorbed the values of $\partial q_{st}/\partial T$ were reported as follows[27]:

Li-faujasite	$-\partial q_{st}/\partial T \approx 63$ (JK^{-1} mol^{-1})
Na-faujasite	≈ 63
K-faujasite	≈ 29.

Thus the differential heat capacity $\bar{C}_s$ of the sorbed nitrogen must be considerable, that of the gaseous nitrogen, $\tilde{C}_g$, being ≈ 29 JK^{-1} mol^{-1}. For these heat capacities one has

$$-(\partial q_{st}/\partial T) = \bar{C}_s - \tilde{C}_g \qquad (40)$$

so that for K-faujasite $\bar{C}_s \approx 58$ and for the other two cationic forms $\bar{C}_s \approx 92$ $JK^{-1}\,mol^{-1}$.

Calculations of fields and field gradients within zeolite crystals have been made using more than one approximation to the charge distribution. In the first of these, for zeolites X and *A*, full cationic charges were placed on all exchange ions and the anionic framework charge was distributed appropriately on all framework oxygens.[29, 45, 46, 47] In a second series of calculations relating to zeolite X full charges corresponding with the valence were placed upon all framework atoms[48] (Al^{3+}, Si^{4+}, O^{2-}) as well as upon the exchangeable cations, although Si-O and Al-O in aluminosilicates are not

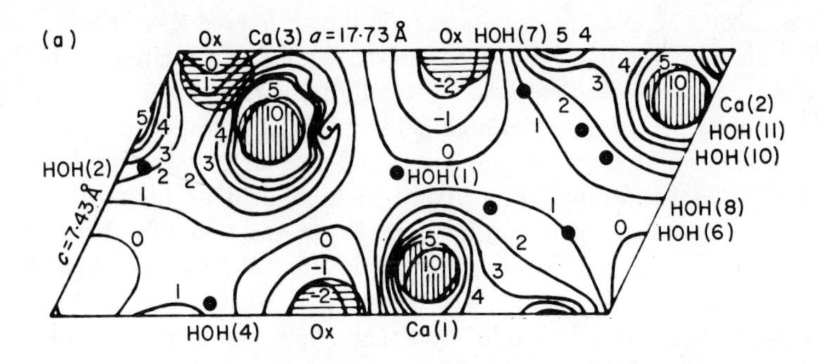

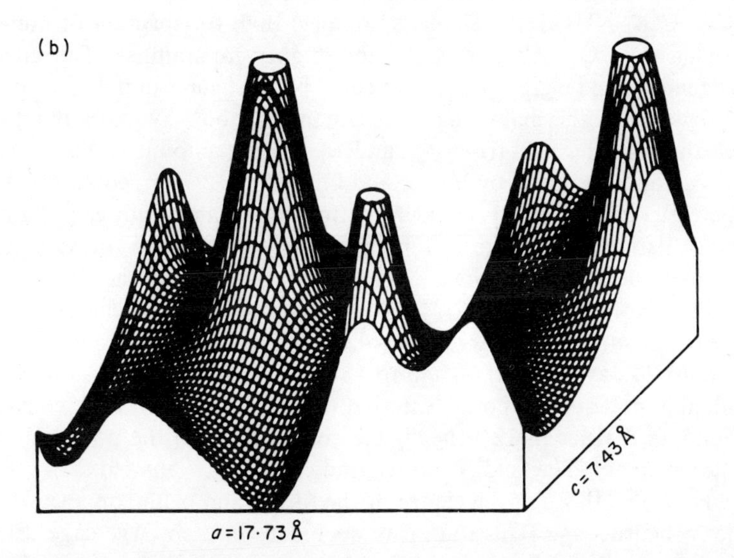

FIG. 11. (a) Equipotential lines in the plane of symmetry of heulandite. Oxygen atoms (Ox), calcium sites (Ca) and the centre of the water (HOH) are represented. (b) Variations of the electrical potential in the plane of symmetry. The lower plane is at −3V and the peaks are truncated above 10V, which corresponds to the size of Ca ions.[50]

fully ionic bonds.[49] In calculations of energy, however, unit negative charge was placed upon each framework Al, with full cationic charge upon each exchangeable cation, an assumption also made by Bonnin and Legrand[50] in calculating electrostatic potentials in heulandite. Figure 11a shows, in projection, equipotential lines in the interlayer plane of symmetry of heulandite. The contours vary from −3V to +10V, the latter immediately around the Ca^{2+} ions. Figure 11b shows, in side elevation, the same equipotential contours peaking at the Ca^{2+} ions. Calculations by all methods agree in showing that electrostatic fields and field gradients in zeolites may be large and can vary extremely rapidly over one or two angstroms ($F = -\text{grad } V$; $\dot{F} = \text{div grad } V$).

Experimental consequences of the rapidly varying values of F and $\dot{F}$ appear to be that ϕ_P is relatively small compared with ϕ_P calculated for molecules interacting with isolated cations. Especially for larger molecules parts of the molecule may be in regions of high field and parts in low field regions. However for the energy terms $\phi_{F\mu}$ and $\phi_{\dot{F}Q}$ the dipolar or quadrupolar parts of the molecules are drawn respectively to high field or high field gradient regions so that the whole of a large molecule is not involved in these two interactions, which may be big. Nevertheless as the sorption increases these special intracrystalline regions are occupied so that $\phi_{F\mu}$ and $\phi_{\dot{F}Q}$ decline rather rapidly. As seen in Table 4, energetic heterogeneity is usually more in evidence for sorption of molecules with permanent electric moments (e.g. N_2, CO, CO_2, NH_3, H_2O, alcohols, amines) than for sorption of non-polar molecules (e.g. O_2, Ar, Kr, CH_4 and higher *n*-paraffins). The energetic heterogeneity can be greater than indicated by the figures in this table because an increase in exothermal sorbate-sorbate interaction may partially offset the decreasing contributions from $\phi_{F\mu}$ and $\phi_{\dot{F}Q}$ as the amount sorbed increases.

The calculations made by Barrer and Gibbons[29, 45] for zeolite X assumed charge distributions of 10 univalent cations associated with each 26-hedral cavity (4 cations of charge + 1 in or against the four 6-rings; 4 cations distributed statistically against the six central 4-rings of the "ribs" of the cage, i.e. six charges of +2/3; and 12 cations each of charge +1/6 in hexagonal prisms). This charge was balanced by allocating anionic charges of −10/48 to 24 oxygens and −5/48 to 48 oxygens. This distribution develops a negligible potential at points external to the large cavity, and, conversely, surrounding cavities make a negligible contribution to the potential within the chosen cavity. The field F and its gradient $\dot{F}$ was evaluated along the *ppp* axis which runs through the centre of the 6-ring and of the opposite 12-ring of the 26-hedral cage. There are four such axes in each large cage. Electrostatic energies $\phi_{F\mu}$ and ϕ_P were then calculated for NH_3 and $\phi_{\dot{F}Q}$ and ϕ_P for CO_2 as these molecules move with their long axes along the *ppp* axis towards the Na^+ or other ion in or against the six-ring. Some results of the calculations are shown in Table 7. The distances in column 1 are those between the centre of the cation and the centre of the molecule moving along the *ppp* axis. The screening effect of the ambient anionic oxygens upon both ϕ_P and $\phi_{\dot{F}Q}$ is well shown. Especially for the polarization energy screening is most important so that even at contact of molecule and cation ϕ_P is usually small in the zeolite (columns 3 to 7) relative to its value for the isolated cation and

TABLE 7

Comparison of ϕ_P and $\phi_{\dot{F}\mu}$ for Interactions of CO_2 with isolated Cations and with Cations in or against the 6-ring[a] in 26-hedra of Zeolite X[(45)]

Distance (Å)	$-\phi_P$(kJ mol^{-1})					
	Isolated Cation	Li-X[a]	Na-X[a]	K-X	Rb-X	Cs-X
3·16	18·8	6·3	—	—	—	—
3·51	12·1	3·3	3·3	—	—	—
3·89	8·8	2·5	2·5	1·3	—	—
4·04	7·1	1·7	1·7	1·0	$0{\cdot}4_2$	—
4·25	5·9	1·3	1·3	0·8	$0{\cdot}2_1$	0
	$-\phi_{\dot{F}Q}$(kJ mol^{-1})					
3·16	38	31	—	—	—	—
3·51	$27{\cdot}_6$	21	21	—	—	—
3·89	20·0	14	14	$17{\cdot}_6$	—	—
4·04	18·0	13	13	$15{\cdot}_9$	$14{\cdot}_6$	—
4·25	15·5	10	10	$12{\cdot}_5$	10	$9{\cdot}_6$

[a] Li^+ and Na^+ have the same position in the 6-ring of the 26-hedral cage. The larger cations are displaced laterally into the cage by distances depending on their size.

molecule (column 2). The relative magnitudes of the calculated energies $\phi_{\dot{F}Q}$ and $\phi_{F\mu}$ for CO_2 and NH_3 at contact of intrazeolitic cation with the guest molecules in kJ mol^{-1} were as follows:[(45)]

	Li-X	Na-X	K-X	Rb-X	Cs-X
$-\phi_{\dot{F}Q}(CO_2)$	31	21	$17{\cdot}_6$	$14{\cdot}_6$	$9{\cdot}_6$
$-\phi_{F\mu}(NH_3)$	51	33	20	$17{\cdot}_6$	$15{\cdot}_0$.

These figures demonstrate the importance of the maximum values of the electrostatic terms but Table 7 shows that such terms decline rapidly as molecules are displaced away from the intracrystalline cations.

6. DISPERSION AND CLOSE-RANGE REPULSION ENERGIES

For non-polar molecules in polar sorbents the universally present energies ϕ_D, ϕ_R and ϕ_P are the only components of the sorption energy. In § 5 it was suggested that ϕ_P is not very large in most cases. The magnitudes of ϕ_D and ϕ_R will be considered in this section. A calculation of $(\phi_D + \phi_R)$ was outlined in Chapter 3 in connection with the evaluation of sorption equilibrium constants.

For an isolated pair of interacting molecules i and j the dispersion energy takes the form

$$\phi_D = -\frac{A_{ij}^{(6)}}{r_{ij}^6} - \frac{A_{ij}^{(8)}}{r_{ij}^8} - \frac{A_{ij}^{(10)}}{r_{ij}^{10}} \tag{41}$$

where the A_{ij} are coefficients that can be estimated from the theory of dispersion energies and r_{ij} is the centre to centre distance of the interacting pair. The first term is due to the coupling of instantaneous dipoles; the second to that of instantaneous dipole and quadrupole; and the third to the coupling of instantaneous quadrupoles.

The exothermic terms of eqn 41 are offset in part by the endothermic close range repulsion energy, ϕ_R. This is often represented as an inverse power of r_{ij} by

$$\phi_R = B_{ij}/r_{ij}^m \tag{42}$$

where m is usually taken to be 12 and B_{ij} is a constant. Equation 42 is empirical, and an exponential form has likewise been used which is also at least semi-empirical.

The next approximation is that all the interactions ϕ_D and ϕ_R between a guest molecule within the crystal and the lattice atoms are additive, so that $\phi_D + \phi_R$, together with the polarization energy ϕ_P, is given by

$$\phi_D + \phi_R + \phi_P = -\sum_{i, r_{ij}} \left(\frac{A_{ij}^{(6)}}{r_{ij}^6} + \frac{A_{ij}^{(8)}}{r_{ij}^8} + \frac{A_{ij}^{(10)}}{r_{ij}^{10}} + \frac{B_{ij}}{r_{ij}^{12}} \right) - \alpha_j F^2/2 \tag{43}$$

where α_j is the polarizability of the guest molecule, j, and F the electrostatic field acting on it at its centre. The assumption that dispersion and repulsion interactions are additive is not fully justifiable[51, 52] but is necessary for the ready calculation of $(\phi_D + \phi_R)$. B_{ij}, the repulsion energy constant between the guest molecule j and the ith kind of lattice atom, follows from $(\phi_D + \phi_R + \phi_P)$ for j and i interacting in isolation from all other atoms. At the equilibrium distance of separation $(r_0)_{ij}$, $\partial(\phi_D + \phi_R + \phi_P)/\partial r_{ij} = 0$. Accordingly

$$B_{ij} = \frac{1}{2}A_{ij}^{(6)}(r_0)_{ij}^6 + \frac{3}{4}A_{ij}^{(8)}(r_0)_{ij}^4 + \frac{5}{6}A_{ij}^{(10)}(r_0)_{ij}^2 - \frac{1}{6}\left\{\frac{\partial(\alpha_j F^2)}{\partial r_{ij}}\right\}_{r_{ij}=(r_0)_{ij}}(r_0)_{ij}^{13}. \quad (44)$$

If i is a neutral atom and without electric moment $F = 0$ and the last term on the right of eqn 44 disappears. If i is an ion of charge ze, where e is the electronic charge, then $F = ze/r_{ij}^2$ and the last term on the right of eqn 46 becomes $+2\alpha_j(ze)^2(r_0)_{ij}^8/3$.

An approximation in the evaluation of ϕ_p arises in the over-simple relation $\phi_P = -\alpha_j F^2/2$. For molecules of finite size near an ion or in a zeolite F cannot be given throughout the molecule by its value at the centre of the molecule and strictly one should not use the mean polarizability, α_j, of the molecule at all points in it.[53,54]

Three approximations are in common use for $A_{ij}^{(6)}$, associated respectively with the names of London,[55, 56] Slater and Kirkwood,[57] and Kirkwood[58] and Muller.[59] They are

$$A_{ij}^{(6)} = \frac{3\alpha_i\alpha_j}{2}\frac{E_iE_j}{(E_i + E_j)} \qquad \text{(London)} \qquad (45)$$

$$A_{ij}^{(6)} = \frac{3}{4\pi}\frac{eh}{\sqrt{m}}\frac{\alpha_i\alpha_j}{\sqrt{\frac{\alpha_i}{n_i}} + \sqrt{\frac{\alpha_j}{n_j}}} \qquad \text{(Slater–Kirkwood)} \qquad (46)$$

$$A_{ij}^{(6)} = -\frac{6mc^2\alpha_i\alpha_j}{\left(\frac{\alpha_i}{X_i} + \frac{\alpha_j}{X_j}\right)}. \qquad \text{(Kirkwood–Muller)} \qquad (47)$$

These expressions are given in c.g.s. units. For SI units $(4\pi\epsilon_0)^{-2}$, $(4\pi\epsilon_0)^{-2}$ and $(16\pi^2\epsilon_0)^{-1}$ (where ϵ_0 is the permittivity of a vacuum) must be introduced into eqn 45, 46 and 47 respectively. In c.g.s. units the expressions for $A_{ij}^{(8)}$ and $A_{ij}^{(10)}$ corresponding with the Kirkwood–Muller relation are:

$$A_{ij}^{(8)} = \frac{45h^2}{32\pi^2 m}\alpha_i\alpha_j\left[\frac{1}{2\left(\frac{\alpha_j}{X_j}\Big/\frac{\alpha_i}{X_i}\right)+1} + \frac{1}{2\left(\frac{\alpha_i}{X_i}\Big/\frac{\alpha_j}{X_j}\right)+1}\right] \qquad (48)$$

$$A_{ij}^{(10)} = -\frac{105h^4}{256\pi^4 mc^2}\alpha_i\alpha_j\left[\frac{\alpha_i/X_i}{3\left(\frac{\alpha_j}{X_j}\Big/\frac{\alpha_i}{X_i}\right)+1} + \frac{3}{4\left(\frac{X_j}{\alpha_j}+\frac{X_i}{\alpha_i}\right)} + \frac{\alpha_j/X_j}{3\left(\frac{\alpha_i}{X_i}\Big/\frac{\alpha_j}{X_j}\right)+1}\right] \qquad (49)$$

In eqn 45 to 49 quantities not already defined are:

α_i = polarizability of lattice atom of type i;

χ_i, χ_j = diamagnetic susceptibilities of i and j;
E_i, E_j = energies characteristic of i and j, often taken as their energies of ionization;
h = Planck's constant; m = mass of the electron;
c = velocity of light;
n_i, n_j = numbers of electrons in the outermost complete electronic shells of i and j respectively.

In calculations of the energy of interaction of long chain n-paraffins in zeolite A Spangenberg *et al.*[47] evaluated $A_{ij}^{(6)}$, $A_{ij}^{(8)}$ and $A_{ij}^{(10)}$ using the Kirkwood–Muller approximation, with the results given in Table 8. The

TABLE 8

Parameters Involved in Estimating the Dispersion Energy for Interaction Between CH_2 and Lattice Atoms in Zeolite A[47]

	CH_2	$O^{-\delta e}$ ($\delta = 0{\cdot}25$)	Na^+	Ca^{2+}	
α_j	18·2	38·9	1·9	5·1	10^{-25} cm^3 ϵ_0 mol^{-1}
χ_j	—19·75	—20·92	—7·0	—14·95	10^{-30} cm^3 μ_0 mol^{-1}
r	2·00	1·437	0·95	0·99	10^{-8} cm
$A_{ij}^{(6)}$	—	1·799	0·204	0·519	10^{-45} kcal cm^6 mol^{-1}
$A_{ij}^{(8)}$	—	0·490	0·026	0·067	10^{-60} kcal cm^8 mol^{-1}
$A_{ij}^{(10)}$	—	0·163	0·004	0·011	10^{-75} kcal cm^{10} mol^{-1}
B_{ij}	—	2·222	0·661	2·801	10^{-90} kcal cm^{12} mol^{-1}

lattice atoms or ions considered were $O^{-\delta e}$, Na^+ and Ca^{2+}; the group with which they were interacting was CH_2 and the long chain paraffin was regarded as being composed of such groups (the terminal CH_3 groups were taken as equivalent to CH_2 groups). If we consider the interaction between CH_2 and $O^{-\delta e}$ as an example, then when $j(=CH_2)$ and $i(=O^{-\delta e})$ are in contact and r_{ij} is taken as $(r_{CH_2} + r_{O^{-\delta e}}) = 3{\cdot}437$ Å, the energies are

$$E_A = A_{ij}^{6}/r_{ij}^{6} \approx -1{\cdot}1 \text{ kcal mol}^{-1}$$
$$E_B = A_{ij}^{(8)}/r_{ij}^{8} \approx -0{\cdot}25 \text{ kcal mol}^{-1}$$
$$E_C = A_{ij}^{(10)}/r_{ij}^{10} \approx -0{\cdot}07 \text{ kcal mol}^{-1}.$$

Although E_B and E_C are not insignificant they are together less than 30% of E_A in the case considered. In most calculations of dispersion energy in the literature they have been omitted.

One may next compare the values of $A_{ij}^{(6)}$ given by the three approximations of eqns 45, 46 and 47 respectively. Among others, Salem[60] has considered their theoretical limitations. The London approximation, using the ionization energies for E_i and E_j, is in effect considering only the optical electron. In a multielectron atom other electrons are also capable of making a contribution so that eqn 45 is expected to give too low a value for ϕ_D, except possibly for very simple molecules such as He or H_2 with few electrons. In the Slater–Kirkwood approximation (eqn 46) the total wave function for the closed shell system is considered as the product of individual electron functions with spherical symmetry and neglects electron correlation. The Kirkwood–Muller

TABLE 9

Interaction Constants $A_{ij}^{(6)}$ for Pairs of Molecules,[(60)] in a.u.
(1 a.u. = $0{\cdot}9571 \times 10^{-60}$ erg cm^6)

Interacting pair	"Experi-mental" 2nd Virial[a] Coefficient	$A_{ij}^{(6)}$ Viscosity[a]	London	Slater–Kirkwood	Kirkwood–Muller
He – He	1·644	1·644	1·31	1·74	1·70
Ne – Ne	10·4	9·7	4·3	8·1	12
Ar – Ar	107·7	114	52·6	67	135
Kr – Kr	214	242	108	125	295
Xe – Xe	606	587	246	259	730
CH_4 – CH_4	265	271	110	155	237
H_2 – H_2	13·5	13·5	12·07	13·04	—
O_2 – O_2	117	114	38·9	—	—
Ar – He	—	12·17[b]	8·03	10·4	15·0

[a] Assuming the Lennard–Jones 12-6 potential.
[b] Assuming the Lennard–Jones 12-6 potential and fitting thermal diffusion data.

approximation (eqn 47) also neglects electron correlation, and as a result neither can be expected to give fully satisfactory dispersion energies. The Kirkwood–Muller approximation tends to give too high values of ϕ_D. To illustrate these observations some results gathered from the literature by Salem are indicated in Table 9. As expected, for the molecules with few electrons (He–He and H_2–H_2) the calculations are close to each other and to the "experimental" values. As the number of electrons increases the London and the Slater–Kirkwood values tend to become too small and the Kirkwood–Muller values too large. The average of either of the first two and the Kirkwood–Muller value would then appear to be a better approximation to the "experimental" values than any one of the three approximations on its own.

Barrer and Ruzicka,[(61)] in calculating the energy of clathration in gas hydrates, also evaluated and compared the London, Slater–Kirkwood and

Kirkwood–Muller approximations for the coefficients $A_{ij}^{(6)}$, this time for simple guest molecules interacting with water molecules. Their values are given in Table 10. For the H_2–H_2O pair the three approximations give rather similar values, but for guest molecules with more and more electrons the London and Kirkwood–Muller values diverge so much that their ratios in certain instances can exceed two (e.g. Kr – H_2O and Xe – H_2O). The divergence is once more in the direction indicated by Salem.[60]

The Kirkwood–Muller approximation to $A_{ij}^{(6)}$ has the great advantage that

TABLE 10

Interaction Constants $A_{ij}^{(6)}$ for Water and Molecule,[61] in cal cm^6 mol^{-1} × 10^{44}

Gas	$(r_0)_{ij}$ in Å	London	Slater–Kirkwood	Kirkwood–Muller
H_2	2·6	36·3	42·3	43·1
Ne	3·0	16·5	32·5	39·9
N_2	3·40	79·0	110·8	117·1
Ar	3·31	60·0	99·7	144
O_2	3·34	66·2	107·9	—
CH_4	3·55	93·5	139·2	138·4
Kr	3·42	86·1	140·5	216
Xe	3·68	130	197·5	340
C_2H_4	4·0	134	222·5	158·1
C_2H_6	3·9	148·5	240	280

group or atom polarizabilities, magnetic susceptibilities and radii may be used for large molecules, so that the interaction energies for parts of large molecules with their intracrystalline environments may be determined, and added to give the energy for the whole molecule. Thus one may use this device to estimate $(\phi_D + \phi_R)$ for chain molecules in different chain configurations and orientations within a cavity. By this means the anisotropy of such molecules may also be allowed for since anisotropy for the individual groups or atoms composing the molecule need not be considered.

7. COMPARISON OF DISPERSION ENERGY FOR POROUS AND NON-POROUS SORBENTS

The universally present dispersion energy contribution to the physical bond between sorbate and sorbent can, other things being equal, vary according to the local environment. The extent to which this is theoretically possible

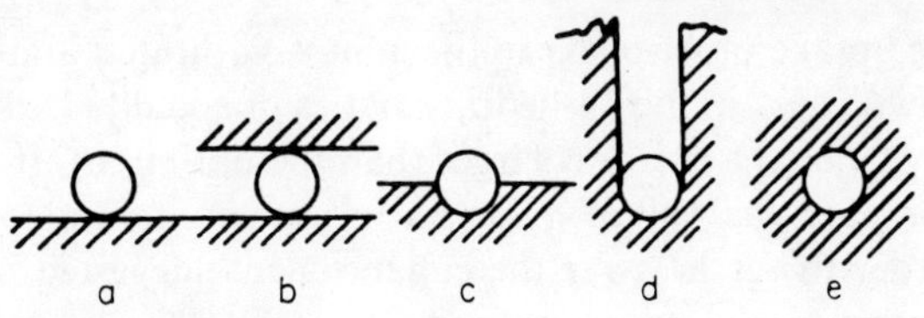

FIG. 12. Adsorbed molecules in various environments on or in a sorbent.[24]

was first clearly indicated by de Boer and Custers.[62] They considered a molecule in each of the five environments a to e in Fig. 12[24], and evaluated ϕ_D by integrating $-A_{ij}^{(6)}/r_{ij}^6$ throughout the volume of the sorbent. For optimum "fit" of the sorbate molecule into the openings, they found the ratios of ϕ_D for the local environments *a* to *e* respectively

$$_a\phi_D = {_b\phi_D}/2 = {_c\phi_D}/4 = {_d\phi_D}/6\cdot 36 = {_e\phi_D}/8.$$

Thus in the optimum hole of environment *e* an eight-fold enhancement in ϕ_D is found as compared with ϕ_D for the same molecule adsorbed on a sheet of the same sorbent. This neat fit into a nearly optimum "hole" applies for

TABLE 11

Some Heats of Sorption of Argon[24] (kcal mol^{-1})[a]

Sorbent	N × 10^{-22} (atoms of sorbent per cm^3)	q_{st} as $\theta \to 0$	as $\theta \to 0\cdot 5$ (or for cm^3 at s.t.p. g^{-1} given in brackets)
β-quinol[63]	2·27 (C) 0·76 (O) 2·27 (H)	4·8	4·8
Chabazite (outgassed at 480°C)[24]	3·0 (O)	6·0	3·0 (50)
Chabazite (outgassed at 300°C)[64]	3·0 (O)	4·6	3·6 (40)
H-mordenite (outgassed at 350°C)[7]	$3\cdot 3_6$ (O)	4·6	3·0
Na-X[27]	$2\cdot 4_7$ (O)	2·7	2·7 (25)
Rutile[65]	10·6 (O) 3·3 (Ti)	3·5	2·5
Carbon			
(a) Spheron heated to 2700°C[66]	11·5 (C) (as in graphite)	2·70	$2\cdot 8_0$
(b) Graphon[67]		2·60	2·75
(c) Saran S84[66]		3·50	2·63
(d) Saran S600H[66]		3·90	3·63

[a] Figures in brackets in the last column indicate cm^3 sorbed at s.t.p. per g to which the heat refers. In these instances θ is not 0·5.

instance to the uptake of Ar or Kr in the quinol clathrates, and is approached for the same two gases in the 14-hedral cavities of sodalite hydrate (although in the latter instance the cavity is larger than the optimum). It is therefore of interest to compare the values of q_{st} in or on zeolite and on non-porous sorbents in order to see how far the enhancement suggested by the calculations of de Boer and Custers is realized. Table 11[24] gives relevant data for Ar in a number of porous and non-porous sorbents.

The heats do not cover the wide range which is indicated by the de Boer and Custers calculation, the main reason for which is the honeycomb-like structure of the porous clathrate and zeolite sorbents. Thus, comparing the carbon-rich porous sorbent, β-quinol, with the graphitic Spheron or Graphon, the density of carbon atoms in the latter two sorbents is five times that in the β-quinol. This at once introduces a major reduction in q_{st} for β-quinol. Among the oxygen-rich sorbents one may similarly compare the molecular-sieve zeolites with non-porous rutile. Again the oxygen atom density in rutile is three to four times as large as that in the zeolites. There is in H-mordenite evidence of enhancement of the initial value of q_{st}, when it is probable that most of the argon is in the side-pockets[7] closely surrounded by oxygens; likewise in β-quinol and in the microporous Saran carbons some enhancement is observed, although it is not large. There is an approximate balance between increases in ϕ_D expected for molecules in cavities and the diminution in N, the atom density, in "honeycomb" structures.

8. CALCULATIONS RELATING TO SORPTION ENERGIES

In § 6 an outline has been given of procedures for calculating $(\phi_D + \phi_R)$, and of the assumptions and limitations to the accuracy of such calculations. One may now consider some examples of the evaluation of $(\phi_D + \phi_R)$ and of electrostatic energy, with a view to comparing observed and calculated sorption heats.

8.1 Inert Gases

The simplest examples occur when inert gases are sorbed in porous crystals because the only contributions to the heats come from ϕ_D, ϕ_R and ϕ_P. One of the first calculations was made for zeolite X in 1958.[27] For the zeolite enriched in various cations the differences between the isosteric heats of sorption of argon for $\theta \to 0$ and $\theta = 0{\cdot}25$ are given in Table 12. It has already been pointed out that ϕ_P is expected to decline rapidly so that the difference between the values of q_{st} for $\theta \to 0$ and $\theta = 0{\cdot}25$ should represent approximately the contribution to q_{st} from ϕ_P, while q_{st} at $\theta = 0{\cdot}25$ should arise approximately from $(\phi_D + \phi_R)$. This estimate of ϕ_P shows it to be considerable only when the exchange ion is the small divalent and highly polarizing Ca^{2+}. As the cation increases in size the contribution to q_{st} from $(\phi_D + \phi_R)$ also increases, as would be expected. This is seen in the series Ca < Sr < Ba and in Li < Na < K. However the changes are not large and most of q_{st} at $\theta = 0{\cdot}25$ arises from interactions of Ar with the framework oxygens.

TABLE 12

q_{st} for $\theta \rightarrow 0$ and for $\theta = 0{\cdot}25$ for Ar in Zeolite X[27]

Exchange form	q_{st} (kJ mol^{-1}) $\theta \rightarrow 0$	$\theta = 0{\cdot}25$	Difference ($\approx -\phi_P$)
Ca	20·9	11·3	9·6
Sr	15·9	13·2	2·7
Ba	13·8	13·8	0
Li	13·8	11·3	2·5
Na	11·7	11·7	0
K	12·5	12·5	0

TABLE 13

Argon-cation Interactions in kJ mol^{-1}

Molecule or Ion	r (Å)	$10^{24}\alpha$ cm^3	$-10^{30}\chi$ (c.g.s. units)	$-(\phi_D + \phi_R)$	$-\phi_P$
Ar	1·92	1·63	31·0	—	—
Li^+	0·78	0·029	0·99	0·210	21·3
Na^+	0·98	0·180	6·95	0·670	16·0
K^+	1·33	0·840	27·54	1·80	10·2
Ca^{2+}	0·99	0·471	22·1	2·13	63·5
Sr^{2+}	1·13	0·863	46·1	3·26	52·7
Ba^{2+}	1·35	1·560	76·4	4·22	40·7

The energy of interaction between isolated cation—argon pairs was calculated using the Kirkwood–Muller approximation and ionic and van der Waals radii to evaluate $(\phi_D + \phi_R)$, while ϕ_P was estimated as $\phi_P = -\frac{1}{2}\alpha F^2 = -\alpha(ze)^2/2(r_0)^4_{ij}$. The results are shown in Table 13.[27] If the cations in the zeolite were wholly unscreened, as in the calculations of Table 13, ϕ_P would be important, especially for the divalent cations. In fact the comparison of Table 13 and Table 12, in accordance also with Table 7, suggests a very strong screening of the cations by the much more numerous anionic oxygens of the framework into which the cations tend to be recessed. Table 13 also confirms that the contribution of $(\phi_D + \phi_R)$ to q_{st} arising from the cations should be quite small, and indicates that this contribution should increase according to the sequences already referred to in the previous paragraph. In the exchange forms there are of course twice as many univalent ions as divalent ones; and the distributions of the various ions among the possible crystallographic sites in zeolite X (and in zeolites in general) are also known to vary according to the cation.[68]

The ratio O : (Al + Si) = 2 in any tectosilicate and the ratio M : Al = 1 where M is a monovalent ion. Also the cationic Al and Si are much less polarizable than the anionic O of the framework, as well as being less numerous, and all are buried in tetrahedra of oxygens so that unlike the oxygens and the exchangeable cations they can never make direct contact with the guest molecules. For these reasons the dominant contribution to $(\phi_D + \phi_R)$ must come from the numerous and polarizable anionic oxygens. An estimate was therefore made of this contribution for argon[27] assuming the large cavities in zeolites X and A to be spherical and of radius R from the centre of the cage to the centre of each oxygen forming the wall of the cavity. If a is the distance from the centre of an argon to the centre of each cavity, then the first approximation is to replace pairwise summation by integration throughout the volume of the crystal external to R, using an average density of oxygens. This gives

$$\phi_D + \phi_R = \pi N A_{ij}^{(6)} [-P(x)/R^3 + (r_0)_{ij}^6 Q(x)/5R^9] \tag{50}$$

where N is the number of oxygens per unit volume and

$$(r_0)_{ij} = \tfrac{1}{2}(r_{\mathrm{Ar}} + r_{\mathrm{O}}).$$

Also

$$\left.\begin{aligned} P(x) &= 4x^6/3(x^2-1)^3 \\ Q(x) &= x^9/2[-(x+1)^{-8}/8 + (x+1)^{-9}/9 + (x-1)^{-8}/8 + \\ &\qquad (x-1)^{-9}/9] \end{aligned}\right\} \tag{51}$$

TABLE 14

$-(\phi_D + \phi_R)$ for Ar-anionic oxygens, according to eqn 50

Zeolite	10^{-22}N (Oxygens cm^{-3})	r_{oxygen} (Å)	$10^{24}\alpha_{\mathrm{oxygen}}$ (cm^3)	$-10^{30}\chi_{\mathrm{oxygen}}$ (c.g.s. unit)	$-(\phi_D + \phi_R)$ (kJ mol^{-1})
X	2·47	1·40	3·89	20·92	6·3
A	2·58				6·7

where $x = R/a$. The Kirkwood–Muller approximation for $A_{ij}^{(6)}$ was used with the values of N, r, α and X given in Table 14, together with the calculated values of $-(\phi_D + \phi_R)$. r, α and χ for Ar are given in Table 13. q_{st} is given by eqn 16 and, regarding sorbed argon as a classical oscillator, $F(T) = 3RT/2$. At 200 K, which is in the temperature range of the isotherms measured,[27] it follows that at the minima of the potential energy curves (for which the values of $-(\phi_D + \phi_R)$ in Table 14 are recorded) $\approx 5 \cdot 7$ and $\approx 6 \cdot 1$ kJ mol^{-1} of q_{st} in zeolites A and X respectively are accounted for. The small zero point energy correction is neglected in this estimate. When these contributions are compared with q_{st} at $\theta = 0 \cdot 25$ in Table 12 it is seen that they account for only about half the observed values.

The low value of the calculated heat, despite using the Kirkwood–Muller value of $A_{ij}^{(6)}$, undoubtedly arises from the integration procedure which assumes a uniform density of oxygens averaged over all space outside the cavity. However the "honeycomb" structure of the crystal means that the oxygen atom density varies from point to point in the structure. In particular in the wall of the cavity in which the argon is located the density of the oxygens is much above the average and it is these oxygens which are nearest the argon and contribute largely to $(\phi_D + \phi_R)$. Clearly summation of argon-oxygen interactions is preferable at least for the oxygens in the wall of the cavity.

The 26-hedral cavities in zeolite A have walls composed of 8-, 6- and 4-rings of (Al, Si)O_4 tetrahedra; and in the 26-hedra of zeolite X the walls are composed of 12-, 6- and 4-rings. The inner peripheries of all such rings are formed by oxygen atoms. By summing pair interactions of the peripheral oxygens and argon in or against isolated rings the maximum values of $-(\phi_D + \phi_R)$ for ideal planar rings were found to be:

12-ring of free diameter 8Å, $-(\phi_D + \phi_R) =$	$9 \cdot 2$ kJ mol^{-1}
8-ring of free diameter $4 \cdot 2$Å	$21 \cdot 3$ kJ mol^{-1}
6-ring of free diameter $3 \cdot 1$Å	$17 \cdot 6$ kJ mol^{-1}
4-ring of free diameter $1 \cdot 98$Å	$11 \cdot 7$ kJ mol^{-1}

Although only a few oxygens are involved the calculated values now span the values of q_{st} at $\theta = 0 \cdot 25$ (Table 12) and may reflect the rather exaggerated estimates expected from the use of the Kirkwood–Muller approximation for $A_{ij}^{(6)}$.

An additional factor concerns the value of α taken for oxygen (Table 14). This is the value for O^{2-} ions. From the refractivity of potash felspar one may estimate $\alpha \approx 1 \cdot 65 \times 10^{-24}$ cm^3. Barrer and Peterson[7] therefore calculated $-(\phi_D + \phi_R)$ and the contributions made by $-(\phi_D + \phi_R)$ to q_{st} for inert gas molecules located in the side pockets of mordenite using both values of α, and summing the pair interactions rather than integrating. For each value of α, X was taken to be $-20 \cdot 9 \times 10^{-3}$ c.g.s. units. Their results

TABLE 15

Calculated q_{st} (kJ mol^{-1}) for Inert Gases in Side Pockets of H-mordenite, and Observed Values of q_{st}[7]

Gas	London		Slater–Kirkwood		Kirkwood–Muller		Initial Experimental Value[a]
	$\alpha=1\cdot65$Å^3	$\alpha=3\cdot89$Å^3	$\alpha=1\cdot65$Å^3	$\alpha=3\cdot89$Å^3	$\alpha=1\cdot65$Å^3	$\alpha=3\cdot89$Å^3	
He	2·5	5·8	4·2	7·5	4·2	5·8	1·7 (2·5)
Ne	5·4	12·5	10·5	17·6	12·5	15·0	5·8 (6·7)
Ar	13·0	30·5	20·9	38·5	29·3	37·2	19·3 (19·3)
Kr	17·6	42	27·2	51	41	54	—
Xe	22·6	54	32·6	64	55	73	—

[a] The bracketed figures are for Na-mordenite.

are summarized in Table 15. The summation was made over 37 oxygen atoms around and composing the side pocket, and including small contributions from two Na^+ ions in the distorted 8-rings at the back of each pocket. The conclusions drawn from these calculations were.

(i) London and Slater–Kirkwood approximations to $A_{ij}^{(6)}$ are much more sensitive to the value of α than is the Kirkwood–Muller approximation.

(ii) For the lower value of α the range in values of $A_{ij}^{(6)}$ for each gas is large as between the three approximations, but the higher values of α lead to similar values for a given gas for all three approximations.

(iii) For $\alpha = 3\cdot89$Å^3 the calculated heats exceed those observed for each of the three approximations for $A_{ij}^{(6)}$ suggesting that this value of α for anionic oxygen in tectosilicates is too large.

(iv) For $\alpha = 1\cdot65$Å^3 London's approximation for $A_{ij}^{(6)}$ gives too low values of q_{st} (except for He) and are smaller than are those calculated using the other two approximations. The Kirkwood–Muller approximation for $A_{ij}^{(6)}$ in particular gives values of q_{st} which are considerably too large.

Conclusion (iv) is in line with Salem's[60] comments (§ 6) upon the limitations of the three approximations for $A_{ij}^{(6)}$, and suggests therefore that $\alpha = 1\cdot65$Å^3 is a reasonable value to take for the polarizability of anionic oxygen in tectosilicates. In mordenite the first molecules sorbed are thought largely to be concentrated in the side pockets of mordenite[7] so that the comparison with initial experimental values of q_{st} using the calculations for molecules at the potential energy minima is then justified.

A further detailed evaluation of $(\phi_D + \phi_R + \phi_P)$ has been made for Ne and Ar in the 26-hedral cavities of zeolite *A*, and summarized by Kiselev and Lopatkin.[46] For this zeolite a charge of 0·25*e* was placed on each oxygen. The polarizability was then interpolated from a plot of polarizability against charge. In this plot the value $\alpha = 3{\cdot}89\text{Å}^3$ was used for O^{2-} and values for O^- and O were calculated by the Kirkwood formula.[69] The value of X was calculated from that of α.[69] Plots of α and X against charge on the oxygen were both linear. For Ne, Ar and Na^+ experimental values of α and X were used. The constants $A_{ij}^{(6)}$ and $A_{ij}^{(8)}$ were calculated according to eqns 49 and 50. The repulsion energy constants B_{ij} were then found as described in § 6.

For evaluating the lattice sums required in eqn 45 the atomic coordinates given by Howell[70] were employed. Summation of pair interactions was made over all atoms comprising the eight 14-hedral (sodalite) cages surrounding a central 26-hedron while integration was performed throughout the remaining volume of the crystal. The contribution from the term $A_{ij}^{(10)}/r_{ij}^{10}$ in eqn 45 was not considered since, as pointed out in § 6, it is small. The calculations were made with the sorbate molecule on each of 40 positions on axes of symmetry of orders II, III and IV. There are respectively 12, 8 and

TABLE 16

Experimental and Calculated Values of Heats of Sorption of Ar and Ne in Zeolite Na-*A* (kJ mol^{-1})[46]

Guest molecule	q_{st} observed	q_{st} (calculated) $r = 1{\cdot}40$Å	$r = 1{\cdot}52$Å	T/K
Ne	4·2	7·1	5·2	84·1
Ar	11·7	13·2	11·7	228·2

6 axes of orders II, III and IV, so that the division of the cavity appears to be detailed. The field F was evaluated assuming full cationic charge on each exchange ion and, as already stated, 0·25*e* on each oxygen, for each of the 40 positions, and $\phi_P = -\frac{1}{2}\alpha F^2$ evaluated. Finally a zero point energy correction was applied assuming that the guest molecule executed simple harmonic motion about each minimum of the curve of energy against distance along each axis. These curves are shown for Ar and Ne in Fig. 13. a_0, the unit cell edge, was taken as 12·31Å. The curves are calculated assuming for oxygen $r = 1{\cdot}40$Å (Fig. 13a) and $r = 1{\cdot}52$Å (Fig. 13b). $m = 1$, 2 and 3

in the diagram refer respectively to the axes of order III, II and IV. The mean energy of binding of guest molecules was then taken to be

$$\langle \Delta E \rangle = \sum_{i=1}^{3} n_i \int_0^{R_m} \phi_i(R_j) \exp(\phi_i(R_j)/kT) R_j^2 \mathrm{d}R_j \Big/ \sum_{i=1}^{3} n_i \int_0^{R_m} \exp(\phi_i(R_j)/kT) R_j^2 \mathrm{d}R_j$$

where R_m is the radius of the idealized spherical cavity, n_i is the number of axes of the ith type with the origin at the centre. From these energies of sorption the heat $\langle \Delta H \rangle$ may be found and compared with the experimental values (Table 16). In zeolite A the experimental heats of sorption of Ne and of Ar were almost independent of amount sorbed over the range of coverage

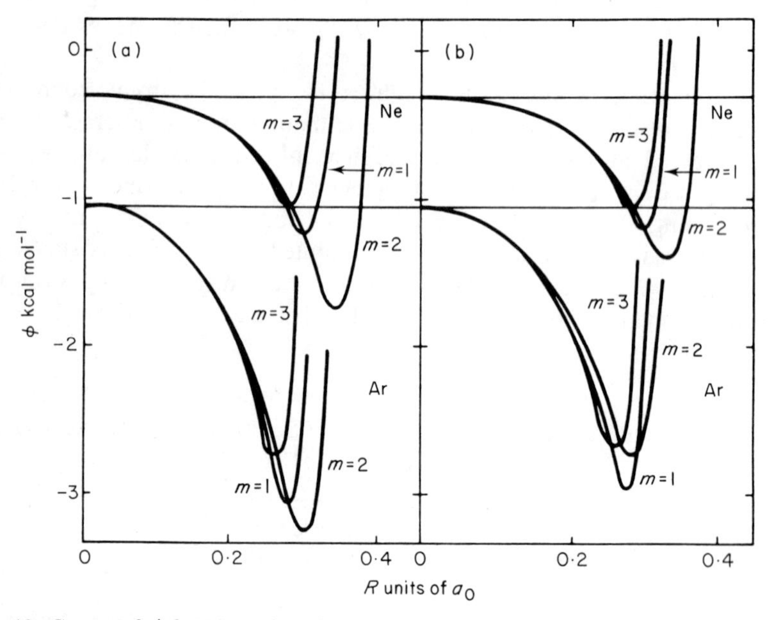

FIG. 13. Curves of ϕ for Ne and Ar in zeolite Na-A plotted against distance along each of the three chosen axes of symmetry, $m = 1$, 2 and 3. R is reckoned from the centre of the large cavity.[46] (a) Oxygen radius = 1·40Å; (b) Oxygen radius = 1·52Å.

studied. For the larger value of $r = 1{\cdot}52$Å the agreement between calculated and observed heats is satisfactory but for the more likely value $r = 1{\cdot}40$Å the calculated heats exceed those observed, as would be expected from the use of the Kirkwood–Muller approximation for $A_{ij}^{(6)}$.

The three examples of calculations of interaction between crystal and inert gases which have been discussed serve to show the degree of success, the problems involved and the limitations in the theory of the physical bond even for the simplest systems. The limitations go back to the theory of dispersion energy and can be reduced only by calculations of a much more elaborate type than anything so far attempted.

8.2 Hydrocarbons

For molecules of considerable size it is necessary to evaluate separately and then to add the contributions to the sorption energy of the several conveniently small groups into which the molecule can be divided. For *n*-paraffins these groups are terminal CH_3— and $>CH_2$ forming the rest of the chain. Polarizabilities and diamagnetic susceptibilities can be ascribed to these groups so that the Kirkwood–Muller approximation can, as noted earlier, be used to evaluate the interaction between the groups and the atoms of the surrounding host lattice. This approach was used by Spangenberg *et al.*[47, 30, 71] in calculations of the sorption energies of hydrocarbons in zeolite *A*. The relevant physical constants have been given in Table 8.

The procedure used was similar to that employed for Ar and Ne in zeolite Na-*A*.[46] The values of $\phi_D + \phi_R + \phi_P$ were calculated for $>CH_2$ at various positions along selected axes traversing the 26-hedron. The sorption energy of the $>CH_2$ group was taken as the weighted mean of the energy minima when the group moved along the selected axes. The weighting factors were the numbers of axes of each kind which traversed the large cavity. Close to the wall this energy was $E^W = -21 \cdot 5$ kJ mol^{-1} ($-5 \cdot 13$ kcal mol^{-1}). The CH_2 groups in a large 26-hedral cavity were divided into those adjacent to the wall—the wall phase—and those nearer the centre of the cavity, termed the volume phase. If the centre of a large cavity is the origin the volume phase comprised all CH_2 groups whose centres lay in the range $0 \leq r \leq (r_{\text{min}} - r_{CH_2})$ where r_{min} are the distances from the cage centre of the minima in the potential energy curves and r_{CH_2} is the radius of the CH_2 group. The higher the temperature and so the more vigorous the vibrations of the CH_2 groups the greater is the proportion of these groups in the volume phase.

The weighted average energy for the volume phase was estimated as $E^B = -7 \cdot 4$ kJ mol^{-1} ($-1 \cdot 77$ kcal mol^{-1}). At any given temperature the binding energy per CH_2 must lie between E^W and E^B. At absolute zero all CH_2 groups will tend to be in the wall phase and the sorption energy for the paraffin considered as a complex of N CH_2 groups will be NE^W. Thus the heat of sorption for *n*-heptane would be $-150 \cdot 5$ kJ mol^{-1}. The high calculated binding energy per CH_2 for the wall phase ($-21 \cdot 5$ kJ mol^{-1}) may in part reflect the use of the Kirkwood–Muller approximation (§ 6, Tables 9 and 10). Also the polarizability of the anionic oxygen was taken to be that of O^{2-} (Table 8) although the assumed anionic charge per oxygen was taken as $0 \cdot 25e$.

The equilibrium between wall and volume phases was considered for the CH_2 groups and estimates made of mean molar energy, entropy and heat capacity for *n*-heptane, considered as a complex of 7 CH_2 groups.* Such

* No distinction was made between CH_3 and CH_2 groups for purposes of evaluating E^W and E^B.

calculations were made for several degrees of filling of the cavities ($\theta = 0 \cdot 025$, $0 \cdot 1$ and $1 \cdot 0$) and over a temperature range of 298 K to 800 K. The molar heat capacities for the model reached very high values with maxima that varied in height and temperature according to the degree of filling. Despite the artificialities and limitations of the model these heat capacity maxima are of considerable interest because maxima have been observed experimentally (Chapter 5, § 4).

In the evaluation of ($\phi_D + \phi_R + \phi_P$) by Fiedler *et al.*[71] only $A_{ij}^{(6)}$ and B_{ij} were used in calculating $\phi_D + \phi_R$ for CH_2 groups in the large cavities of

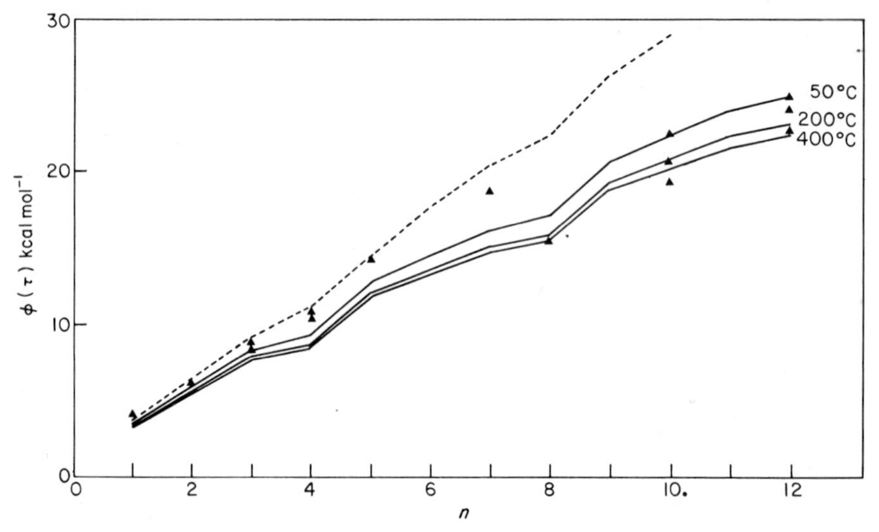

FIG. 14. Sorption heats calculated and observed for *n*-alkanes as functions of carbon number.[71]

▲ = experimental values; .... $N(CH_2)$ complex; ———— *n*-alkane conformation.

(Ca, Na)-*A*. The method of calculation was similar to that of Spangenberg *et al*[47] but because only $A_{ij}^{(6)}$ was used in evaluating θ_D the B_{ij} differed from those in Table 8. Polarization energy, ϕ_P, was rather small compared with ϕ_D but was comparable in size with ϕ_R. When experimental heats and calculated heats were compared, according to the model in which a *n*-paraffin with *N* carbon atoms is represented as *N* CH_2 groups, the correspondence was reasonable when *N* was small. However, the calculated heats were too large for longer chain paraffins (Fig. 14). This led the authors to examine a second model in which dodecane was assigned a specific configuration within the 26-hedral cavity, which paraffins with shorter chains followed as far as possible. This model takes account of the hindered rotation of CH_2 groups found in any paraffin chain. The calculated values of the heats for *n*-paraffins at 50, 200 and 400°C are also shown in Fig. 14 and appear in reasonable agreement with most of the experimental values.

The general form of the relation between q_{st} and the carbon number N indicated by the experimental points in Fig. 14 is also found for paraffins C_1 to n-C_4 in H-chabazite and zeolite H-*L*[72] (Fig. 11, Chapter 3); C_1 to n-C_6 in zeolite (K, Na)-*L*[5] (Fig. 3); and C_1 to n-C_8 in zeolite X[73] (Fig. 15).

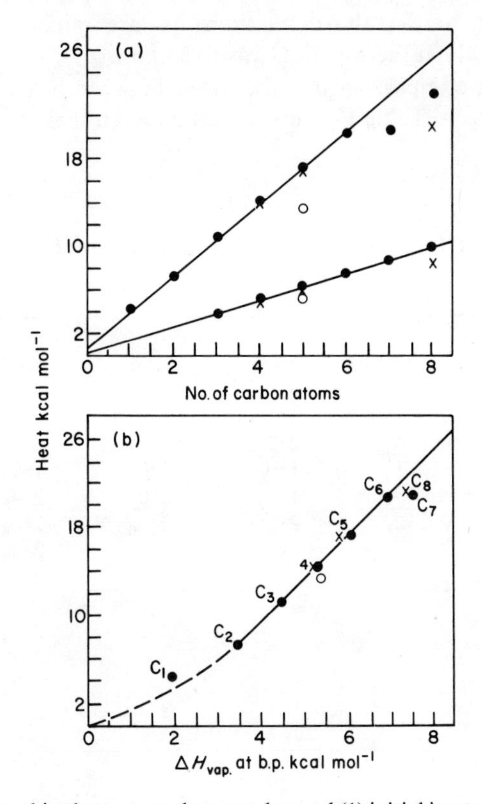

FIG. 15. (a) Relationships between carbon number and (1) initial isosteric heats of sorption and (2) heats of liquefaction at 25°C of some hydrocarbons.[73] Lower curve: heat of liquefaction vs carbon number. (b) Isosteric heats of sorption plotted against heats of liquefaction.

● = *n*-paraffins; X = isoparaffins; ○ = neo-pentane

Some of these relations approach straight lines not only in zeolites, but also on graphitic carbon and in silica gels and other sorbents.[74] This behaviour was also approximated in plots of heat of sorption against carbon number, N, for each of three fixed amounts sorbed for (Na, Ca)-*A* and Mg-*L*.[75] Some equations for the best lines are:

$$q_{st}\ (\text{kcal mol}^{-1}) = 2{\cdot}75 + 2{\cdot}0\, N\ (\text{H-chabazite},^{(72)}\ N = 1 \text{ to } 4)$$

$$q_{st}\ (\text{kcal mol}^{-1}) = 2\cdot5 + 2\cdot0\,N\ (\text{H-}L,^{(72)}\ N = 1 \text{ to } 4)$$
$$q_{st}\ (\text{kcal mol}^{-1}) = 2\cdot6 + 2\cdot2\,N\ (\text{K-}L,^{(5)}\ N = 1 \text{ to } 6)$$
$$q_{st}\ (\text{kcal mol}^{-1}) = 1\cdot3 + 3\cdot0\,N\ (\text{Na-X},^{(73)}\ N = 1 \text{ to } 8)$$
$$q_{st}\ (\text{kcal mol}^{-1}) = 3\cdot9 + 1\cdot45\,N\ ((\text{Na, Ca})\text{-}A,^{(75)}\ N = 1 \text{ to } 18).$$

These lines are averages and over a sufficient range in carbon numbers there is evidence (see Fig. 14) that the slope decreases with increasing values of N.[75] The heats in the above relationships are initial heats, except for (Na, Ca)-A, for which the sorption was 0·2 mmol g^{-1}.

An interesting comparison may be made between heats of sorption and heats of liquefaction. Table 17 shows ratios between these heats for a number

TABLE 17

Ratios of Heats of Sorption in Zeolites and Heats of Liquefaction for Some Hydrocarbons

Hydrocarbon	Ratios for H-chabazite	H-L	K-L	Na-X	(Ca, Na)-A
CH_4	$2\cdot6_3$	$2\cdot3_6$	$2\cdot4_7$	$2\cdot2$	$2\cdot0_5$
C_2H_6	$2\cdot1_0$	$1\cdot9_8$	$2\cdot0_8$	$1\cdot9_4$	$1\cdot7_5$
C_3H_8	$1\cdot9_8$	$1\cdot9_5$	$2\cdot0_5$	$2\cdot4_8$	$1\cdot9_3$
n-C_4H_{10}	$2\cdot0_8$	$2\cdot0_5$	$2\cdot1_0$	$2\cdot6_6$	$1\cdot9_9$
n-C_5H_{12}			—	$2\cdot7_8$	$2\cdot3_7$
n-C_6H_{14}			—	$2\cdot9_6$	—
n-C_7H_{16}			—	$2\cdot6_9$	$2\cdot4_9$
iso-C_4H_{10}			$2\cdot2_5$	$2\cdot8_4$	—
iso-C_5H_{12}			—	$2\cdot9_0$	—
iso-C_8H_{18}			—	$2\cdot8_4$	—
neo-C_5H_{12}			$2\cdot1_9$	$2\cdot4_1$	—

of hydrocarbons and zeolites. The heats of sorption involved in obtaining these ratios are not all fully comparable in that the heats tend to be functions of amount sorbed. Those for H-chabazite and H-L are standard heats, $\Delta H^{\ominus}$; the remainder are differential heats for small uptakes (hydrocarbons for which N is small) or for larger uptakes (bigger values of N). Absolute comparisons are not intended, but the ratios are seen to be high for CH_4 and go through a minimum for C_2H_6 or C_3H_8. The ratios average two or more and suggest a substantial enhancement of binding energy in the zeolites as compared with that of the same molecule in its liquid. For perfluoro compounds in zeolite Na-X Barrer and Reucroft[6] found similar ratios to those in Table 17:

Compound	$\Delta\bar{H}_{\theta\to 0}/\Delta H_{liq.}$	$\Delta\bar{H}_{max}/\Delta H_{liq.}$
CF_4	1·8	2·4
SF_6	1·5	2·1
C_2F_6	1·9	2·6
C_3F_8	2·0	2·7
n-C_4F_{10}	2·6	3·5 .

The heat $\Delta\bar{H}$ was initially nearly independent of amount sorbed but rose to a maximum near saturation of the zeolite. The ratios are given above both for initial and for maximum heats of sorption.

It has been suggested that the high ratio of $\Delta\bar{H}/\Delta H_{liq}$ may arise in part from different mechanisms of vaporization from a liquid and of desorption from a zeolite.[76,5] If a molecule were removed from within a liquid leaving a permanent hole, z bonds would be broken between the molecule and ambient liquid, where z is the coordination number. This is the nature of the desorption process from a zeolite. In a liquid however the hole collapses and $z/2$ bonds reform. Thus the ratio of the energy of vaporization of liquid for the first mechanism (permanent hole) to that for the second (collapsing hole) would be 2. However the permanent holes in zeolites are often much larger than the molecule, so that desorption from a zeolite is not strictly analogous to the first mechanism.

8.3 Polar Molecules and the Role of Exchange Cation

Calculations and observations on the energetics of sorption have been made for molecules having permanent dipole moments (NH_3 and H_2O) and permanent quadrupole moments (N_2, CO, CO_2). The first demonstration of the important role of quadrupole energy, $\phi_{\dot{F}Q}$, in sorption by zeolites was provided by Kington and McLeod.[3] They considered the measured calorimetric heats of sorption of O_2, H_2, N_2, CO and CO_2 in chabazite, a series of molecules for which the quadrupole moments increase in the sequence given. Carbon monoxide has in addition a small dipole moment the contribution of which was neglected. It was assumed that the initial energetic heterogeneity was dominated by the polarization and quadrupole terms, ϕ_P and $\phi_{\dot{F}Q}$, and that ϕ_D and ϕ_R were independent of θ, at least for smaller values of θ. If $\Delta\bar{H}_{\theta_2}$ and $\Delta\bar{H}_{\theta_1}$ are the values of the heats of sorption when $\theta = \theta_2$ and θ_1 respectively then

$$\Delta\bar{H}_{\theta_2} - \Delta\bar{H}_{\theta_1} = \frac{\alpha}{2}(F_1^2 - F_2^2) + \frac{Q}{4}\left(\frac{dF_2}{dt} - \frac{dF_1}{dt}\right). \qquad (53)$$

In this expression F_1 and F_2 are the electrostatic fields appropriate for addition of a molecule at θ_1 and θ_2 respectively and dF_1/dt and dF_2/dt are the gradients of these fields along the quadrupole axes of the guest molecules. The expression above can be rewritten as

$$\Delta\bar{H}_{\theta_2} - \Delta\bar{H}_{\theta_1} - a\alpha = cQ \tag{54}$$

where a and c are taken as constant for the series of guest molecules. Since argon has no quadrupole moment the value of a was determined using this molecule, for the heats at $\theta_1 = 0{\cdot}01$ and $\theta_2 = 0{\cdot}30$. The left hand side of eqn 54 was then plotted against $q = Q/e$ with the result shown in Fig. 16. The straight line supports the view that energetic heterogeneity is

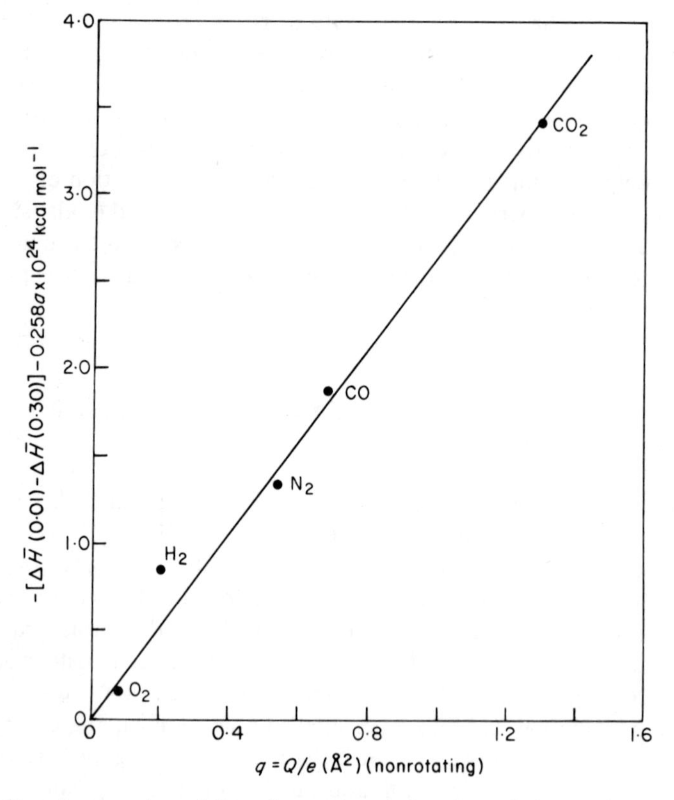

FIG. 16. Plot showing the validity of eqn 54 and the effect of the quadrupole moment upon the energetic heterogeneity of chabazite.[3]

largely electrostatic in origin as well as the tacit assumption in eqn 54 that the guest molecules, at least up to $\theta = 0{\cdot}3$, must preferentially be entering the same elements of interstitial volume in the same order. A similar behaviour was found when Ar, N_2 and CO_2 were sorbed in zeolite Na-X.[28] For argon in this zeolite, a in eqn 54 proved to be nearly zero for $\theta_1 = 0 \cdot 07$ and $\theta_2 = 0 \cdot 26$, so that it was taken as zero for the other two molecules. The plot of $\Delta\bar{H}_{\theta_2} - \Delta\bar{H}_{\theta_1}$ for the above values of θ against $q = Q/e$ was again a straight line.

Calculations have been made[29, 45] of the electrostatic fields and field gradients along the axes running through the centre of each 12-ring window and the centre of the 6-ring window diametrically opposite in the faujasite-type 26-hedral cages of zeolite X. The assumed charge distribution was that

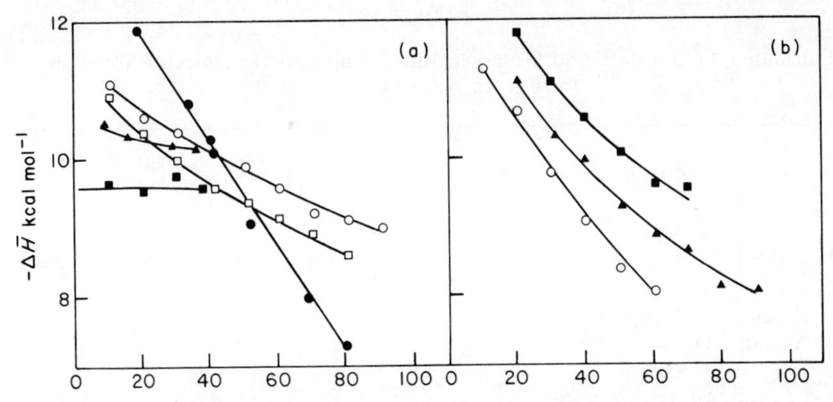

FIG. 17. Heats $q_{st} = -\Delta\bar{H}$ as functions of amount of CO_2 sorbed.[45]
(a) ● = Li-X; ○ = Na-X; □ = K-X; ▲ = (Rb, Na)-X; ■ = (Cs, Na)-X.
(b) ○ = Ca-X; ▲ = Sr-X; ■ = Ba-X.

already described in § 5, one of the cations being located in or against the 6-ring windows. There are four of these *ppp* axes per cage. Values of ϕ_P and $\phi_{\dot{F}Q}$ have been given earlier for CO_2 as functions of distance along the *ppp* axis and at contact with the cation (Table 7). These confirm that the maximum value of $\phi_{\dot{F}Q}$ is large and always outweighs ϕ_P when the linear quadrupole axis lies along the *ppp* axis.

The isosteric heats for carbon dioxide enriched in different exchange ions are shown as functions of amount sorbed in Fig. 17.[45] Both in their initial magnitudes and in the degree of energetic heterogeneity the strong influence of the exchange ion can be seen. The larger the univalent ion the smaller are the initial heats and the extent of the energetic heterogeneity. For divalent ions the heats follow the sequence $Ba^{2+} > Sr^{2+} > Ca^{2+}$ and the energetic heterogeneity is very marked in each case. Calculations were made of $-\phi = -(\phi_D + \phi_R + \phi_P + \phi_{\dot{F}Q} - \phi_0)$, where ϕ_0 denotes estimated zero point energy, for carbon dioxide at the deepest point of the curve of ϕ against distance as the molecule moves along the *ppp* axis. These are compared for the different cationic forms with the energy of sorption, ΔE in Table 18. The comparison of calculations numbered 2 and 4 with each other and with ΔE

suggests that there is a strong tendency for CO_2 to be oriented along *ppp* in the vicinity of the cation. Comparison of 2 with 5 and with ΔE suggests that the kinetic theory radius is not suitable for energy calculations involving the three smallest cations. When the CO_2 is freely rotating an average polarizability is used both for $A_{ij}^{(6)}$ and for ϕ_P. The individual components of ϕ for calculation 2 are summarized in Table 19. The importance of the quadrupole energy is again emphasized, as is that of the nature of the exchange ion present in the zeolite structure.

TABLE 18

Calculated Values of ϕ and Observed Initial Values of Energies of Sorption, ΔE for CO_2 in Na-X (kJ mol^{-1})[45]

ΔE or ϕ	Cationic form				
	Li	Na	K	Rb	Cs
1. $-\Delta E$	51·4	45·1	43·9	42·2	36·8
2. $-\phi$ Kirkwood–Muller $A_{ij}^{(6)}$, Axis of CO_2 along *ppp*	55·2	38·9	28·8	23·4	22·2
3. $-\phi$ London $A_{ij}^{(6)}$, Axis of CO_2 along *ppp*	48·9	33·0	23·0	19·3	15·5
4. $-\phi$ Kirkwood–Muller $A_{ij}^{(6)}$, Axis of CO_2 ⊥ to *ppp*	7·1	9·6	2·5	8·8	16·3
5. $-\phi$ Kirkwood–Muller $A_{ij}^{(6)}$, Kinetic radius[a], CO_2 rotating	123	62·7	55·6	43·5	38·5

[a] The kinetic radius based on viscosity was taken as 1·65Å (5). The half-length of CO_2 along its axis was 2·56Å (2 and 3); the cross-sectional radius was 1·40Å (4).

The strong influence of the exchange ion upon the initial heat of sorption and upon energetic heterogeneity has been demonstrated even for the sorption of argon.[27] It is extremely clear when the guest molecule is polar (N_2, CO, CO_2, NH_3, H_2O), as is shown for water in Fig. 8, and for CO_2 in Fig. 17. With water and ammonia the major electrostatic term is due to the field-dipole energy, $\phi_{F\mu}$. For NH_3 moving along the *ppp* axis the calculated energy terms contributing to the overall energy are summarized in Table 20 together with the initial experimental energy of sorption.[29] The energies are those at the minimum of the potential energy curves along the above axis. Agreement between the calculated ϕ and $-\Delta E$ is not good suggesting that while the first molecules of ammonia sorbed may be in

TABLE 19

Components of ϕ in kJ mol^{-1} of CO_2 for Calculation 2 of Table 18[45]

Component	Cationic form				
	Li	Na	K	Rb	Cs
$-\phi_D$ (oxygens)	15·9	13·0	7·1	4·6	4·6
$-\phi_D$ (cations)	0·4$_2$	0·8$_4$	3·3	4·6	9·2
$-\phi_P$	9·6	5·0	2·1	0·8	≈0
$-\phi_{\dot{F}Q}$	30·$_9$	21·$_3$	17·$_6$	14·$_6$	9·6
ϕ_0	1·7	1·3	1·3	1·3	1·3

contact with the cations, they are also nearer the anionic oxygens and so to one side of the *ppp* axis. However there is no doubt of the considerable size of the electrostatic terms.

The effect of the radius of the exchange ion upon the heat of hydration of these ions, and upon the heat of sorption in zeolite X when 0·5, 2 and 6 water molecules are sorbed per large cavity is compared in Fig. 18, based upon a calorimetric study.[10] For the largest uptake, the cations are less strongly involved and the heat of sorption actually increases in the direction Li < Na < K. The process of filling the large cages was considered on the basis of the calorimetric and infra-red evidence to take place in more than one stage, the first of these involving the exchange ions very directly.[10, 77]

The effect of the cation for still greater water uptakes than those considered in Fig. 18b can be demonstrated by measuring heats of immersion of previously outgassed zeolite. The contribution to the heat of immersion arising from wetting the external crystallite surfaces is negligible compared with the heat liberated on filling the intracrystalline pore volume. Thus from heats of

TABLE 20

Energy Terms in kJ mol^{-1} calculated for NH_3 in ion-exchanged Forms of Zeolite X[29]

Cation form	$-\phi_D$ (oxygens)	$-\phi_D$ (cations)	$-\phi_R$	$-\phi_{F\mu}$	$-\phi_P$	$+\phi_0$	Total ϕ	ΔE experimental
Li	47·2	1·7	41·3	50·6	23·0	3·3	−77·7	−76·5
Na	33·8	3·3	23·4	33·0	9·6	2·5	−53·9	−72·3
K	9·6	6·7	14·6	20·1	3·8	1·7	−23·8	−59·8
Rb	7·9	11·7	18·0	17·6	2·9	1·7	−20·5	−55·6
Cs	7·5	16·3	19·6	15·0	2·1	1·7	−19·6	−47·2

immersion one obtains integral heats of saturation of the zeolite with water (or other guest molecules). Figure 19[13] shows for zeolites X and Y the heats q_D and Q_I as functions of r_c^{-2} where r_c is the cation radius, q_D is the heat of immersion per g of dehydrated crystal and Q_I is the heat evolved per mol of water taken up. r_c^{-2} is proportional to the field strength for the isolated cations. The integral heat per mol of water taken up from the vapour phase is obtained by adding the heat of condensation of water vapour: $\Delta\tilde{H} = -(Q_I + 10{\cdot}5)$ kcal mol^{-1}. The figure shows that q_D and Q_I are

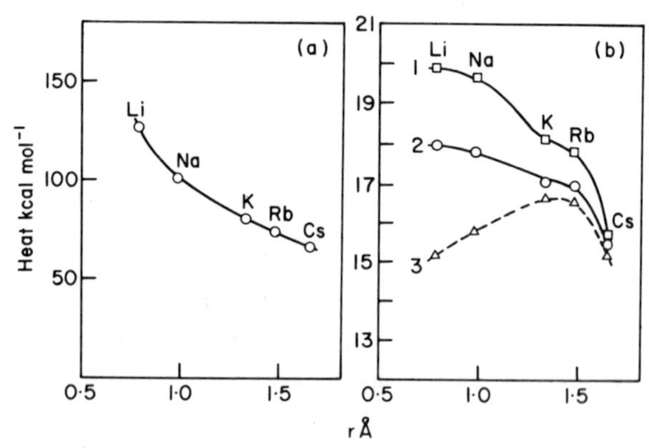

FIG. 18. Dependence on the radius of ions of alkali metals of
(a) heats of hydration of the gaseous ions and
(b) heats of sorption for (1) 0·5, (2) 2 and (3) 6 molecules of water per large cavity in zeolite X.[10]

greater in zeolite X than in zeolite Y, so that these heats are considerably influenced by the total charge density. The heats also depend in a complex way upon the nature of the cations present and are normally larger for divalent than for univalent ions. For univalent ions and for Ba, Sr and Ca ions the heats q_D increase with decreasing ionic radius. The way in which Q_I changes with r_c^{-2} shows that the cations of different kinds are on average variously screened by anionic oxygens.

Coughlan and Carroll[77a] observed that, when the heats of wetting per equivalent of cations were plotted against the reciprocals of the cation radii, the points often fell close to a straight line. There were exceptions to this behaviour, but it was demonstrated, using the results of Barrer and Cram,[13] for Tl^+, Cs^+, Rb^+, K^+ and Na^+ in chabazite, and for Cs^+, Rb^+, K^+, Na^+ and Ag^+ in zeolite Y. It was also shown to be true of their own results for the divalent ions Cd^{2+}, Pd^{2+}, Mn^{2+}, Zn^{2+}, Cu^{2+}, Ca^{2+} and Ni^{2+} in zeolite Y. To explain this result they assumed that immersing the outgassed zeolite in water was analogous to immersing gaseous cations in water. This view of the wetting process is however an approximation and their observation seems best regarded as empirical.

Heats of immersion of zeolites have also been employed to estimate mean field strengths within the structures.[78, 79] Calcium and sodium forms of zeolites X and Y were used with different degrees of exchange and different SiO_2/Al_2O_3 ratios. The wetting liquids included 1-nitropropane, *n*-butanol, *n*-octane and *n*-hexane. As with water, heats of immersion were greater for Ca-rich than for Na-rich zeolite. The larger the dipole in the molecule the more extreme was the change in heat with increasing Ca content. From the heat of immersion in cal g^{-1} heats of immersion in erg cm^{-2} of mono-

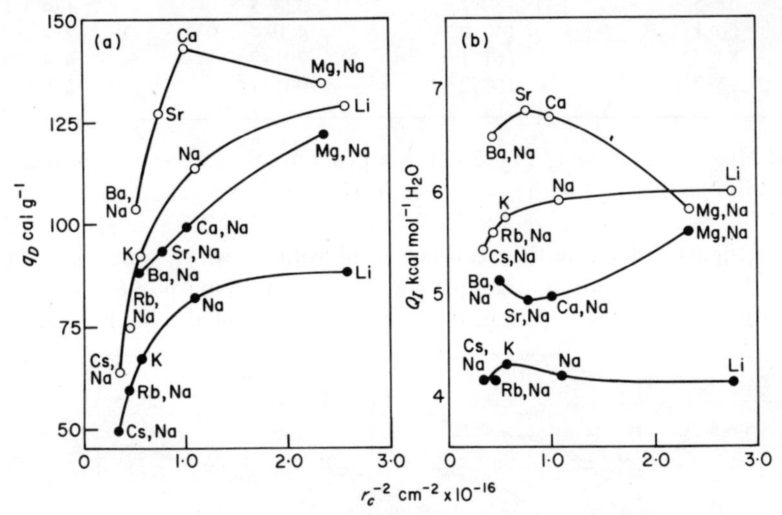

FIG. 19. (a) q_D and (b) Q_I plotted against $r_c{}^{-2}$ for zeolites X and Y.[13]
Open circles = zeolite X; filled circles = zeolite Y.
q_D is the heat of wetting per g of dehydrated zeolite;
Q_I is the heat of wetting per mol of water imbibed;
r_0 is the cation radius.

layer equivalent area were derived. It was then assumed that the difference between these latter heats for 1-nitropropane and *n*-hexane was given by $\phi_{F\mu}$ (erg cm^{-2}) $= -N_s\mu F_m$ where N_s is the number of molecules sorbed per unit area, μ is the dipole moment of 1-nitropropane (3·57 Debye) and F_m is the mean electrostatic field. Results are given in Table 21.[78] Although the above assumption may be over-simple, Table 21 confirms the very large value of F_m, in agreement with calculated values and with the values of the field-dipole energies discussed in § 5 and earlier in this section. The rapid increase in F_m with Ca-content as exchange exceeds 50% indicates that Ca^{2+} introduced at this stage is more accessible to the sorbate than Ca^{2+} in the less fully exchanged zeolite.

As the final example of guest molecules with electric moments the olefines may be considered. Such molecules as well as aromatic molecules can have quadrupole moments as illustrated in Table 1 compiled by Stogryn and Stogryn[80] for ethylene and benzene. Whereas the polarizabilities of ethylene

TABLE 21

Heats of Immersion (erg cm^{-2}) and Mean Fields in (Ca, Na)–Y[78]

Ca Exchange %	Heats of immersion *n*-hexane	1-nitropropane	$10^{-8}F_m$ (e.s.u. cm^{-2})
0	143·8	334·8	1·3
22·5	130·9	298·3	1·2
41·5	130·2	323·5	1·4
60·1	140·9	480·4	2·4
87·8	161·1	791·2	4·4
95·7	168·2	1001·2	5·8

and ethane are similar, Fig. 20 shows pronounced differences between the sorption isotherms for the two gases in (Li, Na)-, Na-, (K, Na)-, (Rb, Na)- and (Cs, Na)- forms of zeolite X.[4] The two (Li, Na)-X samples referred to in the figure differed in their cationic compositions, which were $Li_{0\cdot54}Na_{0\cdot33}$ with 13% decationation in sample 1 and $Li_{0\cdot91}Na_{0\cdot08}$ in sample 2. The heats, shown in Fig. 2, demonstrate the strong influence of the double bond. As

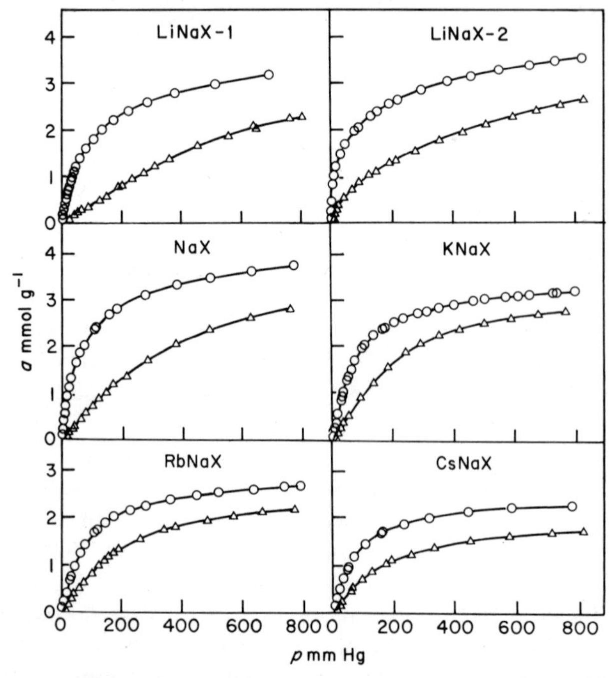

FIG 20. Sorption isotherms of C_2H_6 (Δ) and C_2H_4 (○) on various cation exchanged forms of zeolite X at 50°C (open points refer to adsorption and black points refer to desorption).[4]

observed for the polar molecules, such as CO_2 and NH_3 already considered, energetic heterogeneity for ethylene is more pronounced the smaller the exchange ion and so is the difference between the initial heats for C_2H_4 and C_2H_6. This difference in heats was shown to decline rather smoothly but less and less rapidly with the ionic radius of the exchange ion from ≈ 19 kJ mol^{-1} for Li-enriched to ≈ 4 kJ mol^{-1} for Cs-enriched zeolite X. The electric moment of the charge distribution of π-bonds in alkenes interacts in a specific way not only with zeolites but with other polar surfaces such as hydroxylated silica[81] or barium sulphate.[82]

The discussion in §§ 8.1, 8.2 and 8.3 indicates the progress that has been made in interpreting the energetics of sorption in molecular sieves. The examples chosen are only a few of those which could have been considered, but they illustrate in general both how the problems of calculation have been approached and some of the experimentally established properties. One may now consider further how zeolite modification alters the energetics of intracrystalline sorption.

9. ZEOLITE MODIFICATION AND HEATS OF SORPTION

Zeolites may be modified chemically in a number of ways which include the following:

(i) by changing the exchange ions
(ii) by changing the Si/Al ratios by direct synthesis under different conditions
(iii) by decationation to produce "hydrogen" zeolites
(iv) by dealumination through acid treatment of the zeolite.

The effects of introducing cations of different radius and valency have been described in the previous section in particular and so will not be considered further. It has also been shown in the previous section, by using faujasite-type zeolites with varying Al/Si ratios, that for polar molecules, the greater the charge density in the structure, the greater tends to be the energy of binding of such molecules. This is further illustrated in Table 22[28] which gives the isosteric heats for CO_2 in zeolites *A*, X and Y for different amounts sorbed, and also relates these heats to the exchange capacity (charge density) of the zeolite. The greater the charge density the greater is the heat for all uptakes and the more marked the energetic heterogeneity. The behaviour must be determined in the case of CO_2, with its large linear quadrupole moment, by field gradients which are bigger the greater the intracrystalline charge density.

Decationation of zeolites is normally effected by exchanging the metallic

TABLE 22

Isosteric Heats for CO_2 and Cation Density[28]

cm^3 of CO_2 at s.t.p. per cm^3 intracrystalline pore volume	Na-*A*	Na-X	Na-Y
	$10^{-21} \times Na^+$ ions cm^{-3}		
	$6\cdot4$	$5\cdot6$	$3\cdot7$
	q_{st} (kJ mol^{-1})		
20	$51\cdot_0$	$48\cdot_5$	$33\cdot_8$
40	$48\cdot_9$	$46\cdot_8$	$30\cdot_9$
60	$47\cdot_7$	$45\cdot_1$	$30\cdot_5$
80	$46\cdot_0$	$44\cdot_3$	$30\cdot_5$
100	$45\cdot_1$	$43\cdot_4$	$30\cdot_5$
150	$42\cdot_2$	$41\cdot_4$	$30\cdot_9$
200	$41\cdot_4$	$40\cdot_5$	$31\cdot_6$
250	$40\cdot_5$	$39\cdot_3$	$31\cdot_6$
300	$39\cdot_3$	$38\cdot_0$	$31\cdot_6$

ions with ammonium and then heating the ammonium form in air or under a vacuum to give the initial reaction

$$NH_4^+ \; \overset{\ominus}{Al}{-}O{-}Si \rightarrow Al \;\; HO{-}Si + NH_3$$

(with the possibility of some side reactions[20]). Thus the decationated zeolite is rich in silanol hydroxyls which, though polar, would be expected to produce smaller fields and field gradients than the cationic forms of the parent zeolite. Dealumination of silica rich zeolites (mordenite, clinoptilolite and ferrierite) can be effected to different extents by treatment with mineral acids of varying normalities and for regulated times and temperatures. Less siliceous zeolites can sometimes be partially dealuminated using milder acids such as HEDTA.[83, 84] Such treatments both decationate and dealuminate the zeolite. The initial step of dealumination may be represented as

$$4HCl + M^+\,\overset{\ominus}{Al}(O{-}Si)_4 \rightarrow 4\,Si{-}OH + MCl + AlCl_3.$$

and outgassing may cause dehydroxylation of the "nest" of —OH groups to give two Si–O–Si bonds. In this event, outgassed, decationated and dealuminated zeolite may be produced which is poorer and poorer in hydroxyl groups. Such sorbents derived from mordenite and clinoptilolite were investigated[14, 85] using Kr, CO_2 and H_2O as probe molecules. The SiO_2/Al_2O_3 ratios in the parent zeolite and the treated products were as follows:

Sorbent	SiO_2/Al_2O_3
Parent Na-mordenite	9·5/1
1N HCl	9·5/0·83
2N HCl	9·5/0·48
6N HCl	9·5/0·32
12N HCl	9·5/0·27
Parent Clinoptilolite	9·3/1
0·25N HCl	9·3/0·58
0·5N HCl	9·3/0·33
1N HCl	9·3/0·07
2N HCl	9·3/0·0.

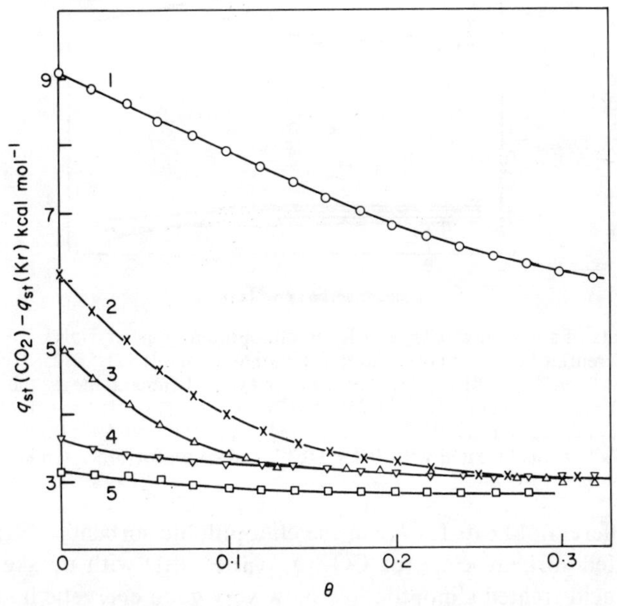

FIG. 21. The difference between isosteric heats of CO_2 and Kr in mordenites as functions of degree of filling, θ.[14]

1. Parent Na-mordenite 2. H-mordenite
3. "2N" mordenite 4. "6N" mordenite
5. "12N" mordenite.

In Figs 21 and 22 the figures in inverted commas denote the strengths of the HCl aq with which the zeolite was treated. The SiO2/Al2O3 ratios of the acid-treated zeolites are given in the text.

The products were still crystalline to X-rays after 4 h refluxing with the acid solutions to give the above sorbents. The initial heats for carbon dioxide and energetic heterogeneity changed substantially and progressively among the sorbents. Figure 21 shows for the parent and modified mordenites the differences in the isosteric heats for CO_2 and Kr as functions of the amount sorbed. The less polar the sorbents become (i.e. the more fully dealuminated) the less the initial differences and the smaller their dependence upon the amount sorbed. Figure 22 shows the differential and integral heats for CO_2

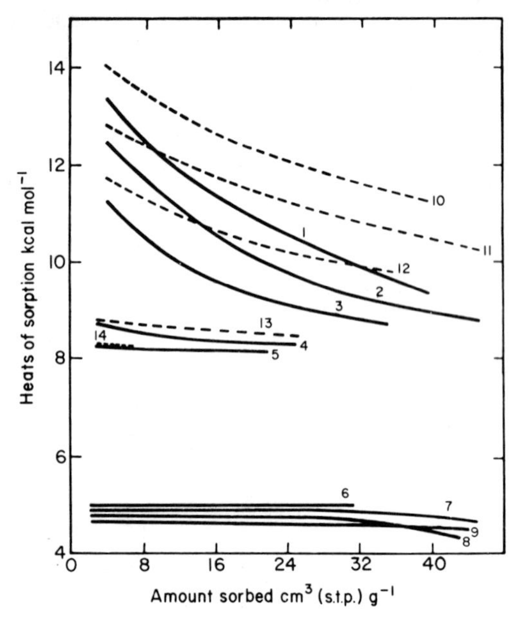

FIG. 22. Heats of sorption of CO_2 and Kr in clinoptilolite-type crystals.[85] For CO_2 and differential heats the curves are: (1) parent mineral; (2) "0·25N" clinoptiloite; (3) "0·5N"; (4) "1N"; (5) "2N". For CO_2 and integral heats the curves are: (10) parent mineral; (11) "0·25N"; (12) "0·5N"; (13) "1N"; (14) "2N". For Kr and differential heats the curves are: (6) "0·25N"; (7) "0·5N"; (8) "1N"; (9) "2N". 0·5N" means "treated with 0·5N HCl", and so on for other symbols in inverted commas.

and the differential heats for Kr in the clinoptilolite sorbents.[85] In the fully dealuminated sorbent, even for CO_2, q_{st} varies little with uptake. Towards Kr all the acid-treated clinoptilolites show very good energetic homogeneity, although the actual value of q_{st} decreases somewhat as Al is progressively extracted.

A study of the influence of progressive decationation of the Na-form of zeolite Y upon the heats of adsorption of a number of probe molecules was made by Patzelova.[86] Gas chromatography was used to obtain retention times at each of five temperatures, and the heats of sorption were derived

TABLE 23

Heats of Sorption (kcal mol^{-1}) for Guest Molecules in Progressively Decationated Na-Y[86]

Sorbate	Heat for % Decationation of Na-Y					
	0%	19%	39%	57%	87%	99%
CH_4	−4·33 ± 0·14	−4·39 ± 0·11	−4·12 ± 0·10	−4·11 ± 0·16	−3·75 ± 0·08	−4·12 ± 0·17
C_2H_6	−6·16 ± 0·35	−6·17 ± 0·21	−5·72 ± 0·18	−5·77 ± 0·16	−6·20 ± 0·26	−5·69 ± 0·15
C_3H_8	−8·12 ± 0·30	−8·15 ± 0·27	−7·84 ± 0·29	−7·44 ± 0·28	−7·92 ± 0·14	−7·60 ± 0·32
CO	−5·79 ± 0·13	−5·47 ± 0·23	−4·92 ± 0·11	−5·19 ± 0·20	−4·85 ± 0·20	−4·76 ± 0·10
C_2H_4	−11·26 ± 0·37	−9·64 ± 0·22	−9·03 ± 0·18	−9·29 ± 0·25	−9·15 ± 0·12	−8·87 ± 0·27

from these times. The changes in the heats with percentage decationation are given in Table 23. The hydrogen zeolite is less polar than its Na-form, but for the saturated hydrocarbons CH_4, C_2H_6 or C_3H_8 the heats are not very dependent upon the percentage decationation. For CO and C_2H_4, with permanent electric moments, the differences in heat between Na-Y and H-Y are considerably larger. Also for H-Y the heat for C_2H_4 is still 2·17 kcal mol^{-1} above the heat for C_2H_6, so that the electrostatic components of the heat in H-Y remain important for ethylene. For Na-Y the heat difference between C_2H_4 and C_2H_6 is 5·1 kcal mol^{-1}.

By suitably strong heating, the 70% decationated Y-type zeolite was partially dehydroxylated and the heats of sorption of the guest molecules were then also determined chromatographically. The heats are compared in Table 24 for the parent $(Na_{0\cdot3}H_{0\cdot7})$-Y and the partially dehydroxylated form.

TABLE 24
Heats of Sorption (kcal mol^{-1}) in Zeolite Y[86]

Sorbate	Original	$(Na_{0\cdot3}H_{0\cdot7})$-Y Partially Dehydroxylated
CH_4	−4·17 ± 0·11	−3·93 ± 0·10
C_2H_6	−6·33 ± 0·35	−5·88 ± 0·28
C_3H_8	−8·08 ± 0·20	−6·97 ± 0·31
CO	−5·26 ± 0·18	−4·85 ± 0·18
C_2H_4	−9·50 ± 0·21	−8·62 ± 0·20

The polarity of the sorbent is evidently reduced by elimination of some hydroxyl water, with an appreciable decrease in the magnitude of the heats.

It is seen that much can be done to control and modify local intracrystalline fields and field gradients, so that by decationation, dealumination and dehydroxylation relatively non-polar molecular sieve sorbents can be made from appropriate stable zeolites. In particular the affinities of these sorbents for polar molecules can be strongly changed in comparison with affinities shown by the parent zeolites.

10. CONCLUDING REMARKS

Of all physical methods of investigating the sorption properties of molecular sieve sorbents the study of energetics using selected probe molecules has proved one of the most powerful. Although there are both theoretical and practical limitations in the evaluation of dispersion and electrostatic energies these energies can be calculated to a reasonable degree of approximation. The analyses presented here show that polarization energies, while not inconsiderable, usually tend to be less than dispersion energies and comparable with close range repulsion energies. The "specific" components of the bond are the field-dipole and field gradient-quadrupole terms. They can be very large and are responsible for much of the energetic heterogeneity and hence dependence of the heats upon degree of saturation of the zeolite. Selectivity shown by zeolite sorbents for polar molecules can be ascribed primarily to these electrostatic components.

REFERENCES

1. R. M. Barrer and R. Papadopoulos, *Proc. Roy. Soc.* (1972) A **326**, 315.
2. R. M. Barrer, R. Papadopoulos and D. J. F. Ramsay, *Proc. Roy. Soc.* (1972) A **326**, 331.
3. G. L. Kington and A. C. McLeod, *Trans. Farday Soc.* (1959) **55**, 1799.
4. A. G. Bezus, A. V. Kiselev, Z. Sedlacek and Pham Quang Du, *Trans. Faraday Soc.* (1971) **67**, 468.
5. R. M. Barrer and J. A. Lee, *Surface Science* (1968) **12**, 341.
6. R. M. Barrer and P. J. Reucroft, *Proc. Roy. Soc.* (1960) A **258**, 449.
7. R. M. Barrer and D. L. Peterson, *Proc. Roy. Soc.* (1964) A **280**, 466.
8. M. J. Sparnaay, Surface Sci. (1968) **9**, 100.
9. M. J. Sparnaay, "Clean Surfaces", Ed. G. Goldfinger, Marcel Dekker (1970) p. 153.
10. O. M. Dzhigit, A. V. Kiselev, K. N. Mikos, G. G. Muttik and T. A. Rahmanova, *Trans. Faraday Soc.* (1971) **67**, 458.
11. W. Schirmer, K-H. Sichhart, M. Bulow and A. Grossmann, Chem, Tech., (1971) **23**, 476.
12. R. M. Barrer and G. C. Bratt, *J. Phys. Chem. Solids* (1959) **12**, 146.
13. R. M. Barrer and P. J. Cram, *in* "Molecular Sieve Zeolites-II", Advances in Chem. Series No. 102, Amer. Chem. Soc. (1971) p. 105.
14. R. M. Barrer and E. V. T. Murphy, *J. Chem. Soc.* A (1970) 2506.
15. H. W. Habgood, *Chem. Eng. Symp. Series* (1967) **63**, 45.
16. L. Bertsch and H. W. Habgood, *J. Phys. Chem.* (1963) **67**, 1621.
17. N. G. Parsonage, *in* "Proc. of 3rd Internat. Conference on Molecular Sieves", Ed. J. B. Uytterhoeven, Sept. 3rd–7th, Zurich (1973) p. 296.
18. J. A. Gard, *in* "Molecular Sieves", Soc. Chem. Ind. (1968) p. 26.
19. R. M. Barrer and J. Klinowski, *J. Chem. Soc., Faraday I* (1975) **71**, 690.
20. G. T. Kerr, *in* "Molecular Sieves", Advances in Chem. Series No. 121, Amer. Chem. Soc. (1973) p. 219.
21. H. W. Habgood, *Can. J. Chem.* (1964) **42**, 2340.
22. T. A. Egerton and F. S. Stone, *Trans. Faraday Soc.* (1970) **66**, 2364.
23. T. A. Egerton and F. S. Stone, *J. Chem. Soc., Faraday I* (1973) **69**, 22.
24. R. M. Barrer, *J. Coll. and Interface Sci.* (1966) **21**, 415.
25. R. M. Barrer, *Proc. Roy. Soc.* (1938) A **167**, 392.
26. W. E. Addison and R. M. Barrer, *J. Chem. Soc.* (1955) 757.
27. R. M. Barrer and W. I. Stuart, *Proc. Roy. Soc.* (1959) A **249**, 464.
28. R. M. Barrer and B. Coughlan, *in* "Molecular Sieves", Soc. Chem. Ind., (1968) p. 241.
29. R. M. Barrer and R. M. Gibbons, *Trans. Faraday Soc.* (1963) **59**, 2569.

30. W. Schirmer, G. Meinert and A. Grossmann, *Monatsberichte* (1969) **11**, 886.
31. R. A. Beebe, B. Millard and J. Cynarski, *J. Amer. Chem. Soc.* (1953) **75**, 839.
32. A. V. Kiselev, *Quart. Rev.* (London) (1961) **15**, 99.
33. R. M. Dell and R. A. Beebe, *J. Phys. Chem.* (1955) **59**, 754.
34. B. Millard, R. A. Beebe and J. Cynarski, *J. Phys. Chem.* (1954) **58**, 468.
35. R. A. Beebe and R. M. Dell, *J. Phys. Chem.* (1955) **59**, 746.
36. R. M. Barrer and J. A. Davies, *Proc. Roy. Soc.* (1970) A **320**, 289.
37. N. N. Avgul, A. G. Bezus, E. S. Dobrova and A. V. Kiselev, *J. Coll. Interface Sci.* (1973) **42**, 486.
38. R. M. Barrer and S. Wasilewski, *Trans. Faraday Soc.* (1961) **57**, 1140.
39. N. G. Parsonage, *Trans. Faraday Soc.* (1970) **66**, 723.
40. N. G. Parsonage, *J. Chem. Soc.* A (1970) 2859.
41. H. Peters and E. Tappe, *Monatsber.* (1970) **12**, 29.
42. L. E. Drain, *Trans. Faraday Soc.* (1953) **49**, 650.
43. H. S. Carslaw and J. C. Jaeger, "Conduction of Heat in Solids", Oxford University Press (1959) p. 484, and Table III, p. 487.
44. J. O. Hirschfelder, C. F. Curtiss and R. B. Bird, "Molecular Theory of Gases and Liquids" Wiley (1954) pp. 28 and 840.
45. R. M. Barrer and R. M. Gibbons, *Trans. Faraday Soc.* (1965) **61**, 948.
46. A. V. Kiselev and A. A. Lopatkin, *in* "Molecular Sieves", Soc. Chem. Ind., London (1968) p. 252.
47. von H-J. Spangenberg, K. Fielder, H. J. Ortlieb and W. Schirmer, *Zeit. phys. Chem. Lpzg* (1971) **248**, 49.
48. E. Dempsey, *in* "Molecular Sieves", Soc. Chem. Ind., London (1968) p. 293.
49. L. Pauling, "The Nature of the Chemical Bond" Cornell University Press (1940) Chap. II.
50. D. Bonnin and A. P. Legrand, *Chem. Phys. Letters* (1975) **30**, 296.
51. M. J. Sparnaay, *Physica* (1959) **25**, 217.
52. H. Margenau and J. Stamper, Advances in Quantum Chemistry (1967) **3**, 129.
53. S. Brunauer, "The Adsorption of Gases and Vapours" Oxford University Press (1944) p. 208.
54. F. V. Lenel, *Zeit. phys. Chem.* (1933) B **33**, 379.
55. F. London, *Zeit. Phys.* (1930) **63**, 245.
56. F. London, *Zeit. Phys. Chem.* (1940) B **11**, 222.
57. J. C. Slater and J. G. Kirkwood, *Phys. Rev.* (1931) **37**, 682.
58. J. G. Kirkwood, *Phys. Zeit.* (1932) **33**, 57.
59. A. Muller, *Proc. Roy. Soc.* (1936) A **154**, 624.
60. L. Salem, *Molecular Phys.* (1960) **3**, 441.
61. R. M. Barrer and D. J. Ruzicka, *Trans. Faraday Soc.* (1962) **58**, 2253.
62. J. H. de Boer and J. F. H. Custers, *Zeit. phys. Chem.* (1934) B **25**, 225.
63. J. H. van der Waals, *Trans. Faraday Soc.* (1956) **52**, 184.
64. L. Garden, G. L. Kington and W. Laing, *Trans. Faraday Soc.* (1955) **51**, 1558.
65. L. E. Drain and J. A. Morrison, *Trans. Faraday Soc.* (1952) **48**, 840.
66. R. A. Beebe and D. M. Young, *J. Phys. Chem.* (1954) **58**, 93.
67. J. G. Aston and J. Greyson, Proc. 2nd Internat. Cong. Surface Activity Butterworth, London (1957) Vol. 3 "Solid/Gas Interface", CC 199.
68. J. V. Smith *in* "Molecular Sieve Zeolites-I", Advances in Chemistry Series, No. 101, Amer. Chem. Soc. (1971) p. 171.
69. see P. G. Gombas "Theorie und Lesungsmethoden des Mehrteilchenproblems der Wellenmechanik", Basel (1950).

70. P. Howell, *Acta Cryst.* (1960) **13**, 737.
71. K. Fiedler, H-J. Spangenberg and W. Schirmer, Monatsber. (1967) **9**, 516.
72. R. M. Barrer and J. A. Davies, *Proc. Roy. Soc.* (1971) A **322**, 1.
73. R. M. Barrer and J. W. Sutherland, *Proc. Roy. Soc.* (1956) A **237**, 439.
74. A. V. Kiselev and K. D. Shcherbakova, *in* "Molecular Sieves", Soc. Chem. Ind. (1968) p. 289.
75. W. Schirmer, G. Fiedrich, A. Grossmann and H. Stach *in* "Molecular Sieves", Soc. Chem. Ind. (1968) p. 276.
76. R. M. Barrer and D. A. Langley, *J. Chem. Soc.* (1958) 3817.
77. A. V. Kiselev, V. I. Lygin and R. V. Starodubceva, *J. Chem. Soc., Faraday I* (1972) **68**, 1793.
77a. B. Coughlan and W. M. Carroll, *J. Chem. Soc., Faraday I* (1976) **72**, 2016.
78. K. Tsutsumi and H. Takahashi, *J. Phys. Chem.* (1970) **74**, 2710.
79. K. Tsutsumi and H. Takahashi, *J. Pys. Chem.* (1972) **76**, 110.
80. D. E. Stogryn and A. P. Stogryn, *Molecular Phys.* (1966) **11**, 371.
81. A. G. Bezus and A. V. Kiselev, *Zhur. Fiz. Khim.* (1966) **40**, 580.
82. L. D. Belyakova, A. V. Kiselev and G. A. Soloyan, *Chromatographia* (1970) **3**, 254.
83. G. T. Kerr, *J. Phys. Chem.* (1968) **72**, 2594.
84. G. T. Kerr, *J. Phys. Chem.* (1969) **73**, 2780.
85. R. M. Barrer and B. Coughlan, *in* "Molecular Sieves", Soc. Chem. Ind., (1968) p. 141.
86. V. Patzelova, *Chem. zvesti* (1975) **29**, 331.

5
Entropy and Heat Capacity

An important contribution to the understanding of sorption equilibrium comes from the study of the entropy changes involved. For a homogeneous sorbent the calculation of the entropy of the sorbate is straightforward for certain limiting cases of localized or of non-localized sorption. These limiting situations can then be used as reference states with which real systems can be compared. Complications arise *inter alia* because, like other sorbents, zeolites and porous crystals may be heterogeneous. Further information can be expected from measurements of the heat capacity of the intracrystalline guest molecules, again with complications arising from heterogeneity of the sorbents. Some of these aspects of sorption equilibria will now be considered.

1. THE ENTROPY OF SORBED MOLECULES

For N_s molecules localized upon N identical sites and without interaction between the sorbed molecules their total entropy is

$$S_s = N_s(S_v + S_R + S_I + S_c) \tag{1}$$

where S_v, S_R, S_I and S_c denote vibrational, rotational, internal and configurational entropies per molecule. These component entropies are related to the orresponding partition functions, q_v, q_R and q_I by

$$S = k \ln q + kT\frac{\mathrm{d} \ln q}{\mathrm{d}T} \tag{2}$$

and the configurational entropy is

$$S_c = kN_s \ln\frac{N - N_s}{N_s} - kN \ln\frac{N - N_s}{N}. \tag{3}$$

If there is one- or two-dimensional translational freedom S_v is reduced from its value for a three-dimensional oscillator to that for a molecule executing two or one vibrations respectively. The thermal translational entropy for a molecule in a one- or two-dimensional gas must be added, and the configurational entropy for localized sorption must be replaced by that for the gas. Especially for localized sorption it would also be possible for rotations of sorbed molecules to be wholly or partially converted to additional vibrations.

Equation 19 in Chapter 3 indicates how the differential entropy, $\bar{S}_s$, may be evaluated from the differential heat of sorption and the entropy of the gaseous molecules. $\bar{S}_s$ is a finite, positive quantity which is the sum of a configurational terms $\bar{S}_c$ and a thermal entropy $\bar{S}_{T_h}$. The analysis of $\bar{S}_s$ in this way was first made for zeolites by Kington and his coworkers.[1, 2, 3] His method has been developed, for example, in studies of simple gases in chabazite[2, 3, 4] faujasite[5] and mordenite,[6,7] of fluorocarbons and hydrocarbons in faujasite[8,9] of hydrocarbons in zeolite L,[10] of ammonia and carbon dioxide in ion-exchanged faujasites,[11, 12] of carbon dioxide and krypton in dealuminated clinoptilolites[13] and of carbon dioxide in various other zeolites.[14]

As noted above one may write

$$\bar{S}_s = \bar{S}_c + \bar{S}_{Th} = \tilde{S}_g + \Delta\bar{H}/T \tag{4}$$

and $\bar{S}_s$ may then be plotted as a function of amount sorbed. Kington *et al.* sought expressions for $\bar{S}_c$, which, subtracted from $\bar{S}_s$, would give $\bar{S}_{Th}$. They considered the model isotherms of ideal localized sorption (Langmuir equation) localized sorption with interaction (Lacher equation) and the Volmer or van der Waals isotherm equations assuming one- or two-dimensional gases. For the first two one has

$$\bar{S}_c = R \ln [(1-\theta)/\theta] \tag{5}$$

$$\bar{S}_c = R \ln \left[\frac{(1-\theta)}{\theta}\left(\frac{2-2\theta}{\beta+1-2\theta}\right)^z\right] + \frac{\omega}{T}\frac{(\beta-1+2\theta)}{\beta} \tag{6}$$

where z, β and ω are as already defined in § 2, Chapter 3. For the non-localized sorption isotherms of Volmer or van der Waals the translational plus configurational entropies, ${}_1\bar{S}_T$ and ${}_2\bar{S}_T$ for one- and two-dimensional gases are respectively

$${}_1\bar{S}_T = R \ln \left[\frac{(2\pi mkT)^{1/2}}{h} N_{\text{sat}}^{1/3} \frac{(1-\theta)}{\theta}\right] + R\left(\frac{1}{2} - \frac{\theta}{1-\theta}\right) \tag{7}$$

$${}_2\bar{S}_T = R \ln \left[\frac{2\pi mkT}{h^2} N_{\text{sat}}^{2/3} \frac{(1-\theta)}{\theta}\right] + R\left(\frac{\theta}{1-\theta}\right). \tag{8}$$

In eqns 7 and 8, N_{sat} denotes the number of molecules saturating unit intracrystalline volume, and the other symbols have their usual significance. For these situations we have

$$\left.\begin{aligned}\bar{S}_s &= \bar{S}_{\mathrm{Th}} + \bar{S}_c \text{ (localized sorption)}\\ \bar{S}_s &= \bar{S}'_{\mathrm{Th}} + {}_1\bar{S}_T \text{ (one-dimensional gas)}\\ \bar{S}_s &= \bar{S}''_{\mathrm{Th}} + {}_2\bar{S}_T \text{ (two-dimensional gas)}\end{aligned}\right\} . \tag{9}$$

In eqns 7 and 8, part of the thermal entropy is retained in ${}_1\bar{S}_T$ and ${}_2\bar{S}_T$ so that $\bar{S}'_{\mathrm{Th}}$ and $\bar{S}''_{\mathrm{Th}}$ in eqn 9 are the corresponding residual thermal entropies. The calculation of thermal entropy thus depends upon models whereas that of $\bar{S}_s$ involves only a completely general argument. Nevertheless the thermal entropies should obey certain criteria if the isotherm models are appropriate:

(i) they must be independent of θ;
(ii) they must be positive quantities of reasonable physical magnitude, yielding vibration frequencies of the correct order of magnitude, as discussed in § 2.

Reasons are given in § 2 as to why a two-dimensional gas is highly improbable inside a zeolite, and in Chapter 3 we have seen that Volmer and van der Waals sorption isotherms usually describe experimental isotherms poorly. On the other hand there is sometimes correspondence with the models for localized sorption over a range in θ. Over this range, with $\bar{S}_c$ given by one of eqn 5 or 6, criterion (i) is satisfied.

$\tilde{S}_g$ in eqn 5 is calculated for monatomic gases from

$$\tilde{S}_g = R\ln(kT/p) + R\ln\left[\frac{(2\pi mkT)^{3/2}}{h^3}\right] + (5/2)R. \tag{10}$$

For linear molecules the expression is

$$\tilde{S}_g = R\ln(kT/p) + R\ln\left[\frac{(2\pi mkT)^{3/2}}{h^3}\,\frac{8\pi^2 IkT}{\sigma h^2}\right] + (7/2)R; \tag{11}$$

and for non-linear molecules

$$\tilde{S}_g = R\ln(kT/p) + R\ln\left[\frac{(2\pi mkT)^{3/2}}{h^3}\cdot\frac{8\pi^2(2\pi kT)^{3/2}(I_1I_2I_3)^{1/2}}{\sigma h^3}\right] + 4R. \tag{12}$$

In eqns 11 and 12, I is the moment of inertia of a linear molecule and I_1, I_2 and I_3 are the principal moments of inertia of a non-linear molecule; σ is the symmetry number. It can reasonably be assumed in physical sorption that intramolecular vibrations and states of simple molecules are not significantly changed upon sorption, so that internal entropy is the same in the gas phase and after sorption. The symmetry numbers, σ, have the following values in some typical instances:

$\sigma =$ 1 NO, CO, HD
2 H_2, D_2, N_2, O_2, CO_2
2 H_2O (isosceles triangle)
3 NH_3 (triangular pyramid)
4 C_2H_4 (rectangle)
12 CH_4 (regular tetrahedron)
12 C_6H_6 (regular hexagon).

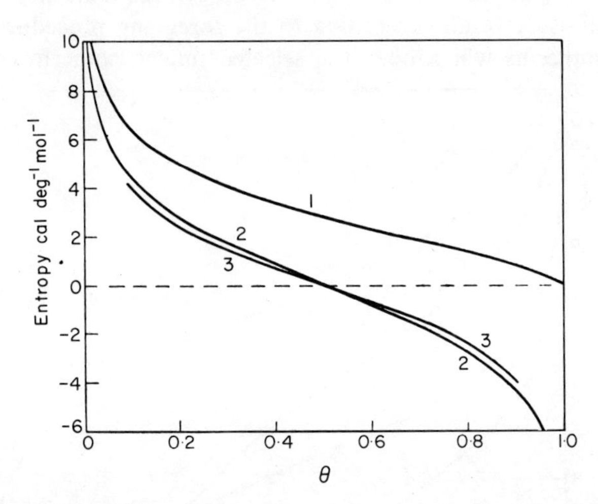

FIG. 1. Configurational entropies.[1] 1. $\tilde{S}_c$ of ideal phase. 2. $\bar{S}_c$ of ideal phase. 3. $\bar{S}_c$ of phase with sorbate-sorbate interaction for $2\omega/z = -240$ cal mol^{-1} and $z = 2$ (eqn 6).

The configurational entropy $\bar{S}_c$ given by eqn 5 is shown in Fig.1[1] curve 2. Its integral value, $\tilde{S}_c$, is

$$\tilde{S}_c = R\left[\frac{1}{\theta}\ln\frac{1}{\theta} - \left(\frac{1}{\theta} - 1\right)\ln\left(\frac{1}{\theta} - 1\right)\right] \tag{13}$$

and is shown in curve 1. The figure also shows $\bar{S}_c$ for argon calculated according to eqn 6. The site coordination number, z, was taken as 2, and $2\omega/z$ as 240 cal mol^{-1} (10 003 J mol^{-1}); $2\omega/z$, being the interaction energy for a pair of sorbed molecules, can be estimated from a knowledge of the potential energy curve of these molecules. When $\theta = \frac{1}{2}$ $\bar{S}_c = 0$ for eqn 5 and $\tilde{S}_c = 2R \ln 2$ for eqn 13 so that for this model one may readily find the thermal entropy, $\bar{S}_{Th}$, of eqn 9.

Figure 2[1] shows ${}_2\bar{S}_T$ and ${}_1\bar{S}_T$ for argon at 90·2K (curves 2 and 4 respectively) and the corresponding integral entropies, ${}_2\tilde{S}_T$ (curve 1) and ${}_1\tilde{S}_T$ (curve 3). The integral entropies are given respectively by

$$_2\tilde{S}_T = R \ln \left[\frac{2\pi mkT}{h^2} \cdot N_{sat}^{2/3} \frac{(1-\theta)}{\theta} \right] + 2R \quad (14)$$

and

$$_1\tilde{S}_T = R \ln \left[\frac{(2\pi mkT)^{1/2}}{h} N_{sat}^{1/3} \frac{(1-\theta)}{\theta} \right] + (3/2)R. \quad (15)$$

Again when $\theta = \frac{1}{2}$, the expressions 7, 8, 14 and 15 simplify and serve to derive the thermal parts of the respective differential and integral entropies.

The analysis of results according to the foregoing procedure is possible only for isotherms which obey the selected model isotherm equation and

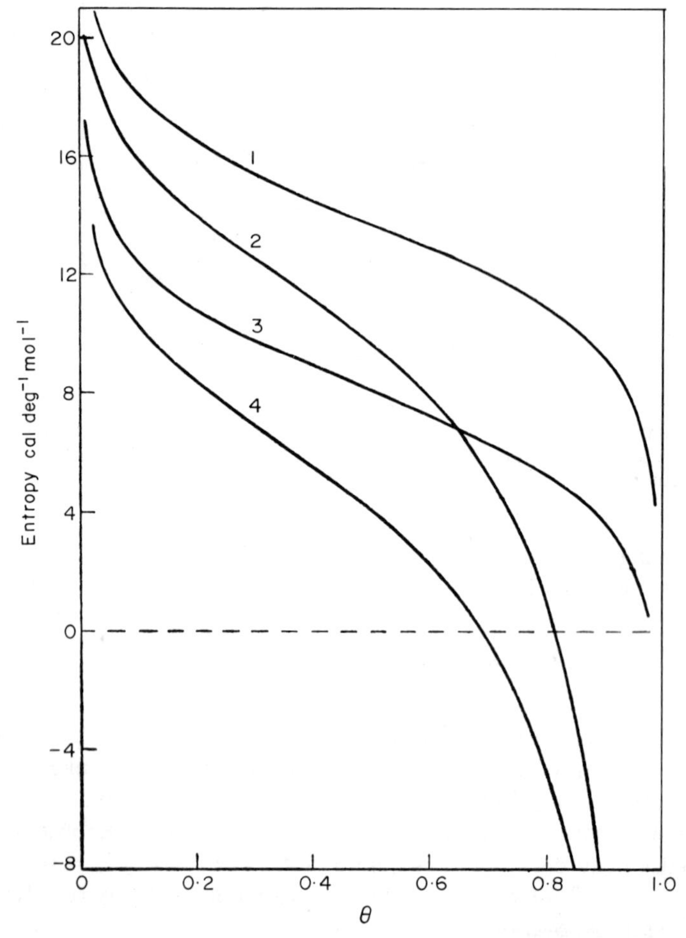

FIG. 2. Translational entropy of argon at 90·19 K.[1] For two translational degrees of freedom curve 1 is the molal entropy $_2\tilde{S}_T$ and curve 2 is the differential entropy $_2\bar{S}_T$. For one translational degree of freedom curve 3 is $_1\tilde{S}_T$ and curve 4 is $_1\bar{S}_T$. The cross-sectional area of argon was taken as 11·4Å².

this constitutes a severe limitation. Few isotherms can be represented adequately by Volmer or van der Waals isotherm equations (Chapter 3) and although the Langmuir isotherm equation is considerably more successful it is also of limited validity. Experimentally based values of $\bar{S}_s$ as functions of amount sorbed[5] are given in Fig. 3 for Ar in ion-exchanged faujasites (zeolite X), together with the isosteric heats $q_{st} = -\Delta\bar{H}$. For those systems (Ar in Li-, Na-, K-, Sr- and Ba-forms of zeolite X) where q_{st} changed little with amount sorbed $\bar{S}_s$ decreased monotonically with uptake, as expected from Figs 1 and 2. However when energetic heterogeneity became more marked (Ar in Ca-X) $\bar{S}_s$ changed much less with amount sorbed. For still more energetically heterogeneous systems (I_2 in zeolites Na-X, Ca-*A* and chabazite) $\bar{S}_s$ may actually increase with coverage and pass through a maximum (Fig. 12, Chapter 3). For NH_3 in (Rb, Na)-, Na- and Li-forms of zeolite X, $\bar{S}_s$ behaved in an equally complex way[11] (Fig. 4). The Rb-enriched form is the least energetically heterogeneous of these sorbents towards NH_3 (Fig. 5) and $\bar{S}_s$ decreased monotonically with uptake. In Na-X q_{st} decreased more strongly with amount sorbed and $\bar{S}_s$ passed through a maximum, as with I_2. The most energetically heterogeneous sorbent was Li-X and the complex form of the curve of $\bar{S}_s$ vs uptake is shown in Fig. 4c, and reflects the sigmoid shape of the curve of q_{st} against uptake, as seen from Fig. 5.[15] Dubinin, Kadlec and Zukal[16] plotted the function $(\bar{S}_s - \tilde{S}_l)$ for sorption from liquid water into Na-X against the amount sorbed for each of many temperatures between 20 and 220°C. The curves shown in Fig. 6 also show a minimum followed by a maximum, and reflect the known heterogeneity of the H_2O + Na-X system.

The inference from the above results is that for energetically heterogeneous sorbents, the entropies $\bar{S}_s$ of the guest molecules bear little resemblance to those calculated for idealized model isotherms. Instead they are related to the variations of q_{st} with coverage, and tend to compensate for these changes as already noted for the thermal entropy in § 3 Chapter 3. The configurational part of $\bar{S}_s$ was considered theoretically by Sparnaay[17] for energetically heterogeneous sorbents. For ideal localized sorption on a homotattic sorbent with *n* parts, arranged in order of increasing affinity, one has at equilibrium for sorbate distributed over the parts

$$\mu_1 = \mu_2 = \ldots = \mu_i \ldots = \mu_n = \mu_g$$

and

$$\theta_1 < \theta_2 < \ldots < \theta_i < \ldots < \theta_n.$$

For such a surface Spaarnay obtained

$$\bar{S}_c = \bar{S}_s - \bar{S}_{\mathrm{Th}} = -R \ln \frac{\theta_1}{1-\theta_1} - R \sum_{i=2}^{n} \frac{\mu_1^\ominus - \mu_i^\ominus}{RT} \; \frac{N_i\theta_i(1-\theta_i)}{\sum\limits_{i=1}^{n} N_i\theta_i(1-\theta_i)}. \tag{16}$$

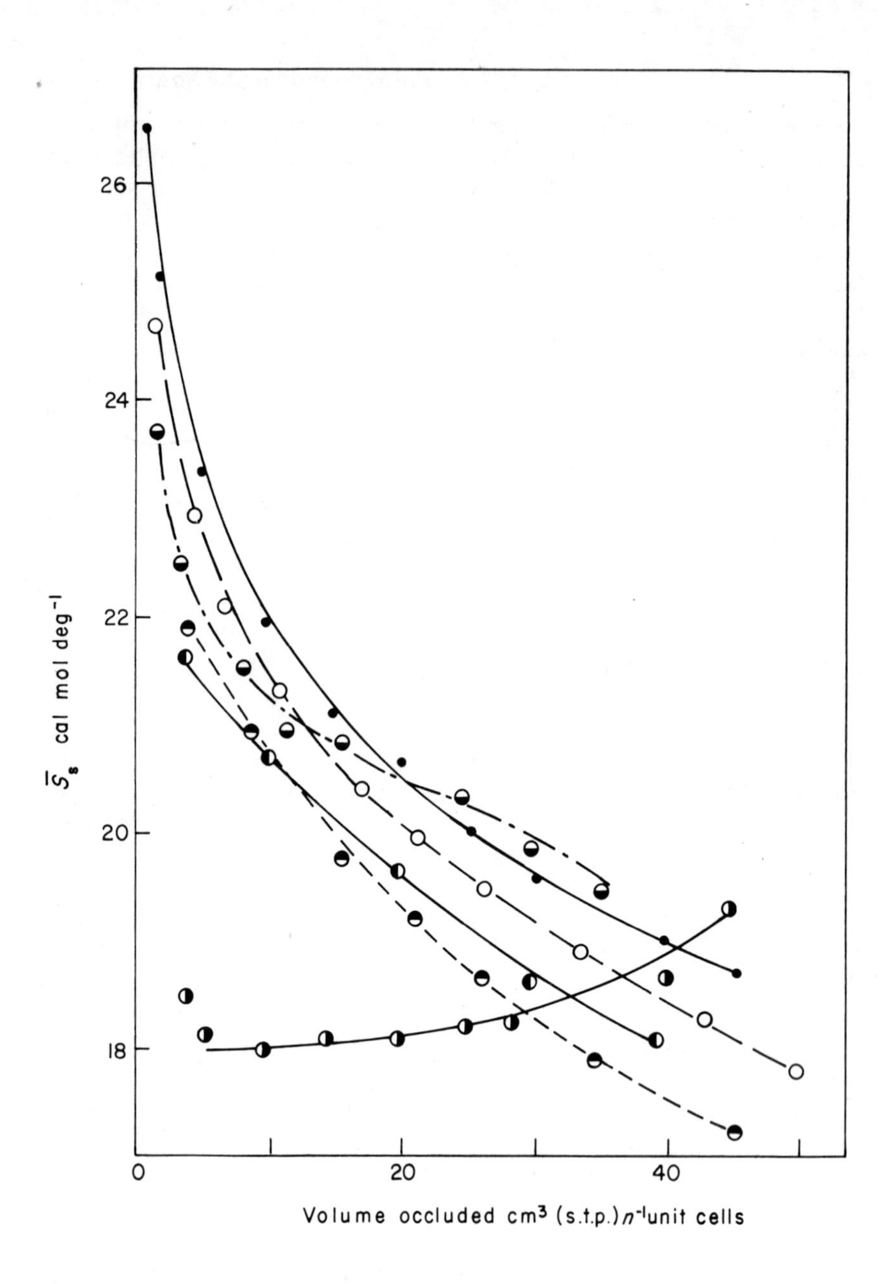

FIG. 3 (a) Differential entropies of argon sorbed by faujasite-type crystals with different exchangeable cations. n denotes the number of unit cells per g of hydrated Na-X having 26% by weight of water.[5]

◒ = L-iX; ● = Na-X; ○ = K-X; ◑ = Ca-X; ◓ = Sr-X; ◐ = Ba-X.

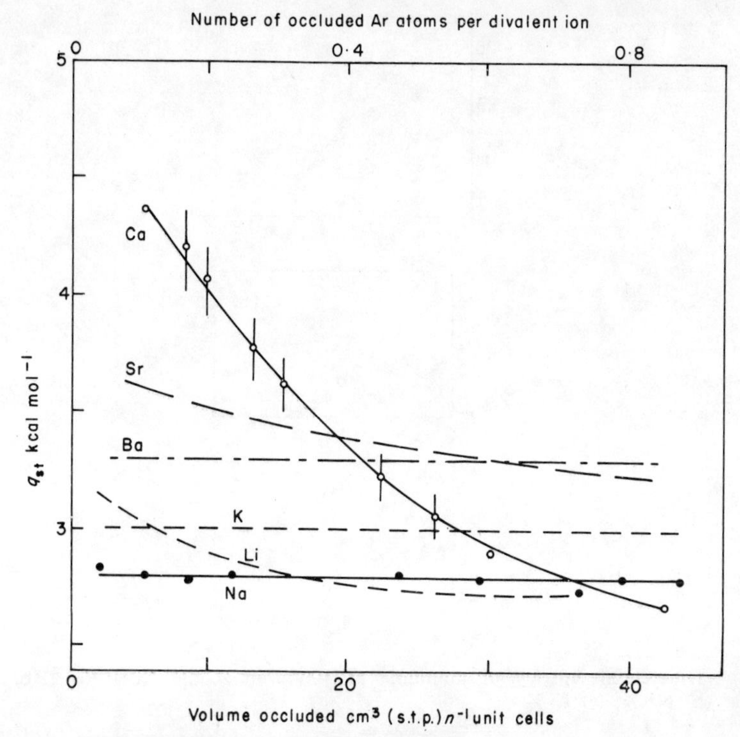

FIG. 3. (*b*) Isosteric heats of argon in the above faujasites, as functions of amount sorbed.[27]

The terms $\bar{S}_{\text{Th}}$ collects all the non-configurational parts of $\bar{S}_s$. For homogeneous surfaces $(\bar{S}_s - \bar{S}_{\text{Th}})$ would have been equal to $-R \ln [\theta/1 - \theta]$. For low coverages most of the adsorbed atoms are on the relatively few sites with large sorption affinity. If m_i is the amount sorbed on the ith part and m is the total uptake, Sparnaay showed that

$$\left(\frac{\partial m_i}{\partial m}\right)_{T,\,A} = \frac{N_i \theta_i (1 - \theta_i)}{\sum\limits_{i=1}^{n} N_i \theta_i (1 - \theta_i)} \tag{17}$$

and for $i > 2$ this differential is relatively large. A is the total area of the surface considered. In spite of the fact that $\theta_1 \ll \theta$ the overall configurational entropy is less than that for an ideal energetically uniform surface for small uptakes. For larger uptakes however $(\partial m_i/\partial m)_{T,\,A}$ decreases. The sorbed molecules now have a wide choice among the numerous remaining sites with low sorption affinity, which leads to an $\bar{S}_c$ higher than that for the uniform surface. The result is that $\bar{S}_s$ is less dependent upon coverage than expected for the homogeneous surface (cf. Fig. 6, Chapter 4). This, however, is not the complete picture since the thermal entropy, $\bar{S}_{\text{Th}}$, also tends to vary

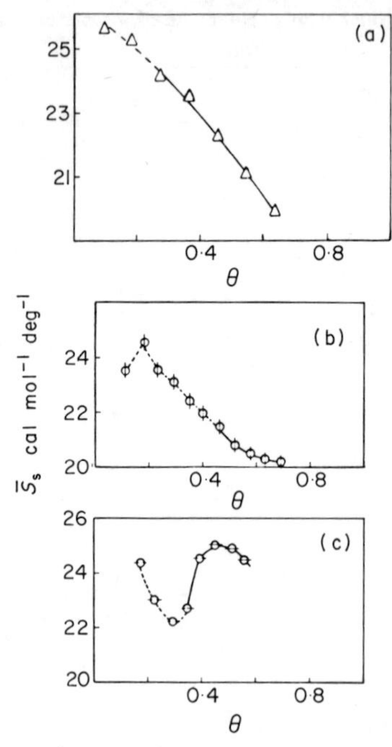

FIG. 4. Differential entropy of ammonia at 100°C in zeolite X.[11] (a) (Rb, Na)-X; (b) Na-X; (c) Li-X.

in the opposite way to the energy of binding of the sorbed molecules (§3 Chapter 3). The combination of the configurational and thermal entropies of heterogeneous sorbents is required to explain the complex behaviour recorded as illustrations in Figs 3, 4 and 6 and in Fig. 12 of Chapter 3.

2. THE STANDARD ENTROPY OF SORPTION

In Chapter 3 equilibrium constants were considered in terms of three conventions which can be compared for the Henry's law limit as follows:

$$\left.\begin{aligned} K_c &= C_s/C_g \\ K_p &= C_s/p \\ K_i &= \theta/p \end{aligned}\right\} \qquad (18)$$

($i = 1, 2, 3$ or 4 for Langmuir, Lacher, Volmer and van der Waals isotherm equations respectively). K_p and K_i may or may not have dimensions according as one or each of C_s, C_g and p is or is not divided respectively by unit concentration or pressure (§ 1, Chapter 3). The relations between these equilibrium constants in the Henry's law limit when K_p and K_i have dimensions are

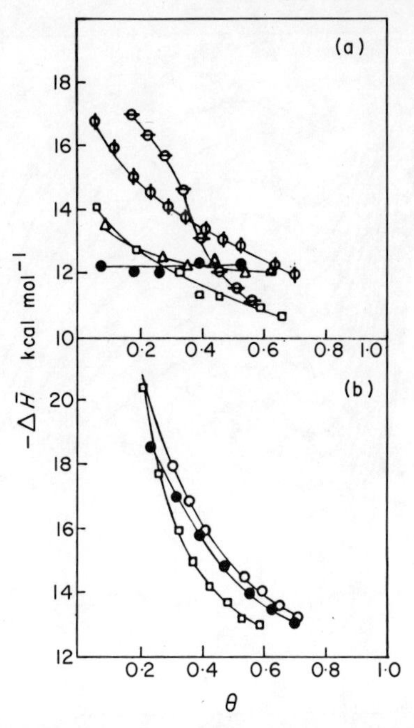

FIG. 5. Isosteric heats of NH_3 as functions of θ.(15)

(a) ⊖ = Li-X, θ for 110°C
ϕ = Na-X
□ = K-X
△ = (Rb, Na)-X
● = (Cs, Na)-X
(Na-X to (Cs, Na)-X: θ for 100°C)

(b) ○ = Sr-X
□ = Ca-X
● = (Ba, Na)-X
(θ for 100°C)

$$K_c = RT\,K_p = C_{\text{sat}}\,RT\,K_i \tag{19}$$

since $\theta = C_s/C_{\text{sat}}$ where C_{sat} denotes the number of moles which completely fills unit volume of the intracrystalline channels and cavities. Statistical thermodynamic models based upon localized sorption lead naturally to expressions relating p or C_g with θ. There is a choice of considering guest molecules and host crystal as a solution, in which the intracrystalline concentration, C, is expressed as mol per unit volume of crystal; or of considering sorption as a volume filling process with concentration, C_s, per unit volume of accessible intracrystalline pore space. The numerical value of K_c will change accordingly. The relation between C and C_s is

$$C = C_s\,\rho\,V \tag{20}$$

where ρ is the density of hydrated zeolite and V the accessible pore volume per unit mass of hydrated crystal.

It is always possible in comparing observed quantities such as $\Delta S^\ominus$ with those expected from statistical thermodynamic models to adjust the values

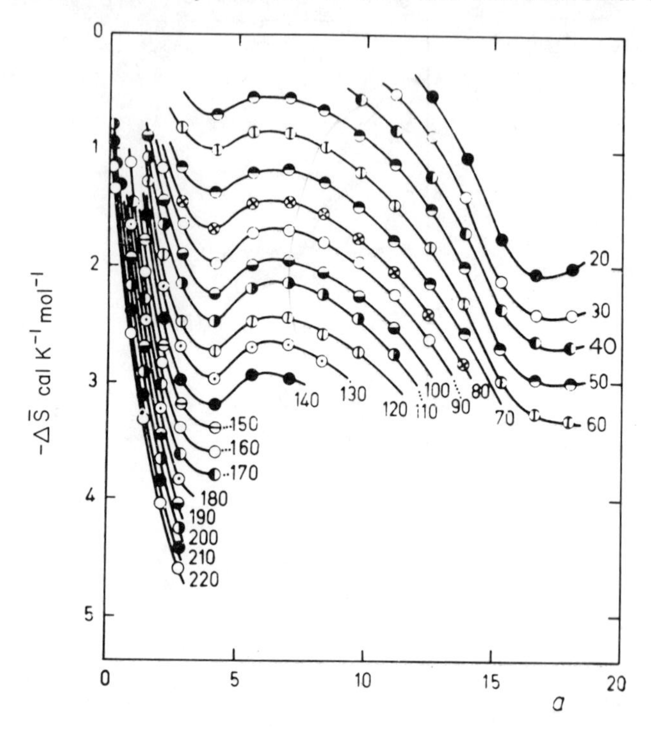

FIG. 6. Entropy of sorption (cal K^{-1}mol^{-1}) of liquid water into zeolite Na-X as a function of amount sorbed (mmol g^{-1}) at the temperature (°C) indicated by the numbers on the curves.[16]

appropriately to the equilibrium constant being considered. Thus for $\Delta S^{\ominus}_{K_c}$ and $\Delta S^{\ominus}_{K_p}$ corresponding to K_c and K_p in eqn 19 one has

$$\Delta S^{\ominus}_{K_c} = \Delta S^{\ominus}_{K_p} + R \ln RT + R. \tag{21}$$

If one wishes to obtain the most direct connection with sorption in general it may be best to consider sorption as volume filling rather than as solution. The standard entropy of sorption is the difference between the entropy of the sorbed guest molecules in their standard state and the entropy of these molecules in their standard state in the gas phase. In the gas phase the molecules have three translational degrees of freedom ($3t$). Linear molecules have an additional two rotational degrees of freedom ($2r$) with moment of inertia, I, and non-linear molecules have three rotations ($3r$) with principal moments I_1, I_2 and I_3. When the molecules are sorbed various (t, v, r) states may arise, six of which are exemplified in Table 1,[18] together with corresponding values of $\Delta S^{\ominus}_{K_c}$ and $\Delta C^{\ominus}_{K_c}$. In Table 1, σ is the symmetry number of the guest molecules. It is assumed that internal entropy, S_I, is not changed

TABLE 1

Some limiting values of $\Delta S^{\ominus}_{K_c}$ and $\Delta C^{\ominus}_{K_c}$ for homogeneous sorbents

(t, v, r) States in sorption volume	$\Delta S^{\ominus}_{K_c}$	$\Delta C^{\ominus}_{K_c}$
Monatomic, linear and non-linear molecules, respectively:		
$3t$; $(3t, 2r)$; or $(3t, 3r)$	0	0
$(2t, v)$; $(2t, v, 2r)$; or $(2t, v, 3r)$	$R \ln \left[\left(\frac{kT}{2\pi m} \right)^{1/2} \frac{N_{sat}^{1/3}}{\nu} \right] + R/2$	$R/2$
$(t, 2v)$; $(t, 2v, 2r)$; or $(t, 2v, 3r)$	$R \ln \left[\frac{kT}{2\pi m} \frac{N_{sat}^{2/3}}{\nu^2} \right] + R$	R
$3v$; $(3v, 2r)$; or $(3v, 3r)$	$R \ln \left[\left(\frac{kT}{2\pi m} \right)^{3/2} \frac{N_{sat}}{\nu^3} \right] + 3R/2$	$3R/2$
Linear molecules		
$5v$	$R \ln \left[\left(\frac{kT}{2\pi m} \right)^{3/2} \frac{N_{sat}}{\nu^5} \frac{kT\sigma}{8\pi^2 I} \right] + 5R/2$	$5R/2$
Non-linear molecules		
$6v$	$R \ln \left[\left(\frac{kT}{2\pi m} \right)^{3/2} \frac{N_{sat}}{\nu^6} \frac{(kT)^{3/2}\sigma}{8\pi^2 (2\pi)^{3/2} (I_1 I_2 I_3)^{1/2}} \right] + 3R$	$3R$

by sorption. The frequency, ν, is a mean value for the vibrations of the guest molecule relative to its intracrystalline environment. These vibrations are taken to be simple harmonic and it is assumed that $h\nu \ll kT$. For $h\nu \approx kT$ one would, for example, have instead of the relation in Table 1 for 6ν the expression,

$$\Delta S^{\ominus}_{K_c} = R \ln \left[\frac{h^3}{(2\pi mkT)^{3/2}} \left(1 - \exp - \frac{h\nu}{kT} \right)^{-6} \frac{h^3 \sigma N_{sat}}{8\pi^2 (2\pi kT)^{3/2} (I_1 I_2 I_3)^{1/2}} \right]. \tag{22}$$

Before using the expressions in Table 1 to interpret $\Delta S^{\ominus}_{K_c}$ or $\Delta S^{\ominus}_{K_p}$ of Tables 8 and 9 of Chapter 3 one may consider how realistic they may be in the intrazeolitic cavities and channels. Firstly for translational freedom the cavities or channels are idealized as boxes or tubes of rectangular cross-section within which there is uniform sorption potential and which have dimensions a, b and c. The energy levels are then

$$\epsilon_{l,m,n} = \frac{h^2}{8\pi m}\left[\frac{l^2}{a^2} + \frac{m^2}{b^2} + \frac{n^2}{c^2}\right]. \quad (23)$$

In eqn 9, l, m and n are quantum numbers, each of which may have values 1, 2, 3 . . . . If a is the smallest dimension and l changes from l to $(l + 1)$ with m and n constant, the change in energy is

$$\Delta\epsilon = \frac{h^2}{8\pi m} \cdot \frac{2l+1}{a^2}. \quad (24)$$

The smallest edge of the rectangular cross-section, a, is then the smallest free diameter of the channel or cavity less the van der Waals diameter of

TABLE 2

Values of $\Delta\epsilon_{l=1}/kT$ for the a dimension at 78 and 273 K

Zeolite	Guest molecule (and diameter)	a-dimension (Å)	$\Delta\epsilon_{l=1}/kT$ 78 K	273 K
Mordenite	H_2	3·7	2·6	0·75
Offretite	(3·0 Å if	3·3	3·3	0·94
Zeolite L	rotating)	4·1	2·1	0·61
Erionite		3·3	3·3	0·94
Chabazite		3·5	2·9	0·84
Mordenite	N_2	2·6	0·38	0·11
Offretite	(4·1 Å if	2·2	0·53	0·15
Zeolite L	rotating)	3·0	0·29	$0{\cdot}08_3$
Erionite		2·2	0·53	0·15
Chabazite		2·4	0·45	0·13

the guest molecule. Some values of $\Delta\epsilon_{l=1}/kT$ are given in Table 2 and show that neither H_2 (at 78 and 273K) nor N_2 (at 78K) will behave as a classical gas so far as movement across channels or cavities is concerned. Especially since the energy contour across the cages or channels is not flat (cf. Fig. 20 of Chapter 3) translations must be replaced by soft vibrations for transverse motions. These considerations appear to preclude $3t$ and $2t$ for the translational degrees of freedom of guest molecules inside porous crystals. In structures where the channels are continuous and unobstructed (H-mordenite, offretite, zeolite L and zeolite Ω) the remaining possibilities for monatomic gases are $(t, 2v)$ or $3v$, the latter if there is energy periodicity along the channels with barriers appreciably in excess of RT. In extremely open three-dimensional channel networks such as those in faujasite it is just possible that $\Delta S^{\ominus}$ could approximate to that for $(2t, v)$ but the tendency of guest molecules to stick to the walls of the cavities due to the non-uniformity of the sorption potential within the cavities still suggests that it would be more probable to find soft vibrations.

A second approach to the degrees of freedom of guest molecules can be made by considering $\Delta S^{\ominus}_{K_c}$. Values of $\Delta S^{\ominus}_{K_c}$ are given for Kr for several zeolites in Table 9 of Chapter 3 and additional values in ref. 19. $\Delta S^{\ominus}_{K_c}$ is never zero, confirming the total unsuitability of $3t$. For soft vibrations of physically sorbed molecules one expects the vibrational energy quanta to be small compared with kT. This may be used as the criterion to check the reasonableness of (t, v) states, and on such a basis one has the following quanta:

ν (s^{-1})	$N_0 h\nu$ (kJ mol^{-1})
10^{13}	3·97
5×10^{12}	1·99
10^{12}	0·397
5×10^{11}	0·199
10^{11}	0·0397.

At 188 K, $RT = 1{\cdot}57$ kJ mol^{-1} so that for $N_0 h\nu \ll RT$ (N_0 = Avogadro's number) the reasonable frequencies would for binding energies around or less than 20 kJ mol^{-1} be less than 10^{12} and probably not much above 5×10^{11}s^{-1}. Frequencies calculated from $\Delta S^{\ominus}_{K_c}$ for the states $(2t, v)$, $(t, 2v)$ and $3v$ are given in Table 3 for Kr at 188 K in several zeolites and in Table 4 for a number of gases in H-chabazite. On the basis of the foregoing criterion, $(2t, v)$ can be eliminated for all the systems and $(t, 2v)$ for the first four. $3v$ is reasonable for all the systems, but $(t, 2v)$ is not ruled out for the last five all of which involve zeolite *L* which has wide relatively unobstructed channels parallel with the *c*-direction.

From Table 4 the $3v$ state is again seen to be reasonable for Ar, Kr and Xe in H-chabazite, but the frequencies calculated for the linear molecules H_2, O_2, N_2 and CO_2 are high for the state $(3v, 2r)$ and therefore suggest considerable interference with free rotation. This interference is probably enhanced in the case of N_2 and CO_2 by their permanent quadrupole moments which interact with electrostatic field gradients so as to orient these molecules. The results in Table 4 are based on the $\Delta S^{\ominus}_{K_p}$ values of Table 8 Chapter 3 but values of $\Delta S^{\ominus}_{K_p}$ for K_p were first recalculated to give $\Delta S^{\ominus}_{K_c}$ (see eqn 21). The concentration, C, in mol m^{-3} of crystal had also to be converted to C_s in mol m^{-3} of intracrystalline pore space (see eqn 20). The conversion factor was 2·2 ($C_s = 2{\cdot}2C$).

Because of the importance of zeolites in separating hydrocarbons and as catalysts for reactions involving hydrocarbons there is particular interest in hydrocarbon sorption equilibria. In one of these investigations[(20)] C_1 to

TABLE 3

Frequencies ν (s^{-1}) from $\Delta S^{\ominus}_{K_c}$ for Kr in zeolites at 188 K

Zeolite	Models $2t, v$	$t, 2v$	$3v$
La-L	$1{\cdot}7 \times 10^{13}$	$1{\cdot}9 \times 10^{12}$	$9{\cdot}4 \times 10^{11}$
H-offretite	$7{\cdot}4 \times 10^{12}$	$1{\cdot}3 \times 10^{12}$	$7{\cdot}2 \times 10^{11}$
H-erionite	$6{\cdot}3 \times 10^{12}$	$1{\cdot}2 \times 10^{12}$	$6{\cdot}8 \times 10^{11}$
H-L	$5{\cdot}4 \times 10^{12}$	$1{\cdot}1 \times 10^{12}$	$6{\cdot}5 \times 10^{11}$
Li-L	$2{\cdot}6 \times 10^{12}$	$7{\cdot}4 \times 10^{11}$	$5{\cdot}1 \times 10^{11}$
Na-L	$1{\cdot}7 \times 10^{12}$	$6{\cdot}2 \times 10^{11}$	$4{\cdot}4 \times 10^{11}$
Ba-L	$1{\cdot}6 \times 10^{12}$	$5{\cdot}9 \times 10^{11}$	$4{\cdot}3 \times 10^{11}$
Cs-L	$1{\cdot}4 \times 10^{12}$	$5{\cdot}6 \times 10^{11}$	$4{\cdot}1 \times 10^{11}$
K-L	$8{\cdot}1 \times 10^{11}$	$4{\cdot}2 \times 10^{11}$	$3{\cdot}4 \times 10^{11}$

TABLE 4

Frequencies ν (s^{-1}) from $\Delta S^{\ominus}_{K_c}$ for gases in H-chabazite

Gas	T/K	State considered	ν
Ar	161	$3v$	$8{\cdot}9 \times 10^{11}$
Kr	188	$3v$	$5{\cdot}6 \times 10^{11}$
Xe	253	$3v$	$4{\cdot}2 \times 10^{11}$
H_2	90·2	$(3v, 2r)$	$3{\cdot}1 \times 10^{12}$
O_2	165	$(3v, 2r)$	$1{\cdot}03 \times 10^{12}$
N_2	167	$(3v, 2r)$	$1{\cdot}8 \times 10^{12}$
CO_2	299	$(3v, 2r)$	$3{\cdot}3 \times 10^{12}$

C_4 equilibria of *n*-paraffins in H-chabazite and in H-L were analysed by the method of § 8, Chapter 3 to give $-RT\ln K_p$, $\Delta S^{\ominus}_{K_p}$ and $\Delta H^{\ominus}$. Some results are given in Table 5, K_p being given in atm^{-1} mol dm^{-3} of crystal; $-\Delta S^{\ominus}_{K_p}$ increases steadily with increasing carbon number. Not only is it likely that the larger paraffins lose the rotational freedom involving the three principal moments of inertia but also with *n*-butane the restrictions imposed by the environment must increase hindrance to intramolecular segmental rotations. The longer the *n*-paraffin chain, the greater the expected loss of such rotational entropy. A similar thermodynamic analysis was made by Eberly[21] for equilibria of *n*-hexane, cyclohexane and benzene in Na- and H-mordenites. He tabulated the half-standard values of $\Delta\bar{G}$ and $\Delta\bar{S}$. Thus $\Delta\bar{G} = \Delta\bar{G}^0$ where $\Delta\bar{G}^0$ is the molar free energy of transfer from the gas phase in its standard state to the zeolite over which the equilibrium pressure is p. Langmuir's isotherm was assumed valid so that at $\theta = 0\cdot5$, $\bar{S}_s = \bar{S}_{Th}$. The analysis then followed the procedure of ref. 9.

For very long chain molecules in channel systems such as those in mordenite, offretite or zeolite L the loss of rotational freedom will be synonymous with loss of many molecular configurations possible in the gas phase. The long chains will be in a stretched configuration with only limited opportunity to pucker or coil. The loss of configurational entropy of this kind was estimated approximately some time ago.[22, 18] One may let

TABLE 5

Standard thermodynamic functions for some paraffins in zeolite H-L[20]

Guest molecule	T/K	$10^{-3}C^{\ominus}$ mol m^{-3} of crystal	$-10^{-3}RT\ln K_p$ J mol^{-1}	$10^{-3}\Delta H^{\ominus}$ J mol^{-1}	$\Delta S^{\ominus}_{Kp}$ J K^{-1}mol^{-1}
CH_4	154·6	0·65	−7·57		−69·9
	176·1	0·67	−5·90		−71·1
	189·1	0·68	−4·97	−18·4	−71·1
	203·3	0·70	−4·10		−70·3
	217·4	0·70	−3·18		−69·9
	237·4	—	−2·59		−70·3
C_2H_6	217·9	—	−10·9		−82·4
	233·9		−9·75	−28·9	−81·3
	253·2	0·65	−7·78		−83·6
	273·2	0·64	−6·53		−82·0
	293·2	$0\cdot64_5$	−4·73		−82·4
	333·2	0·64	−1·42		−82·4
C_3H_8	253·2	0·55	−14·5		−87·0
	273·2	0·58	−12·7	−36·6	−87·4
	293·2	$0\cdot57_5$	−10·8		87·9
	323·2	0·58	−8·2		−87·9
	348·2	$0\cdot59_5$	−5·23		−90·0
	373·2	0·55	−3·72		−89·1
	398·2	—	−1·97		−87·0
n-C_4H_{10}	273·2	—	−18·7		−91·2
	298·2	—	−17·0	−44·9	−93·7
	323·2	0·45	−15·1		−92·5
	348·2	0·46	−12·2		−94·1
	373·2	$0\cdot47_5$	−9·87		−93·7
	398·2	$0\cdot48_5$	−7·28		−94·6
	423·2	$0\cdot47_5$	−5·10		−94·1
	448	—	−2·93		−93·7

r_{max} = the maximum possible distance between the two terminal carbons of a fully stretched chain in gas phase or zeolite channel;

r_{min} = the smallest possible distance between the terminal carbons of the chain in the gas phase;

r_z = the smallest possible distance between these carbons in the zeolite.

In general $r_{min} < r_z < r_{max}$. When $P(r)$ denotes the probability of finding r-values between r and $r + dr$ one has

$$\left.\begin{aligned} \int_{r_{min}}^{r_{max}} P(r)\,dr = 1 \\ \Delta S_{con} = R \ln \int_{r_z}^{r_{max}} P(r)\,dr - R \ln \int_{r_{min}}^{r_{max}} P(r)\,dr = R \ln \int_{r_z}^{r_{max}} P(r) \end{aligned}\right\} \tag{25}$$

where ΔS_{con} is the loss of configurational entropy of the long chain guest molecule in the zeolite channel due to the environmental restrictions. Values of $P(r)$ are known which are appropriate for the long chains in the gas or liquid phase[23] and these were used to estimate $-\Delta S_{con}$. This quantity may be a lower limit only since some configurations available to the chain in gas or liquid, for the given separation of the two terminal carbons, may not be possible within the crystal. In Fig. 7, $-\Delta S_{con}$ is plotted as a function of carbon number and of the quantity ρ where

$$\rho = 100 \times \frac{\text{Range of allowed lengths for chain configurations in channel}}{\text{Maximum range in lengths for gas phase}}.$$

The effect described commences with C_4 and, for a given ρ, increases strongly with carbon number. For a given carbon number $-\Delta S_{con}$ decreases rapidly as ρ increases. As the chain increases in length so also with both $-\Delta \bar{H}$ and $-T\Delta S_{con}$, thus accounting in part for the compensation between $\Delta \bar{H}$ and $T\Delta \bar{S}$ reported by Schirmer *et al.*[24] for hydrocarbons in zeolite Ca-*A* (see § 3, Chapter 3).

3. THERMAL ENTROPY OF INTRACRYSTALLINE FLUIDS

In systems where the quotient $\theta/p(1 - \theta)$ is constant over a range of values of θ(§§ 2 and 3, Chapter 3) one may in this range take the differential configurational entropy as $\bar{S}_c = R \ln [(1 - \theta)/\theta]$ and so evaluate $\bar{S}_{Th} = (\bar{S}_s - \bar{S}_c)$. $\bar{S}_{Th}$ does not then depend on θ over this range so that $\bar{S}_{Th} \approx \tilde{S}_{Th}$. For inert gases where rotations are not involved one has for simple harmonic oscillators (state $3v$)

$${}_3\tilde{S}_v = 3R\{u(\exp u - 1)^{-1} - \ln(1 - \exp - u)\} \tag{26}$$

where $u = \frac{h\nu}{kT}$, ν being the oscillator frequency. Thus for argon at 173 K in Na-faujasite[5] eqn 26 gave $\nu = 6 \times 10^{11} s^{-1}$ so that the criterion $h\nu \ll kT$ is satisfied. The saturation capacity was taken as 185 cm^3 at s.t.p. g^{-1} of hydrated Na-faujasite in which there are n unit cells. If instead $\bar{S}'_{Th}$ and $\bar{S}''_{Th}$ of eqn 9 were determined ν was respectively $1{\cdot}0 \times 10^{12} s^{-1}$ and $2{\cdot}5 \times 10^{12} s^{-1}$

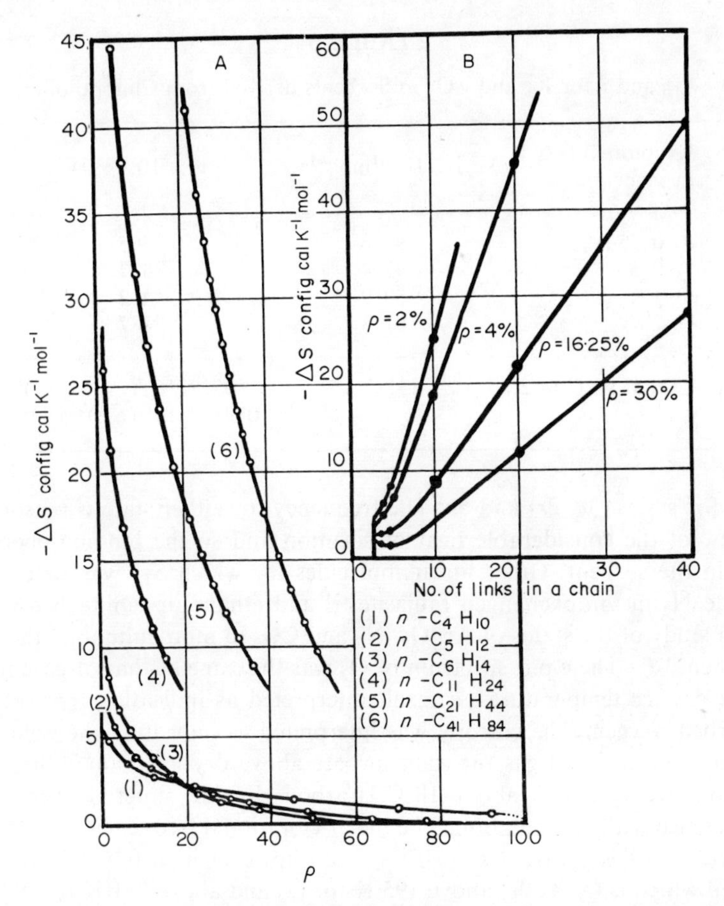

FIG. 7. ΔS_{config} as a function of ρ and of the number of links in the carbon chain.[22,18]

values which are higher than expected for small heats of sorption according to the criterion that $h\nu \ll kT$. The state ($2t$, v) can certainly be ruled out and ($1t$, $2v$) is improbable.

As a second example involving monatomic gases, one may consider Kr in sorbents prepared by progressive dealumination of clinoptilolite using 0·25, 0·5, 1, and 2N HCl solutions. For these systems $\theta/p(1-\theta)$ was constant over substantial ranges in θ (Fig. 4, Chapter 3). Proceeding as in the previous paragraph one obtains the thermal entropies and frequencies in Table 6.[13]

The "soft" mean frequency accords with values expected from that given above for Ar in Na-faujasite and the frequencies for Ar, Kr and Xe given in Tables 3 and 4. The table also contains a result for the linear molecule of

TABLE 6

$\bar{S}_{Th}$ and ν for Kr and CO_2 in Sorbents derived from Clinoptilolite

(a) Krypton at 0°C in sorbent	$\bar{S}_{Th}(JK^{-1}mol^{-1})$	$\nu \times 10^{11}(s^{-1})$
0·25 N	93·6	3·5
0·5 N	92·4	3·7
1 N	89·9	4·2
2 N	87·8	4·7
(b) CO_2 at 90°C in 2N	131·7	$1{\cdot}05 \times 10^{12}({}_3\tilde{S}_v + {}_2\tilde{S}_R)$ $0{\cdot}93 \times 10^{12}(\tilde{S}_{5v})$

CO_2, for states ($3v$, $2r$) and $5v$. The frequency for either state is reasonable, in view of the considerable heat of sorption and so the binding energy of CO_2 in the sorbent. Other linear molecules for which $\bar{S}_{Th}$ was determined include N_2 in ion exchanged faujasites[5] and ethane in zeolite Na-X.[9] A recent study of the states of Ar, O_2, N_2 and CO_2 in mordenite used the same argument.[6, 7] The e.p.r. spectrum of O_2 was the same as that of gaseous O_2 above dry ice temperatures, a result interpreted as indicating free rotation of sorbed oxygen. The entropy was interpreted as indicating the behaviour of a one-dimensional gas for each sorbate above dry ice temperature, with rotation of N_2 above about 18°C. Carbon dioxide suffered even larger interference with free rotation. The mean vibrational frequencies for O_2 and N_2 of $2 \times 10^{12}s^{-1}$ and $2{\cdot}6 \times 10^{12}s^{-1}$ are however high for the state considered which is (t, $2v$, $2r$) above 195 K for O_2 and above 291 K for N_2. The side-pockets.[25] along the wide mordenite channels should provide energetically sorbing sites for all the above sorbates and in these pockets guest molecules should have states ($3v$, $2r$) to $5v$. Only guest molecules remaining in the wide channels with free diameters $5{\cdot}8 \times 7{\cdot}0$Å could possibly have translational freedom.

Where the quotient $\theta/p(1 - \theta)$ is reasonably constant the integral configurational entropy, $\tilde{S}_c$, given by eqn 13 can be subtracted from the integral entropy $\tilde{S}_s$, where

$$\tilde{S}_s = \frac{1}{\theta}\int_0^\theta \bar{S}_s \mathrm{d}\theta \tag{27}$$

to give the corresponding thermal entropy, $\tilde{S}_{Th}$. This entropy can be compared with that of the liquid sorbate at the same temperature. Some examples are given in Table 7 for carbon dioxide in erionite, Na-Y and a cellulose carbon.[14] The entropies for free rotation are also given. For all sorbents $S_{liq} > \tilde{S}_{Th}$ so that, since $\tilde{S}_c$ for liquids is believed to be small, CO_2 is more localized in the zeolites than in the liquid state. The difference is however very small for the carbon sorbent. It is also seen that $\tilde{S}_{Th}$ is smaller the larger the value of $-\Delta\tilde{H}$, where $\Delta\tilde{H}$ is the integral heat of sorption; and that $\tilde{S}_{Th}$ is nearly independent of θ.

TABLE 7

$\tilde{S}_s$ and $\bar{S}_{Th}$ for CO_2 in several sorbents ($JK^{-1}mol^{-1}$)[14]

Sorbent	$-\Delta\tilde{H}$ kJ mol^{-1} for 30 cm^3 at s.t.p. g^{-1}	T(°C)	θ	$\tilde{S}_s$	$\tilde{S}_{Th}$	S_{liq}	S_{rot}
Erionite	45·1	30	0·2	122	101	152	54·5
			0·3	120	103		
			0·4	117	103		
			0·5	114	102		
			0·6	110	101		
			0·7	107	100		
			0·8	104	99		
Na-Y	31·0	30	0·1	143	115	152	54·5
			0·2	143	121		
			0·3	140	123		
			0·4	137	123		
			0·5	134	123		
			0·6	132	122		
Cellulose Carbon	24·5	−40	0·1	143	115	122	52·3
			0·2	139	117		
			0·3	136	120		
			0·4	133	120		
			0·5	131	120		
			0·5$_5$	130	119		
		−20	0·1	146	118	129	53·0
			0·2	142	121		
			0·3	140	123		
			0·4	137	123		
			0·5	135	123		
			0·6	132	123		

The ammonia-faujasite system studied by Barrer and Gibbons[11] may be submitted to the same analysis. For (Cs, Na)-faujasite, which was energetically rather homogeneous, one obtains at 100°C for example, the results below.

θ	$-\Delta\tilde{H}$ (kJ mol^{-1} at $\theta = 0{\cdot}4$)	$\tilde{S}_s$ (JK^{-1}mol^{-1})	$\tilde{S}_{Th}$ (JK^{-1}mol^{-1})	S_{liq} (JK^{-1}mol^{-1})
0·09		121	91	
0·18	50	115	92	122
0·36		106	91	
0·55		100	89	
0·64		97	89	

The substantial difference between S_{liq} and $\tilde{S}_{Th}$ shows that zeolitic ammonia even at 100°C is more localized than in the bulk liquid. If the states of the sorbed NH_3 are taken as $\tilde{S}_{Th} = {}_3\tilde{S}_v + {}_3\tilde{S}_R$ to $\tilde{S}_{Th} = \tilde{S}_{6v}$ the frequencies ν are in the range $3{\cdot}3$ to $3{\cdot}5 \times 10^{12}\text{s}^{-1}$. The frequencies for CO_2 with $\tilde{S}_{Th} = {}_3\tilde{S}_v + {}_2\tilde{S}_R$ to ${}_5\tilde{S}_v$ lay in the range $2{\cdot}2$ to $1{\cdot}4 \times 10^{12}\text{s}^{-1}$ in erionite; $1{\cdot}2$ to $0{\cdot}95 \times 10^{12}\text{s}^{-1}$ in zeolite Na-Y; and $0{\cdot}85$ to $0{\cdot}71 \times 10^{12}\text{s}^{-1}$ in the cellulose carbon. The frequencies decline with decrease in binding energy.

4. HEAT CAPACITIES OF INTRACRYSTALLINE FLUIDS

An important property of $\bar{S}_s$ or $\tilde{S}_s$ is its positive temperature coefficient, as illustrated in Fig. 12, Chapter 3. Figure 8 for N_2 in K-faujasite (zeolite K-X) further demonstrates this property. If the mean value of $\bar{S}_s$ in the temperature interval $(T_1 - T_2) = \delta T$ is $\bar{S}_m$, and the mean temperature is $T_m = (T_1 + T_2)/2$, then the mean heat capacity, $\bar{C}_m$ is given by

$$\bar{C}_m = T_m \frac{\delta \bar{S}_m}{\delta T} \tag{28}$$

as an approximation to $\bar{C}_s = T\dfrac{\partial \bar{S}_s}{\partial T}$, the differentiation being for a fixed amount sorbed. Some values of $\bar{C}_m$ for Kr and Xe in various zeolites, are given in Table 8.[26, 19] If the inert gas atoms are supposed to behave as a set of identical classical simple harmonic oscillators in an inert medium* their heat capacities should be $24{\cdot}9$ JK^{-1} mol^{-1}, while for the states $(t, 2v)$

* For random occupation of identical sites and for a fixed uptake of guest molecules $\partial\bar{S}_s/\partial T = \partial\bar{S}_{Th}/\partial T$, and so $\bar{C}_s = \bar{C}_{Th}$ where $\bar{C}_{Th}$ is the contribution to the heat capacity from the thermal entropy.

and $(2t, v)$ the values should be respectively $20 \cdot 7$ and $16 \cdot 5$ JK^{-1} mol^{-1}. Krypton and xenon behave in general as if their states lay near to $(t, 2v)$ with some systems tending to $3v$ or toward $(2t, v)$. However reasons have been given previously (§ 2) for believing that $(2t, v)$ is unlikely. In directions across the channels soft vibrations which may be anharmonic will replace

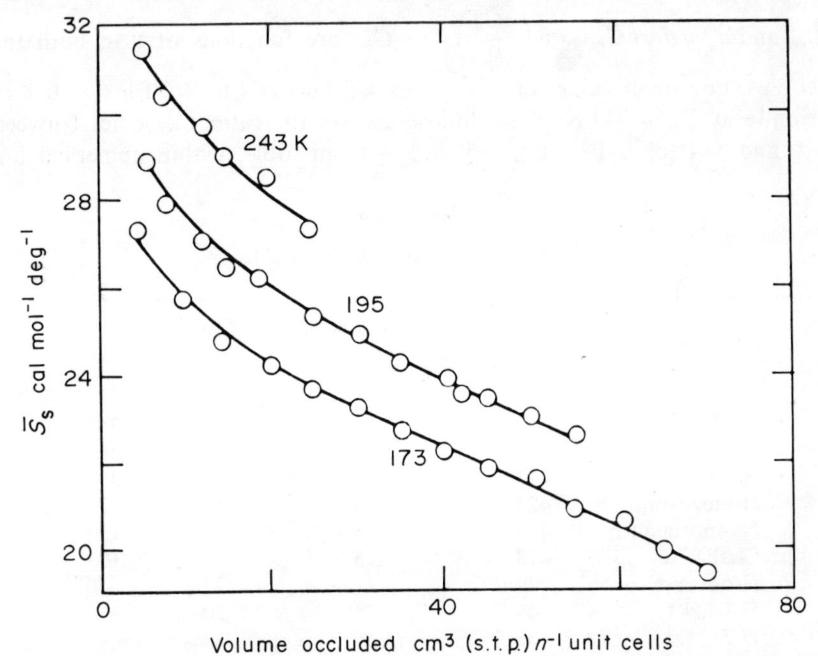

FIG. 8. Differential entropy of sorption of nitrogen in K-faujasite (K-X) at several temperatures.[5]

translations. As with the evaluation of $\bar{S}_s$, that of $\bar{C}_m$ according to eqn 28 depends only on a general argument and not upon specific isotherm models.

As examples of linear molecules, nitrogen and carbon dioxide may be considered. For nitrogen in zeolite K-X between 173 and 195 K (Fig. 8) and for an uptake of 20 cm^3 at s.t.p. in an amount of anhydrous K-X containing the same number, n, of unit cells as in 1 g of anhydrous Na-X,[5] $\bar{C}_m$ was 62 JK^{-1} mol^{-1}, and when this uptake was 40, $\bar{C}_m$ was 54 JK^{-1} mol^{-1}. These values are in excess of $5R$, the maximum to be expected for the ideal classical oscillator when the two rotations of nitrogen have also become vibrations. There are several possible reasons for such deviations. Firstly in a heterogeneous sorbent molecules will redistribute themselves among sites as the temperature changes. Secondly if there is a change of rotational freedom as the temperature alters, the quadrupole energy of interaction of nitrogen with intracrystalline field gradients will change.[27] Thirdly, the assumption that the zeolite is an inert medium cannot always be justified.

It is known for example that the siting of cations in a zeolite is influenced by their interactions with the guest molecules. The possible effect of site heterogeneity is considered later. Table 9 gives some integral values of the heat capacity obtained from the expression $\tilde{C}_m = T_m \frac{\delta \tilde{S}_s}{\delta T}$ for carbon dioxide. $\tilde{C}_m$, and *a fortiori* $\bar{C}_m = \theta \left(\frac{\partial \tilde{C}_m}{\partial \theta}\right)_{T_m} + \tilde{C}_m$, are functions of θ in both the zeolites. For small values of θ, $\tilde{C}_m/R$ is ≈ 5 but rises to 9·6 for $\theta = 0·8$ in erionite at $T_m = 333$ K. On cellulose carbon the ratio $\tilde{C}_m/R$ lies between 4·9 and 5·4 for $0·1 \leq \theta \leq 0·5$ and is compatible, within numerical un-

TABLE 8

$\bar{C}_m$ for Kr and Xe in some zeolites

Sorbent	T_m/K	Amount sorbed (cm³ at s.t.p. g⁻¹)	$\bar{C}_m$(JK⁻¹mol⁻¹) Kr	Xe
Na-Y	473	15	22	24
Ca-X	473	15	20	21
Ca-*A*	473	15	20	25
H-mordenite	473	15	18	22
Na-mordenite	473	15	20	19
Chabazite	473	15	21	19
H-offretite	179·2	2[a]	$22·_6$	—
H-erionite	164·7	4[a]	$20·_0$	—
H-*L*	166·2	3[a]	$22·_2$	—

[a] molecules per unit cell.

certainty, with the state 5*v*. In the zeolites for small θ, $\tilde{C}_m$ is compatible with this state, but for larger uptakes one or more of the factors referred to in the previous paragraph must be involved.

Ammonia may be considered as an example of a non-linear molecule. Table 10 gives values of $\bar{C}_m$ in several ion exchanged faujasites, derived from the work of Barrer and Gibbons,[11] which, at the rather high temperatures involved, represent a simpler behaviour. $\bar{C}_m$ does not depend strongly upon amounts sorbed over the ranges given in the table and the ratios $\bar{C}_m/R$ lie between extremes of 4·3 and 5·7. While the (Cs, Na)-X behaves as a rather homogeneous sorbent the other two are energetically heterogeneous and redistribution effects could play a part in influencing $\bar{C}_m$. However, as a whole the results suggest vibrations with considerable rotational freedom. If the states were (3*v*, 3*r*), (4*v*, 2*r*), (5*v*, *r*) and 6*v* respectively $\bar{C}_m/R$ would be 4·5, 5, 5·5 and 6 respectively. For gaseous ammonia at 300 K $C_v/R = 3·25$

TABLE 9

$\tilde{C}_m$(JK^{-1}mol^{-1}) for CO_2 in several sorbents

Sorbent	Range in T(°C)	T_m/K	θ	$\tilde{C}_m$
Erionite	30 to 90	333	0·2	42
			0·3	46
			0·4	58
			0·5	65
			0·6	74
			0·7	76
			0·8	81
	90 to 150	395	0·2	44
			0·3	47
			0·4	52
			0·5	58
			0·6	63
Na-Y	30 to 90	333	0·1	44
			0·3	56
			0·4	65
			0·5	72
			0·6	76
Cellulose Carbon	−40 to −20	243	0·1	41
			0·2	41
			0·3	45
			0·4	45
			0·5	45

instead of 3·00 for (3t, 3r) so that internal vibrational modes are contributing a little to C_v. At the higher temperatures of Table 10 this contribution will be enhanced so that there may be rather more rotational freedom than is suggested when internal modes are ignored.

As further examples of polyatomic guest molecules hydrocarbons can be regarded as among the more important. Table 11 gives values of $\bar{C}_m$ for several n- and isoparaffins and for cyclo-pentane, benzene and toluene. $\bar{C}_m$ is always considerably less than $C_p^\ominus$, the standard heat capacity of the liquid hydrocarbon (298 K, 1 atm). Also $\bar{C}_m$ increases with carbon number and all values of $\bar{C}_m/R$ are substantial, ranging from a lowest figure of 10 for cyclopentane and isobutane to a maximum of 25 for n-heptane. For these hydrocarbons there are usually internal segmental librations to contribute to $\bar{C}_m$, but other factors such as redistribution among sites as the temperature changes are also possible.

Sichart, Koelsch and Schirmer[28] and Schirmer, Koelsch, Peters and

Stach[29] reported direct calorimetric and sorption studies of the systems (Ca, Na)-*A* plus *n*-heptane and (Ca, Na)-*A* plus ammonia, in the respective temperature ranges 25 to 240°C and 23 to 300°C, from which heat capacities of zeolitic heptane and ammonia were obtained. The results shown in Fig. 9a for n-C_7H_{16} are those obtained by direct calorimetry. The curves of

TABLE 10

$\bar{C}_m(JK^{-1}mol^{-1})$ for NH_3 in ion-exchanged faujasites (zeolite X)

Zeolite	NH_3 sorbed (mg g^{-1})	Range in T (°C)	T_m/K	$\bar{C}_m$
Sr-X	60	200 to 290	518	43
	80	110 to 230	443	40
	100	110 to 230	428	40
Na-X	40	200 to 290	518	48
	70	140 to 230	458	45
	100	140 to 180	433	41
(Cs, Na)-X	20	100 to 200	423	41
	50	100 to 160	403	36

$\tilde{C}_s$ all show maxima. When $\tilde{C}_s$ was obtained via the temperature dependence of the heat of adsorption, similar peaks were observed. The amounts sorbed in the curves of Fig. 9a range between $0\cdot164$ cm³ at s.t.p. g^{-1} to $26\cdot7$ cm³ at s.t.p. g^{-1}. For the former extremely small uptake the maximum in $\tilde{C}_s/R$ is about 250, a value so large as to suggest a chemical process. For the latter uptake $\tilde{C}_s/R$ is ≈ 37 at the maximum and is beginning to approach the value of $\bar{C}_m$ recorded in Table 10. Figure 9b shows values of $\tilde{C}_s$ for NH_3, obtained from the temperature dependence of the isosteric heat of sorption. The extremes in amounts sorbed are $3\cdot34$ and $84\cdot6$ cm³ at s.t.p. g^{-1}. Maxima again appear and that for the smallest uptake is so high as again to suggest a chemical process. However for the largest amount sorbed $\tilde{C}_s$ is again approaching the values recorded for $\bar{C}_m$ in Table 10. The authors were of the opinion that at least in part their results could be accounted for by redistributions among sites as the temperature changed, although for the smallest uptakes chemisorption was not excluded.

Stroud and Parsonage[30] and Richards, Smith and Parsonage[31] determined the heat capacities of CH_4 and C_2H_6 sorbed in zeolite Ca-*A*, their results for CH_4 being shown in Fig. 10. Again low maxima were recorded. For methane, curves of C_s/R against temperature do not differ very much for loadings of $0\cdot8933$, $1\cdot808$ and $4\cdot099$ molecules per cavity, the maxima in the curves being between 8 and 9. With ethane the maximum was in the vicinity of 9 at a temperature of ≈230 K. The maxima may be ascribed to redistributions as the temperature changes of molecules between sites having different affinities for these molecules. The mixed (Ca, Na)-*A* was considered to contain cavities with 6Ca, 5Ca + 2Na, 4Ca + 4Na etc. Each such cavity thus provides a different local environment for the guest molecules. With various simplifying assumptions model calculations were made[32] to see how nearly the experimental curves of C_s/R could be simulated for CH_4 and C_2H_6. The calculations led to maxima in curves of heat capacity against temperature having the same general contours as those observed experimentally, although quantitative agreement was not found.

TABLE 11

Values of $\bar{C}_m$(JK^{-1}mol^{-1}) for hydrocarbons in zeolite Na-X

Hydrocarbon	Temperature Range (in K)	Amount sorbed (cm³ at s.t.p. g⁻¹)	$\bar{C}_m$	$C_p^\ominus$ for liquid[a] (298K, 1 atm)
n-C_4H_{10}	323 to 343	37	105	
	298 to 318	46	103	
n-C_5H_{12}	323 to 343	37	111	
	298 to 323	41	114	171·5
n-C_6H_{14}	323 to 343	34	153	
	298 to 318	37	142	195·0
n-C_7H_{16}	323 to 343	32·5	209	
	298 to 323	34	176	224·7
iso-C_4H_{10}	313 to 343	39	96	
	298 to 318	45	85	
iso-C_5H_{12}	318 to 343	37	166	
	298 to 318	40	116	164·9
iso-C_8H_{16}	323 to 343	26	153	
	313 to 338	26·5	185	
neo-C_5H_{12}	318 to 343	26	133	
	298 to 318	31	129	
cyclopentane	328 to 343	41	94	
	298 to 313	53	85	126·8
benzene	323 to 343	49	90	
	298 to 313	52·5	120	136·1
toluene	333 to 353	43·5	93	
	313 to 343	44·5	92	166

[a] From Landolt–Bornstein "Zahlenwerte u. Funktionen", 6th Edition, 2nd volume, Part 4 (Springer Verlag) 1961, p. 261.

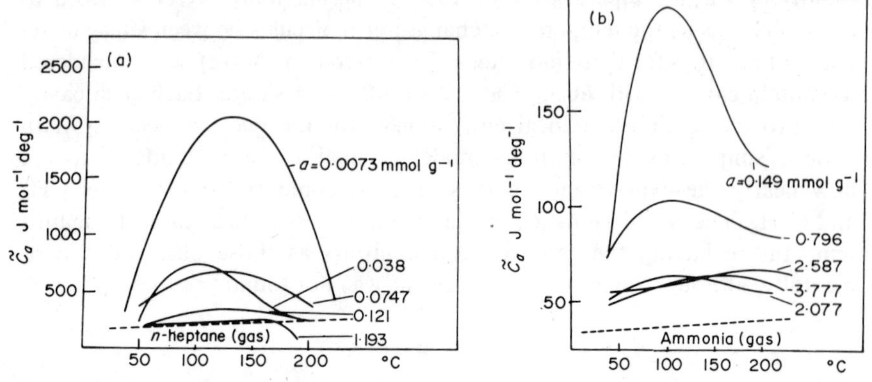

FIG 9. Heat capacities of (a) n-C_7H_{16} and (b) NH_3 in zeolite (Ca, Na)-A as functions of temperature, for several loadings of the crystals(28,29). $\tilde{C}_a$ is $\tilde{C}_s$ for the amount sorbed, a, as indicated on each curve.

Even in a single cavity containing a number of molecules all the molecules may not be able to occupy locations of equal sorption affinity. One may assume that the total sorption volume V_s (i.e. the accessible intracrystalline pore volume) is composed of volume elements totalling $V_1, V_2, \ldots V_i \ldots V_n$ where $\sum_{i=1}^{n} V = V_s$ and that in any chosen volume V_i the energy of sorption $\Delta\bar{E}_i$ and standard sorption affinity $\Delta\mu_i^{\ominus}$ are constant but differ from these quantities in any of the other volumes. The molar energy change, $\Delta\tilde{E}$, for reversible isothermal sorption of n_s mols is then

$$\Delta\tilde{E} = \sum_{i=1}^{n} X_i \, \Delta\tilde{E}_i \tag{29}$$

where $X_i = n_i/n_s$. Then keeping n_s constant one has

$$\Delta\tilde{C}_s = \sum_{i=1}^{n} \Delta\tilde{E}_i \left(\frac{\partial X_i}{\partial T}\right)_{n_s} + \sum_{i=1}^{n} X_i \, \Delta\tilde{C}_i \tag{30}$$

so that

$$\tilde{C}_s = \sum_{i=1}^{n} \Delta E_i \left(\frac{\partial X_i}{\partial T}\right)_{n_s} + \sum_{i=1}^{n} X_i \, \tilde{C}_i \tag{31}$$

because $\Delta\tilde{C}_s = (\tilde{C}_s - \tilde{C}_g)$ and $\Delta\tilde{C}_i = (\tilde{C}_i - \tilde{C}_g)$ where $\tilde{C}_g$ is the molar heat capacity of the gas. The condition for a maximum or minimum in the curve of $\Delta\tilde{C}_s$ against T is

$$2\sum_{i=1}^{n} \Delta\tilde{C}_i \left(\frac{\partial X}{\partial T}\right)_{n_s} + \sum_{i=1}^{n} \Delta\tilde{E}_i \left(\frac{\partial^2 X_i}{\partial T^2}\right)_{n_s} + \sum_{i=1}^{n} X_i \left(\frac{\partial \Delta\tilde{C}_i}{\partial T}\right)_{n_s} = 0. \tag{32}$$

The first term in eqn 31 contains a contribution to $\tilde{C}_s$ due to redistributions as T changes at constant n_s. Some terms $(\partial X_i/\partial T)n_s$ will necessarily be positive and some negative. Only if $(\Delta \bar{E} - T\Delta \bar{S}_{Th})$ was the same for all volume elements would the $(\partial X_i/\partial T)_{n_s}$ all be zero (cf § 3, Chapter 3).

Several investigations have been made of the heat capacity of zeolitic

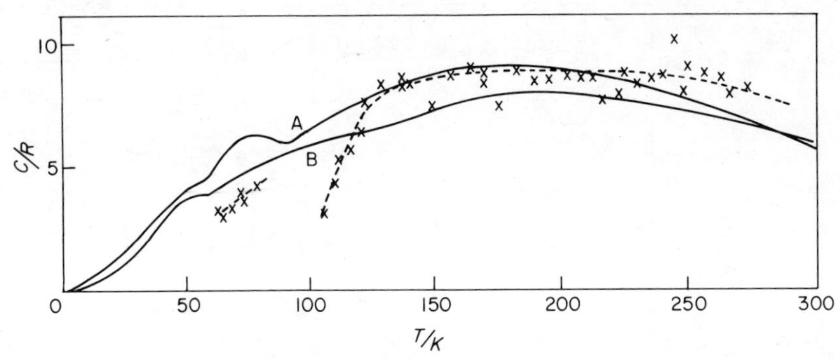

FIG. 10. Heat capacities of methane sorbed in zeolite Ca-*A* as a function of temperature for three loadings. Curves A and B are smoothed curves for sieves with 1·808 and 0·8933 molecules per cavity respectively. Points X and dashed curve: 4·099 molecules per cavity.[31]

water. As with all direct calorimetric studies the heat capacity is found for the container plus zeolite free of sorbate (here water) and for container plus zeolite plus sorbate. It is then assumed that the difference is the heat capacity of the sorbate. This argument may be an over-simplification since the guest molecules may perturb the host crystals, for example by resiting the cations. Such perturbations will be less significant with weakly bonded molecules like CH_4 or C_2H_6 but with polar molecules like ammonia or water some cation resiting is likely. Basler and Lechert[33] reported the apparent heat capacities for water in zeolite Na-X which are shown in Fig. 11. These capacities are shown also for ice and liquid water as reference lines. The heat capacities of the zeolitic water for different constant uptakes by the zeolite are somewhat above those of ice at low temperatures (100 K). They increase with temperature but do not reach the values for liquid water for uptakes of 46 and 93 mg/g of zeolite. For the larger uptakes (140, 202 and 293 mg/g) the values for liquid water are reached by 300 K. Such results support the view that there is no sharp melting point for zeolite water in faujasite-type crystals (Na-X), a conclusion also reached by Ducros[34] on the basis of NMR measurements of the mobility of water in some zeolites. Instead the water loses fluidity continuously as the temperature falls.

This view of the measurements by Basler and Lechert appears to contrast with the interpretation by Haly[35] who measured heat capacities at different uptakes of water by zeolite Ca-*A* as functions of temperature. For larger uptakes (0·293 and 0·248 g/g) flat maxima between −30°C and −20°C were interpreted as indicating a melting process for water. However re-distribution of water molecules or cation-resiting could also provide explanations. Indeed in a further study by Berezin, Kiselev and Sinitsyn[36] where the heat capacity of water was measured in (K, Na)-X at 34°C as a function

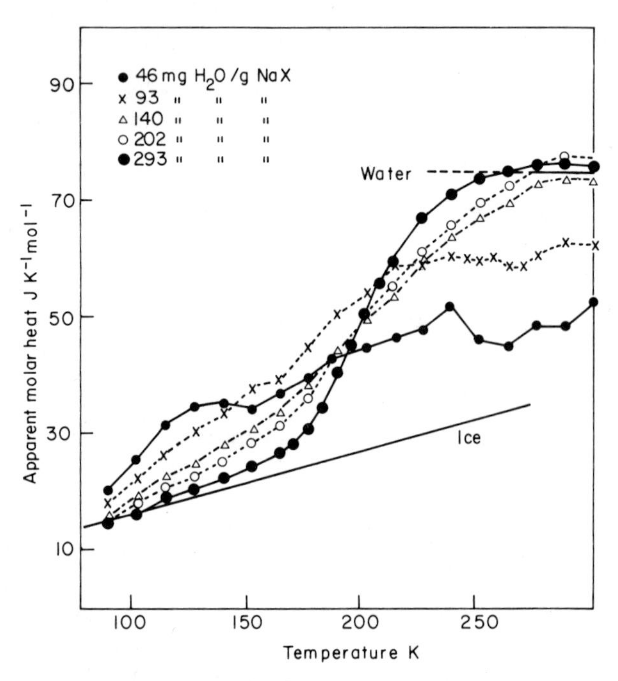

FIG. 11. Apparent molar heat capacities for water in zeolite Na-X as functions of temperature for a number of loadings.[33]

of amount sorbed two peaks were reported in the average molar heat capacity, C_m, as shown in Fig. 12. These peaks were interpreted as due to distributions of water at certain critical degrees of filling. The contribution to C_m due to this effect was given as

$$\Delta C_{ij} = \frac{1}{a} \cdot \frac{\mathrm{d}a_j}{\mathrm{d}T} (q^i_{st} - q^j_{st})$$

where q^i_{st} and q^j_{st} are isosteric heats for water sorbed in the more exo-energetic state and in the less exo-energetic state respectively; a_j is the amount of water sorbed in the less exo-energetic state when the total uptake is a. Thus $\frac{1}{a}\left(\frac{\mathrm{d}a_j}{\mathrm{d}T}\right)$ is the relative repopulation of the adsorbate molecules from the level q^i_{st} to the level q^j_{st} under conditions of isosteric heating of the system by 1K.

Basler and Lechert[33] also reported heat capacity data for SO_2 in Na-X, for 453 mg g^{-1}. The curve of apparent molar heat capacity against temperature lay slightly below that of solid SO_2 in the range 100 to 200 K. It increased continuously and above 270 K exceeded that of liquid SO_2. The results can be interpreted, as with water, to mean a continuous loss of

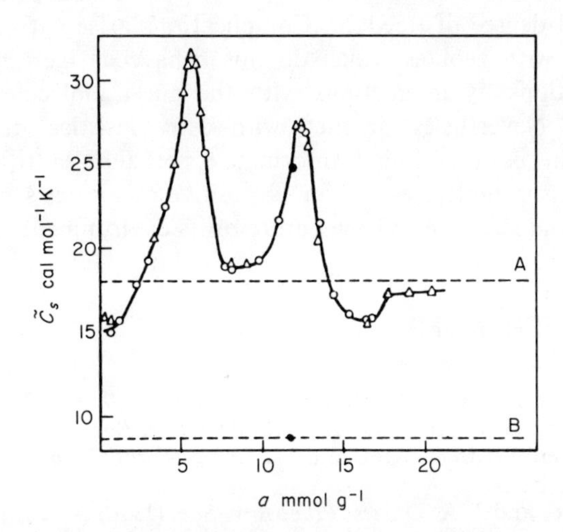

FIG. 12. Heat capacity of water in (K, Na)-X at 34°C as a function of amount sorbed.[36] Lines A and B give the respective heat capacities for liquid water, and water vapour at (constant volume).

mobility without a sharp freezing point as temperature falls. For benzene the heat capacity was obtained for uptakes of 79, 157, 211 and 221 mg g^{-1} of Na-X. The heat capacities declined as uptake increased, and as temperature fell. The values clustered around that of bulk benzene at 100 K, but tended to drift increasingly below this value at temperatures in excess of 200 K. The intracrystalline benzene can therefore be considered above 200 K to have fewer degrees of freedom than benzene in bulk, especially for larger uptakes.

5. CONCLUDING REMARKS

The entropy and thermal entropy of guest molecules, the standard entropy of sorption and the heat capacity of zeolitic fluids all serve to throw light upon their physical state, by helping to define the balance between their vibrations, rotations and translations relative to the host crystal or giving evidence of redistributions among heterogeneous sites resulting from temperature changes. While unambiguous allocations cannot usually be made the (t, v, r) state for simple molecules can be assessed within reasonable limits. If heats of intercalation are not large (less than about 20 kJ mol^{-1}) vibrational frequencies for such molecules normally lie in the range below 10^{12} and above $10^{11}s^{-1}$, although for strongly bonded molecules such as NH_3 or CO_2 the frequencies may rise above $10^{12}s^{-1}$. In certain zeolites, especially those with wide parallel channels, some results can be interpreted as involving a translational degree of freedom. Complications arise with heterogeneous sorbents, and with zeolites which do not behave as inert media, due to resiting of cations by interactions with the guest molecules or to other perturbations. Nevertheless if used with care statistical thermodynamic procedures can be useful and informative, because entropies and heat capacities of guest molecules are finite, positive quantities sensitive both to the nature of the guest and to the nature of its environment.

REFERENCES

1. L. A. Garden and G. L. Kington, *Proc. Roy. Soc.* (1956) A **234**, 24.
2. L. A. Garden, G. L. Kington and W. Laing, *Proc. Roy. Soc.* (1956) A **234**, 35.
3. L. A. Garden, G. L. Kington and W. Laing, *Trans. Faraday Soc.* (1955) **51**, 1558.
4. R. M. Barrer and J. A. Davies, *Proc. Roy. Soc.* (1970) A **320**, 289.
5. R. M. Barrer and W. I. Stuart, *Proc. Roy. Soc.* (1959) A **249**, 484.
6. T. Takaishi, A. Yusa and F. Amakasu, *Trans. Faraday Soc.* (1971) **67**, 3565.
7. T. Takaishi, A. Yusa, Y. Ogino and S. Ozawa, *Japan. J. App. Phys.* (1974) Suppl. 2, Part 2, 279.
8. R. M. Barrer and P. J. Reucroft, *Proc. Roy. Soc.* (1960) A **258**, 431.
9. R. M. Barrer, F. W. Bultitude and J. W. Sutherland, *Trans. Faraday Soc.* (1957) **53**, 1111.
10. R. M. Barrer and J. A. Lee, *Surface Sci* (1968) **12**, 354.
11. R. M. Barrer and R. M. Gibbons, *Trans. Faraday Soc.* (1963) **59**, 2875.
12. R. M. Barrer and R. M. Gibbons, *Trans. Faraday Soc.* (1965) **61**, 948.
13. R. M. Barrer and B. Coughlan, *in* "Molecular Sieves", Soc. Chem. Ind. (1968) p. 141.
14. R. M. Barrer and B. Coughlan, *in* "Molecular Sieves", Soc. Chem. Ind., London (1968) p. 241.
15. R. M. Barrer and R. M. Gibbons, *Trans. Faraday Soc.*, 1963, **59**, 2569.
16. M. M. Dubinin, O. Kadlec and A. Zukal, *Collection Czechoslov. Chem. Comm.* (1966) **31**, 406.
17. M. J. Spaarnay, *Surface Sci.* (1968) **9**, 100.
18. R. M. Barrer *in* Colston Papers, Vol. X, Butterworth (1958) p. 6.
19. R. M. Barrer and I. M. Galabova, *in* "Molecular Sieves", Eds W. M. Meier and J. B. Uytterhoven, Advances in Chemistry Series, Amer. Chem. Soc. (1973) No. 121, p. 356.
20. R. M. Barrer and J. A. Davies, *Proc. Roy. Soc.* (1971) A **322**, 1.
21. P. E. Eberly Jr, *J. Phys. Chem.* (1963) **67**, 2404.
22. R. M. Barrer, *Trans. Faraday Soc.* (1944) **40**, 374.
23. L. R. G. Treloar, *Proc. Phys. Soc.* (1943) **55**, 23.

24. W. Schirmer, G. Fiedrich, A. Grossmann and H. Stach, *in* "Molecular Sieves", Soc. Chem. Ind., London (1968) 276.
25. R. M. Barrer and D. L. Peterson, *Proc. Roy. Soc.* (1964) A **280**, 466.
26. R. M. Barrer R. Papadopoulos and J. D. F. Ramsay, *Proc. Roy. Soc.* (1972) A **326**, 331.
27. R. M. Barrer and W. I. Stuart, *Proc. Roy. Soc.* (1959) A **249**, 464.
28. K. H. Sichart, P. Kolsch and W. Schirmer, *in* "Molecular Sieve Zeolites", Advances in Chemistry Series, Amer. Chem. Soc. (1971) No. 102, p. 132.
29. W. Schirmer, P. Kolsch, H. Peters and H. Stach *in* Proc. 3rd Int. Conference on Molecular Sieves, Ed. J. B. Uytterhoven, Zurich, Sept. 3–7th (1973) p. 285.
30. H. J. F. Stroud and N. G. Parsonage, *in* "Molecular Sieve Zeolites", Advances in Chemistry Series, Amer. Chem. Soc., No. 102 (1971) p. 138.
31. E. Richards, J. H. Smith and N. G. Parsonage in Proc. 3rd Int. Conference on Molecular Sieves, Ed. J. B. Uytterhoven, Zurich, Sept. 3–7th (1973) p. 292.
32. E. L. Richards, Ph.D. Thesis, University of London (1973) p. 100 *et seq.*
33. W. D. Basler and H. Lechert, *Zeit. Phys. Chem. Neue Folge*, Frankfurt (1972) **78**, 199; Proc. 3rd Internat. Conference on Molecular Sieves, Ed. J. B. Uytterhoeven, Zurich, Sept. 3rd–7th (1973) p. 298.
34. P. Ducros, *Bull. Soc. Franc. Mineralog. Crist.* (1960) **83**, 85.
35. A. R. Haly, *J. Phys. Chem. Solids* (1972) **33**, 129.
36. G. I. Berezin, A. V. Kiselev and V. A. Sinitsyn, *J. Chem. Soc. Faraday Trans. I.* (1973) **69**, 3.

6

Diffusion in Zeolites

1. INTRODUCTION

In many important uses of porous crystalline zeolites (selective sorption, desiccation, molecule sieving, and catalysis) one stage involves migration of sorbates within the crystals. Accordingly there is much interest in factors which control this migration, and in relationships between molecular size and shape and the topologies of the open frameworks within which the molecules move. In catalysis Weisz[1] has estimated that a useful reactor must transform about 10^{-6} mol s^{-1} for every cm^3 invested in catalytic material. This sets a lower limit to useful flow rates into and out of catalyst beads. There are two situations involving transport in zeolites. In the first, diffusion within the crystals is alone the object of study. In the second the zeolite crystals form a component in a larger compact and flow involves transport additional to intracrystalline diffusion. Pure intrazeolitic flow can involve migration of a single compound in the zeolite or it may involve co- and counter-diffusion of two or more compounds. The simplest example of co- or counter-diffusion is tracer diffusion.

2. METHODS OF STUDYING THE MOBILITY OF SORBED MOLECULES

The purpose of investigating the mobilities of sorbed molecules is often to derive and interpret diffusion or other useful rate coëfficients. For this purpose direct measurements of mass transport into or out of the porous crystals are usually employed. This transport can be obtained from

(i) the change with time of the volume of gas or vapour around the sorbent, keeping the pressure constant;
(ii) the change with time of the pressure of gas or vapour around the sorbent, keeping the volume constant;
(iii) the change with time in the weight of sorbent bathed in a constant pressure of sorbate vapour;
(iv) the volume change with time of a liquid sorbate in a sorption pipette connected to the sorbent via the vapour phase;
(v) the transfer with time of tracer between gas or liquid phase and intracrystalline sorbed phase.

In a second group of methods, the mobilities or diffusion coefficients are inferred by examining properties which are related to sorbate content or mobility. These include

(vi) birefringence changes in single crystal plates which parallel changes of sorbate content;

(vii) sorbate mobility inferred from jump times derived by NMR;
(viii) sorbate mobility inferred from dielectric relaxation processes;
(ix) molecular motion evaluated by neutron scattering spectroscopy;
(x) changes with time, as the amount sorbed changes, in the intensity of specific infrared absorptions characteristic of the sorbate interacting with the sorbent.

In addition, intracrystalline diffusion coefficients have been evaluated from studies of gas chromatography,[2, 3] and of flow of a single gaseous component through compacts used as membranes[4] (see §§ 10.1 to 10.5). The methods (i), (ii) and (iii) involve equipment and procedures typical of volumetric and gravimetric equilibrium sorption studies. Limitations are set by the rates at which pressure, volume or weight changes can be accurately recorded. With the exception of counter-diffusion between traced and untraced sorbate of a given molecular species mass transport measurements are not strictly isothermal since heat is evolved or absorbed when sorption or desorption occurs. If for example sorption is extremely rapid the kinetics may be influenced, and even controlled over the later stage, by the temperature drop of the sorbent as the heat evolved is dissipated. The equilibrium uptake drifts accordingly, with accompanying time-dependent pressure changes in the ambient sorbate vapour. If however the molar heat of sorption is small, or the uptake of sorbate is small the heating of the sorbent may be negligible and, especially if sorption occurs only slowly, does not interfere with the kinetics. Criteria for gauging the importance of heat of sorption upon the kinetics have been considered by Chihara *et al.*[5]

2.1 Direct Measurements of Mass Transport

An example of apparatus for the volumetric study of desorption of gas from microporous and zeolite sorbents is that developed by Nelson and Walker.[6] It was designed for accurate measurements with gases above about 300°C when the amounts sorbed are small. The crystals were charged with gas up to a pressure in excess of one atmosphere and the gas released was measured when the pressure was suddenly lowered to one atmosphere. The pressure was measured with a micromanometer, in which the movement of a bubble of air was observed. The bubble was trapped in a small-bore capillary which connected two U-tube reservoirs containing a low boiling intermediate fluid, and mercury. Because the reference pressure side was exposed to the same unsteady temperatures as the measuring side, very small rates of gas outflow could be determined even in the presence of fluctuations in furnace and room temperatures. After outgassing the sorbent the sample tube on the right of Fig. 1 was exposed to the gas at pressure above 1 atm through tap S1. The remainder of the apparatus including the second sample tube was exposed to

1 atm of the same gas through tap S3. After equilibration the pressure in the right-hand sample tube was quickly lowered to 1 atm. Taps S3 and S4 were closed, S6, S7 and S10 opened, and reservoir R2 lowered to displace the air bubble in the micromanometer from its equilibrium position. Time and volumeter recordings were then noted when the bubble regained its equilibrium position. The volumeter had a capacity of $1 \cdot 5\ cm^3$.

In a series of experiments involving low boiling gases (He, Ne, Ar, Kr) and crystalline sorbents with frameworks of less open types (tridymite and cristobalite,[7] sodalite hydrate and cancrinite hydrate,[8] zeolites K-M and phillipsite[9] and stilbite and heulandite[10]) the crystals were charged with the sorbate at known high pressures and temperatures. After equilibrium was established, the crystals were first quenched, still at the high pressures, the gas sorbed at the high temperature being thereby "frozen" in the frameworks.

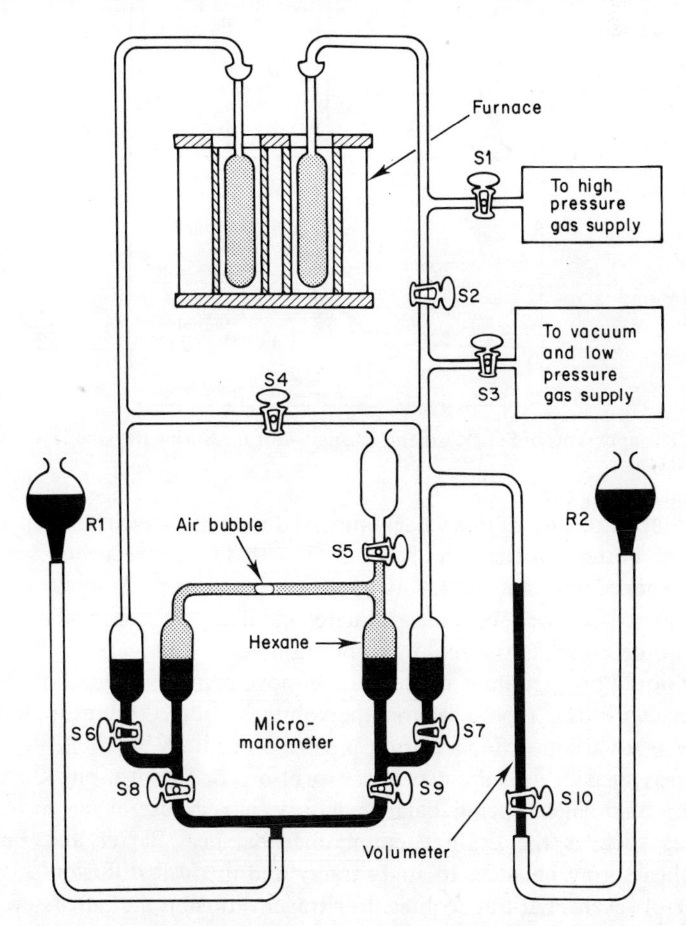

FIG. 1. The volumetric apparatus of Nelson and Walker[6] for studying desorption kinetics.

The pressures were then released, the crystals kept at 78 K were transferred, still in the steel sorption vessel, to a vacuum apparatus. The rates of release of trapped gas were subsequently measured after first raising the temperature of the sorbent to values at which convenient rates of desorption occurred.

Another volumetric procedure suitable for studying the kinetics of imbibition of liquids by zeolites was devised by Aleksandrov *et al.*[11], and was used also by Satterfield and Cheng.[12] The principle is as follows. A known weight of outgassed zeolite is contained in the vessel shown in Fig. 2. To start a run,

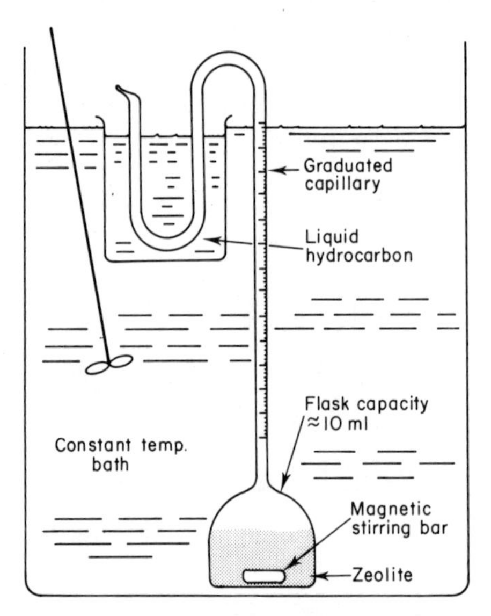

FIG. 2. The apparatus of Satterfield and Chang[12] for measuring the kinetic of sorption of liquids.

the U-shaped section of this vessel immersed in the hydrocarbon was snapped off at the bottom of the U and at time $t = 0$. The hydrocarbon was immediately sucked into the flask and graduated capillary. At precisely 0·1 or 0·2 minutes later the top of the graduated capillary was snapped off. As soon as the liquid contacts the zeolite, imbibition starts and so the volume of the system liquid and zeolite is reduced. The movement of the level of the liquid meniscus with time then measures the volumes imbibed after any time. This method is a variant of the sorption pipette procedure.[13]

In gravimetric measurements of sorption or desorption kinetics the McBain Bakr silica spring balance has been very useful, as have microbalances such as the Cahn vacuum microbalance. Barrer and Fender[14] used silica spring balances to study tracer and intrinsic diffusion of water in crystals of several natural zeolites. For tracer diffusion the outgassed crystals were allowed to sorb D_2O to equilibrium at a relative pressure p/p_0. The crystals were suspended from the springs in glass-cloth buckets. A stream of a non-sorbed gas carrying water vapour also at a partial pressure p/p_0 flowed at a regulated velocity past and through the crystals and the meshes of the glass cloth bucket containing the crystals. The exchange diffusion process was then followed through the decrease in weight of the crystals.

2.2 The Optical Procedure

Birefringence changes with amount sorbed were used by Tiselius[15, 16] to study diffusion of water in heulandite. Single crystals cut in the form of thin plates were employed for this purpose and examined under a polarization microscope. Under crossed nicols the progress of dark bands parallel to the crystal edges served to monitor the diffusion process. The laminated structure of heulandite is such that diffusion normal to the layers is negligible compared with diffusion parallel to the layers. For diffusion from the edge $x = 0$ of a plate and for the boundary conditions for the concentration C

$$C = C_0 \text{ at } x = 0 \tag{1}$$

$$C = C_1 \text{ for } x > 0 \text{ and } t = 0 \tag{2}$$

$$C = C_1 \text{ as } x \to \infty \text{ for all } t \tag{3}$$

the solution of the equation $\partial C/\partial t = D\partial^2 C/\partial x^2$ is

$$\frac{C_0 - C_x}{C_0 - C_1} = \operatorname{erf}[x/2(Dt)^{\frac{1}{2}}] \tag{4}$$

where C_x is the concentration at the plane x.
This equation provides three methods of finding D:

(i) measure x as a function of t for a constant known C
(ii) measure C as a function of t for a given x
(iii) measure C as a function of x for a given t.

Tiselius examined all three of these procedures. For $C_0 = 13{\cdot}45\%$ $C_1 = 19{\cdot}67\%$ and $C_x = 13{\cdot}87\%$ by weight x^2 was a linear function of t and D at 20°C in the 010 plane was $3{\cdot}4 \times 10^{-7}$ cm^{2}s^{-1}. From the curve of C_x as a function of t at constant x this value of D was found to be $3{\cdot}9 \times 10^{-7}$ cm^{2}s^{-1}. When C was plotted as a function of x at constant t the curve was not however that expected from eqn 4, so that D was a function of C, and the numerical values of D given above are mean values. For concentration dependent diffusion coefficients the Fick equation is

$$\frac{\partial C}{\partial t} = \frac{\partial}{\partial x}\left(D\frac{\partial C}{\partial x}\right). \tag{5}$$

From the concentration–distance curve at time t one may evaluate differential diffusion coefficients $\bar{D}(C)$ for each C using the Boltzmann-Matano method[17, 18]

$$\bar{D}_{C=C_x} = \frac{1}{2t}\frac{dx}{dC}\int_{C=C_x}^{C=C_0} x\,dC. \tag{6}$$

The integral is obtained graphically for each C_x from the curve of C vs x. Diffusion coefficients so obtained by Tiselius are given in Table 1a.

When the crystal plate of heulandite was cut with edges parallel to the 201 and the 001 faces the movement of the dark bands showed that the smallest rate of diffusion parallel with the 010 plane was normal to the 001 face. The diffusion anisotropy measured at 20°C on heulandites of different origins ranged from $\bar{D}_{201}/\bar{D}_{001} = 13\cdot6$ to 20. The corresponding activation energies were found to be 5·4 and $9\cdot1_4$ kcal mol^{-1}.

TABLE 1

(a) $\bar{D}(c)$ for H_2O at 20°C in heulandite[15] in 010 plane, normal to 201 face
$C_1 = 19\cdot67\%$ by weight

C (wt %)	$\bar{D} \times 10^7$ (cm^2 s^{-1}) $C_0 = 8\cdot3\%$	$\bar{D} \times 10^7$ (cm^2 s^{-1}) $C_0 = 13\cdot20\%$	$\bar{D} \times 10^7$ (cm^2 s^{-1}) $C_0 = 16\cdot20\%$
10	0·04[a]	—	—
11	0·2[b]	—	—
12	0·7	—	—
13	1·3	—	—
14	2·0	2·1	—
15	2·7	2·6	—
16	3·0	3·5	—
17	4·0	4·2	4·0
18	4·0	4·1	4·2
19	3·3	3·5	4·1

(b) $\tilde{D}$ for NH_3 in analcime,[16] for $C_0 = 0$

T (°C)	C_1 (cm^3 at s.t.p. g^{-1})	$\tilde{D} \times 10^8$ (cm^2 s^{-1})
302	27·5	1·20
323	18·6	1·68
351	10·7	3·15
389	5·3	5·04
412	3·5	$8\cdot6_7$
437	2·3	$11\cdot6_3$
478	1·3	19·4

[a] Very uncertain. [b] Uncertain.

Tiselius[16] extended the method to the diffusion of ammonia in analcime, using powder and also plates of analcime cut parallel to cube and to diagonal faces. Analcime is an isotropic crystal [19] and so, as expected, the value of the diffusivity was the same for each plate. The kinetics of uptake obeyed the $\sqrt{t}$ law well (uptake proportional to $\sqrt{t}$) and the integral diffusivity over the concentration ranges $C_0 = 0$ to C_1 where C_1 has the values given in Table 1(b). The isotherms obeyed Langmuir's equation, as might be expected for this zeolite, where sorption occurs into identical interstitial positions, each one capable of containing one ammonia molecule only. For a Langmuir isotherm and interstitial diffusion $\tilde{D}$ should be independent of $(C_1 - C_0)$, as discussed in § 6·4, so that $\tilde{D} = \bar{D} = D$. The heat of sorption was 16·6 kcal mol^{-1} and the energy of activation for diffusion was 11·5 kcal mol^{-1}.

2.3 NMR Measurements

Bloembergen *et al.*[20] showed that nuclear magnetic resonance relaxation times are often closely connected with the motions of the molecules which contain the nuclei. These nuclei have magnetic moments which result in a local magnetic field at the site of a given near-neighbour nucleus. The rotations or translations of the molecules cause the local magnetic field to fluctuate with time. This fluctuating field is able to bring about the transitions between the nuclear magnetic energy levels which must occur if the nuclear spin system is to relax towards thermal equilibrium within itself. If a system of nuclear magnets is at equilibrium in a constant magnetic field of strength H_0 the magnetization M_0 for the system of nuclear spins $hI/2\pi$ with magnetic moments $\gamma hI/2\pi$ is

$$M_0 = \frac{\gamma^2 h^2 I(I+1)H_0}{(2\pi)^2 3kT} \tag{7}$$

where γ is the nuclear gyromagnetic ratio.

The spin system is then perturbed so as to leave $M_x = M_0$ and $M_y, M_z = 0$ (as is often the case in practice). M_x, M_y and M_z are the components of M_0. The simplest expressions for the relaxation kinetics are[21]

$$\frac{M_0 - M_z}{M_0} = \exp(-t/T_1) \tag{8}$$

$$\frac{M_x}{M_0} = \cos(-\gamma H_0 t) \exp(-t/T_2) \tag{9}$$

$$\frac{M_y}{M_0} = \sin(-\gamma H_0 t) \exp(-t/T_2) \tag{10}$$

although the relaxations do not always follow these expressions. T_1 is termed the spin-lattice or longitudinal relaxation time since it refers to relaxation along the field. T_2 is the spin-spin or transverse relaxation time.

T_1 and T_2 have been determined as functions of temperature for various guest molecules sorbed in certain zeolites.[21, 22, 23] The variation of T_1 and T_2 for water in zeolite Na-X is shown in Fig. 3. There is a minimum in T_1

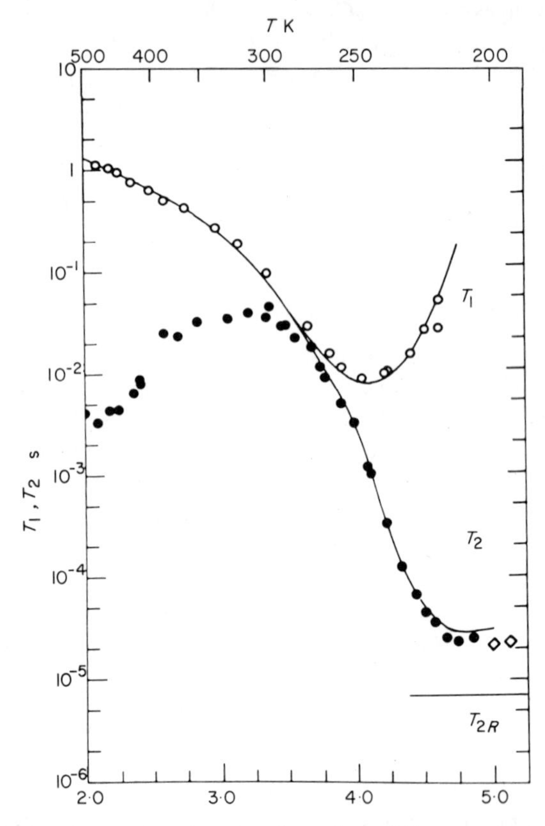

FIG. 3. Spin-lattice and spin-spin relaxation times, T_1 and T_2 respectively, plotted against reciprocal temperature for water in zeolite Na-X.[23] $\omega = 12$ MHz.

but not in T_2. The spin-lattice relaxation time, T_1, is the resultant of translational and rotational terms. If the times, τ, between translational and rotational jumps are comparable a single minimum occurs; if these are sufficiently different, two minima may be expected. For translational jumps at the minimum[21]

$$\gamma H_0 \tau \approx 0{\cdot}62 \tag{11}$$

so that τ can be determined. For the diffusion coefficient D in an isotropic medium it is often assumed that

$$D = l^2/6\tau \tag{12}$$

where l is the jump distance per unit translation. For water in Na-X Resing and Thompson[23] reported a jumping rate at room temperature which was about 100-fold smaller than for bulk water or for water in a charcoal with pores of ≈25Å. On the other hand, in the intrazeolitic fluid the jumps were at least 1000-fold more frequent than in ice.

For the H_2O– chabazite system Ducros[22] used the formula of Kubo and Tomita[24] to relate T_2 and τ:

$$\left(\frac{1}{T_2}\right)^2 = \left(\frac{1}{T_2}\right)^2_{rig} \frac{\pi}{2} \mathrm{Tan}^{-1}\left(\frac{\pi\tau}{4T_2 \ln 2}\right) \tag{13}$$

where $(T_2)_{rig}$ is the spin-spin relaxation time when $\tau \to \infty$. In this system τ so calculated obeyed the Arrhenius equation $\tau = \tau_0 \exp(-E/RT)$ with $\tau_0 = 5 \times 10^{-11}$ s and $E = 5{\cdot}8$ kcal. It is of interest that there was no sharp "freezing" of the water in chabazite at a particular temperature but that the intracrystalline fluid became increasingly and continuously more "viscous" as the temperature was reduced. This behaviour is probably typical of other sorbates in porous crystals. The correlation times, τ, referred to above are also probably mean values. There may be a spectrum of relaxation times for which the distribution function is $P(\tau)$. The more complex relations between τ and T_1 and T_2 have been considered for this situation by Resing[21] and by Resing and Thompson.[23]

Pulsed field gradient NMR has also been used to measure effective diffusion coefficients, D_{eff}, in beds of crystalline zeolite powder. The spin echo amplitude, $X(\sigma g)$ is given by[25]

$$\ln X(\sigma g) = \ln X(o) - \gamma^2\sigma^2 g^2 \Delta t \left\{ D^* + \frac{pD_g^*}{\gamma^2\sigma^2 g^2 \tau p D_g^* + 1} \right\} \tag{13a}$$

where γ is the gyromagnetic ratio; σ, g and Δt are respectively the width and amplitude of and the time interval between the gradient pulses; τ is the mean life time of molecules within the intracrystalline space; p is the fraction of the sorbate molecules in the intercrystalline space; and D^*, D_g^* are respectively the intra- and intercrystalline self-diffusion coefficients. The quantity in the curly bracket is D_{eff} and is $\approx \overline{d^2}/6\Delta t$ where $\overline{d^2}$ is the mean square displacement of a molecule within the bed of powder over the interval Δt. Two extreme situations then arise:

(i) $\tau \gg \Delta t$, i.e. $(\overline{d^2})^{1/2} \ll$ crystal radius. For this condition D_{eff} approaches D^*.

(ii) $\tau \ll \Delta T$, i.e. $(\overline{d^2})^{1/2} \gg$ crystal radius. In this case D_{eff} approaches pD_g^*.

Either extreme gives a linear plot of ln $X(\sigma g)$ against $\sigma^2 g^2$ at constant Δt and p. For the intermediate case when $\tau \approx \Delta t$ this plot becomes curved.

Examples of pulsed field gradient NMR studies include CH_4 in (Ca, Na)-*A* and water, n-C_4H_{10} and n-C_7H_{16} in zeolite Na-X.[25, 26] Under conditions considered to represent case (i) D_{eff} for water increased slightly when the fractional saturation of the crystals, θ, rose from 0·22 to 0·90. The activation energy was 4·5 kcal mol^{-1}. As regards the relative mobility of water in liquid water and in the zeolite the results are compatible with those of Resing and Thompson[23] already referred to. For n-C_4H_{10} and n-C_7H_{16} under the conditions of case (i) D_{eff} decreased with increasing chain length and with θ. At 20°C and $\theta = 0{\cdot}8$ D_{eff} was $\approx 1{\cdot}4 \times 10^{-5}$ and $\approx 0{\cdot}63 \times 10^{-5}$ cm^2 s^{-1} for n-butane and n-heptane respectively, and in the range $-90°$ to $+20$°C D_0 and E in the relation $D_{eff} = D_0 \exp -E/RT$ were

$$D_0\,(\text{cm}^2\text{s}^{-1})\text{: } 6 \times 10^{-4}\,(n\text{-}C_4);\ 15 \times 10^{-4}\,(n\text{-}C_7)$$

$$E\,(\text{kcal mol}^{-1})\text{: } 2{\cdot}2 \pm 0{\cdot}5\,(n\text{-}C_4);\ 3{\cdot}4 \pm 0{\cdot}5\,(n\text{-}C_7).$$

For this open zeolite the D_{eff} in Na-X near room temperature are similar to values expected for the bulk liquids.

2.4 Dielectric Relaxation

When a dielectric is polarized in a field, electronic polarization occurs in $\ll 10^{-13}$ s. The atomic nuclei also respond in times of $\approx 10^{-13}$ s. In a field which varies periodically with time, at frequencies less than 10^{11} cycles per second, the electronic and atomic polarizations are therefore always in phase with the field, and there corresponds with them a permittivity ϵ_∞. At very low frequencies, other polarization mechanisms may be involved and the permittivity approaches the static value $\epsilon_0 > \epsilon_\infty$. In zeolites, polarizations may include ion jumps or rotations of dipolar molecules, which may not be in phase with the field. One then has the complex frequency dependent permittivity

$$\epsilon = \epsilon' - i\epsilon''. \tag{14}$$

If polarization decays exponentially with time when the field is removed it can be shown that[27]

$$\epsilon = \epsilon_\infty + \frac{\epsilon_0 - \epsilon_\infty}{1 + i\omega T} \tag{15}$$

where the relaxation rate is such that polarization decays to e^{-1} of its original value in time T, and ω is the frequency of the field. Thus in the ϵ' ϵ'' plane the complex dielectric constant is a semicircle on the abscissa, which it intersects at ϵ_∞ and ϵ_0. The frequency at which ϵ'' has its maximum value is then $\omega = 1/T$.

When there is a spectrum of relaxation times, T, the complex permittivity can be formally expressed as

$$\epsilon = \epsilon_\infty + (\epsilon_0 - \epsilon_\infty) \int_0^\infty \frac{G(T_0)\mathrm{d}T}{1 + i\omega T} \tag{16}$$

where the function $G(T)$ gives the distribution of relaxation times, normalized so that $\int_0^\infty G(T)\mathrm{d}T = 1$. Figure 4[22] shows the plot of ϵ'' against ϵ' for a monocrystalline plate of chabazite cut parallel to the (100) plane. When there is such a spectrum of times T these plots can approximate to the arc of a circle. However Ducros reported that the curve in Fig. 4 was not exactly an

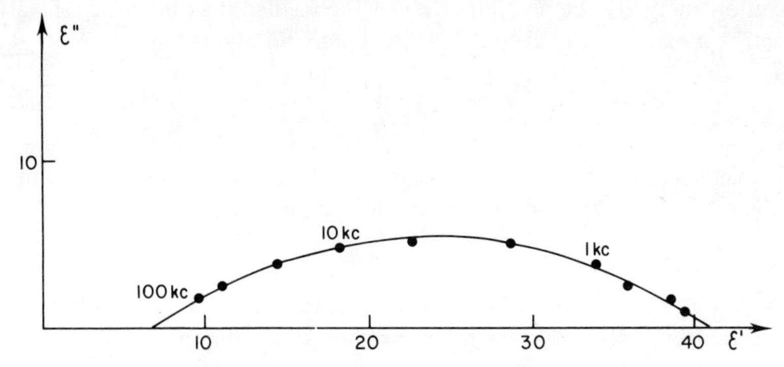

FIG. 4. Cole–Cole diagram at 46°C for a monocrystalline plate of chabazite[22] cut parallel with the plane (100).

arc, and also that it could not be represented by a Gaussian distribution for $G(T)$.

If the plot of ϵ'' against ϵ' is an arc, Cole and Cole[28] suggested in place of eqn 16 the semi-empirical expression

$$\epsilon = \epsilon_\infty + \frac{\epsilon_0 - \epsilon_\infty}{1 + (i\omega T)^{(1-\alpha)}} \tag{17}$$

where $\alpha < 1$. The points where the arc intersects the axis of ϵ' give ϵ_0 and ϵ_∞. The circle of which the arc forms part has its centre at $\left(\frac{(\epsilon_0 + \epsilon_\infty)}{2}, -\frac{(\epsilon_0 - \epsilon_\infty)}{2} \text{Cot}\,(1-\alpha)\frac{\pi}{2}\right)$ and radius $(\epsilon_0 - \epsilon_\infty)\,\text{Cosec}\,(1-\alpha)\frac{\pi}{2}$ so that α may be found. T_0 is a relaxation time, the value of which is found from $\omega = 1/T_0$ where ω is the frequency at which ϵ'' has its maximum value. The distribution function to which eqn 17 corresponds is $G(T) = \frac{1}{2\pi}\frac{\text{Sin}\,\alpha\pi}{\text{Cosh}\,(1-\alpha)s - \text{Cos}\,\alpha\pi}$ where $s = \ln T/T_0$. Other relationships analogous to eqn 17 have been suggested, for example by Davidson and Cole[29] and Fuoss and Kirkwood[30].

Even when a single T is involved it remains to decide what is the physical process to which it corresponds. In zeolites this may be a cation jump, or it may be a rotational jump of a sorbed dipolar molecule. It may also represent the Maxwell-Wagner relaxation of polarization either at the crystal-electrode interface or around the points of contact between crystal grains if the dielectric is (as is usual) a powder. The powder is normally compacted to reduce its porosity, but the compaction pressures needed to reduce porosity to low values, and so to increase the crystal-crystal contact area, may also damage the crystals and produce layers of amorphous material at the points where the pressure has ground the crystals together. In such layers, dielectric relaxations may also be possible. Debye[31] considered an intrinsic relaxation time, T^*, which he related to T by the expression

$$T = T^* \frac{(\epsilon_0 + 2)}{(\epsilon_\infty + 2)}. \tag{18}$$

For molecular rotations he then connected T^* with the inner friction hindering the rotation, so that for a pure liquid such as water

$$T^* = 4\pi\eta a^3/kT \tag{19}$$

where η is the viscosity coefficient at temperature T, and a the radius of the dipole molecule. His approach exemplifies one attempt to interpret the relaxation time T^* in terms of a transport property, η. In order for dielectric relaxation to give information about translational diffusion coefficients it would be necessary for jump times and energy barriers for rotation to be closely similar to those for translation. For this reason dielectric studies are of more direct use when considering translational jumps of ions than in the case of dipolar molecules such as water. For long chain molecules segmental rotations may result in changes in the position of the centre of gravity of the chain and hence can contribute more directly to translational diffusion.

2.5 Neutron Scattering Spectroscopy

Aims of work on radiation scattering are to obtain information about atomic structure on the one hand and about atomic and molecular motions in the scattering medium on the other. Neutron scattering provides such a method for studying molecular vibrational, rotational and diffusive motions on the 10^{-11} to 10^{-13} second time scale. In crystalline solids the structure and the thermal motions can be considered to be separated in the scattering effects they produce. For glasses and liquids this is not the case and a picture of the microscopic behaviour is needed to develop the theory. In the case of liquids several such models of molecular motion are available. In large cavity zeolites nearly saturated with sorbate the behaviour of the guest molecules can approximate that of dense fluids, while zeolites only lightly charged with sorbate might behave more like clathrates. Neutron scattering time-of-flight spectra have been recorded for instance for zeolite 3*A* containing H_2O, D_2O, NH_3, CH_3OH and CH_3CN.[32] These spectra can be divided into inelastic and quasi-elastic scattering regions. For neutrons with wave-lengths of 5·3Å these regions are compared in Fig. 5 with those for liquid water for the zeolite containing H_2O and NH_4^+.

The background theory relating these spectra to the momentum and energy transferred in scattering collisions is summarized in the paper of Egelstaff *et al.*[32] The time, τ, between jumps, the diffusion coefficient D, the momen-

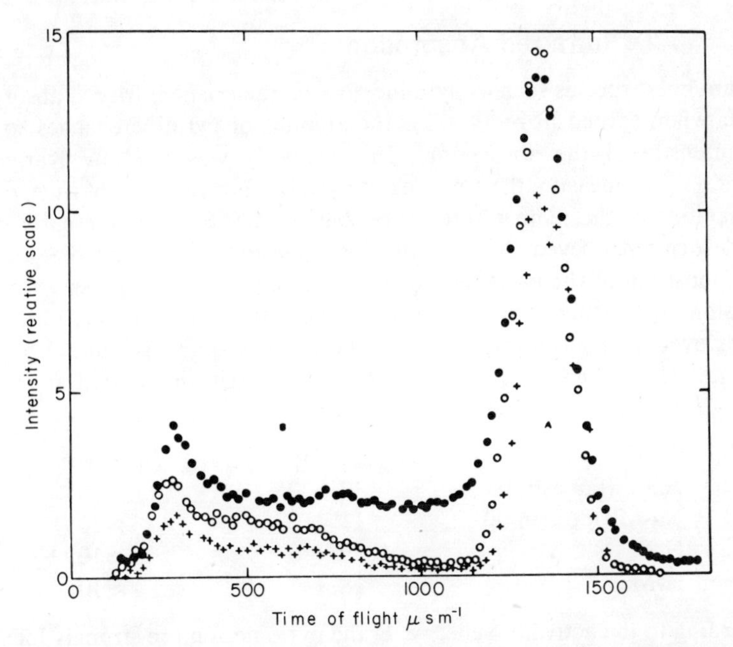

FIG. 5. Neutron and time-of-flight spectrum for the scattering of 5·3Å wavelength neutrons at 60° by zeolite K-*A* containing water, or NH_4-*A*, compared to similar data on liquid water.[32]
● = liquid water; ○ = NH_4^+ sieve; + = H_2O sieve.

tum transfer on collision Q, and the width ΔE of the quasi-elastic peak could be related by

$$\Delta E = \frac{h}{\tau\pi}\left[1 - \frac{1}{(1 + Q^2 D\tau)}\right]. \tag{20}$$

Values of τ and of $D = l^2/6\tau$ at 20°C and for $l = 3{\cdot}8$Å were estimated as:

CH_3CN	$\tau = 1{\cdot}5 \pm 0{\cdot}5 \times 10^{-11}$ s	$D \approx 2 \times 10^{-5}$ cm² s⁻¹
CH_3OH	$1{\cdot}0 \pm 0{\cdot}5 \times 10^{-11}$	$\approx 3 \times 10^{-5}$
NH_4^+	$1{\cdot}3 \pm 0{\cdot}4 \times 10^{-11}$	$\approx 2 \times 10^{-5}$.

It does not follow that these diffusion coefficients represent mass transport from one cavity to another. They may represent random motions of guest molecules, including rotations, within the cavities, the D for transfer between cavities then being smaller. In this connection it is of interest that at 35°C tracer diffusion coefficients of water in chabazite, gmelinite and heulandite(14) were in the range 10^{-8} to 10^{-7} $cm^2\ s^{-1}$, two orders of magnitude less than those given above.

2.6 Infra-red Absorption

Certain guest species such as pyridine give characteristic infra-red absorption bands when sorbed in zeolites. As the amount of pyridine changes so does the intensity of the absorption. This property was used by Kárge and Klose(33,34) to measure the amounts of pyridine as a function of time, during desorptions in the temperature range 250° to 400°C. Over the early stages the desorption obeyed the $\sqrt{t}$ diffusion equation, suggesting that thermal decomposition of the pyridine was not the rate process being measured. The diffusion coefficients derived from the $\sqrt{t}$ diffusion equation (§ 3) are mean values over the concentration interval between zero and the initial pyridine content. The values of D_0 and of E in the Arrhenius equation $D = D_0 \exp(-E/RT)$ for several zeolites are given below:

Zeolite	E (kcal mol^{-1})	D_0 ($cm^2\ s^{-1}$)
Na-Y (faujasite)	12 ± 2	$\approx 10^{-7}$
Na-M (mordenite)	11 ± 2	$\approx 10^{-6}$
H-Y	29 ± 3	$\approx 10^{-3}$
H-M	18 ± 3	$\approx 10^{-5}$.

According to the activation energy, E, the pyridine is more strongly localized in the hydrogen zeolites than in the Na-forms, possibly because with the hydrogen zeolites pyridine form pyridinium ions by proton transfer.

This survey of some methods of examining molecular mobilities in zeolite sorbents shows that the mobility is not always related to mass transport between cavities. It may instead relate to rotations or random motions within a cavity. This may be true of the results obtained with such indirect methods as NMR relaxation, dielectric relaxation and neutron scattering spectroscopy. However it is not the case with the optical procedure used by Tiselius, the infra-red absorption method of Karge and Klose or the pulsed field gradient NMR method.

3. INTRACRYSTALLINE DIFFUSIVITIES: SINGLE DIFFUSANT

In a two-component solution of species A and B, five diffusion coefficients have been considered:(35) an interdiffusion coefficient, D_{AB}; two intrinsic diffusion coefficients, D_A and D_B; and two tracer diffusion coefficients D_A^* and D_B^*. If component B is a zeolite, D_B and D_B^* are effectively zero because the crystals do not swell appreciably as sorption occurs and so the lattice-forming units are immobile. In this situation $D_{AB} = D_A$ and one need consider only D_A and D_A^*. The relation between them is often assumed to be given by the Darken relation

$$D = D^* \frac{\partial \ln a}{\partial \ln C} \tag{21}$$

where a is the activity of diffusant at a point where its concentration is C. The subscript "A" has been omitted from eqn 21. However the Darken relation is incomplete, the full expression being[36,37]

$$D = D^* \frac{\partial \ln a}{\partial \ln C} \frac{1}{\left(1 - \dfrac{CL_{AA^*}}{C^*L_{AA}}\right)} \tag{22}$$

where L_{AA^*} is the cross-coefficient between untraced and traced molecules of A and L_{AA} is the straight coefficient, in the irreversible thermodynamic formulation of diffusion. C^* is the concentration of isotopically labelled A at a point where the concentration of this untraced A is C. $(C + C^*)$ is assumed constant throughout the crystal and D in eqn 22 is defined as

$$D = L_{AA}RT \frac{\partial \ln a}{\partial C} \tag{23}$$

Barrer and Fender[14,38] studied sorption and intrinsic and tracer diffusion of water in chabazite, heulandite and gmelinite using heavy water as tracer. The activity correction

$$\frac{\partial \ln a}{\partial \ln C} = \frac{\partial \ln p}{\partial \ln C} \tag{24}$$

was derived from the slopes of the water sorption isotherms. Their data were later used to estimate the term CL_{AA^*}/C^*L_{AA}, with the results shown in Table 2.[39] These results suggest that, although the correction to Darken's relation arising from this term is here much smaller than $\partial \ln a/\partial \ln C$, the correction is not negligible. Similar conclusions were reached from intrinsic and tracer diffusion coefficients for surface flow of SO_2 through microporous carbon compacts.[36] Cross-coefficients L_{AA^*} must involve direct interaction between A and A^* and should be significant whenever sorbate concentrations are sufficient for collisions between A and A^* to become frequent.

When the diffusion coefficient for a single component is a function of concentration, one often needs to consider separately integral and differential

TABLE 2
Relation between D^* and D for water in several zeolites near saturation of the zeolite[14]

Zeolite	T°C	$D^* \times 10^8$ cm^2 s^{-1}	$D \times 10^6$ cm^2 s^{-1}	$\frac{\partial \ln a}{\partial \ln C}$	$\frac{CL_{AA*}}{C^*L_{AA}}$
Chabazite B	75	46	10·7	23·0	0·09$_1$
	65	32	7·6	24·0	0·15
	55	21	5·5	25·5	0·17
	45	14	3·8	27·0	0·21
	35	9·0	2·5	28·0	0·27
Heulandite	75	9·8	3·0	30·0	0·39
	65	6·1	2·0	32·0	0·44
	55	3·7	1·26	34·0	0·48
	45	2·2	0·78	35·5	0·51
	35	1·24	0·47	37·0	0·56
Gmelinite	55	7·3	2·0	26·5	0·15
	45	5·0	1·40	28·0	0·06$_5$
	35	3·3	0·97	29·5	0·030

coefficients which can be denoted by $\tilde{D}$ and $\bar{D}$ respectively and may be related by

$$\tilde{D} = \frac{1}{C}\int_0^C \bar{D}(C)\,\mathrm{d}C. \tag{25}$$

For constant diffusion coefficients $\tilde{D} = \bar{D} = D$.

4. INTERPRETATION OF SORPTION KINETICS

Sorption kinetics are usually presented as plots of $(Q_t - Q_0)/(Q_\infty - Q_0)$ against time t or against $\sqrt{t}$ where Q_t, Q_0 and Q_∞ are the amounts sorbed at time t, when $t = 0$ and at equilibrium $(t \rightarrow \infty)$ respectively. From these curves, one seeks to obtain rate coefficients which characterize the relaxation of the system towards equilibrium, such as the diffusion coefficient D. Some examples of the procedures involved follow.

4.1 Single Diffusant, D is a Constant

Most diffusion studies made in this regime have involved zeolite powders and weakly sorbed gases. Sorption occurs either at constant pressure, or at constant volume with variable pressure. Suppose sorption takes place into a powder of isotropic crystals which can be approximated as spheres all with the same radius r_0. The diffusion equation $\partial C/\partial t = D$ div grad C has to be solved for the boundary conditions

$$\left.\begin{aligned} C &= C_\infty \text{ at } r = r_0 \text{ for } t > 0 \\ C &= C_0 \text{ for } 0 \le r < r_0 \text{ at } t = 0 \end{aligned}\right\}. \tag{26}$$

The solution is

$$\frac{M_t}{M_\infty} = \frac{Q_t - Q_0}{Q_\infty - Q_0} = 1 - \frac{6}{\pi^2} \sum_{n=1}^{\infty} \frac{1}{n^2} \exp\left(-\frac{Dn^2\pi^2 t}{r_0^2}\right) \tag{27}$$

or alternatively

$$\frac{M_t}{M_\infty} = \frac{6}{r_0}\left(\frac{Dt}{\pi}\right)^{1/2} \left\{1 + 2\pi^{1/2} \sum_{n=1}^{\infty} \text{ierfc} \frac{nr_0}{\sqrt{Dt}}\right\} - \frac{3Dt}{r_0^2}. \tag{28}$$

When t is large, eqn 27 approaches the expression

$$\ln \frac{Q_\infty - Q_t}{Q_\infty - Q_0} = \ln (6/\pi^2) - \frac{D\pi^2 t}{r_0^2} \tag{29}$$

while eqn 28 for small t approaches the expression

$$\frac{M_t}{M_\infty} = \frac{6}{r_0}\left(\frac{Dt}{\pi}\right)^{1/2} = \frac{2A}{V}\left(\frac{Dt}{\pi}\right)^{1/2} \tag{30}$$

where A is the total external surface and V the total volume of the crystallites.

Alternatively, sorption occurs under constant volume variable pressure conditions. At $t = 0$ an extra amount of gas is introduced into the gas volume and allowed to equilibrate with the sorbent. Then when Henry's law governs the equilibrium

$$\frac{M_t}{M_\infty} = 1 - 6K(K+1) \sum_{\alpha_n} \exp\left(-\frac{\alpha_n^2 Dt}{r_0^2}\right) \Big/ [9(K+1) + \alpha_n^2 K^2] \tag{31}$$

where K = ratio of amount of sorbate in the gas phase to amount in the sorbent at equilibrium and α_n is the nth positive root of

$$\tan \alpha = 3\alpha/(3 + K\alpha^2). \tag{32}$$

An alternative solution, suitable for short times[40] is

$$\left.\begin{aligned} \frac{M_t}{M_\infty} = (1+K)\Bigg[1 &- \left(\frac{\gamma_1}{\gamma_1+\gamma_2}\right) \text{e erfc}\left\{\frac{3\gamma_1}{\text{K}}\left(\frac{Dt}{r_0^2}\right)^{1/2}\right\} \\ &- \left(\frac{\gamma_2}{\gamma_1+\gamma_2}\right) \text{e erfc}\left\{-\frac{3\gamma_2}{K}\left(\frac{Dt}{r_0^2}\right)^{1/2}\right\}\Bigg] \\ &+ \text{higher terms} \end{aligned}\right\} \tag{33}$$

Here e erfc Z $\equiv$ exp Z^2 erfc Z; $\gamma_2 = (\gamma_1 - 1)$ and $\gamma_1 = \frac{1}{2}\left\{\left(1 + \frac{4K}{3}\right)^{1/2} + 1\right\}$

For large times eqn 31 approaches

$$\ln\left(\frac{Q_\infty - Q_t}{Q_\infty - Q_0}\right) = \ln\left[\frac{6K(K+1)}{9(K+1) + \alpha_1^2 K^2}\right] - \frac{\alpha_1 Dt}{r_0^2} \tag{34}$$

while for small enough t eqn 33 reduces to

$$\frac{M_t}{M_\infty} = \frac{2A}{V}\frac{(1+K)}{K}\left(\frac{Dt}{\pi}\right)^{1/2} \tag{35}$$

which differs from eqn 30 by the factor $(1 + K)/K$.

For rectangular parallelepipeds with edges $2a$, $2b$ and $2c$ the expressions corresponding with eqns 29 and 30 are respectively

$$\ln\left(\frac{Q_\infty - Q_t}{Q_\infty - Q_0}\right) = \ln\frac{512}{\pi^6} - \frac{D\pi^2 t}{4}\left(\frac{1}{a^2} + \frac{1}{b^2} + \frac{1}{c^2}\right) \tag{36}$$

and

$$\frac{M_t}{M_\infty} = 2\left(\frac{Dt}{\pi}\right)^{1/2}\left(\frac{1}{a} + \frac{1}{b} + \frac{1}{c}\right) = \frac{2A}{V}\left(\frac{Dt}{\pi}\right)^{1/2}. \tag{37}$$

Equations to parallel those for spheres and parallelepipeds already given are also available for cylinders.[41] It appears from a comparison of the $\sqrt{t}$ laws (e.g. eqns 30 and 37) that the limiting slopes of plots of M_t/M_∞ vs $\sqrt{t}$ are expressible as slope $= \frac{2A}{V}\left(\frac{D}{\pi}\right)^{1/2}$, whatever the shape and size distributions of the component crystals. This is one of the very useful properties of the $\sqrt{t}$ law, which is not shown by the long-time approximations (eqns 29, 34 or 36), or by the full equations (eqns 27, 28 and 31). The slopes of the $\sqrt{t}$ law plots then serve to give D, once A and V have been separately evaluated.

It is seen from eqns 29, 34 and 36 that for large enough times, the plot of $\ln\left(\frac{Q_\infty - Q_t}{Q_\infty - Q_0}\right)$ against t approaches a straight line, the negative slope of which is proportional to D. That is, at large t, diffusion kinetics for sorption into crystallites all of one shape and size approach more and more nearly the form of first order reaction kinetics. The respective slopes ($-D\pi^2/r_0^2$ for spheres at constant pressure; $-D\alpha_1^2/r_0^2$ for spheres at constant volume, variable pressure, and for Henry's law; $-\frac{D\pi^2}{4}\left(\frac{1}{a^2} + \frac{1}{b^2} + \frac{1}{c^2}\right)$ for rectangular parallelepipeds at constant pressure) then serve to evaluate D if the crystal shape and size is known.

Finally, one may obtain D by using the eqn 27 or 28 (constant pressure) or eqn 31 or 33 (constant volume, variable pressure and Henry's law).

M_t/M_∞ is calculated in terms of, and plotted against the variable $Z = Dt/r_0^2$. The experimental curve of M_t/M_∞ against t is then adjusted by a change in the scale of t until it coincides with the calculated curve. The factor by which the time scale must be adjusted to bring the two curves into coincidence is D/r_0^2.

It is not usually possible to obtain crystallites all of one shape and size, although uniformity can be improved by fractionation using a sedimentation column. A distribution of sizes among the crystallites will distort the experimental plots of M_t/M_∞ against t as compared with the plot for uniform crystallites. M_t/M_∞ will initially increase more rapidly (an effect due to the smaller than average crystallites) and finally will increase less rapidly (because of the larger than average crystallites), than with crystallites all of the same size. This situation was considered by Ruthven and Loughlin[42] in a study of sorption of n-butane in cubic crystallites of zeolite Ca-A. For sorption at constant pressure in crystallites of one size

$$\frac{M_t}{M_\infty} = 1 - \sum_{l=1}^{\infty}\sum_{m=1}^{\infty}\sum_{n=1}^{\infty} \frac{512}{\pi^6} \frac{\exp\left[-\dfrac{\pi^2 Dt}{4a^2}\{(2l-1)^2+(2m-1)^2+(2n-1)^2\}\right]}{(2l-1)^2(2m-1)^2(2n-1)^2} \tag{38}$$

where $2a$ is the cube edge. The diffusion coefficient of n-butane is expected to be concentration dependent, but over small increments of uptake D can be assumed constant. Then for a normal distribution of crystal sizes the contributions of each size group may be summed and so eqn 38 was replaced by

$$\frac{M_t}{M_\infty} = 1 - \frac{512S}{\pi^6\sqrt{\pi}} \sum_{l=1}^{\infty}\sum_{m=1}^{\infty}\sum_{n=1}^{\infty} \int_0^{\infty} \frac{\exp\left[-(\tfrac{1}{2})S^2(y-1) - (\pi^2 Dt/4y^2\mu^2)\{(2l-1)^2+(2m-1)^2+(2n-1)^2\}\right]\mathrm{d}y}{(2l-1)^2(2m-1)^2(2n-1)^2} \tag{39}$$

where $S = \mu/\sigma$; σ = standard deviation of the cube half edge; μ = weight mean cube half side; y = dummy variable. For any given value of S the sorption curve of M_t/M_∞ against Dt/μ^2 may be calculated from eqn 39. The value of D/μ^2 can then be found by matching the experimental data to the appropriate theoretical curve.

Examples of sorption kinetics are illustrated in Fig. 6 for water diffusing in chabazite, heulandite and gmelinite under constant pressure conditions.[14] Figure 7 gives corresponding kinetic curves for Ne and Kr in several exchange forms of mordenite for constant volume variable pressure conditions when sorption was governed approximately by Henry's law.[43] Both these figures show the initial validity of the $\sqrt{t}$ law.

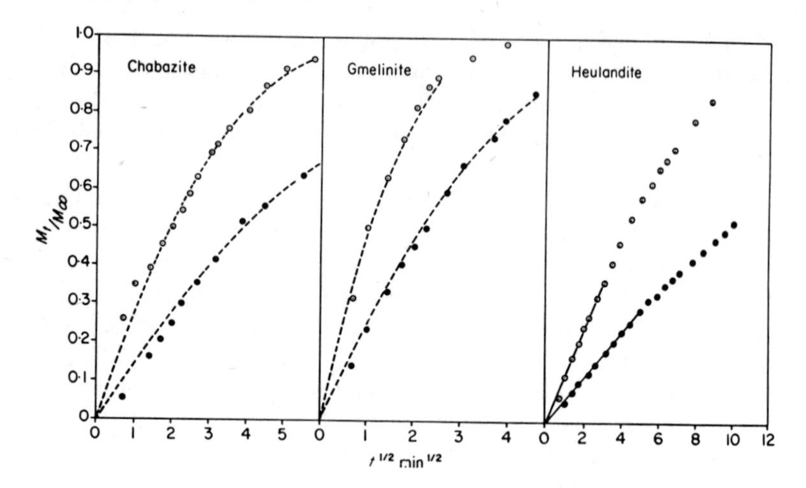

FIG. 6. Kinetics of diffusion of water at constant pressure into chabazite, gmelinite and heulandite.[14] The dashed lines are calculated curves. The experimental points are as follows: Chabazite: ⊙, 75·4°C; ●, 30·8°C. Gmelinite: ⊙, 62·5°C, ●, 31·7°C. Heulandite: ⊙, 77·8°C, ●, 37·4°C.

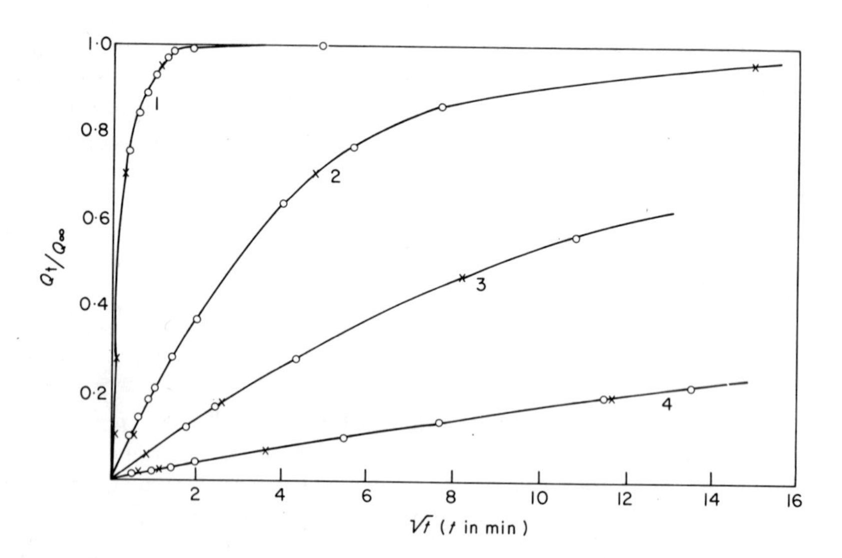

FIG. 7. Kinetics of diffusion of gases into mordenites and levynite at constant volume, variable pressure, when sorption is near the Henry's law range.[43]

Curve 1: Ne in Li-mordenite at −185°C; 1/K = 2·50, $(D/r_0^2)^{1/2}$ = 0·285
Curve 2: Ne in Ca-mordenite at −185°C; 1/K = 2·54, $(D/r_0^2)^{1/2}$ = 0·0214
Curve 3: Kr in Ba-mordenite at 24°C; 1/K = 0·86, $(D/r_0^2)^{1/2}$ = 0·0122
Curve 4: Kr in levynite at 0°C; 1/K = 0·96, $(D/r_0^2)^{1/2}$ = 0·00273
⊙ = experimental points; × = points calculated from eqn 13a of ref. 43.

4.2 Diffusion Coefficients from $\int_0^\infty \left(1 - \frac{M_t}{M_\infty}\right) dt$

The usual plots of M_t/M_∞ against t approach the value one for large enough values of t. In such a curve as that shown in Fig. 8a the cross-hatched area up to time t is

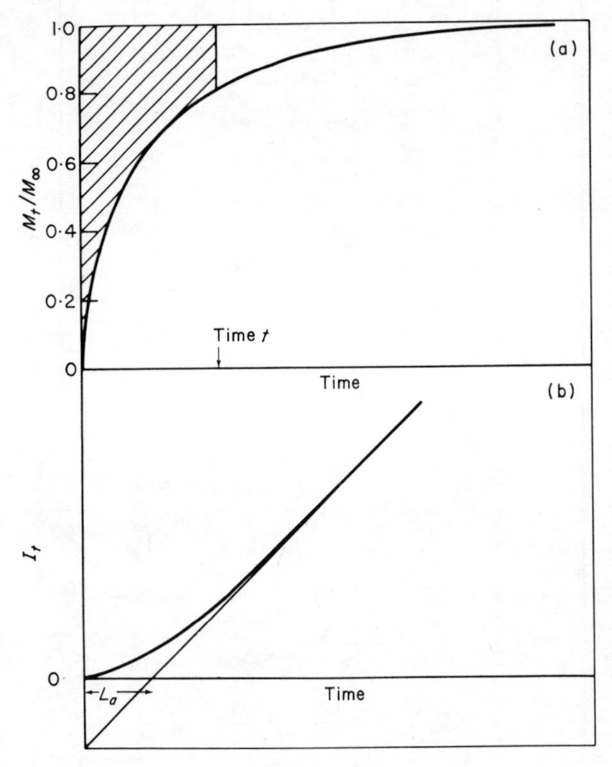

FIG. 8(a) Plot of M_t/M_∞ against t showing cross-hatched the area I_t of eqn. 40. (b) Plot of I_t against t, showing the intercept L_a on the axis of t.

$$I_t = \int_0^t \left(1 - \frac{M_t}{M_\infty}\right) dt = t - \int_0^t \frac{M_t}{M_\infty} dt. \tag{40}$$

Accordingly if I_t is plotted against t, a straight line of unit slope is approached asymptotically, the intercepts on the axis of t or of I_t being equal. This intercept, L_a, is the total cross-hatched area when $t \to \infty$. As M_t/M_∞ is available as a function of time for various situations, rather simple and explicit expressions can be found for measuring D.[44] These are illustrated in Table 3. In these expressions, V is the volume of the sorbent and A its external area. The expressions apply when D is constant and when all sorbent

TABLE 3

The total areas of Fig. 8a, or intercepts L_a of Fig. 8b for different geometries[44]

Diffusion Medium	L_a	
	Constant pressure sorption	Constant volume, variable pressure, Henry's law
Sphere (radius b)	$\frac{b^2}{15D} = \left(\frac{V}{A}\right)^2 \frac{3}{5D}$	$\frac{b^2}{15D} \cdot \frac{K}{(K+1)} = \left(\frac{V}{A}\right)^2 \frac{3}{5D} \frac{K}{(K+1)}$
Cylinder (radius b)	$\frac{b^2}{8D} = \left(\frac{V}{A}\right)^2 \frac{1}{2D}$	$\frac{b^2}{8D} \cdot \frac{K}{(K+1)} = \left(\frac{V}{A}\right)^2 \frac{1}{2D} \frac{K}{(K+1)}$
Sheet (thickness $2l$)	$\frac{l^2}{3D} = \left(\frac{V}{A}\right)^2 \frac{1}{3D}$	$\frac{l^2}{3D} \cdot \frac{K}{(K+1)} = \left(\frac{V}{A}\right)^2 \frac{1}{3D} \frac{K}{(K+1)}$
Cube (edge $2a$)	$\frac{4a^2}{D\pi^2}\left(\frac{8}{\pi^2}\right)^3 \sum_{l=0}^{\infty} \sum_{m=0}^{\infty} \sum_{n=0}^{\infty} \frac{1}{(2l+1)^2(2m+1)^2(2n+1)^2[(2l+1)^2+(2m+1)^2+(2n+1)^2]}$	
Rectangular parallelepiped (edges $2a$, $2b$ and $2c$)	$\frac{4}{D\pi^2}\left(\frac{8}{\pi^2}\right)^3 \sum_{l=0}^{\infty} \sum_{m=0}^{\infty} \sum_{n=0}^{\infty} \frac{1}{(2l+1)^2(2m+1)^2(2n+1)^2\left[\frac{(2l+1)^2}{a^2}+\frac{(2m+1)^2}{b^2}+\frac{(2n+1)^2}{c^2}\right]}$	

crystallites in a given sample of the sorbent have the same shape and size. If D is a function of concentration, by carrying out the sorptions at constant pressure and in small increments, it will be satisfactory to assume that D is constant over each small concentration interval, so that values of D for each concentration could be obtained. Over bigger concentration intervals D determined from L_a will be a mean or integral value.

A bed of zeolite powder provides both inter- and intracrystalline pore systems. In this case the method of moments gives for the initially sorbate-free bed:[44a]

$$\left.\begin{aligned} E_{w_0} &= \int_0^\infty t\,\frac{\mathrm{d}(M_t/M_\infty)}{\mathrm{d}t}\,.\,\mathrm{d}t \\ E_{w_0} &= \frac{l^2}{pD_g\nu(\nu+2)} + E_{w_1} \end{aligned}\right\} \tag{40a}$$

where E_{w_1} is the part of E_{w_0} which arises from intracrystalline diffusion, p is the fraction of the sorbate molecules present in the intercrystalline space, in which the diffusion coefficient is D_g. ν is 1, 2 or 3 respectively, for the powder compact in the form of sheet, cylinder and sphere. l is a characteristic dimension which for the sheet denotes half the thickness. If E_{w_0} is plotted against l^2 for constant p a line of slope $1/(pD_g\nu(\nu+2))$ and intercept E_{w_1} should result, which will show the relative roles of inter- and intracrystalline flows, and in principle can determine each. However, this description is over-simple because p will be an average throughout the bed and will be a function of time. The nearest approach to constant p would be found for the interval method where the interval $(Q_\infty - Q_0)$ is small compared with Q_0, the amount of sorbate initially taken up by the zeolite.

4.3 Energy of Activation for Diffusion

Diffusion processes within porous crystals normally obey the Arrhenius equation $D = D_0 \exp(-E/RT)$. If D is determined over a range of temperatures then E and D_0 are readily calculated. However E can be found without prior determination of D. If one considers the plots of M_t/M_∞ against time at each of two different temperatures T_1 and T_2, and measures the times t_1 and t_2 taken for M_t/M_∞ to reach the same value at each of the two temperatures then one must have

$$D_1t_1 = D_2t_2. \tag{41}$$

Accordingly

$$\ln t_1/t_2 = \ln D_2/D_1 = \frac{E}{R}\left[\frac{1}{T_1} - \frac{1}{T_2}\right]. \tag{42}$$

This method is applicable for any distribution of crystal shapes and sizes.

4.4 Diffusion Coefficient depends on Diffusant Concentration

When the diffusion coefficient is a function of concentration it is not possible to obtain solutions of the kinds discussed in § 4.1. The Boltzmann-Matano method referred to in § 2.2 allows one to evaluate differential diffusion coefficients from concentration profiles, but these can be found only for single crystals. Usually powders of sorbent crystallites have to be considered. For these powders perhaps the simplest procedure is the interval method, in which, after loading the crystals to the level Q_0 a little extra diffusant is introduced to give Q_∞ where $(Q_\infty - Q_0)$ is sufficiently small to take D as a constant in this interval. The procedure is repeated for each of a series of different Q_0 values, to give approximate differential diffusion coefficients as functions of Q_0 (or C_0). For the proper use of the interval method the extra increment should be added at constant pressure. This condition has not always been observed. If the isotherm is rectangular and the pressure alters even slightly $\partial \ln p/\partial \ln C$ near the ingoing surfaces may change substantially and since this is also the activity correction, $\partial \ln a/\partial \ln C$, the assumed constancy of D could be significantly in error.

From experimental observations on sorption kinetics, it has been found that even when D is a function of concentration the $\sqrt{t}$ law is usually valid for small t, in the form

$$\frac{M_t}{M_\infty} = k\sqrt{t}. \tag{43}$$

The validity of eqn 43, and the impossibility of using the interval method under constant pressure conditions for rectangular isotherms where even at exceedingly small pressures the crystals are almost saturated, led Barrer and Clarke[45] to consider the $\sqrt{t}$ region, and to modify the interval method. In a study of the sorption of n-C_4, n-C_6 and n-C_9 paraffins by zeolites (Ca, Na)-A they kept the pressure of sorbate constant by having a reservoir of the liquid paraffin suitably thermostated. Q_0 was changed systematically from 0 to values approaching Q_∞; and Q_∞, by keeping constant the pressure of sorbate vapour, was always the same, and because of the rectangular form of the isotherms, was near to saturation of the zeolite. The initial validity of the $\sqrt{t}$ law is shown in Fig. 9 for typical sorption kinetic runs.

The rate constant, k, in eqn 43 can very well characterize the kinetics. If $J(t)$ is the total flux through the surfaces of all the crystals then

$$(Q_t - Q_0) = \int_0^t J(t)\,\mathrm{d}t \tag{44}$$

so that

$$\frac{\mathrm{d}}{\mathrm{d}t^{\frac{1}{2}}}\left(\frac{M_t}{M_\infty}\right) = \frac{2J(t)t^{1/2}}{Q_\infty - Q_0} = k\,. \tag{45}$$

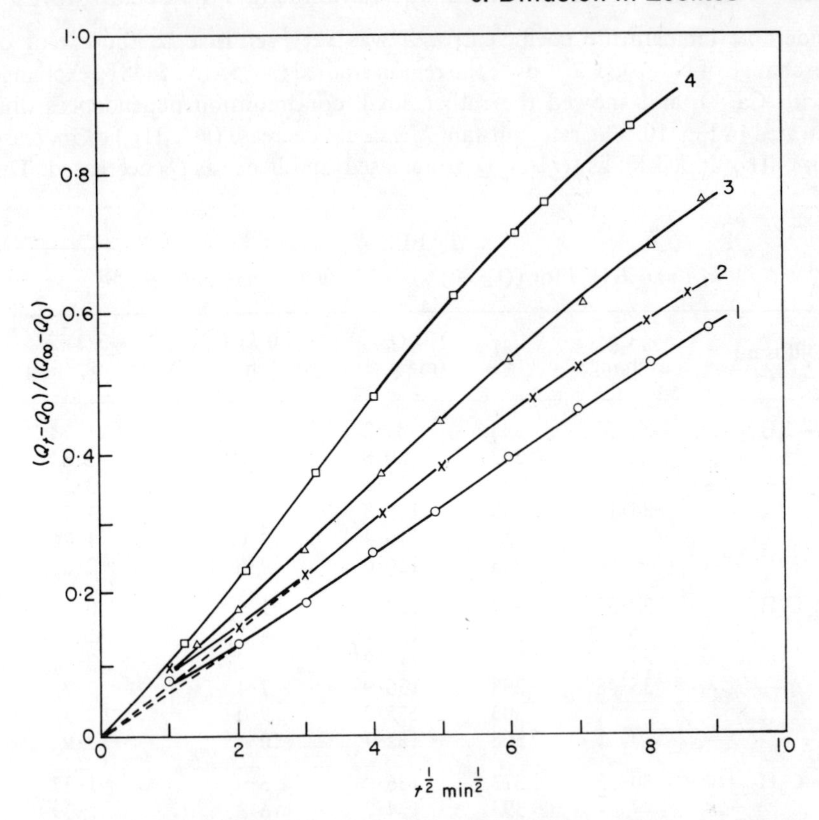

FIG. 9. Plots of $M_t/M_\infty = (Q_t - Q_0)/(Q_\infty - Q_0)$ against $\sqrt{t}$ for n-C_6H_{14} in 30·55% Ca-exchanged zeolite (Ca, Na)-*A* at constant pressure and at 348 K.[45]

$Q_\infty = 120{\cdot}05$ mg g^{-1}
Curve 1: $Q_0 = 0$
Curve 2: $Q_0 = 48{\cdot}08$ mg g^{-1}
Curve 3: $Q_0 = 63{\cdot}37$ mg g^{-1}
Curve 4: $Q_0 = 96{\cdot}99$ mg g^{-1}

The total flux is therefore proportional to $(Q_\infty - Q_0)$ and to $t^{-1/2}$ in the range of validity of the $\sqrt{t}$ law, and at $t = 1$,

$$2J(1)/(Q_\infty - Q_0) = k. \tag{46}$$

If both sides of eqn 45 or 46 are divided by the total external surface area, A, then k/A is a characteristic of the system which is independent of the state of subdivision of the crystals and of their shape or size distributions.

Some values of k/A and of $J(1)/A$ are given in Table 4, for n-paraffins in zeolites (Ca, Na)-A. The ratio $J(1)/A$ is the flux per m^2 of surface one minute after starting the sorption run. At constant temperature and for a chosen sorbent (e.g. 30·55% Ca-exchanged zeolite A at 373 K) k/A and $J(1)/A$ decrease with increasing chain length of n-paraffin. Each also has a considerable temperature coefficient, k/A was very sensitive to the extent of exchange of Na^+ by Ca^{2+} over the region studied (30·55 to 35·48% exchange with Ca^{2+}), and showed the rather small concentration dependences illustrated in Fig. 10. The rate constant k/A could decrease (n-C_4H_{10}) or increase (n-C_9H_{20} at 303 K) as $(Q_\infty - Q_0)$ increased and hence as Q_0 decreased. The

TABLE 4

k/A and $J(1)/A$ for $(Q_\infty - Q_0) = 75$ mg g^{-1} in (Ca, Na)-A[45]

Diffusant	% Ca^{2+} exchange	T/K	Q_∞ (mg g^{-1})	$10^2k/A$ (mg m^{-2}min$^{-1/2}$)	$J(1)/A$ (mg m^{-2}min^{-1})
n-C_4H_{10}	30·55	348	87·2	4·6$_4$	1·74
		363	80·8	7·0$_0$	2·63
		373	73·5	8·8$_6$	3·3$_2$
	34·10	248	116·8	3·4$_4$	1·29
		258	116·4	4·6$_8$	1·76
		273	116·0	7·6	2·8$_5$
n-C_6H_{14}	30·55	323	122·5	1·7$_2$	0·65
		348	120·1	4·1$_4$	1·5$_5$
		373	116·9	8·5$_0$	3·1$_9$
	35·48	298	136·9	7·4	2·7$_6$
		303	135·2	9·0	3·3$_8$
		308	132·7	10·4	3·9$_0$
n-C_9H_{20}	30·55	373	136·0	3·6$_4$	1·37
		393	134·2	6·3$_6$	2·39
		408	132·7	9·5	3·5$_2$
	35·48	343	133·3	5·2$_0$	1·9$_5$

concentration dependences were influenced by the temperature of the experiment as well as by the % exchange with Ca^{2+} and by the hydrocarbon studied.

4.5 The Coefficients $\bar{D}$, $\tilde{D}_1$ and $\tilde{D}_2$ for n-paraffins in (Ca, Na)–A

To obtain diffusion coefficients from the values of k one may refer to the relation $M_t/M_\infty = \frac{2A}{V}\left(\frac{Dt}{\pi}\right)^{1/2}$ for constant pressure sorption at small times. This relation is appropriate when D is constant, and gives $k = \frac{2A}{V}\left(\frac{D}{\pi}\right)^{1/2}$. The simplest assumption to make when D depends on concentration is that

$$k = \frac{2A}{V}\left(\frac{\tilde{D}_1}{\pi}\right)^{1/2} \tag{47}$$

where $\tilde{D}_1$ is a mean or integral diffusion coefficient over the interval $(Q_\infty - Q_0)$ and is related to the differential diffusion coefficient, $\bar{D}$, by

$$\tilde{D}_1(Q_\infty - Q_0) = \int_{Q_0}^{Q_\infty} \bar{D}\,\mathrm{d}Q. \tag{48}$$

From eqn 48 $\bar{D}$ is also related to $\tilde{D}_1$ by

$$\bar{D} = -(Q_\infty - Q_0)\left(\frac{\partial \tilde{D}_1}{\partial Q_0}\right)_{Q_\infty} + \tilde{D}_1 \tag{49}$$

in accordance with the experimental conditions of Barrer and Clarke[45] (Q_∞ constant; Q_0 varied). Accordingly from k, $\tilde{D}_1$ was derived and from plots of $\tilde{D}_1$ against Q_0 $\bar{D}$ was obtained. In terms of $\bar{D}$ two other integral diffusion coefficients, $\tilde{D}_2$ or $\tilde{D}_3$ can be defined as

$$\left.\begin{aligned} \tilde{D}_2 Q_0 &= \int_0^{Q_0} \bar{D}\,\mathrm{d}Q \\ \tilde{D}_3 Q_\infty &= \int_0^{Q_\infty} \bar{D}\,\mathrm{d}Q \end{aligned}\right\}. \tag{50}$$

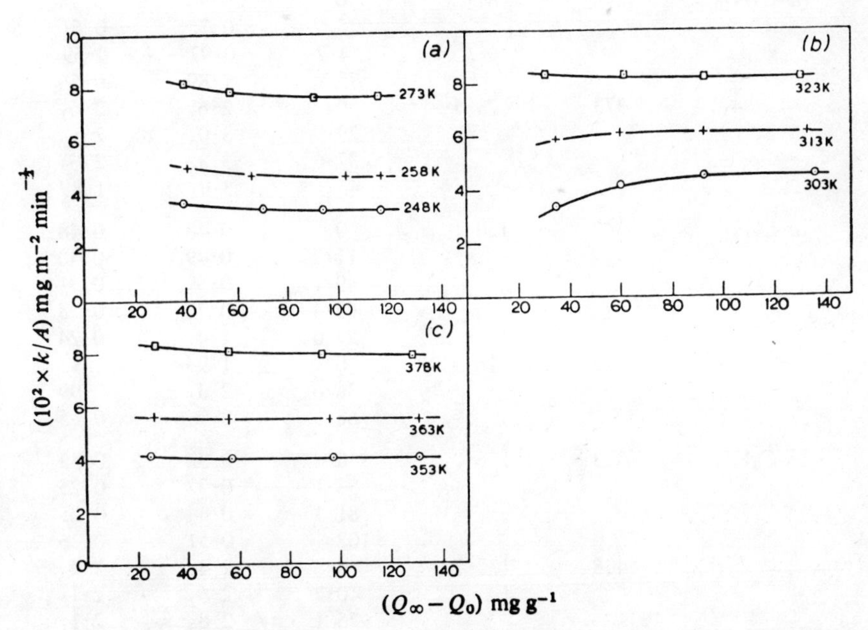

FIG. 10. Examples of plots of k/A, from $M_t/M_\infty = k\sqrt{t}$, as functions of $(Q_\infty - Q_0)$,[45] in 34·10% Ca-exchanged (Ca, Na)-A: (a) n-C_4H_{10}; (b) n-C_6H_{14}; (c) n-C_9H_{20}.

Then $\tilde{D}_1 = \tilde{D}_3$ when $Q_0 = 0$ and $\tilde{D}_2 = \tilde{D}_3$ when $Q_0 = Q_\infty$. From these several relations one finds

$$\tilde{D}_2 = Q_\infty/Q_0[(\tilde{D}_1)_{Q_0=0} - \tilde{D}_1] + \tilde{D}_1. \tag{51}$$

Table 5 gives some values of $\tilde{D}_1$ and $\tilde{D}_2$ for C_4, C_6 and C_9 *n*-paraffins in 30·55% Ca-exchanged (Ca, Na)-*A* at each of two temperatures and Table 6 gives values of $\bar{D}$ for *n*-C_4H_{10} in several (Ca, Na)-*A* zeolites. In most cases $\tilde{D}_1$ increases as $(Q_\infty - Q_0)$ decreases (i.e. as Q_0 increases). On the other hand $\tilde{D}_2$ behaves like $\bar{D}$ in that it decreases with decreasing $(Q_\infty - Q_0)$ (i.e. as Q_0 increases). The concentration dependences of $\tilde{D}_1$ and $\tilde{D}_2$ thus depend on the way they are defined. The values of $\tilde{D}_1$ and $\tilde{D}_2$ both fall with increasing chain length of the *n*-paraffin diffusant. The values of $\tilde{D}_1$ are proportional to k^2 (eqn 47). Therefore if one writes $k = k_0 \exp(-E/RT)$ the activation energy for $\tilde{D}_1$ will be $2E$. Table 7 contains activation energies

TABLE 5

Values of $\tilde{D}_1$ and $\tilde{D}_2$ in zeolite *A* 30·55% exchanged by Ca^{2+} [45]

Hydrocarbon	T/K	Q_∞ (mg g^{-1})	Q_0 (mg g^{-1})	$10^{17}\tilde{D}_1$ m^2 s^{-1}	$10^{17}\tilde{D}_2$ m^2 s^{-1}
n–C_4H_{10}	348	87·22	0	0·74	0·74
			25	0·75	0·61
			54·2	0·97	0·59
			75	1·89	0·56
	373	73·5	0	$2{\cdot}6_7$	$2{\cdot}6_7$
			29·9	$3{\cdot}0_5$	$2{\cdot}1_2$
			37·6	$3{\cdot}3_5$	$2{\cdot}0_2$
			45	$3{\cdot}9_1$	1·89
n–C_6H_{14}	348	120·1	0	0·48	0·48
			11·7	0·49	0·40
			49·1	0·59	0·31
			63·4	0·70	0·28
			97·0	1·47	0·24
	373	116·9	0	1·93	1·93
			36·6	$2{\cdot}3_3$	0·99
			68·5	$3{\cdot}9_8$	0·55
n–C_9H_{20}	373	136·0	0	0·32	0·32
			53·1	0·37	0·25
			81·1	0·44	0·25
			102	0·51	0·26
	408	132·7	0	$2{\cdot}4_7$	$2{\cdot}4_7$
			40·2	$2{\cdot}6_2$	$2{\cdot}1_2$
			75·1	$2{\cdot}8_0$	$2{\cdot}1_1$
			96·5	$2{\cdot}9_4$	$2{\cdot}2_0$

TABLE 6

$\bar{D}$ for n-C_4H_{10} in 30·55, 32·54 and 34·10% Ca-exchanged zeolite A[45]

% Ca^{2+}	T/K	Q_∞ (mg g^{-1})	Q_0 (mg g^{-1})	$10^{17}\bar{D}$ (m^2 s^{-1})
30·55	348	$87{\cdot}2_2$	0	0·74
			5	0·74
			20	0·65
			40	0·55
			60	0·44
			70	0·40
32·54	323	$106{\cdot}5_3$	0	$1{\cdot}8_2$
			5	$1{\cdot}7_5$
			20	$1{\cdot}5_2$
			40	1·42
			60	1·29
			70	1·30
34·10	273	$116{\cdot}0_3$	0	$1{\cdot}9_9$
			5	$1{\cdot}9_9$
			20	$1{\cdot}9_9$
			40	$1{\cdot}9_4$
			60	$1{\cdot}7_9$
			70	$1{\cdot}5_8$

for k and also gives these energies and the pre-exponential constant D_0 in the expression $\bar{D} = D_0 \exp(-E/RT)$. If some fluctuations among the values of E and D_0, attributed to experimental uncertainties, are discounted, the following trends appear:

(i) E decreases as the % Ca^{2+} increases;
(ii) E increases with the chain length of the n-paraffins;
(iii) D_0 increases as E increases (see also Table 8).

The third of these trends indicates that the numerical values of the heat and entropy of activation, ΔH^* and ΔS^*, change in such a way as partially to compensate in their influence upon D and hence upon the free energy of activation $\Delta G^* = \Delta H^* - T\Delta S^*$.

When considering diffusion in certain polymer-penetrant systems it was shown empirically[46] that, for $\bar{D}$ increasing with concentration, $\tilde{D}_2$ and $\bar{D}$ are better related by the empirical relations

$$\tilde{D}_2 = \frac{1{\cdot}67}{Q_0^{1{\cdot}67}} \int_0^{Q_0} Q^{0{\cdot}67} \bar{D} \mathrm{d}Q \tag{52}$$

TABLE 7

Activation energies (kJ mol^{-1}) for k and $\bar{D}$ and coefficients D_0 (m^{2}s^{-1}) in $\bar{D} = D_0 \exp(-E/RT)$[45]

Diffusant	$(Q_\infty - Q_0)$	% Ca-exchanged (Ca, Na)-A			
	(mg g^{-1})	30·55	32·5	34·10	35·48
(a) E in $k = k_0 \exp(-E/RT)$ at constant $(Q_\infty - Q_0)$					
n–C_4H_{10}	25	$27{\cdot}_2$	$21{\cdot}_0$	—	—
	50	$27{\cdot}_9$	$24{\cdot}_2$	$18{\cdot}_0$	—
	75	$28{\cdot}_1$	$23{\cdot}_8$	$17{\cdot}_9$	—
	100	—	$23{\cdot}_6$	$18{\cdot}_1$	—
n–C_6H_{14}	50	$35{\cdot}_5$	$32{\cdot}_6$	30·7	$32{\cdot}_8$
	75	$32{\cdot}_2$	$31{\cdot}_3$	$26{\cdot}_1$	$26{\cdot}_7$
	100	$31{\cdot}_3$	$30{\cdot}_5$	$24{\cdot}_6$	$22{\cdot}_8$
	125	—	$30{\cdot}_2$	$24{\cdot}_4$	$20{\cdot}_6$
n–C_9H_{20}	25	—	$33{\cdot}_4$	$30{\cdot}_7$	$32{\cdot}_0$
	50	$32{\cdot}_9$	$33{\cdot}_6$	$31{\cdot}_3$	$30{\cdot}_7$
	75	$34{\cdot}_9$	$34{\cdot}_0$	$30{\cdot}_9$	$29{\cdot}_5$
	100	$35{\cdot}_6$	$33{\cdot}_8$	$30{\cdot}_7$	$28{\cdot}_8$
	125	$36{\cdot}_1$	$33{\cdot}_6$	$30{\cdot}_5$	$27{\cdot}_3$
(b) E in $\bar{D} = D_0 \exp(-E/RT)$ for $Q_0 = 0$					
n–C_4H_{10}		$40{\cdot}_7$	$47{\cdot}_3$	$36{\cdot}_4$	—
n–C_6H_{14}		$61{\cdot}_1$	$58{\cdot}_2$	$48{\cdot}_5$	$38{\cdot}_9$
n–C_9H_{20}		$74{\cdot}_5$	$67{\cdot}_4$	$60{\cdot}_7$	$53{\cdot}_1$
(c) D_0 in $\bar{D} = D_0 \exp(-E/RT)$ for $Q_0 = 0$					
n–C_4H_{10}		$1{\cdot}8\times10^{-9}$	$7{\cdot}3\times10^{-10}$	$3{\cdot}3\times10^{-10}$	—
n–C_6H_{14}		$6{\cdot}1\times10^{-9}$	$1{\cdot}3\times10^{-8}$	$1{\cdot}5\times10^{-9}$	$2{\cdot}1\times10^{-11}$
n–C_9H_{20}		$3{\cdot}8\times10^{-8}$	$2{\cdot}3\times10^{-8}$	$4{\cdot}8\times10^{-9}$	$9{\cdot}2\times10^{-10}$

than by eqn 50 for adsorption kinetics; and by

$$\tilde{D}_2 = \frac{1{\cdot}85}{Q_0^{1{\cdot}58}} \int_0^{Q_0} (Q_0 - Q)^{0{\cdot}85} \bar{D} \mathrm{d}Q \tag{53}$$

for desorption. When D decreases with concentration 1·67 and 0·67 are respectively replaced in eqn 52 by 1·85 and 0·85; and 1·85 and 0·85 in eqn 53 are replaced by 1·67 and 0·67. It is possible, if not proven, that like relations are satisfactory for diffusion in zeolites. In polymers it has also been found that when $\bar{D}$ is a function of concentration the curves of M_t/M_∞ vs t do not coincide for sorption and desorption runs. It has been suggested that a better approximation to $\tilde{D}_1$ could be

$$\tfrac{1}{2}(k_{\text{ads}} + k_{\text{des}}) = \frac{2A}{V}\left(\frac{\tilde{D}_1}{\pi}\right)^{1/2}. \tag{54}$$

In zeolites the evidence about differences in diffusivities derived from sorption and desorption kinetics is conflicting. For *n*-paraffins in zeolite Ca-*A* no evidence of differences was found.[47,41] On the other hand Satterfield and Frabetti[48] reported for C_1 to C_4 paraffins in single crystals of mordenite that diffusivities for desorption were from 3- to 60-fold smaller than for sorption. Similarly for C_7 and C_8 *n*-paraffins in erionite at 93°C Eberly[49] reported desorption diffusivities 10- to 70-fold less than the corresponding adsorption diffusivities. With sorption in zeolites, complicating factors may occur which are not apparent for diffusion in polymers. Satterfield *et al.*[50] reported that the rate of desorption of cumene from H-mordenite at 25°C decreased by two orders of magnitude as the time of the previous saturation was increased to six days. The decrease was attributed to the slow formation within the zeolite of radical ions and diisopropylbenzene, which are less mobile than and impede the diffusion of cumene. Lack of reproducibility in sorption and desorption runs may also arise from the deposition of strongly sorbed impurity molecules or from different amounts of residual immobile water molecules.[50] Such species may cause large changes in diffusivities as considered more quantitatively in § 7.

5. DETERMINATION OF THE AREA *A*

The BET adsorption method can sometimes be applied directly for measuring the external area of zeolite powders, by using a sorbate molecule too large to enter the crystals. Neo-pentane or isobutane serve for zeolites such as chabazite, zeolite *A* or zeolite *ZK-5*, for which the channel openings are controlled by 8-ring windows, because they are too large to pass through these windows. For still more compact zeolites (analcime, the natrolite group, sodalite and cancrinite hydrates) krypton or nitrogen may also be used, because at the temperatures involved (78 or 90K) neither molecule will penetrate the crystals.

For very open zeolites such as faujasites and zeolites *L*, Ω and offretite, even neo-pentane rapidly penetrates the crystals. However, by first filling the intracrystalline pore volume with a strongly sorbed molecule such as water which is immobile and has no measurable equilibrium pressure at the subsequent experimental temperature (78 or 90K); one may again use a gas such as krypton to evaluate the external area, *A*.

Another method can be used whenever intracrystalline sorption is slow.[43] The amount sorbed, Q_t, may be plotted against $\sqrt{t}$. These plots are initially linear, as considered in §§ 4.1 and 4.4 and are readily extrapolated to zero time. At appropriate temperatures this extrapolation gives positive intercepts, which measure the quantity, $F(p_0)$, very rapidly adsorbed upon external surfaces of the crystallites, at the initial pressure p_0. Accordingly from a succession of kinetic runs, each on initially sorbate-free crystals, in which p_0 is systematically increased, the uptake, $F(p_0)$, can be evaluated as a function of p_0. The BET procedure then serves to find the area A. In a constant pressure sorption run the total uptake $Q_t' = F(p_0) + Q_t$ so that the measured quantity Q_t'/Q_∞ is given by

$$\frac{Q_t'}{Q_\infty} = \frac{F(p_0)}{Q_\infty} + \frac{Q_t}{Q_\infty} = \frac{F(p_0)}{Q_\infty} + \frac{2A}{V}\left(\frac{Dt}{\pi}\right)^{1/2}. \tag{55}$$

Thus, the slope still has the value it would have in the absence of external adsorption. Since A has been found and since V can also be found from the weight and density of the crystallites, D can be determined.

Under constant volume, variable pressure conditions near the Henry's law region one can find the isotherm as before. Hence the amount adsorbed externally can be obtained as a function of time and subtracted from the measured Q_t' to give Q_t. The plot of Q_t/Q_∞ then serves, through eqn 35, to give D.

Flow of a gas or liquid through a column of powder of known extra-crystalline pore volume, ϵ, is often used to evaluate surface areas.[51] The steady state flow rate gives the permeability, from which a smoothed external area is derived. This area does not take account of the fine-scale roughness, cracks, and blind pore character of the column of powder, and can be expected to give only a lower limit to the total external area. When intra-crystalline diffusion is occurring in a subsequent sorption kinetics experiment, as the sorbate penetrates more and more deeply into each crystallite the influence of fine scale roughnesses will appear less and less in the contours of equal concentration. It is thus possible that the smoothed area will become more appropriate in the later stages. The use of both values of A would give upper and lower bounds to the value of D. However these aspects have not been studied experimentally.

6. FACTORS INFLUENCING INTRACRYSTALLINE DIFFUSION

In §§ 2 to 4, the emphasis has been upon experimental methods and their limitations, and upon the interpretation of sorption kinetics. At the same time some properties of diffusivities, D, and of the rate coefficients k (eqn 43) have also been illustrated. Further consideration will be given to some of the variables which influence intracrystalline diffusion. These include

1. intracrystalline channel geometry and dimensions;
2. shape, size and polarity of the diffusing molecules;
3. cation distributions, size, charge and number;

4. concentration of diffusant within the crystals;
5. temperature;
6. lattice defects such as stacking faults;
7. presence of impurity molecules in the diffusion pathways;
8. structural changes brought about by penetrants;
9. structural changes associated with physical and chemical treatments.

The first three of these factors are closely interconnected. They largely determine the total or partial molecule sieving property of zeolites. In total molecular sieving, one component of a mixture, having molecules of the wrong shape and size to enter the crystals, is excluded; while another, somewhat smaller or of the correct shape, is freely absorbed. The crystal surface

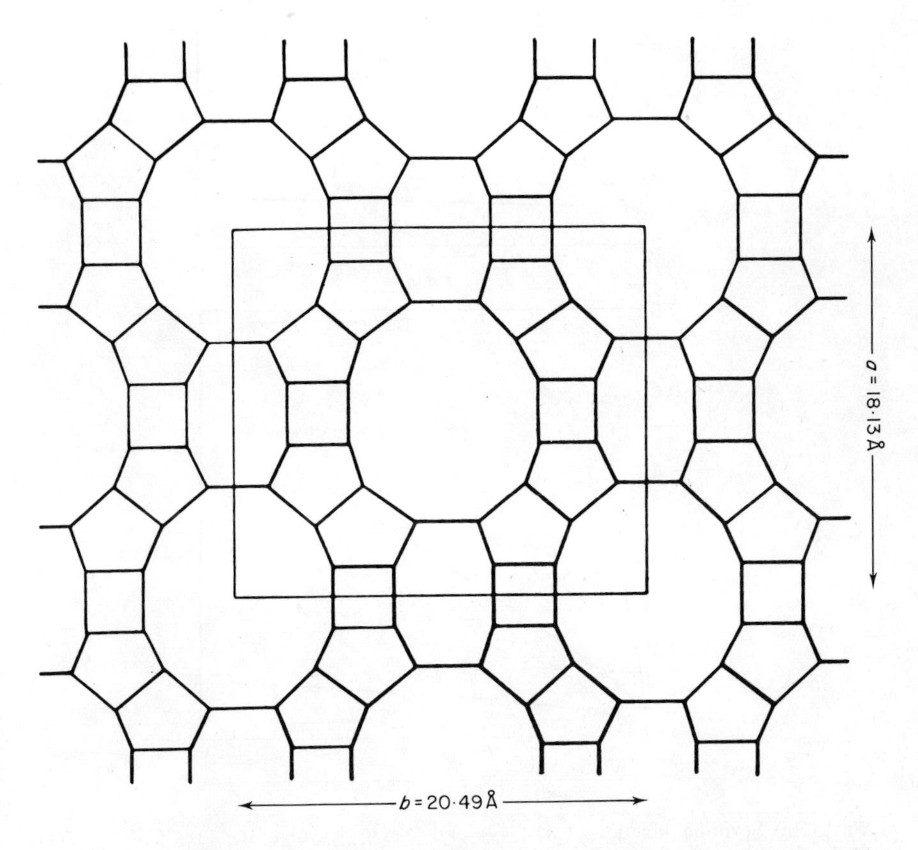

FIG. 11. The framework structure of mordenite in cross-section normal to the wide channels. Al or Si atoms are at each corner and oxygens near the mid point of each edge.[52] The cations are not shown.

functions like a sieve in turning back the first kind of molecule but allowing passage of the second. Figure 11 illustrates the sieve-like aspect of the structure of mordenite.[52]

6.1 Molecule and Channel Dimensions

On the basis of calibrations by molecules of known dimensions, the first classification of three kinds of molecular sieve was made,[53,54] and later extended to five.[55] It is always possible however to have gradations between the different types of sieve, or to convert one type to another as a result of

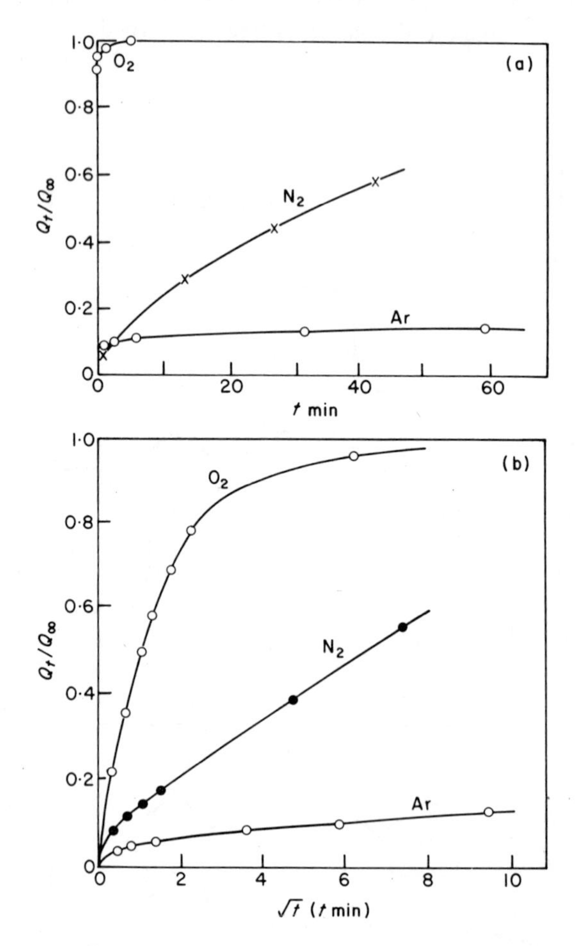

FIG. 12(a) Sorption kinetics of O_2, N_2 and Ar in levynite at $-184°C$.[57] For O_2 $Q_\infty = 10{\cdot}02$; for N_2, $Q_\infty = 9{\cdot}77$ and for Ar, $Q_\infty = 10{\cdot}01$, all in cm^3 at s.t.p. g^{-1}.

(b) Sorption kinetics of N_2, Ar and O_2 in Ca-mordenite at $-78°C$.[43]

modification by ion-exchange and other physico-chemical treatments, or by accelerating the diffusion by suitably raising temperatures. At very low temperatures, so sensitive can the sieving process become, that dramatic differences in sorption rates occur between such pairs of molecules as Ar and O_2, Ar and N_2 and even O_2 and N_2. The van der Waals dimensions of these molecules are:[56]

$$\text{Ar} \approx 3{\cdot}8_3\text{Å}; \ N_2 \approx 3{\cdot}0 \times 4{\cdot}1\text{Å}; \ O_2 \approx 2{\cdot}8 \times 3{\cdot}9\text{Å}.$$

Figure 12a shows the different rates of sorption of these three gases in levynite at −184°C[57] and Fig. 12b shows the differences in Ca-mordenite at −78°C.[43] One of the first large scale applications of molecular sieve sorbents was the use of zeolite Na-*A* in the separation of Ar and O_2.[58]

The interrelation between the critical van der Waals dimension of diffusant molecules and the energy of activation for diffusion in a series of different porous crystals is shown in Fig. 13.[39] In α-tridymite and sodalite hydrate, diffusion is governed at least in part by windows with free diameters of

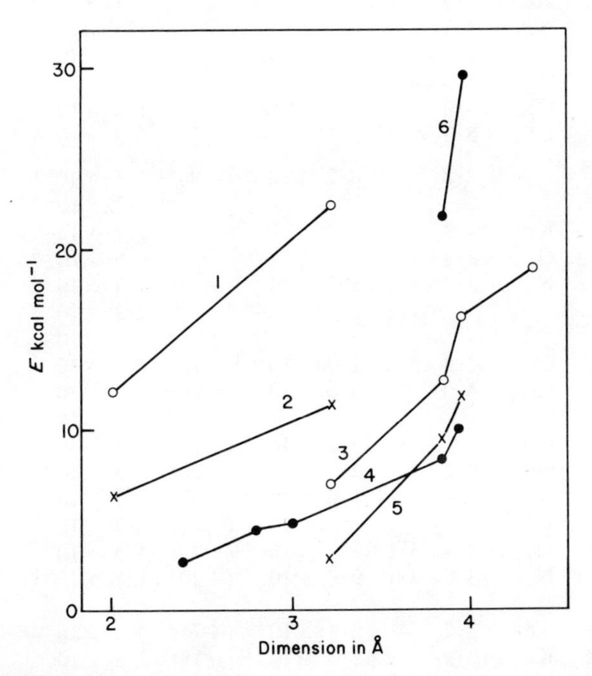

FIG. 13. Dependence of E in $D = D_0 \exp(-E/RT)$ in some porous crystals upon the critical dimension of the diffusing molecule, and upon the open-ness of the porous crystal.[39]
1: tridymite 2: silica glass 3: Zeolite K-*A* 4: K-mordenite 5: Levynite 6: Sodalite hydrate
Reading from left to right the diffusing molecules are: He, H_2, O_2, N_2, Ne, Ar, Kr and Xe.

≈2·2 to 2·6Å. In tridymite the energy barriers are considerable even for helium and neon (line 1). These barriers for argon and krypton in basic sodalite are shown in line 6. In silica glass the network is irregular and some more open diffusion paths result in barriers lower than those in tridymite (line 2). For the potassium form of zeolite *A* the 8-ring windows giving access to the large 26-hedral cavities have free diameters of ±4·3 Å but are partially obstructed by K^+ ions, giving the energy barriers of line 3. Those of line 4

TABLE 8

Relation between diffusivity and molecular dimensions

Zeolite	Mole-cule	Dimen-sions (Å)	D or D^* cm^2s^{-1} (and T°C)	D or $D^* = D_0 \exp(-E/RT)$ D_0 ($cm^2\,s^{-1}$)	 E (kcal mol^{-1})
Ca-*A*[59, 60]	Ar	$3{\cdot}8_3$	—	$\approx 1\times10^{-9}$	<1[a]
	Kr	$3{\cdot}9_4$	—	$7{\cdot}7\times10^{-9}$	2·0[a]
	Xe	$4{\cdot}3_7$	—	$1{\cdot}5\times10^{-9}$	3·0[a]
	O_2	2·8×3·9	—	$2{\cdot}6\times10^{-9}$	1·0[a]
	N_2	3·0×4·1	—	$5{\cdot}2\times10^{-9}$	1·5[a]
	CH_4	4·0	—	$7{\cdot}2\times10^{-8}$	3·0[a]
	CF_4	$5{\cdot}3_3$	—	$2{\cdot}5\times10^{-6}$	$9{\cdot}1_5$[a]
Na-*A*[59, 61, 62]	Ar	$3{\cdot}8_3$	$4{\cdot}2\times10^{-12}(-90)$	$1{\cdot}6\times10^{-6}$	4·7
			—	$1{\cdot}22\times10^{-6}$	5·8[a]
	Kr	$3{\cdot}9_4$	—	$9{\cdot}7\times10^{-8}$	8·1[a]
	O_2	2·8×3·9	—	$6{\cdot}6\times10^{-6}$	4·5[a]
	N_2	3·0×4·1	$5{\cdot}7\times10^{-12}(-70)$	$4\ 3\times10^{-3}$	8·3
			—	$2{\cdot}3\times10^{-9}$	4·1
			—	$9{\cdot}6\times10^{-7}$	6·1[a]
	CO	2·8×5·2	$1{\cdot}04\times10^{-11}(-70)$	$4{\cdot}7\times10^{-5}$	6·2
	CH_4	4·0	$5{\cdot}0\times10^{-12}(-10)$	$7{\cdot}9\times10^{-7}$	6·3
			—	$5{\cdot}8\times10^{-8}$	7·4
	C_2H_6		$3{\cdot}9\times10^{-12}(+10)$	$1{\cdot}1\times10^{-5}$	7·8
K-*A*[63]	Ne	3·2	$2{\cdot}9\times10^{-13}(+20)$	$4{\cdot}8\times10^{-8}$	7·0
	Ar	$3{\cdot}8_3$	$1{\cdot}6\times10^{-17}(+20)$	$3{\cdot}7\times10^{-8}$	12·6
	Kr	$3{\cdot}9_4$	$1{\cdot}8\times10^{-18}(+20)$	$6{\cdot}7\times10^{-7}$	16·4
	H_2	2·4×3·1	$2{\cdot}0\times10^{-13}(+20)$	$4{\cdot}5\times10^{-6}$	9·9
	N_2	3·0×4·1	$9{\cdot}8\times10^{-18}(+20)$	$1{\cdot}5\times10^{-5}$	16·2
K-mor-denite[43]	Ar	$3{\cdot}8_3$	$2{\cdot}4\times10^{-16}(-78)$	$5{\cdot}6\times10^{-7}$	8·4
	Kr	$3{\cdot}9_4$	$1{\cdot}8\times10^{-18}(-78)$	$2{\cdot}5_6\times10^{-7}$	10·0
	H_2	2·4×3·1	$2{\cdot}7\times10^{-13}(-78)$	$1{\cdot}64\times10^{-10}$	2·5
	O_2	2·8×3·9	$2{\cdot}0\times10^{-15}(-78)$	$1{\cdot}59\times10^{-10}$	4·4
	N_2	3·0×4·1	$9{\cdot}2\times10^{-16}(-78)$	$2{\cdot}0\times10^{-10}$	4·8

[a] These values of E and D_0 are for the diffusivity D^* in $D = D^* \dfrac{\partial \ln p}{\partial \ln C}$

refer to K-mordenite and in line 5 to levynite. Each framework topology contributes its specific resistance to diffusion, a resistance often strongly moderated by the cations present, and always very sensitive to the dimensions of the diffusing molecules.

The relation between molecular dimensions of the diffusant and its diffusion coefficient in a given porous crystal is illustrated in Table 8. For the linear molecules (H_2, O_2, CO_2, C_2H_6) the length of the molecule and the cross-sectional diameter are given. Normally it is the cross-sectional diameter which is critical in governing passage through the narrow points of the diffusion paths. The diffusivities for K-*A* and for K-mordenite show that at a given temperature D may change by many powers of ten for small changes in the critical dimension of the diffusing molecule. Increases in this dimension are related also to rapidly increasing activation energies. The diffusivities in Na-*A*(61) for Ar at −90°C, for N_2 and CO at −70°C for CH_4 at −10°C and for C_2H_6 at +10°C refer respectively to the intervals 0 to 20, 40, 90, 20 and

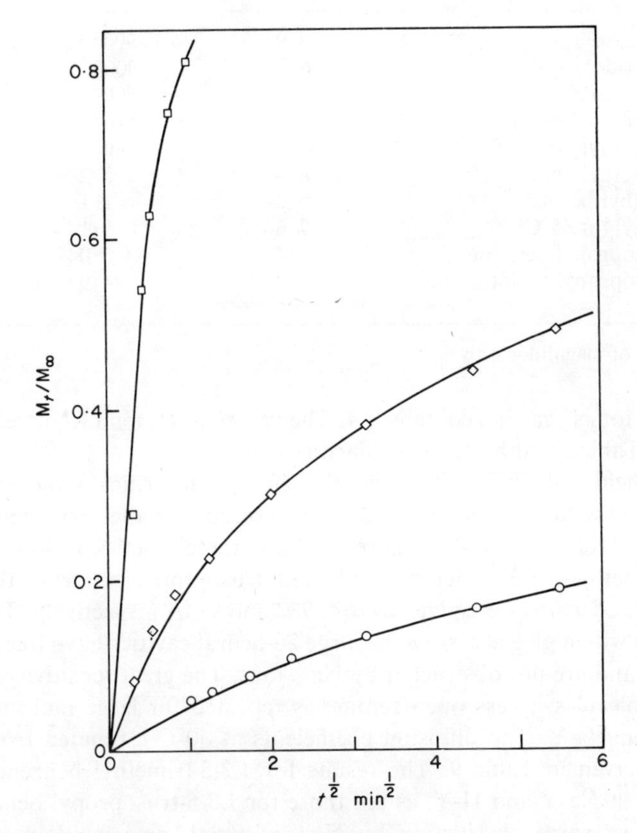

FIG. 14. Sorption kinetics of liquid 1,3,5-trimethyl-(□), 1,3,5-triethyl-(◇) and 1,3,5-triisopropyl benzene (○) in Na-Y at 30°C.(12)

50 cm^3 at s.t.p. g^{-1}, and so are average or integral values over these intervals. For the results obtained in K-*A*[63] and K-mordenite[43] the amounts sorbed are small and are near the Henry's law condition where D should be nearly independent of concentration. The values of D_0 and E for D^* in $D = D^* \partial \ln p / \partial \ln C$ do not involve the temperature coefficients of $\partial \ln p / \partial \ln C$ which, however, is unity in the Henry's law range (when $D = D^*$). Variations are seen among results of different authors when values of E are compared, for

TABLE 9

$\tilde{D}$ at 30°C estimated from the $\sqrt{t}$ law for sorption of liquids into zeolites Na–Y and H–Y[12]

Molecule	Critical dimension (Å)	$10^{13} \times \tilde{D}$ ($cm^2 s^{-1}$) Na–Y	H–Y
n-hexadecane	4·9	>900[a]	—
l-phenyl tridecane	6·8	>900[a]	—
isopropyl benzene	6·8	>700[a]	—
1,2,4-trimethyl benzene	—	>500[a]	—
1,3,5-trimethyl benzene	8·4	60	>500[a]
2,4,6-trimethyl aniline	8·4	5·2	4·1
1,3,5-triethyl benzene	9·2	1·1	—
o-terphenyl (at 65°C)	9·6	1·1	—
1,3,5-triisopropyl benzene	9·4	0·047	9·3
1,3,5-triisopropyl cyclohexane	9·8	4·9	8·7

[a] Order of magnitude only.

example for N_2 gas in zeolite Na-*A*. The values of D_0 tend to increase with E (cf also Table 6), although irregularities occur.

Satterfield and Cheng[12] studied the effect of the critical molecular dimension upon the kinetics of uptake of large molecules in the very open structure of zeolites Na-Y and H-Y. Figure 14 shows these kinetics at 30°C for liquid 1,3,5-trimethyl-, 1,3,5-triethyl- and 1,3,5-triisopropyl benzene, the critical diameters of which were given as 8·4, 9·2 and 9·4Å respectively. The 12-ring windows which give access to the large 26-hedral cavities have free diameters of ≈8Å and are not obstructed by Na^+ ions. The great sensitivity found for small molecules in less open zeolites is repeated for large molecules in the very open Na-Y. The diffusion coefficients at 30°C estimated from the $\sqrt{t}$ law are given in Table 9. The results for 1,3,5-trimethyl benzene are very different in Na-Y and H-Y, as are those for 1,3,5-triisopropyl benzene. This may reflect some obstruction by Na^+ of the 12-ring window, or greater localization through a higher sorption heat in the Na-form. For the polar

2,4,6-trimethyl aniline this difference in $\tilde{D}$ for Na-Y and H-Y is not apparent, nor is it very marked for 1,3,5-triisopropyl cyclohexane. However overall the great importance of molecular dimensions is well illustrated. The presence of a polar group very much reduces $\tilde{D}$, as seen by comparing the results for 1,3,5-trimethyl benzene and 2,4,6-trimethyl aniline, which have the same critical dimension. By its specific interactions with the polar environment, the $-NH_2$ group may enhance the friction between diffusant and medium.

6.2 Exchangeable Cations

The great part played by the exchangeable cations in moderating molecule-sieving properties was first demonstrated with Na-, Ca- and Ba-exchanged mordenites in 1945[53] and was more fully investigated in 1949[43] using mordenites exchanged with Li, Na, K, NH_4, Ca and Ba. From the sorption kinetics, values of $D/\bar{r}_0^2$ were found, where $\bar{r}_0$ is a mean crystal radius, the crystals being idealized for simplicity as spheres. Some values for argon at −78°C are given in Table 10. A range of 10^5 is covered in the values given,

TABLE 10

$D/\bar{r}_0^2$ (min^{-1}) for argon at −78°C[43]

Crystal	$D/\bar{r}_0^2$
Ca-mordenite	$1 \cdot 51 \times 10^{-8}$
Levynite	$7 \cdot 6 \times 10^{-7}$
K-mordenite	$2 \cdot 8 \times 10^{-6}$
Ba-mordenite	$5 \cdot 5 \times 10^{-6}$
Na-mordenite	$3 \cdot 8 \times 10^{-6}$
Li-mordenite	$4 \cdot 0 \times 10^{-4}$
NH_4-mordenite	$1 \cdot 35 \times 10^{-3}$

according to the exchange ion present. The NH_4-mordenite, in the light of more recent experiments, was probably largely converted to H-mordenite by the outgassing at 300–340°C. Thus for the monovalent cations in mordenite $D/\bar{r}_0^2$ changes in the sequence K $<$ Na $<$ Li $<$ H. A comparison of the data in Table 8 for the K and Na forms of zeolite *A* also emphasizes the effect of ion size upon both *D* and *E*.

Other researches have further demonstrated the role of the exchange ion in modifying diffusion kinetics. Reference has been made in § 4.5 to the sensitive dependence of the diffusivity of *n*-paraffins upon the Ca/Na ratio in zeolite *A*,[45] and how *E* decreases as the % exchange of Na by Ca increases. Takaishi *et al.*[64] exchanged Na- and K-forms of zeolite *A* to different extents with Ca and Zn, and studied the effects of this upon the sorption of nitrogen, argon, carbon dioxide, *n*-butane, silane, diborane, phosphine, arsine, but-1-ene, and *trans*- and *cis*-but-2-ene, each at a chosen pressure and temperature. The results in Fig. 15 exemplify the behaviour. The sorption capacities of non-polar molecules underwent strong decreases as the K-content of (K, Zn)-*A* reached values which differed for different sorbates and tempera-

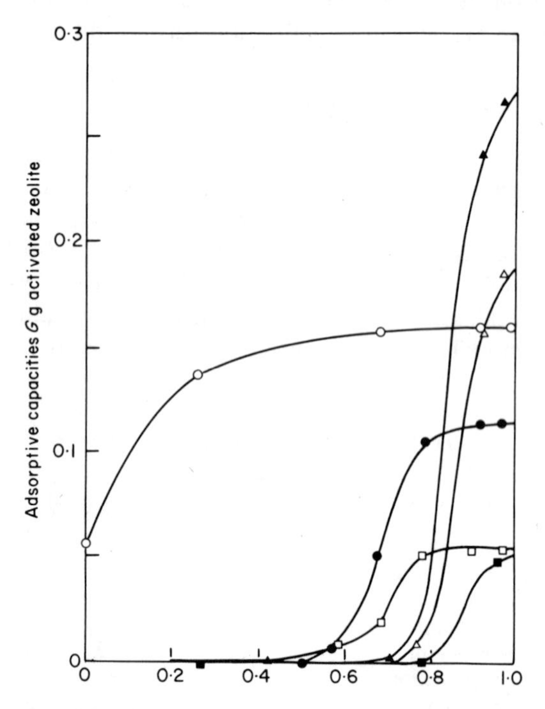

FIG. 15. Sorption capacities of some gases in (K, Zn)-*A* as a function of the cation fraction of Zn^{2+} in the crystal.[64]

△, N_2 at 200 torr and −195°C
▲, Ar at 50 torr and −195°C
○, CO_2 at 210 torr and 0°C
●, *n*-C_4H_{10} at 200 torr, 0°C
■, B_2H_6 at 50 torr, 0°C.
□, SiH_4 at 160 torr, 0°C

tures. This implies that sorption rates have become so small that beyond the critical K-content the potential sorption equilibrium is not achieved in any time available for experiment. If the molecules were polar (CO_2, NH_3, H_2O) the cut-off points were no longer so clear and occurred at higher K-contents. This is seen in the figure for CO_2 at 0°C. Water is freely sorbed in various ion-exchanged forms of chabazite,[65] or zeolite *A* in which the cations are too large or too numerous to allow thc sorption of non-polar molecules. The polarity of such diffusants and their consequent high affinity for the ionic zeolite sorbents ensures their penetration and copiòus uptake at room temperature, with ion solvation and probably with resiting of some of the exchange ions from the sites they occupy in the outgassed zeolite. Such perturbations are not expected to occur appreciably with small non-polar molecules.

Barrer and Baynham[66] synthesized a series of chabazite-type zeolites in which the ratios Si/Al varied between about 1·15 and 2·08. The number of exchangeable cations varied accordingly. The Ca^{2+}- and Na^{+}-exchanged forms were examined as sorbents of oxygen at −183°C. The uptakes at 20 cm Hg pressure are shown in Fig. 16. The sorption capacity changed with

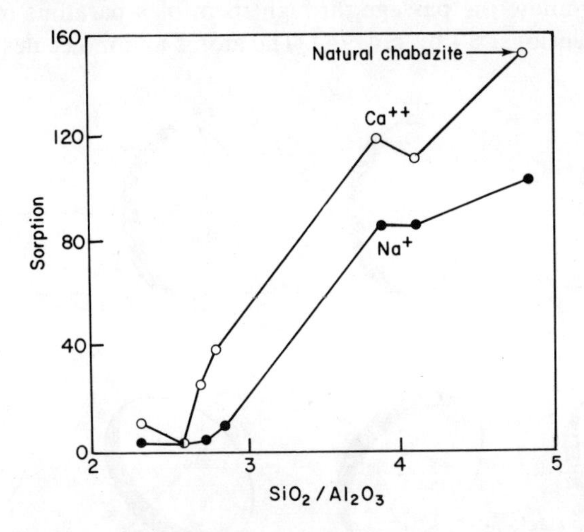

Fig. 16. The influence of the replacements ($\frac{1}{2}$Ca, Al) ⇄ Si and (Na, Al) ⇄ Si on the sorption capacities for O_2 at −183°C and 20 cm pressure in chabazite-type zeolites.[66] The ordinate gives the sorption in cm^3 at s.t.p. for the number of unit cells which are present in 1 g of the hydrated K-zeolite in which $SiO_2 : Al_2O_3$ = 4.15 : 1.

the number of cations from a high value with rapid sorption at the silica-rich end of the series to a very small value at the aluminous end. Here it is not the size of the cations which is altering but their numbers.

6.3 Chain Length of *n*-paraffins

In chabazite for the *n*-paraffins C_3, C_4, C_5 and C_7;[67] in erionite for C_5 to C_8;[50] in zeolite (Ca, Na)-*A* for C_4, C_6 and C_9;[45] in Na-mordenite for C_1 to C_4;[48] and in H-mordenite for C_6, C_7 and C_8[68] initial sorption or desorption rates have been found to decrease with increasing chain length. For C_5 to C_8 in Ca-*A* sorption rates were all similar.[50] In mordenite the main channels are circumscribed by 12-rings and so in the H-form are wide enough to admit straight or branched chain paraffins, cycloparaffins or aromatics. The Na-form may have less open channels, due *inter alia* to the location of

Na ions or possibly of adventitious impurities in these channels. In these straight channels the friction between hydrocarbon and the anionic framework would be expected to increase with the carbon number of the *n*-paraffin. In chabazite, erionite and zeolite *A* access to intracrystalline channels is through 8-ring windows, somewhat distorted in the first two zeolites and nearly planar in zeolite *A*. The maximum and minimum free dimensions are 3·7 and 4·1Å (chabazite), 3·6 and 5·2Å (erionite) and 4·3Å (zeolite *A*). In Ca-rich forms of these three zeolites sufficient 8-ring windows are unblocked by cations to allow the passage through them of *n*-paraffins (critical cross-sectional dimensions ≈4 by ≈4·9Å). The atoms and molecules are not hard

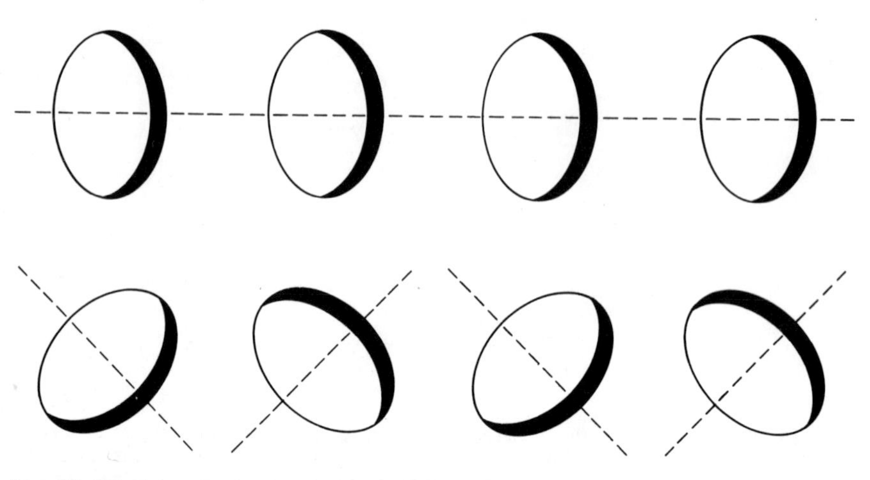

FIG. 17. Windows of equal dimensions, all parallel or oriented at an angle to each other.[39]

spheres, the windows may be slightly deformable, and lattice vibrations can also assist so that without a large activation energy the *n*-paraffins can pass through.

On the basis of the above free dimensions the most ready diffusion of *n*-paraffins might be into erionite and zeolite *A*. However, an additional factor is the mutual orientation of the 8-ring windows, as illustrated formally in Fig. 17.[39] In zeolite *A* the 8-rings are parallel along any one channel and these channels intersect each other in three directions at right angles. In chabazite and erionite the 8-rings are not parallel; six 8-rings form the windows of any one of the cylindrical cavities and so a given central cavity provides access to each of six other identical cavities. For long-chain molecules, the chain conformation required to present one end of a molecule to a window in the most favourable configuration for transit through that window, especially when the other end is anchored in a second window, may be more readily achieved if the windows are parallel and each normal to the same axis than it is in chabazite or erionite, so that resistance to diffusion could be bigger in these two zeolites than in zeolite *A*.

Gorring[69] reported an unexpected behaviour for the diffusion of some longer *n*-paraffins in the K-form of zeolite *T* which is a synthetic zeolite consisting primarily of offretite but with intergrowths of erionite. The diffusion coefficients at first diminished with increasing carbon number, went through a minimum at C_8, rose steeply to a maximum at C_{12} and then again

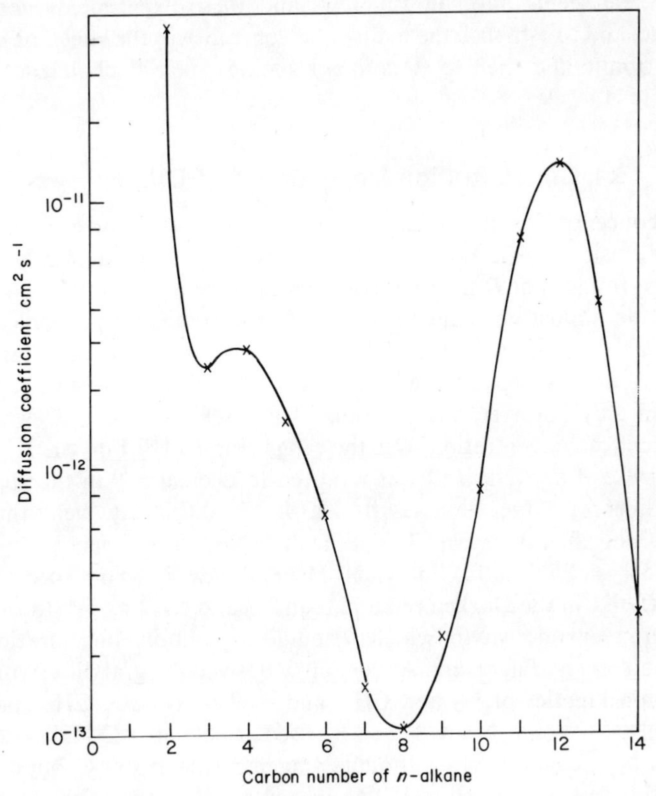

FIG. 18. Diffusion coefficients of *n*-alkanes in the potassium form of zeolite *T* at 300°C,[69] plotted as a function of carbon number.

declined (Fig. 18). Some values of D_0 and E were as given below ($D = D_0 \exp -E/RT$):

	D_0 (cm^2 s^{-1})	E (kcal mol^{-1})
n-nonane	$8 \cdot 5 \times 10^{-8}$	14·8
n-decane	$5 \cdot 7 \times 10^{-7}$	15·4
n-undecane	$4 \cdot 4 \times 10^{-9}$	7·3
n-dodecane	$3 \cdot 3 \times 10^{-8}$	8·0

For *n*-tridecane, E was reported as being almost zero. In partial explanation of the above behaviour it was considered that all molecules in the erionite intergrowths larger than *n*-octane would be too long to align themselves fully stretched within the cylindrical 23-hedral erionite cavities. A molecule such as dodecane will have to extend through two 8-ring windows at once. Thus the ends are necessarily already in orientations and locations which give the molecule an advantage in migrating from one cage to another. Nevertheless E for tridecane seems anomalously low, and these experiments could with advantage, and to establish the reality and generality of the effect, be extended first to erionite and then to some other zeolites such as chabazite, levynite and zeolite A.

6.4 Concentration Dependence of Diffusivities

Diverse concentration dependences have been reported for differential and integral diffusion coefficients, $\bar{D}$ and $\tilde{D}$ respectively. For water in heulandite, Tiselius[15] found that $\bar{D}$ first increased with degree of filling but near saturation became almost constant (Table 1). For *n*-butane or propane in Ca-A Ruthven and Loughlin[70] found that $\bar{D}$ increased with concentration and this behaviour was also reported for ethane, ethylene and *n*-pentane in Ca-A and erionite by Ruthven and Derrah.[71] For CH_4 and CF_4 $\bar{D}$ was nearly independent of concentration over the range studied.[60] For Ar, Kr, Xe, O_2 and N_2 in Ca-A and Na-A $\bar{D}$ was reported to decrease[59] in the Henry law range, while for larger uptakes in Na-A this diffusivity went through a minimum and then increased. It is difficult to understand decreases in $\bar{D}$ for true intracrystalline diffusion when Henry's law is valid, because then $\partial \ln p / \partial \ln C = 1$ in the Darken relation (eqn 21) and so $\bar{D} \approx D^*$. In this dilute range there seems no way in which D^* could depend on concentration.

Habgood[62] and Eagan and Anderson[61] derived integral diffusivities from the sorption kinetics of N_2 and CH_4, and of Ar, N_2, CO, CH_4 and C_2H_6, respectively, in zeolite Na-A. When he used pellets at 0°C Habgood found that $\tilde{D}$ for N_2 increased with C if outgassing occurred at 400°C, but decreased if outgassing was performed at 350°C. In powder $\tilde{D}$ was not dependent on C. Eagan and Anderson found that $\tilde{D}$ increased with C, slightly for Ar and more strongly with CH_4, C_2H_6, N_2 and CO. On the other hand Eberly[50] found in erionite that $\tilde{D}$ determined by a similar experimental method to that used by Barrer and Clarke,[45] decreased with C for C_5, C_6, C_7 and C_8 *n*-paraffins (i.e. keeping Q_∞ constant and varying Q_0). Brandt and Rudloff[72] using chabazite as diffusion medium, found integral diffusivities that sometimes increased with concentration (CO_2) and sometimes, after constant values for small concentrations, decreased for larger C (e.g. C_3H_8, CH_2Cl_2).

Some light may be shed on the above complex pattern of reported concentration dependences by considering the behaviour of the diffusivities $\tilde{D}_1$,

$\tilde{D}_2$ and $\bar{D}$ derived by Barrer and Clarke[45] (§ 4.5). $\tilde{D}_1$ was estimated from the slopes, k, of linear plots of M_t/M_∞ against $\sqrt{t}\left(k = \frac{2A}{V}\left(\frac{\tilde{D}_1}{\pi}\right)^{1/2}\right)$, over the interval $M_\infty = (Q_\infty - Q_0)$, in which Q_0 was varied and Q_∞ kept constant. It was considered to be related to the differential diffusivity, $\bar{D}$, by eqn 48. The second integral diffusivity, $\tilde{D}_2$, was related to $\bar{D}$ by eqn 50. As Q_0 increased towards Q_∞ it was found that $\tilde{D}_2$ decreased and $\tilde{D}_1$ increased (Table 6). Thus the concentration dependence is, at least in part, related to the ways in which the $\tilde{D}$ are defined and determined.

The $\bar{D}$ decreased to some extent for C_4, C_6 and C_9 *n*-paraffins in (Na, Ca)-*A*,[45] contrary to observations of Ruthven and coworkers[70, 71] for several hydrocarbons in Ca-*A* and erionite. Such differences may be real, or may be connected with the experimental conditions. In the interval method used by Ruthven a small additional amount of sorbate is added to the zeolite, already loaded to the extent Q_0, by introducing an extra pressure of vapour and allowing equilibration to occur. The additional uptake could occur at constant pressure, or at variable pressure. If the pressure is variable, even though the increment in Q_0 is small, because of the rectangular nature of the isotherms $\partial \ln p/\partial \ln C$ can be very large (cf Table 2) and can change rapidly with the pressure rise and decline on admitting the dose of vapour even when the equilibrium uptake changes only a little. Thus, in the expression $\bar{D} = D^* \partial \ln p/\partial \ln C$ it may be incorrect in a variable pressure run to assume for the increment in Q_0 that $\bar{D}$ is constant unless there is a strong compensating concentration dependence in D^*. The best way to use the small incremental step in Q_0 to determine $\bar{D}$ is to allow the extra sorption to occur strictly at constant pressure (cf § 4.4). Unfortunately, for rectangular isotherms the equilibrium pressures can be so small over most of the range in Q_0 that the constant pressure procedure is not practicable. The only alternative constant pressure procedure for rectangular isotherms is that adopted by Barrer and Clarke (Q_∞ constant, Q_0 varied systematically). Here the evaluation of integral and differential diffusion coefficients depends upon the initial assumption of eqn 47 $\left(k = \frac{2A}{V}\left(\frac{\tilde{D}_1}{\pi}\right)^{1/2}\right)$. It is possible, if this assumption is not sufficiently accurate, that the concentration dependence of the $\bar{D}$ is influenced.

It was pointed out in § 3 that the expression $\bar{D} = D^* \partial \ln p/\partial \ln C$ does not give the complete relation between $\bar{D}$ and the tracer diffusivity D^*. However for strongly curving isotherms the term $\partial \ln p/\partial \ln \theta$ for a number of model isotherm equations is given below.[39] In these equations K is the equilibrium constant; β and α are as defined in Chapter 3; and the a_i are coefficients which may depend on temperature but not on θ. For $\theta \ll 1$ all these model isotherms give $\partial \ln p/\partial \ln \theta = 1$ (Henry's law). To proceed further one must

Model	Isotherm Equation	$\partial \ln p/\partial \ln \theta$
Ideal localized	$K = \dfrac{\theta}{p(1-\theta)}$	$(1-\theta)^{-1}$
Localized with interaction	$K = \dfrac{\theta}{p(1-\theta)} \exp \beta\theta$	$(1-\theta)^{-1} + \beta\theta$
Volmer eqn of state	$K = \dfrac{\theta}{p(1-\theta)} \exp \dfrac{\theta}{(1-\theta)}$	$(1-\theta)^{-2}$
van der Waals eqn of state	$K = \dfrac{\theta}{p(1-\theta)} \exp \left[\dfrac{\theta}{1-\theta} - \alpha\theta\right]$	$(1-\theta)^{-2} - \alpha\theta$
virial eqn of state	$K = \dfrac{\theta}{p} \exp [2a_1\theta + (3/2)a_2\theta^2 + (4/3)a_3\theta^3 + \ldots]$	$1 + 2a_1\theta + 3a_2\theta^2 + 4a_3\theta^3 + \ldots$

consider the concentration dependence of D^*.

Barrer and Jost[73] considered the situation in which a molecule can migrate from one site to another only when the receiving site is vacant. The chance of this is proportional to $(1 - \theta)$, so that $D^* = D_0^*(1 - \theta)$. According to this model, for Langmuir's isotherm $\bar{D}$ should be independent of θ and $\bar{D} \approx D_0^*$. For localized sorption with interaction $\bar{D} \approx D_0^*[1 + \beta\theta(1 - \theta)]$ so that $\bar{D}$ becomes a function of θ. For Volmer's and van der Waals equations of state the isotherm equations, if combined with the expression $D^* = D_0^*$ $(1 - \theta)$, give respectively

$$\bar{D} \approx D_0^*(1 - \theta)^{-1} \tag{56}$$

$$\bar{D} \approx D_0^*[(1 - \theta)^{-1} - \alpha\theta(1 - \theta)] \tag{57}$$

so that as $\theta \to 1$ D should increase rapidly.

It is also possible to envisage multiple jumps through a succession of vacant sites before the activation energy, E, is dissipated by vibrational collisions with the surroundings.[73] If l is the distance moved per unit diffusion process, ν is the vibration frequency, and if a molecule could jump in any one of q directions, then for a single jump

$$D^* = \frac{\nu}{q} l^2(1 - \theta) \exp -E_1/RT \tag{58}$$

while for a single activation followed by an n-fold jump along a single direction

$$D^* = \frac{\nu}{q} (nl)^2(1 - \theta)^n \exp -E_n/RT. \tag{59}$$

If all possibilities are taken into account, the expression for $\bar{D}$ becomes

$$\bar{D} = \frac{\nu l^2}{q} RT \frac{\partial \ln p}{\partial \ln C} \sum_{n=1,2,\ldots} n^2(1-\theta)^n \exp -E_n/RT. \tag{60}$$

Unless E_n increased strongly with n, multiple jumps in a given direction could, where geometrically possible, contribute to $\bar{D}$, as can be seen by comparing values of $n^2(1-\theta)^n$ for different values of n and θ. Figure 19 gives values of $(1-\theta)^n \partial \ln p/\partial \ln \theta$ for several isotherm models and values of n.[39] $\bar{D}$ could on this basis exhibit a variety of concentration dependences, such as those observed experimentally. However, in the Henry's

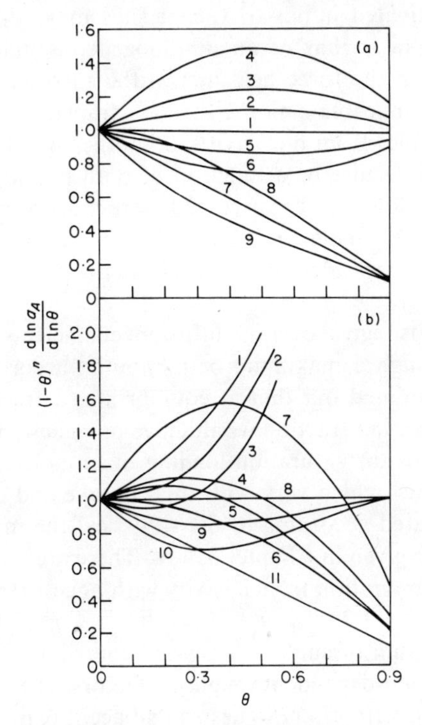

FIG. 19. $(1-\theta)^n \frac{\mathrm{dln}p}{\mathrm{dln}c}$, for $n = 1$ and 2; plotted as functions of θ.[39]
(a) Localized sorption with interaction. For curves 1 to 6, $n = 1$ and $\beta = 0$, 0·5, 1·0, 2·0, −0·5 and −1·0 respectively. For curves 7, 8 and 9, $n = 2$ and $\beta = 0$, +1 and −1 respectively.
(b) Isotherms based on equations of state. For curves 1, 2 and 3 $n = 1$ and equations respectively are that of Volmer; and that of van der Waals with $\alpha = 1$ and 2. Curves 4, 5, 6 and 7 are for $n = 1$ and the virial isotherms with $a_1 = 1·0$, $a_2 = 0$; $a_1 = 1·0$, $a_2 = -0·33$; $a_1 = 0·50$, $a_2 = 0$; and $a_1 = 2·0$, $a_2 = 0$. Curves 8, 9 and 10 are for $n = 2$ and isotherm equations of Volmer, and of van der Waals with $\alpha = 1$ and 2 respectively. Curve 11 is for Volmer's isotherm with $n = 3$.

law range ($\theta \ll 1$; $\partial \ln p / \partial \ln \theta = 1$) $\bar{D}$ will always become independent of concentration and the values of $\bar{D}$ and D^* will converge.

In intrazeolitic fluids the assumption in eqn 58 that for a single jump $D^* = D_0^*(1 - \theta)$ may not be adequate. It should be correct for diffusion of water or ammonia in analcime[16] or of argon and krypton in sodalite hydrate[8] for which only one guest molecule at a time can occupy the intracrystalline interstices. It should also be true of helium and neon diffusing in the crystalline silicas, tridymite and cristobalite.[7] When the interstices become channels or cavities able to accommodate filaments or clusters of guest molecules the situation is more complex. At low or intermediate values of θ the concepts of localized sorption may still reasonably describe the behaviour, and activated jumps to vacant sites may still control diffusion. The general problem is that of an inhomogeneous fluid of mean density proportional to θ, in the force field due to the surrounding anionic framework and the exchangeable cations. In mass transfer by diffusion it is the flux of guest molecules from one cavity to another which is important. This could for simple molecules be a function of both θ_i and $(1 - \theta_j)$, since if a jump through a window is to be successful there must be room on the far side to receive the jumping molecule. θ_i and θ_j are the local degrees of filling of the space on the two sides of the window, the subscript "j" referring to the receiving side.

This view can also give overall diffusion coefficients that increase, decrease or pass through a maximum or minimum, according to Fiedler and Gelbin.[73a] They pointed out that at equilibrium, a fraction of the cavities would be empty, another fraction would have one guest, another would have two and so on up to the saturation loading of m molecules per cavity. The equilibrium fractions, which vary with total uptake and temperature, can in principle be calculated according to the statistical thermodynamic formulation of the isotherm given in Chapter 3, § 10. There are then various diffusion coefficients D_{ij} for migration from a cavity with i guest molecules to one with j molecules where $i = 1,2 \ldots m$ and $j = 0,1 \ldots m$. The overall diffusion coefficient was the sum of suitably weighted contributions from each of the D_{ij}. The concentration-dependent weighting factors reflected the probabilities of finding each i, j distribution of guests in adjacent pairs of cavities. The D_{ij} can also be concentration-dependent. The resultant expression was complex, but was developed for the simpler situation where $m = 2$ to demonstrate the properties of the model.

For straight chain paraffins the diffusion through a window of a zeolite such as Ca-*A* may occur segment by segment, so that a number of segmental unit diffusion processes would be involved in the complete transfer of the molecule from one cavity to the next. If there are N segments in the chain and if when an activation energy E_{n_1} is received, n_1 segments on average pass through the window, the diffusivity would be expected to have the form

$$\bar{D} = \frac{\nu l^2}{q} \cdot RT \frac{\partial \ln p}{\partial \ln C} \sum_{n=1}^{N} n_1^2 \left(\frac{n_1}{N}\right) \exp - E_{n_1}/RT. \tag{61}$$

The ratio n_1/N is introduced to scale $\bar{D}$ from the value appropriate for a segment to that appropriate for the whole molecule. In this expression l represents the length of a segment and the other quantities have the same meaning as before (eqn 60); n_1 (and so E_{n_1}) would be expected to be functions of θ because the ability of segments to move either way through a window will be moderated by the degree of blocking of the paraffin chain by other paraffin molecules on each side of the window.

Ruthven and Derrah[63] developed an expression comparable with eqns 60 and 61 for the diffusivity, with zeolite Ca-*A* particularly in mind:

$$D = \frac{kTl^2}{6h} \frac{f_z^*}{f_z} \exp\left[\frac{(u_z - u_z^*)}{kT}\right] \frac{\partial \ln a}{\partial \ln C} \tag{62}$$

Here f_z^* and f_z are the relevant partition functions for a molecule in the 8-ring window and in the large 26-hedral cavity respectively, and $(u_z - u_z^*)$ is the difference in energy between these two positions. Ruthven and Derrah considered the term $\frac{kTl^2}{6h} \frac{f_z^*}{f_z} \exp\left[\frac{(u_z - u_z^*)}{kT}\right]$ to be nearly independent of concentration, but this is unlikely except in the Henry's law range. Rather one would expect both f_z^*/f and $(u_z - u_z^*)$ to change as the 26-hedral cavities are progressively filled with guest molecules.

Where the sorbate-zeolite system is energetically heterogeneous, the first molecules are most strongly bound to certain positions and are least mobile. D^* and $\bar{D}$ have then low values, as illustrated by the results of Tiselius for water in heulandite (Table 1). Later additions of sorbate are less strongly bound to particular sites and could be more mobile. Energetic heterogeneity is most pronounced for sorption of polar molecules (Chapter 4) so that $\bar{D}$ for these molecules (H_2O, NH_3, CO_2, SO_2) should in general increase with θ, unless and until steric blocking becomes dominant as θ approaches one (cf Table 1). It follows from all the above discussion that simple behaviour is not to be expected for the concentration dependence of either D^* or $\bar{D}$ except in a limited number of special cases. In the same context more refined measurements of $\bar{D}$ as a function of C or θ are needed, made with due care to minimize the considerable problems in performing the kinetic experiments in the best way and in interpreting the results.

7. HINDERED DIFFUSION

Channel networks in the zeolites may be one-, two- or three-dimensional. One-dimensional channel systems include those in mordenite, offretite.

zeolite L and zeolite Ω. Two-dimensional systems include levynite and heulandite or its siliceous variant clinoptilolite. Three-dimensional networks are found for example in chabazite, faujasite and zeolites *A*, *ZK-5* and *RHO*. In all types of channel system diffusion may be moderated by the presence in them of cations (§ 6.2), anions and neutral molecules which are immobile at the temperature at which the sorption of a second component occurs.

Barrer and Harding[74] moderated the diffusion of *n*-hexane and 2,2-dimethyl butane in a (K, TMA)-offretite (TMA = tetramethylammonium). The TMA^+ in the wide channels parallel with the *c*-axis of the crystals was removed progressively by ion exchange with H^+, Li^+, Na^+, K^+, Cs^+, $CH_3NH_3^+$, $(CH_3)NH_2^+$ and $(CH_3)_3NH^+$. When the TMA^+ in these channels was increased to its maximum value not even *n*-hexane was sorbed; when it was all replaced by hydrogen or the inorganic ions both *n*-hexane and 2,2-dimethylbutane were sorbed, *n*-hexane very rapidly and 2,2-dimethyl butane with only marginal variations in rate. Intermediate contents of TMA led to intermediate rates of uptake of *n*-hexane. The partial replacement of K and TMA by methyl-, dimethyl- and trimethylammonium did not prevent the uptake of *n*-hexane, but the rates of sorption differed *inter se* in the order dimethylammonium > methylammonium > trimethylammonium.

Moderation of the channel systems of offretite, mordenite and zeolite *L* was then investigated,[75] using *n*-hexane, 2-methylpentane, 3-methylpentane, 2,2-dimethylbutane, cyclohexane and benzene as probe molecules, after soaking the crystals in solutions of chloride, bromide, chromate, and sulphate. The salts are imbibed in small amounts, the uptake being governed by a Donnan membrane equilibrium,[76] and so being sensitive to the external concentration of the salt. The inorganic salts are thus introduced in controllable amounts. The zeolite can be outgassed at high temperatures with no decomposition of the salts, unlike the organic cations in the offretite used in the study of Barrer and Harding. With appropriate amounts of salts in the crystals it was possible to produce sorbents from these wide pore zeolites which took up only *n*-hexane, or which were selective in amount sorbed and rate for benzene over cyclohexane or branched chain paraffins. Some differences could also be observed at the appropriate salt loadings between neo- and isopentane. The notable effectiveness of salts in moderating the molecule sieving and sorption capabilities of zeolites suggests that adventitious uptake of aluminate, silicate or hydroxide during synthesis could account, at least in part, for the variable sorption behaviour sometimes observed among different preparations of zeolites such as mordenite. The presence of silicate anions in channels of cancrinite hydrate has been suggested and may be an alternative explanation to stacking faults to explain why the wide channel character of this zeolite has never been realized.

Diffusion, sorption capacity and molecule sieving properties are very effectively moderated by the presorption of regulated amounts of small polar molecules, as first demonstrated in 1954.[77] Water, ammonia or methylamine are examples of moderators of the low temperature sorption of gases such as oxygen, nitrogen, hydrogen and argon in mordenite, chabazite and zeolite *A*.[77, 78] Figure 20 shows the effects of presorbed ammonia upon the diffusivities of these four gases in Na-mordenite at −183°C.[77] After critical

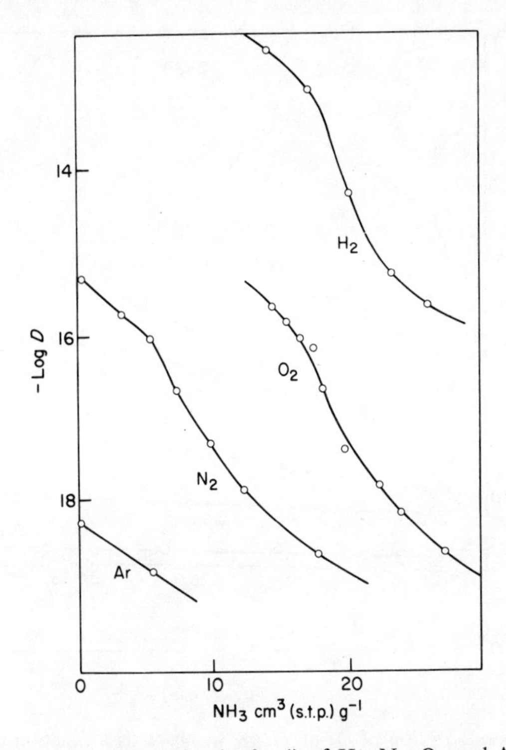

FIG. 20. The change in the diffusivities (cm^2s^{-1}) of H_2, N_2, O_2 and Ar in mordenite at −183°C as a function of the amount of ammonia presorbed.[77]

amounts of presorbed ammonia had been added, the diffusion rates became so low for certain of the gases that sorption equilibrium was not achieved in any reasonable time and so there was a sharp cut-off in the apparent sorption capacity of the zeolite. This cut-off can occur at different ammonia loadings for sorbate molecules of different dimensions, and can thus influence drastically the selectivity of the sorbent. In (Ca, Na)-forms of zeolite *A* the cut-off regions for oxygen depended upon both the extent of exchange of Na^+ by Ca^{2+} and the amounts of presorbed ammonia. The greater the % Ca^{2+} the larger the amount of presorbed ammonia required to effect the cut-off.[78]

The reduction in the rates of sorption of oxygen at 77·4K in zeolite Na-*A* are illustrated in Fig. 21 for different ammonia loadings.[78] In this zeolite Rees and Berry considered the twelve Na^+ ions to be located with one at each of the eight 6-ring windows (Group I), and four at four of the 8-ring windows which link a central large cavity (26-hedron) with each of six other 26-hedra (Group II). Because each 8-ring of free diameter $\approx 4 \cdot 3$Å is shared in this way by two 26-hedra there are eight group II cations for six 8-ring windows. Accordingly two of these are effectively doubly occupied with Na^+

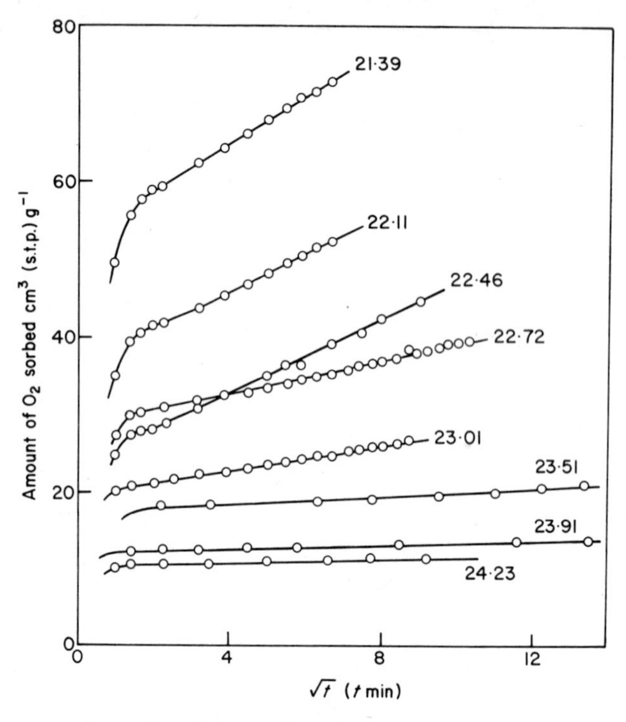

Fig. 21. The rates of sorption of O_2 at 77·4K in Na-*A* as a function of the amounts of presorbed NH_3 (cm^3 at s.t.p. g^{-1}) indicated by the numbers shown with each curve.(78)

on either side. The group II sites were thought to be preferentially solvated by NH_3 and cut-off occurred when each window was effectively blocked. In the absence of NH_3, N_2 was blocked at 77 to 90K, but not O_2 or Ne. When 15% of Na^+ was replaced by K^+ O_2 and Ne were also blocked. For Na-*A* about 24 cm^3 at s.t.p. of presorbed ammonia per g of zeolite blocked the uptake of O_2 (Fig. 21). If blocking is due to NH_3 associated with the group II Na^+ ions then three molecules per unit cell (or six per 26-hedron because of sharing) would completely prevent access of the oxygen; 2·5 molecules would leave only one window open and so would also inhibit diffusion of O_2; and two molecules would leave two windows open per 26-hedron so that O_2 could probably diffuse. The cut-off should therefore lie between 2 and 2·5 NH_3 per unit cell, in reasonable accord with the above critical uptake.

Hammersley[79] considered a similar problem involving penetration of water into a cubic array of pores, i.e. pores disposed in a manner analogous to the channel structure of zeolite *A*. He calculated the degree of wetting, $W(p)$, as a function of the probability, p, that a pore was open between the junction points of pairs of pores. The cut-off of water uptake was sharp for $p \approx 0{\cdot}23$. Thus little sorption will occur when $3 \times 0{\cdot}23 = 0{\cdot}69$ windows per unit cell are open and so $2{\cdot}3$ are closed. Because of the sharing of each 8-ring window in zeolite *A* between two 26-hedra this means $\approx 4{\cdot}6$ of the 8-rings are closed per 26-hedron for $p = 0{\cdot}23$ as compared with $\approx 4{\cdot}8$ for the ammonia uptake at the cut-off point. The agreement with Hammersley's calculation was however not reproduced when K^+ or Ca^{2+} replaced some of the Na^+. The K^+ ion blocks the channels without needing additional solvation by NH_3, and replacement of $2Na^+$ by Ca^{2+} leaves some windows without cation occupants, so that ammonia will no longer tend to be sorbed there.

In three- or two-dimensional channel systems the diffusing molecule can often find its way past an obstruction by an alternative unblocked pathway. In one-dimensional channel systems this is not the case. A block at each end of a channel eliminates the channel as a location where sorption can occur. Accordingly these one-dimensional channels which are parallel and not interconnected will be even more sensitive than three- or two-dimensional channel systems to small numbers of moderating molecules or ions. Again the behaviour to be expected can be formulated.[77]

The immobile moderating species are for simplicity assumed to be distributed at uniform intervals along the parallel channels. Each such molecule constitutes a barrier to pass which the diffusant requires an increased activation energy E_2. In between each pair of barriers diffusion is rapid and involves only the smaller energy of activation E_1, characteristic of the normal energy periodicity experienced by the diffusant moving along the channel in absence of moderating species. The distance between each pair of low energy barriers is l and there are N_s minima (i.e. sites for diffusant) between each pair of the moderating species. The net flow across a low energy barrier (height E_1) is then

$$k_1\theta_1(1 - \theta_2) - k_1\theta_2(1 - \theta_1) = k_1(\theta_1 - \theta_2) \tag{63}$$

where k is the rate constant for crossing this barrier and θ_1 and θ_2 are the degrees of occupancy of sites just before and just after the barrier. Similarly the net flow across a high barrier is

$$k_2(\theta_L - \theta_R)$$

where θ_L and θ_R are the degrees of occupancy of sites to left and right of a high energy barrier. The net flow can be considered to take place from left to right across all barriers. If one assumes a quasi-steady state and equates the two flows, and with $k_1 = \nu_1 \exp - E_1/RT$ and $k_2 = \nu_2 \exp - E_2/RT$ one has

$$\frac{k_1}{k_2} = \frac{\nu_1}{\nu_2} \exp [(E_2 - E_1)/RT] = \frac{\theta_L - \theta_R}{\theta_1 - \theta_2}. \tag{64}$$

Accordingly

$$\frac{\text{Concentration drop across a high barrier}}{\text{Concentration drop between successive high barriers}} = \frac{1}{N_s}\frac{k_1}{k_2} = \frac{\theta_L - \theta_R}{N_s(\theta_1 - \theta_2)}. \tag{65}$$

For many values of N_s, E_2 and E_1 this ratio will be $\gg 1$, so that the concentration drop between high barriers is small compared with that across each high barrier. Then the net rate of filling of the region between two high barriers, the rth and $(r + 1)$th, is

$$\frac{dN_r}{dt} = k_2(\Delta\theta_r - \Delta\theta_{r+1}) \tag{66}$$

where N_r is the number of molecules between the rth and $(r + 1)$th high barriers and

$$\left.\begin{aligned} \Delta\theta_r &= (\theta_L - \theta_R)_r. \\ \Delta\theta_{r+1} &= (\theta_L - \theta_R)_{r+1}. \end{aligned}\right\} \tag{67}$$

The problem is now analogous to that of heat flow in the x direction along a chain of equal radiating thin sheets of very high thermal conductivity, all normal to the x-direction and centred on this axis. The region between two successive high energy barriers is the analogue of the conducting sheet and the high energy barrier is the analogue of the space between each successive pair of conducting sheets. This problem has been solved,[80] and for the analogous material flow problem the ratio of the number N_r of occupied sites between the rth and $(r + 1)$th high barriers to the number, N_0, of these sites occupied at equilibrium is given by

$$\frac{N_r}{N_0} = \frac{(n - r + 1)}{(n + 1)} - \sum_{s=1}^{n} \frac{\text{Sin}\,[rs\pi/(n + 1)] \sin [s\pi/n + 1)]}{(n + 1)[n - \cos \{s\pi/(n + 1)\}]} \times \exp \left\{\frac{tk}{N_s} [1 - \cos s\pi/(n + 1)]\right\} \tag{68}$$

where $2k_2 = k$, n is the number of high energy barriers and can be taken also as equal to the number of regions of low energy barriers; $r = 1,2 \ldots, n$ is the number of the region under consideration and s is an integer such that $1 \leq s \leq n$. When n is very large the above equation can be replaced by

$$\frac{N_r}{N_0} = r \int_0^{tk/N_s} I_r(Z) \exp - Z \frac{dZ}{Z} \tag{69}$$

where $I_r(Z)$ is the Bessel function $\sum_{m=0}^{\infty} [(Z/2)^{r+2m}/\{m!\Gamma(r+m+1)\}]$.

The total number of molecules which has in time t passed the first high energy barrier is then

$$N = \sum_{r=1}^{n} N_r \tag{70}$$

and

$$\frac{Q_t}{Q_\infty} = \frac{1}{n}\sum_{r=1}^{n}\frac{N_r}{N_0} \tag{71}$$

if the first high barrier is just within the channel entry. Taking $n = 9$ calculations showed that Q_t/Q_∞ as a function of tk/N_s was not significantly different from Q_t/Q_∞ for $n \to \infty$ at least up to $tk/N_s = 4$. Moreover, for short times and after an apparent brief initial lag, the $\sqrt{t}$ law was valid.

The problem thus far has been considered as one of hindered diffusion into a single channel. We may now consider the total flow through unit area normal to all the N_c channels per unit area. The total uptake of molecules which in time t has entered the crystal through this face is

$$N_2 = N_c \sum_{r=1}^{n} N_r. \tag{72}$$

In absence of high energy barriers this number would be

$$N_1 = \frac{2N_cN_0}{N_sl}\left(\frac{Dt}{\pi}\right)^{1/2} \tag{73}$$

in the range of validity of the $\sqrt{t}$ law, because $\frac{N_cN_0}{N_sl}$ is the concentration in molecules per unit volume just within the ingoing face. Thus

$$\frac{N_1}{N_2} = \frac{2}{N_sl}\left(\frac{Dt}{\pi}\right)^{1/2}\left\{\sum_{r=1}^{\infty}\left(\frac{N_r}{N_0}\right)\right\}^{-1} \tag{74}$$

if $n \to \infty$. This ratio compares the uptakes N_1 with no high energy barriers with the uptake N_2 in presence of these barriers. Some values of N_1/N_2 are compared in Table 11, assuming that $l = 5 \times 10^{-8}$ cm and that in s^{-1}

$$k = 10^{12} \exp - E_2/RT$$
$$k_1 = 10^{12} \exp - E_1/RT.$$

The very great reductions in the uptakes of sorbate in the region of validity of the $\sqrt{t}$ law are clearly demonstrated, and are greater the smaller N_s and the greater the differences $(E_2 - E_1)/RT$. The calculations parallel the observed behaviour in offretite with different contents of tetramethyl ammonium in the wide channels,[74] and the hindered diffusion of gases in

TABLE 11

N_1/N_2 for several energy barriers E_1 and E_2 and values of N_S
Temperature is $-183°C$[77]

N_S	E_1 Kcal mol^{-1}	E_2	N_1/N_2 for $tk/N_S = 1$	for $tk/N_S = 4$
50	1·8	3·6	59·2	41·2
5	1·8	3·6	187·4	130·7
50	1·8	4·5	721	503
5	1·8	4·5	$2{\cdot}28 \times 10^3$	$1{\cdot}59 \times 10^3$
50	1·8	6·3	$1{\cdot}06 \times 10^5$	$7{\cdot}40 \times 10^4$
5	1·8	6·3	$3{\cdot}36 \times 10^5$	$2{\cdot}34 \times 10^5$
50	2·7	5·4	7·21	503
5	2·7	5·4	$2{\cdot}28 \times 10^3$	$1{\cdot}59 \times 10^3$
50	2·7	6·3	$8{\cdot}77 \times 10^3$	$6{\cdot}12 \times 10^3$
5	2·7	6·3	$2{\cdot}77 \times 10^4$	$1{\cdot}94 \times 10^4$
50	2·7	7·2	$1{\cdot}06 \times 10^5$	$7{\cdot}40 \times 10^4$
5	2·7	7·2	$3{\cdot}36 \times 10^5$	$2{\cdot}34 \times 10^5$

Na-mordenites (Fig. 20). They likewise provide a basis for understanding the often strong influence of adventitiously incorporated impurities upon diffusion kinetics in one-dimensional channels.

8. TRACER DIFFUSION

8.1 Water

Tracer diffusion is the simplest example of counter-diffusion of two species. It is normally studied under the condition $C + C^* = C_0$, a constant, where C and C^* are concentrations of untraced and traced (e.g. isotopic) forms of the diffusant. The differential tracer diffusion coefficient, D^*, and the differential diffusion coefficient, D are related by eqn 22 where $\bar{D}$ is defined by eqn 23. These relations combine to give

$$D^* = \bar{D}\frac{\partial \ln C}{\partial \ln p} - \frac{RT\,L_{AA^*}}{C^*} \tag{75}$$

which becomes Darken's equation (eqn 21) if the second term on the right can be neglected. As discussed in connection with Table 2 Darken's equation provided a reasonable but not an exact connection between $\bar{D}$ and D^* for water diffusing in several zeolites (chabazite, gmelinite and heulandite). When Henry's law describes the sorption isotherm $\bar{D}^*$ and $\bar{D}$ become equal, but outside this range their values will diverge and both become functions of concentration, as shown in Table 1 for $\bar{D}$ when water diffuses in heulandite.

Similarly, over a much more restricted range in the degree of filling, θ, the values of $\bar{D}^*$ for chabazite, gmelinite and heulandite are given in Table 12. $\bar{D}^*$ changes only a little for $0\cdot91 < \theta < 0\cdot98$ in all three zeolites, and in this respect behaves like $\bar{D}$ in for water in heulandite (Table 1). Over a wide range in θ one would expect $\bar{D}^*$ to increase with θ if effects of energetic heterogeneity are dominant.

The results of a number of measurements on tracer diffusion of water are

TABLE 12

$\bar{D}^*$ for water in zeolites as θ approaches one[14]

Zeolite	T(°C)	θ	$\bar{D}^* \times 10^7$(cm^{2}s^{-1})
Chabazite B	75	$0\cdot91_3$	$4\cdot3_9$
		$0\cdot93_9$	$4\cdot1_7$
		$0\cdot97_6$	$4\cdot8_6$
Gmelinite	76·9	$0\cdot91_6$	$1\cdot4_8$
		$0\cdot94_1$	$1\cdot5_1$
		$0\cdot97_8$	$1\cdot6_1$
Heulandite	76·1	$0\cdot91_8$	0·83
		$0\cdot95_2$	0·97
		$0\cdot98_4$	$1\cdot1_4$

compared in Table 13. In analcime and possibly in natrolite the energy barriers may be governed largely by the tightness of the anionic framework, since the van der Waals diameter of water is $\approx 2\cdot8$Å. In the more open zeolites the energy barriers need not be due to window dimensions which are $>2\cdot8$Å. Nevertheless considerable barriers exist which are between those for diffusion in ice and in liquid water respectively. One thus sees the intracrystalline water as intermediate in physical state between ice and water at the temperatures given in the table. In the least open zeolite, analcime, the water occupies particular lattice sites; in the most porous (zeolites X and Y), where the largest clusters of water molecules occur, these clusters are linked into filaments which permeate the whole structure. As porosity and cluster size increase, the water positions become less definite. The energies of activation reported for tracer diffusion of monovalent ions in chabazite[87] do not seem to be very different from E for water in the same zeolite.

Na^+	$E = 6\cdot5_5$ kcal mol^{-1}
K^+	$7\cdot0_1$
Rb^+	$6\cdot7_5$
Cs^+	$7\cdot5_5$
Ca^{2+}	13·8
Sr^{2+}	14·6
Ba^{2+}	8·8

TABLE 13

$\bar{D}^*$, D_0^* and E in $\bar{D}^* = D_0^* \exp(-E/RT)$ for water[81] near saturation of zeolite

Diffusion medium	Free dimensions of windows (Å)	$\bar{D}^*$ (cm^2s^{-1}) (at T°C)	D_0^* (cm^2s^{-1})	E Kcal mol^{-1}	Method
Analcime[81]	2·2–2·4	$1·9_7 \times 10^{-13}$ (46°)	$1·5 \times 10^{-1}$	17·0 ± 0·3	Tracer, HTO
Natrolite[82]	2·6–3·9	—	—	15·0	NMR
Heulandite[14]	2·4–6·1, 3·2–7·8, 3·8–4·5	$2·0_7 \times 10^{-8}$ (45°)	$7·6 \times 10^{-1}$	11·0 ± 0·3	Tracer, D_2O
Chabazite[14]	3·7–4·2	$1·2_6 \times 10^{-7}$ (45°)	$1·2 \times 10^{-1}$	8·7 ± 0·3	Tracer, D_2O
Gmelinite[14]	6·9, 3·4–4·1	$5·8 \times 10^{-8}$ (45°)	$2·0 \times 10^{-2}$	8·1 ± 0·3	Tracer, D_2O
Na-X[83]	≈7·4	$2·1_1 \times 10^{-5}$ (40°)	—	6·9	NMR
Ca-X[83]	≈7·4	$2·4_1 \times 10^{-5}$ (40°)	—	6·8	NMR
Ca-Y[83]	≈7·4	—	—	5·6	NMR
Ice[84, 85]	—	1×10^{-10} (−2°)	—	13·5 ± 1·1	Tracer, $H_2O^{(18)}$ Dielectric relaxation
Water[83, 86]	—	$3·8_7 \times 10^{-5}$ (45°)	$5·6 \times 10^{-2}$	4·6	Tracer, $H_2O^{(18)}$ Dielectric relaxation

Thus the monovalent ions show "stickiness" to preferred sites comparable with that shown by water, which would be expected if cations and water molecules tend to be associated. The divalent ions however show enhanced "stickiness", so that the argument of ion-water association in diffusion must not be taken too far.

For non-polar gases "stickiness" of the kind shown by water is less in evidence and it is in any event most improbable that the gas migrates as ion-gas clusters. Here the topology of the framework, and especially window dimensions, are more important.

8.2 *n*-Alkanes

Quig and Rees[88] measured tracer diffusion coefficients of C_5, C_6, C_7 and C_9 *n*-paraffins in (Na, Ca)-*A* with 32·8 % exchange of Na^+ by Ca^{2+}. This zeolite composition falls within the range of compositions studied by Barrer and Clarke[45] who measured $\bar{D}$ and $\tilde{D}$ for C_4, C_6 and C_9 *n*-paraffins. Thus these two investigations are in part complementary. Both were made under constant pressure conditions. In the tracer measurements the zeolite was equilibrated with *n*-paraffin at pressure p and then at $t = 0$ part of the gas phase was replaced by the corresponding mono-deuterated *n*-alkane, the gas phase, still at pressure p, being circulated through the zeolite bed. The rate of exchange was monitored with a mass spectrometer. Because of the rectangular sorption isotherms all measurements refer to near saturation of the zeolite with *n*-alkane.

The values of $\bar{D}^*$ are given in Table 14. Keeping Q_∞ nearly constant and varying Q_0 the authors measured the initial slope k of the plots of M_t/M_∞ against $\sqrt{t}$ and thence found the values of $\tilde{D}$ $\left(k = \frac{2A}{V}\left(\frac{\tilde{D}}{\pi}\right)^{1/2}\right)$ also given in Table 14. They considered, since the pressure was always high enough to maintain virtual saturation of the crystals, that Q_∞ was nearly constant as noted above. However the variable pressure method used for evaluating $\tilde{D}$ does not ensure constant $\partial \ln p/\partial \ln C$ which may be very large and sensitive to small pressure changes for rectangular isotherms (cf § 6.4). Thus the significance of $\tilde{D}$ is less clear than for truly constant pressure runs. Also, a volumetric rather than a gravimetric procedure was used which is less accurate for condensable vapours. Nevertheless a very limited study of concentration dependence of $\tilde{D}$ for *n*–C_6H_{14} at 106°C and of *n*–C_9H_{20} at 102°C indicated no significant change in $\tilde{D}$ with Q_0. Accordingly $\tilde{D}$ was thought to be close to $\bar{D}$ for all the diffusants.

For tracer diffusion Table 14 shows that D^* decreases and E increases with chain length and that ΔS^* also becomes increasingly and very strongly positive with carbon number. On the other hand, for $\tilde{D}$, E decreases and ΔS^* is increasingly negative as carbon number becomes larger. The authors considered that hole formation was necessary for molecules to diffuse and that this was the origin of positive values of ΔS^*. However hole formation at a given point, since it requires simultaneous cooperative movements of molecular segments away from the point might be expected to give rise to a negative entropy contribution. A large positive entropy of activation could

TABLE 14

Values of D^* and of $\tilde{D}$ for some n-alkanes in 32·8% Ca-exchanged zeolite A[88]

n-alkane	25°C		100°C	
	$\tilde{D}$	$\bar{D}^*$	$\tilde{D}$	$\bar{D}^*$
(a) Diffusion coefficients (m^2s^{-1})				
C_5H_{12}	$2\cdot4 \times 10^{-20}$	$8\cdot5 \times 10^{-20}$	$1\cdot3 \times 10^{-17}$	$4\cdot2 \times 10^{-17}$
C_6H_{14}	$2\cdot0 \times 10^{-20}$	$3\cdot5 \times 10^{-21}$	$7\cdot6 \times 10^{-18}$	$1\cdot2 \times 10^{-17}$
C_7H_{16}	—	$1\cdot15 \times 10^{-21}$	—	$1\cdot2 \times 10^{-17}$
C_9H_{20}[a]	145×10^{-20}	$7\cdot2 \times 10^{-29}$	$3\cdot2 \times 10^{-18}$	$2\cdot4 \times 10^{-24}$
(b) Activation energies E(kJ mol^{-1}) and entropies[b], ΔS^*($JK^{-1}mol^{-1}$)				
	$E_{\tilde{D}}$	$E_{\bar{D}*}$	$\Delta S_{\tilde{D}*}$	$\Delta S^*_{\bar{D}*}$
C_5H_{12}	74	76·5	$-11\cdot_6$	+7·4
C_6H_{14}	72·7	99·0	−17·0	+57·5
C_7H_{16}	—	115·0	—	+101
C_9H_{20}	$64\cdot_4$	183·9	$-46\cdot_1$	+195

[a] Extrapolated values from the experimental range 140 to 180°C.
[b] Assuming a jump distance per unit diffusion process of 6Å.

arise if the required activation energy was distributed among a number of degrees of freedom, as suggested in the zone theory of activation.[89] E and ΔS^* for $\tilde{D}$ are more complex than are these quantities for D^* since they involve temperature coefficients of $\partial \ln p/\partial \ln C$, according to eqns 48 and 22.

A comparison of values of D^* determined from pulsed NMR and from sorption kinetics has revealed the large differences shown in Table 15,[90] which appear to be outside any uncertainties in either method. The values found by NMR are uniformly very large and are similar in magnitude to diffusion coefficients in liquids, whereas the values found from sorption kinetics are sometimes several orders of magnitude less. Also the directions in which diffusion coefficients change with the intracrystalline concentration of the guest is not always the same when determined by the two procedures. The reasons for such differences are of much interest. In order to investigate them Kärger *et al.*[91] studied the sorption kinetics of propane and butane at 23°C and of ethane at −80°C in (Ca, Na)-A. Crystal samples weighed only 15 mg in order to minimize intercrystal diffusion as rate controlling;

and the crystals were selected as uniform in size as possible, but with graded sizes between each sample. For pure intracrystalline diffusion, the sorption kinetics are governed by the parameter D/r_0^2 where r_0 is the radius of the crystals of a given sample, regarded as equal spheres (eqns 27 and 28). To interpret the results it was necessary to make D a function of r_0. As r_0

TABLE 15

Comparison of $\bar{D}^*$ (cm^2s^{-1}) from pulsed NMR and from sorption[a] kinetics[(90)]

System	Temp. °C	Loading in NMR (molecules/ cavity)	$\bar{D}^*$ (NMR)	$\bar{D}^*$ (sorption rates)
CH_4/Ca-rich *A*	23	5	2×10^{-5}	5×10^{-10}
C_2H_6/Ca-rich *A*	23	3	2×10^{-6}	10^{-10}
C_3H_8/Ca-rich *A*	23	4	5×10^{-8}	3×10^{-11}
n-C_7H_{16}/Na-X	164	0·6	5×10^{-5}	3×10^{-9}
cyclo-C_6H_{12}/Na-X	164	1·3	$4{\cdot}5\times10^{-5}$	4×10^{-9}
C_6H_6/Na-X	164	1	2×10^{-6}	10^{-10}
C_2H_4/(Na, Ag)-X	23	3	2×10^{-5} (0% Ag)	5×10^{-7} (0% Ag)
	23	3	7×10^{-6} (10% Ag)	10^{-7} (10% Ag)
	23	3	2×10^{-6} (20% Ag)	5×10^{-8} (20% Ag)
	23	3	2×10^{-7} (50% Ag)	2×10^{-8} (50% Ag)
	23	3	10^{-7} (100% Ag)	10^{-8} (100% Ag)

[a] The values of $\bar{D}^*$ found from sorption kinetics have involved extrapolation to $\theta = 0$, when $\tilde{D} \rightarrow \bar{D} \rightarrow \bar{D}^*$.

increased from ≈ 1 to ≈ 40 μm, D increased by almost two orders of magnitude. This was interpreted to mean that sorption rates were determined by a surface barrier and not by intracrystalline diffusion. Further evidence of the same kind was presented for CH_4 diffusing in chabazite,[(90)] where the NMR values of $\bar{D}^*$ at 0°C were independent of crystallite size at about 10^{-6} cm^2s^{-1}, but those found from sorption kinetics varied from about 2×10^{-10} cm^2s^{-1} for $r_0 \approx 2{\cdot}5$ μm to values approaching that determined by NMR for crystals with $r_0 \geq 1000$ μm. Such behaviour would be expected if there was a surface resistance because as r_0 increases the surface to volume ratio decreases.

For a continuous surface resistance with a transmission coefficient small enough completely to control the sorption rates the $\sqrt{t}$ law of diffusion and other solutions of the diffusion equation appropriate for longer times are replaced by a first order kinetic expression.[(90)] For this case, and also when both barrier transmission and intracrystalline diffusion play a part, the plots of fractional uptake against $\sqrt{t}$ are sigmoid in form.[(92)] This however is contrary to general experience of sorption kinetics, where the $\sqrt{t}$ law in particular has been clearly demonstrated in many circumstances.[45,88] It must also be borne in mind that it is these directly measured sorption kinetics which are technically important because they determine rates in practical applications.

The discrepant values of diffusion coefficients measured by NMR and sorption kinetics are more probably explained as follows. The NMR values of $\bar{D}^*$ in Table 15 are all in the range 10^{-5} to 10^{-8} cm^2 s^{-1}. According to Darken's relation (eqn 21) $\bar{D}$ equals $\bar{D}^*$ in the Henry's law range of sorption equilibrium, or exceeds it for curved type I isotherms. For constant pressure sorption kinetics with diffusivities, D, independent of concentration the approximations of eqns 29 and 30 give half lives, $t_{1/2}$, for the sorption kinetics equal to $2{\cdot}2 \times 10^{-2}\, r_0^2/D$ and $2{\cdot}0 \times 10^{-2}\, r_0^2/D$ respectively. Using the mean of these values one may calculate $t_{1/2}$ for different r_0 and D as below:

	$t_{1/2}$ (s) for				
r_0(cm)	D(cm^{2}s^{-1})$=10^{-6}$	10^{-8}	10^{-10}	10^{-12}	10^{-14}
10^{-5}	$2{\cdot}1\times10^{-6}$	$2{\cdot}1\times10^{-4}$	$2{\cdot}1\times10^{-2}$	$2{\cdot}1$	$2{\cdot}1\times10^{2}$
10^{-4}	$2{\cdot}1\times10^{-4}$	$2{\cdot}1\times10^{-2}$	$2{\cdot}1$	$2{\cdot}1\times10^{2}$	$2{\cdot}1\times10^{4}$
10^{-3}	$2{\cdot}1\times10^{-2}$	$2{\cdot}1$	$2{\cdot}1\times10^{2}$	$2{\cdot}1\times10^{4}$	$2{\cdot}1\times10^{6}$
10^{-2}	$2{\cdot}1$	$2{\cdot}1\times10^{2}$	$2{\cdot}1\times10^{4}$	$2{\cdot}1\times10^{6}$	$2{\cdot}1\times10^{8}$
10^{-1}	$2{\cdot}1\times10^{2}$	$2{\cdot}1\times10^{4}$	$2{\cdot}1\times10^{6}$	$2{\cdot}1\times10^{8}$	$2{\cdot}1\times10^{10}$

Thus, to measure a diffusivity of 10^{-6} cm^2 s^{-1} from sorption rates one would need crystallites with $r_0 \geq 1$ mm. But synthetic zeolite powders mostly comprise crystallites in the size range 10^{-4} to 10^{-3} cm, so that the above tabulation indicates that it would be impossible from sorption kinetics to measure D, unless for the 10^{-4} cm crystals D was $\leq 10^{-12}$ cm^2 s^{-1} or for the 10^{-3} cm crystals D was $\leq 10^{-10}$ cm^2 s^{-1}. Therefore if the NMR values of $\bar{D}^*$ in Table 15 are correct it is hardly surprising that corresponding diffusivities obtained from sorption kinetics are grossly discrepant; such kinetics if controlled by intracrystalline diffusion would require half lives in the range $\approx 10^{-4}$ to ≈ 1 s. It can be concluded that the actual sorption rates were not determined by intracrystalline diffusion but by flow external to the crystals and/or by the rate of dissipation of the heat of sorption.

However many intracrystalline diffusivities are small enough to be measured by sorption kinetics even in synthetic zeolite crystals in the size range 10^{-4} to 10^{-3} cm (1 to 10 μm). Examples are given in Tables 5, 6, 9 and 14; and in Table 8 for Na- and K-*A* and K-mordenite, although the values for Ca-*A* in this table are, in the light of the above discussion, suspect. Also, by using natural zeolites as diffusion media, much larger crystals are available and it is then possible to use sorption kinetics to measure intracrystalline diffusivities as large as 10^{-6} cm^2 s^{-1}. Examples are given in Tables 2, 12 and 13. Some published data giving apparent intracrystalline diffusivities derived from sorption kinetics should be reexamined in terms of the foregoing criteria.

9. CO- AND COUNTER-DIFFUSION

Except for tracer diffusion there have been few studies of co- or counter-diffusion for which quantitative interpretations have been attempted, although this process is of much importance in zeolite catalysts. In one investigation, the kinetics of simultaneous diffusion of benzene and *n*-heptane in zeolite Na-X were investigated, the result being shown in Fig. 22.[93] *N*-heptane was sorbed more rapidly than benzene, but over a longer period was displaced again by the slower moving but more strongly sorbed benzene. Thus the system exhibited both co- and counter-diffusion in the early and later stages respectively.

The quantitative description of these processes is best formulated using the thermodynamics of irreversible processes. Kärger and Bulow[94] wrote for the two fluxes, J_1 and J_2

$$\left.\begin{aligned} J_1 &= -L_{11}\,\mathrm{grad}\,\mu_1 - L_{12}\,\mathrm{grad}\,\mu_2 \\ J_2 &= -L_{22}\,\mathrm{grad}\,\mu_2 - L_{21}\,\mathrm{grad}\,\mu_1 \end{aligned}\right\} \tag{76}$$

where the cross-coefficients satisfy the Onsager relation $L_{12} = L_{21}$. L_{11} and L_{22} are the straight coefficients and μ_1 and μ_2 the chemical potentials of species 1 and 2 respectively. If the two sorbates in the gas phase behave like perfect gases and provided sorption occurs isothermally, then since for the ith component

$$\mu_i = \mu_i^\theta + RT \ln p_i\,(C_1, C_2) \tag{77}$$

one obtains

$$\mathrm{grad}\,\mu_i = \frac{RT}{p_i}\left\{\frac{\partial p_i}{\partial C_1}\,\mathrm{grad}\,C_1 + \frac{\partial p_i}{\partial C_2}\,\mathrm{grad}\,C_2\right\}. \tag{78}$$

The dependence of the partial pressure p_i on the ith sorbed component is determined from $C_i = f(p_1, p_2)$ the isotherm for sorption of the mixture.

The flux equations can now be written as

$$\left.\begin{aligned} J_1 &= -D_{11}\,\mathrm{grad}\,C_1 - D_{12}\,\mathrm{grad}\,C_2 \\ J_2 &= -D_{21}\,\mathrm{grad}\,C_1 - D_{22}\,\mathrm{grad}\,C_2 \end{aligned}\right\} \tag{79}$$

where

$$D_{ij} = RT\left\{L_{11}\,\frac{\partial \ln p_i}{\partial C_j} + L_{12}\,\frac{\partial \ln p_{k(\neq j)}}{\partial C_j}\right\}, \tag{80}$$

so that the corresponding diffusion equations are

$$\left.\begin{aligned}\frac{\partial C_1}{\partial t} &= \operatorname{div}(D_{11}\operatorname{grad} C_1) + \operatorname{div}(D_{12}\operatorname{grad} C_2)\\ \frac{\partial C_2}{\partial t} &= \operatorname{div}(D_{21}\operatorname{grad} C_1) + \operatorname{div}(D_{22}\operatorname{grad} C_2).\end{aligned}\right\} \quad (81)$$

Karger and Bulow then assumed that the cross-coefficients L_{21} and L_{12} were zero and defined a diffusion coefficient D_i by

$$D_i = RT\,L_{ii}/C_i \quad (82)$$

D_i was considered to be independent of concentration, and the above relation with $L_{21} = L_{12} = 0$ enabled eqn 80 to be rewritten as

$$D_{ij} = D_i \frac{C_i}{C_j} \frac{\partial \ln p_i}{\partial \ln C_j}. \quad (83)$$

For $i = j$ this relation reduces to Darken's relation (eqn 21) for the concentration dependence of the diffusivity during sorption of a single pure component, except that it is not an assumption of Darken's relation that D_i is independent of concentration. Further assumptions were that Langmuir's isotherm described the sorption equilibria, that $D_1 \gg D_2$ and that the sorbent particles were equal spheres of radius r_0. The concentrations $C_i\,(r, t)$ at the crystal boundary are the equilibrium ones given by

$$C_i = C_{\text{sat}}\, b_i p_i/(1 + b_1 p_1 + b_2 p_2) \quad (84)$$

where C_{sat} is the concentration of adsorbed component at saturation of the zeolite. It was taken to be the same for both components 1 and 2. Then

$$\frac{\partial C_1}{\partial t} = \frac{1}{r^2}\frac{\partial}{\partial r}\left[r^2\left(D_{11}\frac{\partial C_1}{\partial r} + D_{12}\frac{\partial C_2}{\partial r}\right)\right] \quad (85)$$

$$\frac{\partial C_2}{\partial t} = \frac{1}{r^2}\frac{\partial}{\partial r}\left[r^2\left(D_{21}\frac{\partial C_1}{\partial r} + D_{22}\frac{\partial C_2}{\partial r}\right)\right]. \quad (86)$$

Provided $D_1 \gg D_2$ (as assumed) $\partial C_1/\partial r$ can be replaced by

$$\frac{\partial C_1}{\partial r} = -\frac{D_{12}}{D_{11}}\frac{\partial C_2}{\partial r} \quad (87)$$

which follows from eqn 85 by setting $\partial C_1/\partial t = 0$.

Thus

$$\frac{\partial C_2}{\partial t} = \frac{1}{r^2}\frac{\partial}{\partial r}\left[r^2\left(D_{22} - \frac{D_{21}D_{12}}{D_{11}}\right)\frac{\partial C_2}{\partial r}\right]. \quad (88)$$

The coefficients in eqn 88 are obtained after combining eqns 83 and 84 as

$$\left.\begin{aligned}D_{11} &= D_1\frac{C_{\text{sat}}-C_2}{C_{\text{sat}}-C_1-C_2}\\D_{12} &= D_1\frac{C_1}{C_{\text{sat}}-C_1-C_2}\\D_{21} &= D_2\frac{C_2}{C_{\text{sat}}-C_1-C_2}\\D_{22} &= D_2\frac{C_{\text{sat}}-C_2}{C_{\text{sat}}-C_1-C_2}\end{aligned}\right\} \tag{89}$$

Thus finally for component 2

$$\frac{\partial C_2}{\partial t}=\frac{1}{r^2}\frac{\partial}{\partial r}\left(r^2 D_2\frac{C_{\text{sat}}}{C_{\text{sat}}-C_2}\frac{\partial C_2}{\partial r}\right). \tag{90}$$

This expression is the same as that used by Garg and Ruthven[95] for sorption kinetics of the single pure component 2, so that according to the assumptions made the slower component diffuses into the crystal independently of the faster one.

From eqns 87 and 89 one finds

$$\frac{\partial \ln C_1}{\partial r}=\frac{\partial \ln (C_{\text{sat}}-C_2)}{\partial r} \tag{91}$$

so that

$$C_1(r,t)=k(C_{\text{sat}}-C_2(r,t)) \tag{92}$$

where k is independent of r and t. From eqn 92 with $r=r_0$ and $t=\infty$ one obtains

$$k=\frac{C_1(p_1,p_2)}{C_{\text{sat}}-C_2(p_1,p_2)} \tag{93}$$

or, with eqn 84

$$k=\frac{b_1p_1}{1+b_1p_1}. \tag{94}$$

Since C_{sat} and k do not depend upon r, the form of eqn 92 is preserved during integration and division by the crystal volume so that the relationship between the experimentally observable concentrations $C_i(t)$ is

$$\frac{C_1(t)}{C_{1,\infty}}=\frac{C_{\text{sat}}-C_2(t)}{C_{\text{sat}}-C_{2,\infty}}. \tag{95}$$

The experimental findings[93,94] illustrated in Fig. 22 were discussed in terms of the foregoing treatment, where the indices 1 and 2 denote *n*-heptane and benzene respectively. Thus, in agreement with eqn 90 sorption of the component of lower mobility, benzene, is seen to increase monotonically towards its final equilibrium value. The uptake of *n*-heptane, as expected from eqn 95, initially exceeds the equilibrium value and declines towards its final value in much the same way as is observed according to Fig. 22. The fact that the *n*-heptane uptake does not instantly reach the value given by eqn 95 is a consequence of the finite rate of transport of the *n*-heptane into and within the crystals.

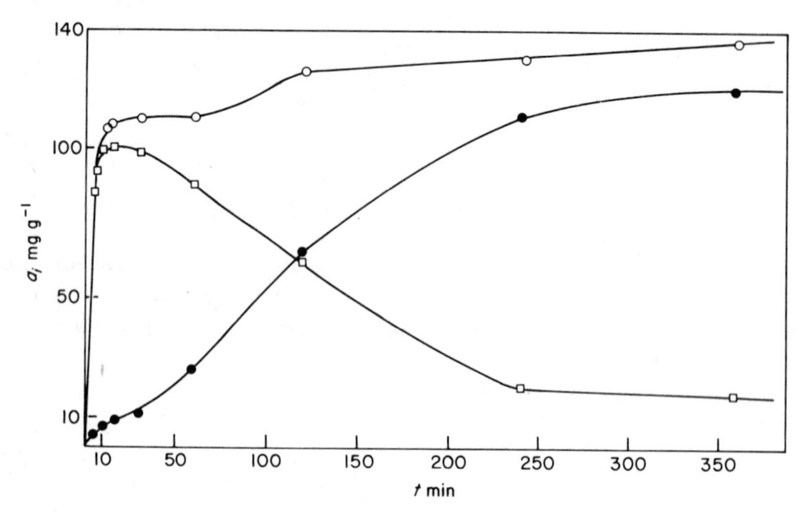

FIG. 22. The kinetics of simultaneous uptake of *n*-heptane and benzene vapours from their azeotropic mixture by zeolite Na-X at 85°C[93]. The saturated vapours from the mixture were continuously circulated through the sorbent with nitrogen as carrier gas.

●, C_6H_6; □, *n*-C_7H_{16}; ○, total uptake.

The zeolites used in the preceding studies were faujasites, which have three-dimensional channel systems. Riekert[96] pointed out that in two- or three-dimensional channel systems, diffusion can occur more readily than in one-dimensional parallel channel systems because in the one-dimensional case, co- or counter-diffusing molecules can no longer by-pass each other by choosing alternative unblocked pathways (§ 7). In a one-dimensional channel system the slower moving molecule may act as a high energy barrier towards the faster moving one, or may limit its migration to a rate little if any faster than that of the slowly moving component when both components are being simultaneously sorbed or desorbed. Examples of this kinetic "poisoning" were first observed in separations of mixtures involving molecules such as acetone in chabazite.[53, 54] Acetone diffused only very slowly in chabazite and greatly reduced the sorption rates of other non-polar components from their mixtures with acetone, even though chabazite provides a three-dimensional channel system. In one-dimensional systems, interference by slow-moving large molecules should be very marked, and the treatment summarized in § 7 illustrates in the limiting case of zero mobility of the blocking component now great the effects on the second diffusant can be.

10. FLOW METHODS OF STUDYING DIFFUSION

The remainder of this chapter will be devoted to considering ways in which flow through columns of pelleted zeolite or through zeolite compacts can give information about diffusion within zeolites. Two procedures will be exemplified: in one there is a carrier gas to which are added pulses of the gas or vapour whose diffusion properties are the subject of investigation,[92,93] in the other there is no carrier gas and the permeation rate and time-lag in establishing the steady state of permeation are to be measured and interpreted.[4]

10.1 The Chromatographic Procedure

The linear velocity of the carrier gas is U and can be varied. The carrier gas passes through a column of pelleted zeolite, so that fairly large and representative samples of the solid can be studied over wide temperature ranges. A pulse of another gas or vapour is injected into the carrier gas and the dispersion of the emergent pulse determined. By doing this at a series of values of U the effective diffusivity of the second gas in the stationary phase (the pellets) can be found. This method is a transient state one where dead-end as well as continuous through channels in the pellets are involved.

In pelleted and extruded zeolites there are macropores connecting with the micropores in the individual crystallites, the micropores having dimensions comparable with those of molecules. Each type of pore system should be reflected in the dispersion of the emergent pulse of the gas being studied. The height, H, of an equivalent theoretical plate is related to U by the expression[97]

$$H = A + \frac{B}{U} + CU \tag{96}$$

where A and B are constants associated with dispersion due to eddy and molecular diffusion respectively in the mobile phase (the gas phase) while C is related to mass transfer effects of the immobile phase (the pellets) forming the chromatographic column; C is found by measuring H as a function of U. It is the limiting slope at high velocities. Similarly A is the intercept on the axis of H of the high velocity asymptote. Once A and C are determined, B

can also be found. Plots of H against U show minima, as illustrated in Fig. 23, at which $B = CU^2$.

In the work of Eberly[2] the relation between H and independently measurable quantities was taken to be[98]

$$H = \frac{LW_e^2}{8t_m t_e}, \tag{97}$$

where L is the length of the chromatographic column, W_e is the width of the pulse at a fraction $1/e = 0{\cdot}368$ of the pulse height; and t_m and t_e are the retention times at the maximum and at a fraction $0{\cdot}368$ of the pulse height respectively.

The constant C is related to diffusivities D_1 and D_2 in the mobile phase (the gas) and in the immobile phase (the pellets) respectively. Eberly used the relationship

$$C = \frac{\dfrac{F_1^2 d_p^2}{75(1-F_1)^2 D_1} + \dfrac{F_1 K d_p^2}{2\pi^2(1-F_1)D_2}}{\left[1 + K\dfrac{F_1}{1-F_1}\right]^2} \tag{98}$$

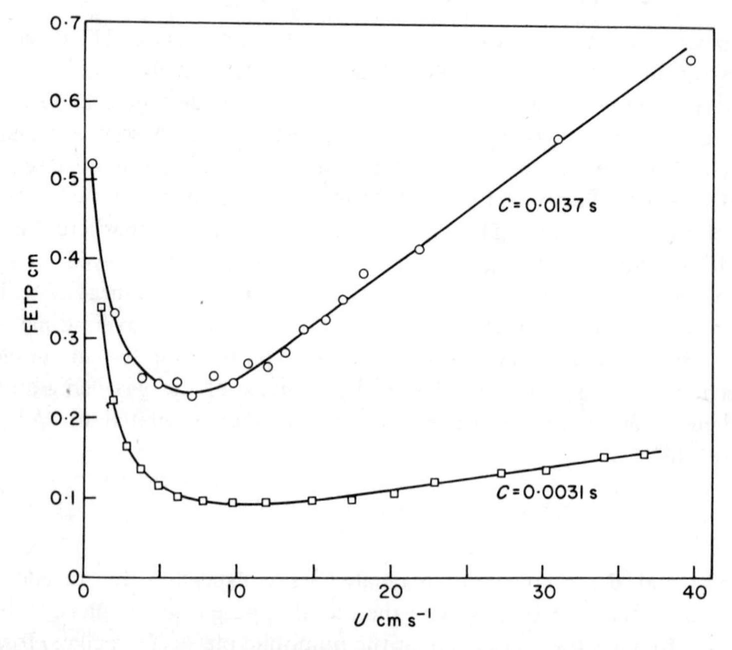

FIG. 23. Height of a theoretical plate as a function of the veolocity, U, for a column in which the immobile phase was pelleted zeolite K-A.[2]
○, He pulses injected into Ar carrier gas
⊡, Ar pulses injected into He carrier gas.

where F_1 is the volume fraction of voids in the packed column, d_p is the diameter of the pellet and K is the distribution coefficient, defined as the equilibrium ratio of the concentration, C_1, in the mobile phase to the sum of the amount, C_p, within the pores of the solid per unit pellet volume plus the amount, C_a, sorbed per unit pellet volume:

$$K = \frac{C_1}{C_2}\,;\; C_2 = C_p + C_a. \tag{99}$$

In absence of sorption ($C_a \ll C_p$), $K = C_1/C_p = 1/\epsilon$ where ϵ is the porosity of the pellet. More generally however

$$1/K = \epsilon + C_a/C_1. \tag{100}$$

The sorption equilibrium constant, K_a, defined as mols sorbed per unit volume of column divided by mols per unit volume of gas phase, is given by

$$K_a = F_2 \frac{C_a}{C_1} \tag{101}$$

where F_2 denotes the volume fraction of pellets in the packed column. It has been shown[99] that

$$K_a = (t - t_0)\, UF_1/L \tag{102}$$

where t and t_0 are the retention times of the adsorbable material and of non-adsorbable material respectively. Thus, combining the last three relations

$$1/K = \epsilon + (t - t_0)\, UF_1/F_2L. \tag{103}$$

This value of K was then used in the expression for C (eqn 101) to evaluate D_2. D_1 was determined as suggested by Satterfield and Sherwood.[100]

D_2 will be influenced by each diffusion process within the pellet, which includes intracrystalline diffusion within each crystallite in the pellet. To demonstrate a role of intracrystalline diffusion zeolite K-A was used, which was assumed at 27°C to admit helium but not argon. Within argon pulses in helium as carrier H was for all U much smaller than with helium pulses in argon. The respective values of C were 0·0031 s (Ar pulses in He) and 0·0137 s (helium pulses in Ar). The experiment indicated that a large part of the pulse dispersion was associated with diffusion in the micropores of the zeolite itself.

Some values of K and D_2 for experiments involving pellets of Na-faujasite are illustrated in Table 16(a) and in Table 16(b) values of heats of sorption and apparent activation energies are also given. The question of how far the properties of D_2 are influenced by intracrystalline diffusion and how far by intrapellet diffusion is not fully clarified. The substantial temperature coefficients in D_2 certainly support the view that intracrystalline diffusion plays a significant part, but the numerical values of D_2 are large even for pellets made from the less open zeolite, mordenite. However this account of Eberly's work serves to illustrate one way in which pulse dispersion has been interpreted.

The methods of moments[101] can also be used in the Henry's law range to give the same kind of information as that obtained above. As before, a

TABLE 16

Diffusion in zeolite pellets studied by pulse dispersion in chromatographic columns
(a) Values of K and of D_2 for gases in Na-faujasite[2]

Gas	T°C	K	D_2(cm^{2}s^{-1})
Argon	27	0·598	0·039
	82	1·18	0·077
	154	1·68	0·180
	227	1·92	0·267
	304	1·92	0·776
Krypton	27	0·149	0·005
	82	0·305	0·019
	154	0·592	0·059
	227	0·886	0·129
	304	1·12	0·207
	427	0·910	0·366
SF_6	154	0·0487	0·001
	227	0·126	0·005
	304	0·260	0·016
	427	0·406	0·055

(b) Heats, ΔH, and apparent energies of activation, E, for diffusion (kcal mol^{-1})

Gas	Solid	$-\Delta H$	E
Argon	Na-Mordenite	4·1	7·6
	H-mordenite	3·6	6·6
	Ca-*A*	—	3·5
	Na-faujasite	—	2·9
	Silica-alumina	—	2·5
Krypton	Na-mordenite	4·3	9·9
	H-mordenite	4·1	9·0
	Ca-*A*	3·6	5·9
	Na-faujasite	3·7	4·5
	Silica-alumina	—	4·0
SF_6	H-mordenite	6·1	14·6
	Na-faujasite	5·0	7·5
	Silica-alumina	4·2	7·2

pulse of sorbable gas is introduced into the carrier gas at the entrance to the diffusion column. The effluent peak is spread and the peak height lowered by the diffusion processes within the column. The nth moment, m_n, the nth absolute moment, μ_n, and the nth central moment μ_n' are defined respectively by

$$\left.\begin{aligned} m_n &= \int_0^\infty C_e t^n \mathrm{d}t \\ \mu_n &= m_n/m_0 \\ \mu_n' &= \int_0^\infty C_e(t-\mu_1)^n \mathrm{d}t \Big/ \int_0^\infty C_e \mathrm{d}t \end{aligned}\right\} \qquad (104)$$

In eqns 104 C_e is conveniently chosen as the concentration in the emergent pulse at the exit to the column. Higher moments are subject to large experimental errors so that μ_1 and μ_1' are of most significance. The expression for μ_1 involves only physical constants of the system and the Henry's law adsorption constant, and serves to give this adsorption constant. The expression for μ_1' contains in addition the relevant diffusion coefficient in the mobile phase and the effective diffusion coefficient within the pellet. The latter in turn may be influenced both by diffusion along intercrystalline channels within the pellet and by intracrystalline diffusion.

10.2 Flow through Compacts with a Zeolite Component

Zeolite crystallites can be compacted under pressure, with or without additional filler to reduce intercrystallite spaces. For flow of sorbable components through such compacts, in absence of any carrier gas, it should be possible to study and separate the several transport processes involved. The basis of the treatment involved follows.[4, 102, 103]

The membrane or compact is bounded by the planes $x = 0$, $x = l$ and flow occurs in the x-direction. The boundary conditions are:

(i) $C = C_0$ at $x = 0$ for $t > 0$
(ii) $C = C_i(x)$ for $0 < x < l$ at $t = 0$
(iii) $C = C_l$ at $x = l$ for $t > 0$.

The total flow through unit cross-section of membrane normal to x is given by

$$J = J_g + J_s + J_i \qquad (105)$$

where "g" "s" and "i" denote "gas phase", "surface-generated" and "intracrystalline" respectively. If it is assumed that local sorption equilibrium is established and maintained at each plane x between gaseous, adsorbed and intracrystalline diffusant then for the chemical potentials we have

$$\mathrm{d}\mu_g = \mathrm{d}\mu_s = \mathrm{d}\mu_i. \tag{106}$$

The Fick expression for eqn 105 can be written as

$$J = -\left[D_g(1 + \alpha C_g) + D_s \frac{\partial C_s}{\partial C_g} + D_i \frac{\partial C_i}{\partial C_g}\right] \frac{\partial C_g}{\partial x} \tag{107}$$

where the D are diffusion coefficients which in order refer to gas phase flow, to the extra flow due to the mobile adsorbed film and to intracrystalline flow, and where for the gas phase flow the term αC_g allows for a possible Poiseuille flow component. C_s, C_g, and C_i are all concentrations in molecules per cm^3 of porous medium.

In the steady state, J is constant through each cross-section and so

$$Jl = \int_{(C_g)_l}^{(C_g)_0} \left[D_{gs}(1 + \alpha C_g) + D_{ss} \frac{\partial C_s}{\partial C_g} + D_{is} \frac{\partial C_i}{\partial C_g}\right] \partial C_g \tag{108}$$

where the additional subscript "s" to the three D indicates steady state values. One may differentiate eqn 108 w.r.t. $(C_g)_0$ keeping $(C_g)_l$ constant (in practice often close to zero) and one obtains

$$l\left(\frac{\partial J}{\partial (C_g)_0}\right)_{(C_g)_l} = \left[D_{gs}(1 + \alpha C_g) + D_{ss} \frac{\partial C_s}{\partial C_g} + D_{is} \frac{\partial C_i}{\partial C_g}\right]. \tag{109}$$

The corresponding expression in terms of the phenomenological coefficients of irreversible thermodynamics is

$$l \frac{\partial J}{\partial (C_g)_0} = \frac{RT}{(C_g)_0} [L_{11}^g + L_{11}^s + L_{11}^i] \tag{110}$$

where L_{11}^g, L_{11}^s and L_{11}^i are the straight coefficients corresponding with $D_g(1 + \alpha C_g)$, D_s and D_i respectively. It follows that from a master plot (Fig. 24) of J against $(C_g)_0$ one may obtain the right-hand sides of eqn 109 and 110 as functions of $(C_g)_0$.

If helium is used as a non-sorbed calibrating gas (i.e. J_s^{He} and J_i^{He} are zero), and, in absence of a viscous flow component in the gas phase, at a given temperature we assume for D_{gs} for any other gas that

$$D_{gs}/D_{gs}^{He} = \sqrt{M_{He}/M} \tag{111}$$

where the M are molecular weights. Next, the intracrystalline flow of a mobile sorbate may be reduced to zero by nearly filling the crystals with a polar strongly sorbed molecule such as water which is immobile and has negligible vapour pressure at the temperature of the flow experiment with the mobile sorbate. This allows one to measure the adsorption upon external

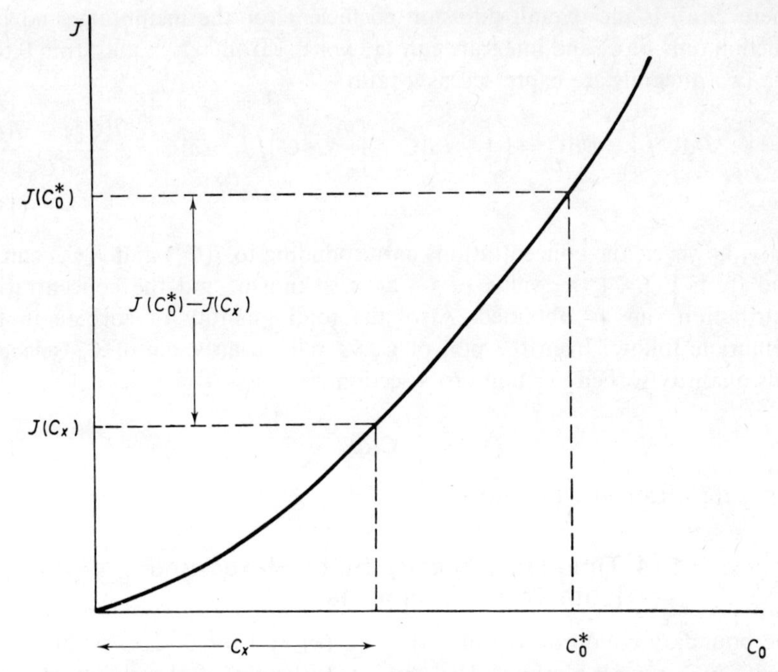

FIG. 24. Plot of the steady-state flux through a compact as a function of concentration at x = 0 for the boundary conditions of eqn. 104.[102]

surfaces only and so obtain $\partial C_s/\partial C_g$. Because $D_{is}\partial C_i/\partial C_g$ is zero and since we know D_{gs} we may estimate D_{ss}.

To complete the experiment the membrane is outgassed, the total sorption measured and thence, subtracting the previously determined external adsorption, the intracrystalline sorption isotherm and its slope σ_i are found. Finally, for the total flow, from the value of $l\partial J/\partial(C_g)_0$ and knowing D_{gs}, $\partial C_s/\partial C_g$, D_{ss} and $\partial C_i/\partial C_g$, the value of D_i may be found.

10.3 Concentration Distribution and Equilbrium Uptake

The master curve of Fig. 24 shows $J = J(C_0)$ plotted as a function of ingoing concentration

$$C_0 = (C_g + C_s + C_i)_{x=0}. \tag{112}$$

In one particular experiment suppose $C_0 = C_0^*$ and in another $C_0 = C_x$. Then the corresponding fluxes are as indicated on the ordinate. We may now take the relation

$$J(C_0^*) = -\frac{D(C)\partial C}{\partial x} \tag{113}$$

where $D(C)$ is the overall diffusion coefficient for the membrane and is a function only of C, and integrate eqn 113 w.r.t. x from 0 to x and from 0 to l. The two integrals are expressed as a ratio:

$$\frac{x}{l} = \int_{C_x}^{C_0^*} D\mathrm{d}C \Big/ \int_0^{C_0^*} D\mathrm{d}C = \left(\int_0^{C_0^*} D\mathrm{d}C - \int_0^{C_x} D\mathrm{d}C\right) \Big/ \int_0^{C_0^*} D\mathrm{d}C = \frac{J(C_0^*) - J(C_x)}{J(C_0^*)} \tag{114}$$

Since, however, the concentrations corresponding to $J(C_0^*)$ and $J(C_x)$ can be read from Fig. 24 the value of C_x at x is known, and the concentration distribution can be obtained. Also, the total quantity of sorbate in the membrane follows from the plot of C_x vs x for each value of C_0^* selected. This quantity sorbed per unit cross-section is

$$Q_\infty = \int_0^l C_x \mathrm{d}x \tag{115}$$

and is the area under the curve.

10.4 Time-lags, Steady State Fluxes and Isotherms in Compacts

The boundary condition (ii) of § 10·2 ($C_i(x) = 0$ for $0 < x < l$ at $t = 0$) results in a time-lag at $x = l$ in the establishment of the steady state of

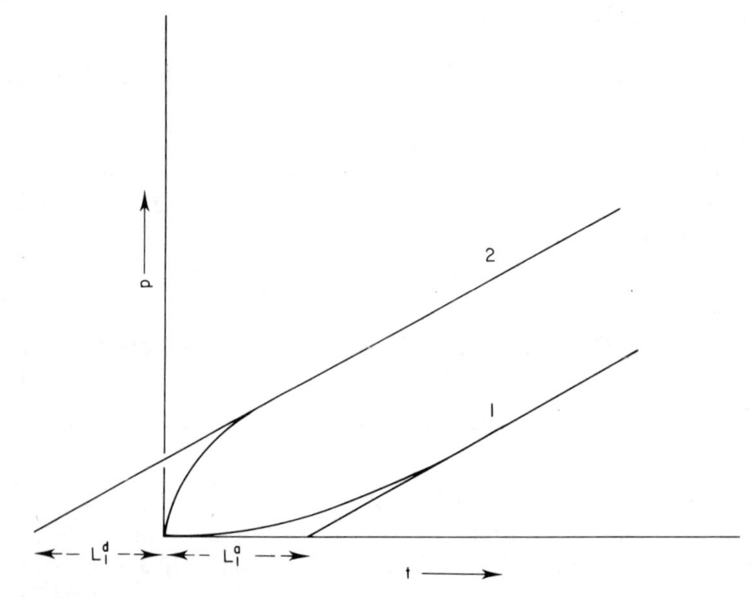

FIG. 25. Pressure changes at the face $x = l$ of a compact as a function of time, for p_0 (at $x = 0$) $\gg p_l$ at ($x = l$) for all t. Curve 1 illustrates the adsorption time-lag and curve 2 the desorption time-lag at, in each case, $x = l$.[104]

flow, which is defined as shown in Fig. 25. The corresponding negative time-lag at $x = 0$ is shown in the same figure.

We now consider, analogous to C_g, C_s and C_i, the concentrations C'_g, C'_s and C'_i denoting concentrations per unit volume of gas phase, per unit area of surface and per unit intracrystalline volume respectively. If the porosity is ϵ (cm^3 cm^{-3} of membrane), the external surface area of the zeolite crystals and binder is A (cm^2 cm^{-3}) and the intracrystalline free volume is V (cm^3 cm^{-3}), one then obtains

$$C_g = \epsilon C'_g \; ; \; C_s = AC'_s \; ; \; C_i = VC'_i. \tag{116}$$

The transient state form of Fick's equation for flow is now written as

$$\frac{\partial C}{\partial t} = \frac{\partial}{\partial x}\left(D_1 \frac{\partial C_g}{\partial x}\right) \tag{117}$$

where $C = (C_g + C_s + C_i)$ and $D_1 = \left(D_g + D_s \dfrac{\partial C_s}{\partial C_g} + D \dfrac{\partial C_i}{\partial C_g}\right)$. With these relations eqn 117 becomes

$$(\epsilon + A\sigma_s + V\sigma_i)\frac{\partial C'_g}{\partial t} = \frac{\partial}{\partial x}\left[(\epsilon D_g + A\sigma_s D_s + V\sigma_i D_i)\frac{\partial C'_g}{\partial x}\right] \tag{118}$$

where $\sigma_s = \mathrm{d}C'_s/\mathrm{d}C'_g$ and $\sigma_i = \mathrm{d}C'_i/\mathrm{d}C'_g$ are the slopes of the isotherms for adsorption upon external surfaces and for intracrystalline sorption respectively.

In the Henry's law range of adsorption and of intracrystalline sorption

$$A\sigma_s = Ak_s \; ; \; V\sigma_i = Vk_i \tag{119}$$

where k_s and k_i are the Henry's law equilibrium constants for adsorption and intracrystalline sorption respectively. Then from eqns 118 and 119 we obtain in this limit

$$\frac{\partial C'_g}{\partial t} = \left(\frac{\epsilon D_g + Ak_s D_s + Vk_i D_i}{\epsilon + Ak_s + Vk_i}\right)\frac{\partial^2 C'_g}{\partial x^2}. \tag{120}$$

For this equation the time lag at $x = l$ for the previously stated boundary condition (ii) of eqn 104 ($C_i(x) = 0$ for $0 < x < l$ at $t = 0$) is

$$L_l = \frac{l^2(\epsilon + Ak_s + Vk_i)}{6(\epsilon D_g + Ak_s D_s + Vk_i D_i}. \tag{121}$$

Proceeding just as before with helium as non-sorbed calibrating gas we have, at constant temperature,

$$L_{\mathrm{He}} = l^2/6D_g^{\mathrm{He}}$$
$$D_g/D_g^{\mathrm{He}} = (M_{\mathrm{He}}/M)^{1/2} \tag{122}$$

so that D_g is found, k_s and k_i are determined for the sorbable gas from the slopes of the Henry's law isotherms, measuring each sorption in the way already described when considering the steady state. A is found from krypton sorption measurements with the crystals filled with water; ϵ is measured separately using the helium method for example; and V is known from the amount of zeolite present. Again there is enough information first to find D_s and then to find D_i.

Four limiting time lags (Fig. 25) are of special interest,[104,105,106] as tabulated below for particular cases of the boundary condition (ii) of eqn 104.

Boundary condition	*Time lag*	*Sign*
(iia) $C_i(x) = 0$ for $0 < x < l$ at $t = 0$	L_l^a at $x = l$	+ve
	L_o^a at $x = 0$	−ve
(iib) $C_i(x) = C_0$, a constant, for	L_l^d at $x = l$	−ve
$0 < x < l$ at $t = 0$	L_o^d at $x = 0$	+ve

For "adsorption time lags", L_l^a and L_0^a, the membrane-diffusant system relaxes to its steady state flow and distribution of diffusant by sorbing diffusant. For "desorption time lags" this relaxation requires desorption of some diffusant.

These time lags can be obtained rigorously[103] using only the conservation of matter condition for flow in the x-direction, i.e.

$$\partial C/\partial t = -\partial J/\partial x. \tag{123}$$

TABLE 17

Time lags and related quantities

$$lJ_\infty L_l^a = \int_0^l xC(x)\mathrm{d}x + I_a$$

$$lJ_\infty L_0^a = \int_0^l (x - l)C(x)\mathrm{d}x + I_a$$

$$J_\infty(L_l^a - L_0^a) = J_\infty \Delta L^a = \int_0^l C(x)\mathrm{d}x = M_\infty$$

$$lJ_\infty L_l^d = \int_0^l xC(x)\mathrm{d}x - lM_0/2 + I_d$$

$$lJ_\infty L_0^d = \int_0^l (x - l)C(x)\mathrm{d}x + lM_0/2 + I_d$$

$$J_\infty(L_l^d - L_0^d) = J_\infty \Delta L^d = \int_0^l (C(x) - C_0)\mathrm{d}x = (M_\infty - M_0)$$

$$lJ_\infty(L_l^a - L_l^d) = lJ_\infty \delta L_l = lM_0/2 + (I_d - I_a)$$

$$lJ_\infty(L_0^a - L_0^d) = lJ_\infty \delta L_0 = -lM_0/2 + (I_d - I_a)$$

$$J_\infty(\delta L_l - \delta L_0) = J_\infty \delta \Delta L = M_0$$

This relation is independent of any mechanism or equation of flow. It obtains for molecular streaming, streamline, turbulent or orifice flow regimes in interparticle spaces, for surface and intracrystalline flows, or for any mixture of two or more of these. If the overall diffusion is expressed in terms of a Fick-type equation ($\partial C/\partial t = \partial(D\partial C/\partial x)/\partial x$) then D may be a function of C, x, and t, or all of these, whether the functions are separable or inseparable. The rigorous expressions in Table 17 can then be obtained. In these expressions J_∞ is as before the steady state flow and Q_∞ and Q_0 are the amounts sorbed at equilibrium and at time $t = 0$ respectively. Also

$$
\begin{aligned}
I_a &= \int_0^\infty \int_0^l [J_\infty - J^a(x, t)] \, \mathrm{d}x\mathrm{d}t \\
I_d &= \int_0^\infty \int_0^l [(J_\infty - J^d(x, t)] \, \mathrm{d}x\mathrm{d}t
\end{aligned} \tag{124}
$$

where $J^a(x, t)$ is the flux through the plane at $x = x$ and at time t under adsorption conditions (iia) and $J^d(x, t)$ is this flux under desorption conditions (iib). $C(x)$ is the steady state concentration distribution, which can be obtained as indicated in § 10.3.

When D in the equation $\partial C/\partial t = \partial(D\partial C/\partial x)/\partial x$ is a function of C only, the integrals I_a and I_d are both zero.[103] This follows because $J(x, t) = - D(C)\mathrm{d}C/\mathrm{d}x$ and on integrating between 0 and l one has

$$\int_0^l J(x, t)\mathrm{d}x = \int_0^{C_0} D(u)\mathrm{d}u = lJ_\infty. \tag{125}$$

However I_a and I_d are not necessarily zero if D is a function of position or of time or both. The expressions in Table 17 afford means of finding both M_∞ and M_0. If M_0 is measured for each of a series of pressures of the gaseous diffusant then the isotherm may be found, with the membrane *in situ*, and in the same run for which by the method of § 10.3 one may find $C(x)$ and M_∞. All the above possibilities exist, but have not yet been fully developed on the experimental side.

10.5 Application of the Membrane Method

For compacts containing zeolites only one study has so far been made employing the analysis given in the previous sections. Barrer and Petropoulos[4] made membranes by compaction of graphite lubricant and zeolites Na-A or Ca-A in a steel tube as membrane holder. The graphite was intended in part to reduce damage to the crystals resulting on compaction and in part to fill interparticle spaces and so reduce gas phase transport. Table 18 illustrates some of the results obtained using nitrogen gas.

TABLE 18

Analysis of mixed flows J_g, J_s and J_i for N_2 in graphite-zeolite compacts[4]

	Membrane	B (graphite + Na-A)				H (graphite + Ca-A)			
	Temperature (°C)	0	23	50	75	0	23	50	75
Steady State	J_{is}/J_{gs}	0·32	0·29	0·19	0·20	0·48	0·37	0·26	0·20
	J_{ss}/J_{gs}	0·36	0·30	0·24	—	0·27	0·22	0·19	—
	$D_{gs} \times 10^3$ (cm²s⁻¹)	0·75	0·72	0·69	0·69	$1{\cdot}1_2$	$1{\cdot}1_0$	$1\ 0_6$	$1{\cdot}0_4$
	$D_{ss} \times 10^4$ (cm²s⁻¹)	0·51	0·75	$1{\cdot}1_1$	—	0·54	0·79	$1{\cdot}2_2$	—
	$D_{is} \times 10^6$ (cm²s⁻¹)	$1{\cdot}9_5$	$3{\cdot}2_2$	$3{\cdot}7_5$	6·5	$1{\cdot}4_6$	$2{\cdot}3_8$	$3{\cdot}5_0$	$5{\cdot}1_0$
	E_s (kcal mol⁻¹)		2·4				2·8		
	E_i (Kcal mol⁻¹)		3·6				3·1		
Time lag	J_i/J_g	$2{\cdot}6_8$	$2{\cdot}5_2$	$2{\cdot}1_2$	—				
	J_s/J_g	0·58	0·27	−0·04	—				
	$D_g \times 10^3$ (cm²s⁻¹)	0·50	0·58	0·64	—				
	$D_i \times 10^6$ (cm²s⁻¹)	10·9	22·3	39·1	—				
	E_i (kcal mol⁻¹)		4·4						

The ratios J_{is}/J_{gs}, J_{ss}/J_{gs}, J_i/J_g and J_s/J_g all decrease with rising temperature because in or near the Henry's law range of sorption the sorbed population decreases faster than the corresponding diffusion coefficient increases. The E_s and E_i are the energies of activation for D_{ss} and D_i or D_{is} respectively. They appear reasonable for N_2 gas diffusing in zeolite A. The steady state value, D_{is}, is considerably smaller than the transient state (time-lag) value, D_i. This is believed to arise from the composite nature of the membrane. Crystallites of zeolite screened behind graphite flakes are all involved in the transient state, as they fill with their equilibrium content of diffusant, but they play little part in the steady state. The steady state flow paths are tortuous because the graphite lubricant flakes tend to orient normal to the direction of compression during the compaction process.

The results in this section show that composite flows may indeed be analysed by the methods described and suggest that further studies of the same kind could give valuable information.

11. Conclusion

It will be seen from this account that diffusion in molecular sieves has made progress, but that problems of measurement and of interpretation of diffusion coefficients and their properties are still considerable. These arise in part because differential integral and tracer diffusion coefficients, except in the Henry's law range, may have different numerical values, temperature coefficients and concentration dependences. Thus comparisons *inter se* may not be appropriate. Very little quantitative work has been performed in certain important areas, such as co- and counter-diffusion, and adventitious impurities in the parent crystals, or in the diffusant may greatly modify the diffusion kinetics. Thus reproducibility can be poor in successive runs, or between sorption and desorption kinetics, or with runs made in different laboratories. Nevertheless many kinetic aspects of molecule sieving are now understood at least semi-quantitatively. The behaviour of a range of zeolite diffusion media has been explored and the kinetics of sorption of single diffusants have been studied and discussed under a variety of conditions. Interest in diffusion seems likely to continue, coupled with efforts to obtain improved diffusivity values and interpretations.

REFERENCES

1. P. B. Weisz, *Chem. Tech.* (1973) **3**, 498.
2. P. E. Eberly Jr, *Ind. Eng. Chem. Fund.* (1969) **8**, 25.
3. A. Zikanova, *J. Chromat. Sci.* (1971) **9**, 248.
4. R. M. Barrer and J. H. Petropoulos, *Surface Sci.* (1965) **3**, 126 and 143.
5. K. Chihara, M. Suzuki and K. Kawazoe, *Chem. Eng. Sci.* (1976) **31**, 505.
6. E. T. Nelson and P. L. Walker Jr, *J. Appl. Chem.* (1961) **11**, 358.
7. R. M. Barrer and D. E. W. Vaughan, *Trans. Faraday Soc.* (1967) **63**, 2275.
8. R. M. Barrer and D. E. W. Vaughan, *J. Phys. and Chem. Solids* (1971) **32**, 731.
9. R. M. Barrer and D. E. W. Vaughan, *Trans. Faraday Soc.* (1971) **67**, 2129.
10. R. M. Barrer and D. E. W. Vaughan, *Surface Sci.* (1969) **14**, 77.
11. G. G. Aleksandrov, O. G. Larinov and A. S. Ponomarev, *Russ. J. Phys. Chem.* (1967) **41**, 489.
12. C. N. Satterfield and C. S. Cheng, *Am. Inst. Chem. Eng.*, 68th National Mtg, Houston 1971, Feb. 28th–Mar. 4th; Adsorption Pt I, Paper 16 f.
13. J. W. McBain "Sorption of Gases by Solids", Routledge (1932) p. 24.
14. R. M. Barrer and B. E. F. Fender, *J. Phys. and Chem. Solids* (1961) **21**, 12.
15. A. Tiselius, *Z. Phys. Chem.* (1934) **169** A, 425.
16. A. Tiselius, *Z. Phys. Chem.* (1935) **174** A, 401.
17. C. Matano, *Jap. J. Phys.* (1933) **8**, 109.
18. C. Matano, *Proc. Phys. Math. Soc. Japan* (1933) **15**, 405.
19. W. M. Meier and D. H. Olson, *in* "Molecular Sieve Zeolites", *Am. Chem. Soc.*, Advances in Chem. Series No. 101 (1971) p. 155.
20. N. Bloembergen, E. M. Purcell and R. V. Pound, *Phys. Rev.* (1948) **73**, 679.
21. H. A. Resing, Advances in Molecular Relaxation Processes Vol. 1 (1967–8) p. 109.
22. P. Ducros, Thesis for Docteur ès Sciences Physique, Université de Paris (1960).
23. H. A. Resing and J. K. Thompson, *in* "Molecular Sieve Zeolites", Am. Chem. Soc., Advances in Chem. Series No. 101 (1971) p. 473.
24. R. Kubo and K. Tomita, *J. Phys. Soc. Japan* (1954) **9**, 888.
25. J. Karger, *Z. Phys. Chem. Leipzig* (1971) **248**, 27.

26. J. Kärger, S. P. Shdanov and A. Walter, *Z. Phys. Chem. Leipzig* (1975) **256**, 319; *also* J. Caro, J. Kärger, H. Pfeifer and R. Schollner, *Z. Phys. Chem. Leipzig* (1975) **256**, 698.
27. C. J. F. Bottcher "Theory of Electric Polarisation", Elsevier (1952) pp. 348 *et seq*.
28. R. H. Cole and K. S. Cole, *J. Chem. Phys.* (1941) **9**, 341.
29. D. W. Davidson and R. H. Cole, *J. Chem. Phys.* (1950) **18**, 1417.
30. R. M. Fuoss and J. G. Kirkwood, *J. Am. Chem. Soc.* (1941) **63**, 385.
31. P. Debye "Polar Molecules", Chemical Catalog Co. New York (1929) Chapter 5.
32. P. A. Egelstaff, J. S. Downes and J. W. White, *in* "Molecular Sieves", Soc. Chem. Ind., London (1968) p. 306.
33. H. G. Karge and K. Klose, *Ber. Buns. Gesell* (1974) **78**, 1263.
34. H. G. Karge and K. Klose, *Ber. Buns. Gesell.* (1975) **79**, 454.
35. J. Crank, "The Mathematics of Diffusion" Oxford (1956) Chapter XI.
36. R. Ash and R. M. Barrer, *Surface Sci.* (1967) **8**, 461.
37. J. Kärger, *Surface Sci.* (1973) **36**, 797.
38. R. M. Barrer and B. E. F. Fender, *J. Phys. Chem. Solids* (1961) **21**, 1.
39. R. M. Barrer, *in* "Molecular Sieve Zeolites", Am. Chem. Soc., Advances in Chem. Series, No. 102 (1971) p. 1.
40. P. C. Carman and R. A. W. Haul, *Proc. Roy. Soc.* (1954) A **222**, 109.
41. J. Crank, "The Mathematics of Diffusion" Oxford (1956) Chapter V.
42. D. M. Ruthven and K. F. Loughlin, *Chem. Eng. Sci.* (1971) **26**, 1145.
43. R. M. Barrer, *Trans. Faraday Soc.* (1949) **45**, 358.
44. R. Ash, R. M. Barrer and R. J. B. Craven, *J. Chem. Soc. Faraday, II*, 1978, **74**, 40.
44a. M. Kocirik and A. Zikanova, *Ind. Eng. Chem. Fund.* (1975) **13**, 347.
45. R. M. Barrer and D. J. Clarke, *J. Chem. Soc. Faraday I* (1974) **70**, 535.
46. J. Crank, "The Mathematics of Diffusion" Oxford (1956) p. 256.
47. W. Schimer, G. Fiedrich, A. Grossman and H. Stach, *in* "Molecular Sieves", Soc. Chem. Ind., London (1968) p. 276.
48. C. N. Satterfield and A. J. Frabetti Jr, *Am. Inst. Chem. Eng.* (1967) **13**, 731.
49. P. E. Eberly Jr, *Ind. Eng. Chem., Prod. Res. and Dev.* (1969) **8**, 140.
50. C. N. Satterfield, J. R. Katzer and W. R. Vieth, *Ind. Eng. Chem.* (1971) **10**, 478.
51. P. C. Carman, "Flow of Gases through porous Media", Butterworth (1956) Chapter 4.
52. R. M. Barrer, *Ber. Bunsengesell Phys. Chem.* (1965) **69**, 786.
53. R. M. Barrer, *J. Soc. Chem. Ind.* (1945) **64**, 130–133.
54. R. M. Barrer and L. Belchetz, *J. Soc. Chem. Ind.* (1945) **64**, 131.
55. R. M. Barrer, *Brennstoff Chemie* (1954) **35**, 325.
56. L. Pauling, "The Nature of the Chemical Bond", Cornell (1940) Chapter V.
57. R. M. Barrer, *Nature* (1947) **159**, 508.
58. R. M. Milton *in* "Molecular Sieves", Soc. of Chem. Ind., London (1968) p. 199.
59. D. M. Ruthven and R. I. Derrah, *J. Chem. Soc. Faraday* I (1975) **71**, 10.
60. D. M. Ruthven and R. I. Derrah, *J. Chem. Soc. Faraday* I (1972) **68**, 2332.
61. J. A. Eagan and R. B. Anderson, *J. Coll. Interface Sci.* (1975) **50**, 419.
62. H. W. Habgood, *Canad. J. Chem.* (1958) **36**, 1384.
63. P. L. Walker Jr, L. G. Austin and S. P. Nandi, *in* "Chemistry and Physics of Carbon" Vol. 2, Dekker (1966) pp. 257–371.

64. T. Takaishi, Y. Yatsurugi, A. Yusa and T. Kuratomi, *J. Chem. Soc. Faraday* I (1974) **70**, 97.
65. R. M. Barrer and D. A. Langley, *J. Chem. Soc.* (1958) 3817.
66. R. M. Barrer and J. W. Baynham, *J. Chem. Soc.* (1956) 2892.
67. R. M. Barrer and D. Ibbitson, *Trans. Faraday Soc.* (1944) **40**, 195 and 206.
68. P. E. Eberly Jr, *Ind. Eng. Chem. Prod. Res. Dev.* (1971) **10**, 433.
69. R. L. Gorring, *J. Catal.* (1973) **31**, 13.
70. D. M. Ruthven and K. F. Loughlin, *Trans. Faraday Soc.* (1971) **67**, 1661.
71. D. M. Ruthven and R. I. Derrah, *J. Coll. Interface Sci.* (1975) **52**, 397.
72. W. W. Brandt and W. Rudloff, *J. Phys. Chem.* (1967) **71**, 3948.
73. R. M. Barrer and W. Jost, *Trans. Faraday Soc.* (1949) **45**, 928.
73a. K. Fiedler and D. Gelbin, *J. Chem. Soc. Faraday* I, in press.
74. R. M. Barrer and D. A. Harding, *Separation Sci.* (1974) **9**, 195.
75. R. M. Barrer, D. A. Harding and A. Sikand, in preparation.
76. R. M. Barrer and A. J. Walker, *Trans. Faraday Soc.* (1964) **60**, 171.
77. R. M. Barrer and L. V. C. Rees, *Trans. Faraday Soc.* (1954) **50**, 852 and 989.
78. L. V. C. Rees and T. Berry *in* "Molecular Sieves", Soc. Chem. Ind., London (1968) p. 149.
79. J. M. Hammersley *in* "Methods in Computational Physics", Vol. I, Academic Press (1963) p. 281.
80. H. S. Carslaw and J. C. Jaeger, "Conduction of Heat in Solids", Oxford (1947) p. 332.
81. A. Dyer and A. Molyneux, *J. Inorg. Nuclear Chem.* (1968) **30**, 829.
82. S. P. Gabuda, *Dokl. Akad. Nauk. SSSR* (1962) **146**, 840.
83. C. Parravano, J. D. Baldeschweiler and M. Boudart, *Science* (1967) **155**, 1535.
84. W. Kuhn and M. Thurkauf, *Helv. Chem. Acta* (1958) **41**, 938.
85. N. Riehl and O. Dengel, Colloq. Phys Ice Crystals, Zurich (1962) L. Onsager and L. K. Runnels, *Proc. National Acad. Sci* (1965) **50**, 208.
86. J. H. Wang, C. V. Robinson and I. S. Edelman, *J. Amer. Chem. Soc.* (1953) **75**, 466.
87. R. M. Barrer, R. F. Bartholomew and L. V. C. Rees, *J. Phys. Chem. Solids* (1963) **24**, 51.
88. A. Quig and L. V. C. Rees *in* Proc. 3rd Internat. Conference on Molecular Sieves, Sept. 3–7th, Zurich (1973) p. 277.
89. R. M. Barrer, *Trans. Faraday Soc.* (1942) **38**, 322.
90. J. Kärger and J. Caro, *J. Chem. Soc. Faraday* I (1977) in press.
91. J. Kärger, J. Caro and M. Bulow, *Z. Chem.* (1976) **16**, 331.
92. J. Crank, "The Mathematics of Diffusion" (Clarendon Press, Oxford) (1975) Fig. 6.5, p. 97.
93. J. Kärger, M. Bulow and W. Schirmer, *Zeit. Phys. Chem., Leipzig* (1975) **256**, 144.
94. J. Kärger and M. Bulow, *Chem. Eng. Sci* (1975) **30**, 893.
95. D. R. Garg and D. M. Ruthven, *Chem. Eng. Sci.* (1973) **28**, 791.
96. L. Riekert, Advances in Catalysis (1970) **21**, 281.
97. J. J. van Deemter, F. J. Zuiderweg and A. Klinkenberg, *Chem. Eng. Sci.* (1956) **5**, 271.
98. H. Purnell, "Gas Chromatography", Wiley (1962) Chapter 7
99. P. E. Eberly Jr and E. H. Spencer, *Trans. Faraday Soc.* (1961) **57**, 289.
100. T. N. Satterfield and T. K. Sherwood "The Role of Diffusion in Catalysis", Addison–Wesley (1963) Chapter 1.

101. M. Suzuki and J. M. Smith, Chapter 5 *in* "Advances in Chromatography", Vol. 13, Ed. J. C. Giddings, Dekker (1975).
102. R. Ash, R. W. Baker and R. M. Barrer, *Proc. Roy. Soc.* (1967) A **299**, 434.
103. R. Ash, R. W. Baker and R. M. Barrer, *Proc. Roy. Soc.* (1968) A **304**, 407.
104. R. M. Barrer *in* "Surface Area Determination", Butterworth (1970) p. 227.
105. J. H. Petropoulos and P. P. Roussis, *J. Chem. Phys.* (1967) **47**, 1491 and 1496.
106. J. H. Petropoulos and P. P. Roussis, *J. Chem. Phys.* (1968) **48**, 4619.

7
Chemisorption and Sorption Complexes

In the voids and channels in zeolites various chemisorption processes, both reversible and irreversible, have been observed. There may also be chemical breakdown of chemisorbed species; these species may initiate lattice changes; intrazeolitic complexes may form between specific ligands and exchange ions; strong interactions may be involved between pairs of sorbed molecules; and irradiation by γ- or X-rays may produce radical ions within the zeolite. Determinations of the spatial arrangements of sorbed species have in some instances also been possible, using X-ray diffraction and other methods. Thus chemical processes within the crystals can involve sorbate molecules and exchange cations, sorbate molecules and the anionic framework or pairs of sorbate molecules. In this last case, the intracrystalline environment may predispose sorbate molecules to interact with each other, even if only by strongly concentrating these molecules in intracrystalline clusters or filaments. It is the purpose of this chapter to give an account of some specific processes and sorption complexes, without, however, considering catalysis which has been discussed elsewhere.[1,2,3,4]

Methods of investigating sorption complexes in zeolites or clay minerals have included infra-red absorption, e.s.r., Mossbauer and Raman spectroscopy, visible and infra-red reflectance spectroscopy and X-ray diffraction. These methods have considerably supplemented information obtained by determinations of isotherms, calorimetry and other basic procedures of surface chemistry.

1. CHEMICAL INTERACTIONS INVOLVING WATER AND HYDROXYL

This important area is concerned with zeolite hydrolysis, formation of hydronium and hydrogen zeolites, secondary processes occurring when hydrogen zeolites are heated and chemisorptions taking place between lattice hydroxyls and appropriate sorbate molecules.

1.1 Hydrolysis

Zeolites containing cations of Groups IA or IIA of the periodic table can be regarded as salts of strong bases with zeolite acids. If the zeolite acid is not very strong, such salts will tend to produce free base and hydronium zeolite (zeolite acid) by hydrolysis and the free base can distribute itself between the intrazeolitic voids and channels and any ambient aqueous phase. In support of such a view, zeolites tend to impose an alkaline pH on water in contact with the crystals. An alkaline reaction can, however, also arise because small amounts of free alkali are entrained within the structures during synthesis. This supply can be expected to become exhausted by successive treatments with distilled water. However, when 10 g of zeolite Na-X were given 40 successive treatments, each with 200 cm^3 of distilled water, the pH of the water sank slowly to 9·3 after 20 treatments and then remained almost at that value.[5] It can be concluded that hydrolysis occurs, which may initially be at terminal crystal faces:

$$[Na^+ \ldots \overset{\ominus}{O}\text{-(Si, Al)}] + H_2O \rightleftarrows NaOH + HO\text{-(Si, Al)} \quad (1)$$

Alkali liberated from terminal faces will also be quickly exhausted whereas the maintenance of high pH in successive treatments indicates a substantial reservoir of alkali resulting from intracrystalline hydrolysis:

$$[Na^+ \ldots > \overset{\ominus}{Al}\text{–O–Si}<] + 2H_2O \rightleftarrows [H_3\overset{+}{O} \ldots > \overset{\ominus}{Al}\text{–O–Si}<] + NaOH \quad (2)$$

$$[H_3\overset{+}{O} \ldots > \overset{\ominus}{Al}\text{–O–Si} <] \rightleftarrows [> Al \; {}^{HO}\diagdown Si <] + H_2O. \quad (3)$$

Reaction 1 differs from 2 plus 3 in that in reaction 1 no > Al–O–Si < bond is broken. In reaction 3 this bond breaks yielding a silanol group adjacent to a tricoordinate Al. This juxtaposition confers different properties on the –OH group which is much more acid and which usually gives a different infra-red absorption band from those –OH groups resulting from reaction 1. Thus in outgassed faujasite-type zeolites X and Y an often observed infra-red absorption band at about 3745 cm^{-1} is considered to arise from terminal –OH groups (reaction 1) or from entrained silicate fragments. Absorptions with maxima around 3650 and 3540 cm^{-1} result from intracrystalline –OH groups in two different sites (reaction 3). Among others[6,7,8] Angell and Schaffer[9] investigated infra-red absorption bands of zeolites X and Y containing cations of Groups IA and IIA and of transition metals. The zeolites were heated to about 500°C to remove molecular water. Absorption bands assigned to –OH groups are summarized in Table 1. The

TABLE 1

Frequencies of structural –OH groups in zeolites X and Y[9]

Zeolite	OH bands in cm^{-1}				OD bands in cm^{-1}		
H-Y	3744		3636	3544	2758	2686	2617
Na-Y	3748		3652				
Li-Y	3744						
Mg-Y	3645	3688	3643	3540	2762	2686	2616
Ca-Y	3746		3645		2762	2690	
Sr-Y	3746	3691					
Ba-Y	3744		3647				
Mn-Y	3748		3644	3545		2685	2616
Co-Y	3748		3646	3540			
Ni-Y	3746	3682	3643	3544			
Zn-Y	3744	3673	3642	3542			
Ag-Y	3745		3634	3550	2762		2610
Na-X	3744						
Ca-X	3744						

assignment was checked by isotopic exchange of H by D, which gave the expected frequency shifts. The absorption at $\approx$3745 cm^{-1} is universally present. The bands near 3690 to 3670 cm^{-1} found for Mg-, Sr-, Ni- and Zn-Y have been assigned to $\gg$ AlOH, but might also result from hydrolysis of cations during outgassing, e.g.

$$[Mg, H_2O]^{2+} \rightarrow [MgOH]^{+} + H^{+} \quad (4)$$

with attack of $\gg \overset{\ominus}{\text{Al}}\text{–O–Si} \ll$ by the proton to give the grouping on the right hand side of reaction 3. Most of the zeolites show the absorptions at 3650 and 3540 cm^{-1} seen in H–Y, especially those containing the transition metal ions, which, being derived from less basigenic elements than those of Groups IA and IIA, would be expected to give zeolite salts that hydrolysed more readily than zeolites exchanged with ions of Groups IA and IIA.

These infra-red measurements show widespread occurrence of hydroxyl groups in cationic forms of zeolites X and Y which must arise from incipient hydrolysis of zeolites during or even before outgassing, by processes which involve cations and framework oxygens (reactions 2 and 3). Little quantitative study has been made of hydrolysis under room temperature conditions and with the zeolite fully hydrated. However the exchange $Na^+ \rightleftarrows H_3O^+$ (reaction 2) has been studied in mordenite at 80°C and at room temperatures.[10] Hydronium mordenite could be reconverted to the Na-form by NaCl, NaOH or sodium acetate, but the forward and reverse isotherms, the completion of which involved the lapse of considerable time, were not coincident although of the same shape. The intracrystalline hydronium ion concentration, if reaction 2 is favoured over reaction 3, will be very high in the fully hydrated zeolite, and a slow dealumination reaction 5 could explain some irreversibility:

$$[\overset{\ominus}{A}lO_4 \ldots H_3^+O] + 2H_2O \rightarrow 4(-OH) + Al(OH)_3. \qquad (5)$$

The four hydroxyl groups must initially be attached one to each of four different but adjacent framework silicons.

1.2 Hydrogen Zeolites

Hydrogen zeolites were first prepared in 1949 by heating ammonium exchanged forms of certain zeolites in air.[11] The reaction can be written as

$$[\geqslant \overset{\ominus}{Al} \underset{\displaystyle O}{\diagdown\!\!\diagup} \overset{+}{NH_4}\; Si \leqslant] \rightarrow [\geqslant Al \quad \overset{HO\diagdown}{Si} \leqslant] + NH_3 \qquad (6)$$

followed by oxidation of ammonia. As already noted the resultant decationated zeolites possess strong Bronsted acidity which is a major factor in many of their catalytic functions. The deammoniation reaction 6 can be the first in a series of additional reactions as calcination is continued at progressively higher temperatures. These eventually result, *inter alia*, in loss of Bronsted acidity and sometimes in the development of Lewis acidity.

Although most studies of framework hydroxyls have been concerned with zeolite Y, other zeolites have also attracted attention. Table 2 summarizes and compares some –OH frequencies after outgassing NH_4-forms of mordenite, clinoptilolite, erionite and zeolite *L* at temperatures which remove all

TABLE 2

Comparison of hydroxyl frequencies for several hydrogen zeolites

Zeolite	SiO_2/Al_2O_3	Heating (°C)	–OH frequencies (cm^{-1})
H-mordenite[12]		350	3740, 3610
H-mordenite[13]	10 to 73	300–450	3740[a], 3650[a], 3600[a]
H-clinoptilolite[14]		>300	3740, 3620, 3560
H-erionite[15]		300–700	3745, 3612, 3565
H-*L* (40% H)[16]		>260	3740, 3630
H-*L* (5% to 66% H)[17]		460	3750, 3640, 3280

[a] Estimated from the corresponding –OD frequencies, 2760, 2700 and 2670 cm^{-1} respectively.

or most zeolitic water and decompose the NH_4-forms to yield H-forms. The band centred at about 3745 cm^{-1} occurs in all the zeolites, including zeolites X and Y (Table 1), and is again assignable to terminal –OH groups at external crystal surfaces or on entrained silica fragments. The remaining frequencies are more structure sensitive. The broadness of the band at 3280 cm^{-1} for H–*L* suggests hydrogen bonding of the relevant –OH groups with framework oxygens.[17]

Acidity studies have been made of zeolite Y,[18,12,19,20,21,22] zeolite X,[18,8,23,24] mordenite,[18,25,26,27] erionite,[18,15] zeolite *L*,[17] and clinoptilolite,[18,14] for various cationic forms and for ammonium forms calcined at particular temperatures or at each of a series of progressively increasing temperatures. The calcined forms can then be treated with bases, for example ammonia, piperidine or pyridine, at temperatures such as 150° or 200°C and at low vapour pressures of the base. Under these conditions the bases are chemisorbed at Bronsted and Lewis acid sites, and colour changes with adsorbed indicators and changes in the infra-red spectra have been used to monitor the reactions. Thus, addition of ammonia to H–Y reverses reaction 6 and therefore causes the disappearance of the bands at 3650 and 3540 cm^{-1} and the appearance of the absorption spectrum of the ammonium zeolite. Analogous changes follow the sorption of pyridine and piperidine except that interaction of pyridine with the less accessible –OH giving absorption at 3540 cm^{-1} is not as strong as with the more accessible –OH absorbing at 3650 cm^{-1}. Under some conditions interaction between the 3540 cm^{-1} band and pyridine is minimal.[12] The pK_a values of piperidine, ammonia and pyridine are respectively 11·2, 9·3 and 5·2 and their interaction with acidic hydroxyls depends both on their basicity and the accessibility of the –OH groups.

Integral heats of sorption of a number of bases and of benzene with

TABLE 3

Integral heats of sorption (kcal mol^{-1}) in some ion-exchanged forms of zeolite Y[28]

Zeolite[a]	Benzene Amount (mmol/dry g)	Benzene Heat	Pyridine Amount (mmol/dry g)	Pyridine Heat	Piperidine Amount (mmol/dry g)	Piperidine Heat	*n*-Butylamine Amount (mmol/dry g)	*n*-Butylamine Heat	Ammonia Amount (mmol/dry g)	Ammonia Heat
Na-Y	2·4	17·9	2·7	24·9	2·5	22·8	2·7	21·3	5·9	11·5
0·16 Ca Na-Y	2·4	17·5	2·7	24·8	2·4	22·7	2·7	21·1	6·1	11·3
0·42 Ca Na-Y	—	—	2·7	24·9	2·5	22·9	—	—	—	—
0·61 Ca Na-Y	—	—	2·7	25·4	—	—	—	—	—	—
0·77 Ca Na-Y	2·4	16·1	2·7	26·9	2·5	24·7	2·7	27·1	6·3	15·4
0·81 Ca Na-Y	2·4	16·3	2·7	27·2	2·5	25·1	2·7	27·8	6·8	15·7
0·19 Cu Na-Y	—	—	3·1	26·1	3·0	23·7	3·4	22·1	6·0	13·1
0·47 Cu Na-Y	—	—	—	—	—	—	—	—	7·4	16·1
0·69 Cu Na-Y	—	—	—	—	—	—	3·5	28·5	10·0	15·9
0·79 Cu Na-Y	2·4	16·4	3·4	28·5	3·2	27·8	3·5	30·1	10·0	16·3
0·79 Cu Na-Y (600°C)	—	—	—	—	—	—	—	—	9·8	16·3
0·08/0·35 H Cu Na-Y	2·4	16·8	—	—	—	—	3·5	28·0	6·5	17·6
0·52 Ce Na-Y	2·4	16·8	—	—	—	—	—	—	—	—
0·58 Ni Na-Y	—	—	3·0	25·2	2·6	25·3	3·0	25·1	8·3	13·1
0·76 Ni Na-Y	2·4	16·1	3·2	26·8	2·8	27·0	3·0	32·1	12·0	14·9
0·97 H Na-Y	2·4	13·4	2·3	33·2	2·4	34·6	2·6	34·7	7·1	16·9

[a] The figures in this column give the equivalent cation fractions of the entering Ca, Cu, Ce, Ni or H in the parent Na-Y.

zeolite Y in various ion-exchanged forms are given in Table 3.[28] The H–Y sorbs the bases more exothermally and benzene less exothermally than do the other cationic forms. For benzene the heats are heats of physisorption; the heats of sorption of the bases have both physi- and chemisorption components. The sorption heats of benzene in Na-Y and 0·97HNa-Y reflect the greater local electrostatic fields in the Na-form. On the other hand the increase in heats for the bases in 0·97HNa-Y compared with Na-*Y* directly reflects the chemisorption component. The exothermal heats are also seen to increase as Ca^{2+}, Cu^{2+} or Ni^{2+} replaces Na^{+} in the zeolite. With Ca^{2+} the increase can be ascribed to larger field-dipole energy; for the transition metal ions there may be an additional chemisorption as the bases coordinate with accessible ions.

A recent study[17] of partial H-forms of zeolite *L* exemplifies acidity in a zeolite of quite different structure from that of zeolite Y. The partial H–*L* samples were made by heating the partial NH_4-forms of zeolite *L* at 460°C. These H-containing samples were then allowed to chemisorb pyridine vapour at 100, 200, 300 and 400°C, followed by evacuation for an hour at the temperature of chemisorption to remove as much physically sorbed excess pyridine as possible. Chemisorption caused the bands due to –OH frequencies to be replaced by other frequencies, some of which were assigned

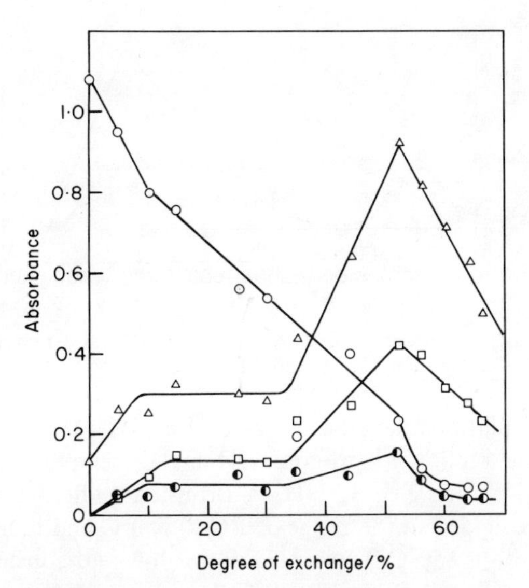

FIG. 1. The influence of the extent of exchange of K^{+} by H^{+} in zeolite *L* upon the intensity of infra-red bands.[17]

○, 1436 cm^{-1} band after evacuation at 100°C
△, 1490 cm^{-1} band after evacuation at 100°C
□, 1540 cm^{-1} band after evacuation at 200°C
◐, 1450 cm^{-1} band after evacuation at 400°C

as follows: pyridine bound to K^+, 1436 cm^{-1}; pyridine on Lewis sites, 1450 cm^{-1}; pyridine probably bound to both K^+ and pyridinium, 1490 cm^{-1}; pyridinium ions (i.e. pyridine chemisorbed on Bronsted sites), 1540 cm^{-1}; pyridine probably bound to K^+, 1580 cm^{-1}. The influence of hydrogen exchange upon the intensities of these bands is shown in Fig. 1. As expected from the above assignment, the 1436 cm^{-1} band decreases almost linearly with the extent to which potassium ions are replaced by hydrogen and reaches zero for $\approx 70\%$ exchange. This suggests that $\approx 30\%$ of the K^+ is located at positions inaccessible to pyridine (e.g. in the cancrinite cages).

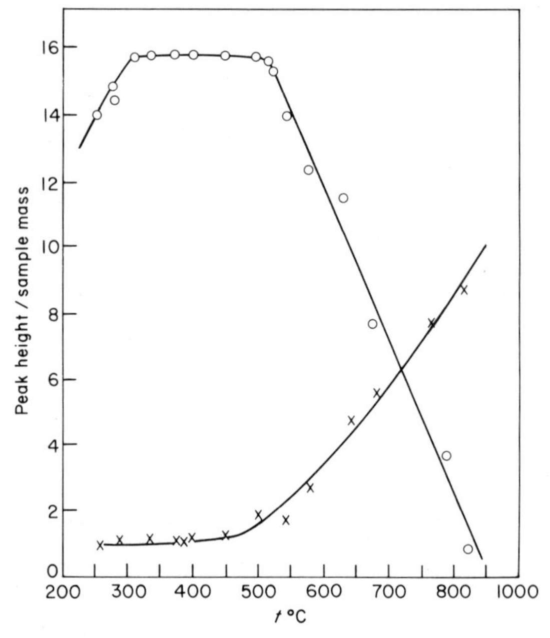

FIG. 2. Changes in populations of Bronsted acid sites (○) and Lewis acid sites (X) as functions of calcination temperature of zeolite H-Y.[12]

Acid site populations have been studied for NH_4–Y and NH_4-mordenite by calcining the zeolites at progressively higher temperatures. Results are shown in Figs 2[12] and 3.[27] In NH_4–Y Bronsted acidity was nearly constant for the zeolite calcined in the range 350°C to 550°C and thereafter declined, reaching zero about 850°C. Lewis acidity was low and constant up to 450°C and then increased up to around 850°C. In calcined NH_4-mordenite on the other hand Lewis and Bronsted acidities were both constant from 300°C to 450°C, but both acidities decreased after still higher calcination temperatures, Bronsted acidity more rapidly. The Lewis sites in the calcined mordenites retained pyridine up to about 450°C but the Bronsted sites began to lose the base at 200°C and were virtually free of it at 450°C. This indicated that in mordenite Lewis acidity was stronger than Bronsted acidity. One stoichiometry which has been proposed to explain development of Lewis acidity at the expense of Bronsted acidity is the reaction 7:

$$2[> \mathrm{Al} \overset{\mathrm{HO}}{\diagdown} \mathrm{Si} <] \rightleftarrows [> \overset{\ominus}{\mathrm{Al}}\text{–O–Si} <] + [> \mathrm{Al}\ \overset{\oplus}{\mathrm{Si}} <] + \mathrm{H_2O}. \quad (7)$$

However with calcined H–Y, although addition of water regenerated Bronsted acidity, the relevant absorption at 3650 cm^{-1} was not observed. Also,

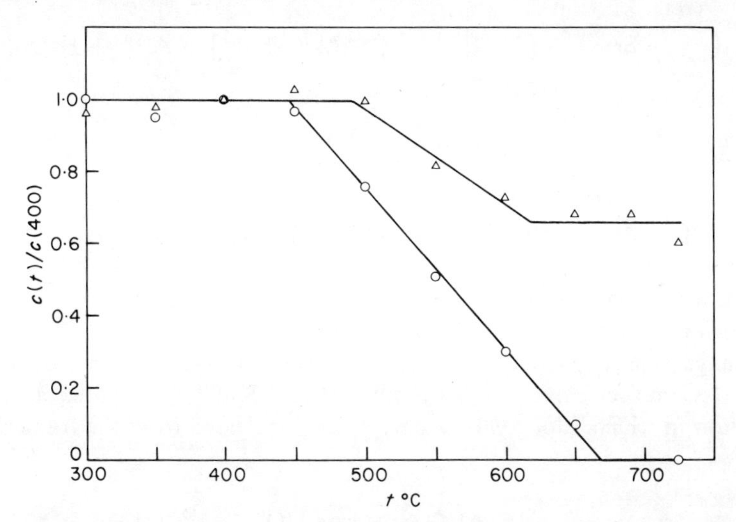

Fig. 3. Changes in populations of Bronsted acid sites (lower curve) and Lewis acid sites (upper curve) as H-mordenite is calcined at different temperatures, relative to these populations on mordenite heated at 400°C.[27]

reaction 7 generates Lewis acidity at the expense of Bronsted acidity and while this appears to be true of calcined H-Y (Fig. 2), in H-mordenite both acidities decline together (Fig. 3). Thus reaction 7 cannot represent all the processes associated with dehydroxylation. Some of these are considered below.

There is evidence that calcining NH_4-zeolites liberates Al-bearing fragments from the aluminosilicate frameworks.[29, 30, 10] Kerr[29] reported that up to 25% of the framework Al could be released from NH_4-Y during calcination in presence of water vapour. The first step can be

$$[> \mathrm{Al} \overset{\mathrm{HO}}{\diagdown} \mathrm{Si} <] + 2\mathrm{H_2O} \rightarrow 4(-\mathrm{OH}) + \mathrm{AlOOH} \quad (8)$$

which is analogous to reaction 5. Further reactions may then occur between

AlOOH and $[>\mathrm{Al}\overset{\mathrm{HO}}{\diagdown}\mathrm{Si}<]$ groups to regenerate normal $[>\overset{\ominus}{\mathrm{Al}}\text{–O–Si}<]$ bonds and form ions $[Al(OH)_2]^+$, $[AlOH]^{2+}$ and Al^{3+}:

$$\left.\begin{aligned}
&[>\mathrm{Al}\overset{\mathrm{HO}}{\diagdown}\mathrm{Si}<] + \mathrm{AlOOH} \rightarrow [>\overset{\ominus}{\mathrm{Al}}\text{–O–Si}<] + [\mathrm{Al(OH)_2}]^+ + \mathrm{H_2O}\\
&[>\mathrm{Al}\overset{\mathrm{HO}}{\diagdown}\mathrm{Si}<] + [\mathrm{Al(OH)_2}]^+ \rightarrow [>\overset{\ominus}{\mathrm{Al}}\text{–O–Si}<] + [\mathrm{AlOH}]^{2+} + \mathrm{H_2O}\\
&[>\mathrm{Al}\overset{\mathrm{HO}}{\diagdown}\mathrm{Si}<] + [\mathrm{AlOH}]^{2+} \rightarrow [>\overset{\ominus}{\mathrm{Al}}\text{–O–Si}<] + \mathrm{Al}^{3+} + \mathrm{H_2O}
\end{aligned}\right\} \quad (9)$$

If reactions 8 and 9 are added the result is

$$4[>\mathrm{Al}\overset{\mathrm{HO}}{\diagdown}\mathrm{Si}<] \rightarrow 3[\overset{\ominus}{\mathrm{Al}}\text{–O–Si}<] + \mathrm{Al}^{3+} + 4(\text{–OH}) \qquad (10)$$

which would account for the observed loss of 25% of the exchange capacity.

However Bolton and Lanewala[31] reported more than 25% loss in exchange capacity. Complete dehydroxylation of H-Y involves the loss of one H_2O molecule per two Al atoms,[32] and Kuhl[30] proposed a further reaction in which, along with water, AlO^+ is released from the framework:

$$2[>\mathrm{Al}\overset{\mathrm{HO}}{\diagdown}\mathrm{Si}<] \rightarrow [>\overset{\ominus}{\mathrm{Al}}\text{–O–Si}<] + \mathrm{AlO}^+ + \left[\begin{smallmatrix}| & |\\ \mathrm{O} & \mathrm{O}\\ | & |\end{smallmatrix}\right] + \mathrm{H_2O} \qquad (11)$$

where $\left[\begin{smallmatrix}| & |\\ \mathrm{O} & \mathrm{O}\\ | & |\end{smallmatrix}\right]$ denotes the two new $>\text{Si–O–Si}<$ bonds. This reaction involves a 50% loss of framework Al and exchange capacity if there is full reaction and if AlO^+ is exchangeable. Barrer and Klinowski[10] indicated two additional reactions:

$$[>\mathrm{Al}\overset{\mathrm{HO}}{\diagdown}\mathrm{Si}<] \rightarrow \left[\begin{smallmatrix}| & |\\ \mathrm{O} & \mathrm{O}\\ | & |\end{smallmatrix}\right] + \mathrm{AlO(OH)} \qquad (12)$$

with total loss of framework Al and of exchange capacity, for full reaction; and

$$3[>\mathrm{Al}\overset{\mathrm{HO}}{\diagdown}\mathrm{Si}<] \rightarrow 2[>\overset{\ominus}{\mathrm{Al}}\text{–O–Si}<] + \mathrm{Al(OH)}^{2+} + \left[\begin{smallmatrix}| & |\\ \mathrm{O} & \mathrm{O}\\ | & |\end{smallmatrix}\right] + \mathrm{H_2O} \qquad (13)$$

This gives $33\frac{1}{3}\%$ loss of framework Al and of exchange capacity for the complete reaction.

Barrer and Klinowski looked for evidence of reactions such as 10, 11, 12 and 13 by investigating differences in the extent of exchange, with Na^+ from dilute aqueous NaOH and NaCl solutions respectively, in H-mordenites calcined to different temperatures. With dilute NaOH or NaCl any $[\geqslant Al \overset{HO}{\diagdown} Si \leqslant]$ groups should react:

$$[\geqslant Al \overset{HO}{\diagdown} Si \leqslant] + NaOH\,(NaCl) \rightarrow [\geqslant \overset{\ominus\;\;Na^+}{Al\text{–}O\text{–}Si} \leqslant] + H_2O(HCl) \quad (14)$$

but in addition it was thought that the AlO^+, $[AlOH]^{2+}$ and Al^{3+} would react readily only with NaOH:

$$\left.\begin{array}{l} AlO^+ + NaOH \rightarrow Na^+ + AlOOH \\ [AlOH]^{2+} + 2NaOH \rightarrow 2Na^+ + AlOOH + H_2O \\ Al^{3+} + 3NaOH \rightarrow 3Na^+ + AlOOH + H_2O \end{array}\right\} \quad (15)$$

Figure 4 shows the extent to which Na was reintroduced into H-mordenite after having heated it to the temperatures recorded on the abscissa. The result with NaCl aq shows that all acidic silanols had gone by about 700°C, and the difference between the uptakes of Na^+ from NaCl and from NaOH solutions should then indicate the weighted sum of the reactions 15. Considerable Na^+ is taken up into the zeolite from dilute NaOH even by H-mordenites heated to close on 1000°C. The final dehydroxylation of the zeolite would involve elimination of water between Al-bearing fragments by such reactions as

$$[AlOH]^{2+} + AlOOH \rightarrow [Al_2O_2]^{2+} + H_2O. \quad (16)$$

How far the observed behaviour in generation and loss of Lewis acidity on heating can for different zeolites be accounted for by the reactions 7 to 16 remains uncertain. Morrow and Cody,[33] who investigated silica dehydroxylated at 600°C observed two new infra-red bands at 908 and 888 cm^{-1} that increased in intensity to a maximum at 1200°C. The absorptions were associated with the formation of active sites which reversibly coordinated with pyridine and trimethylamine. In explanation it was suggested that strained surface siloxane bridges formed on dehydroxylation, in which a Si atom was electron deficient and so functioned as a Lewis acid centre. If this is so, intracrystalline dehydroxylation to give strained siloxane bonds (reactions 11, 12 and 13) may also confer Lewis acidity.

The foregoing considerations indicate various possibilities for reactions involving the hydrogen zeolites. Additional indications of such reactions come from the very different thermal stabilities that can be conferred on zeolite Y, according to the ways in which the ammonium form is calcined. The preparation of zeolite catalysts of maximum stability is a major objective and efforts to confer "ultrastability" have met with notable success in the case of zeolite Y.[34] Three procedures which result in the ultrastable state of the zeolite can be summarized as follows.[35]

Procedure A	Procedure B	Procedure C
Na-Y ↓ NH_4Cl aq $(NH_4)_{(0.75-0.90)}$ $Na_{(0.25-0.10)}$-Y ↓ Wash; heat in static or steam-containing atmos. to 600°–800°C. About 1% unit cell shrinkage ↓ NH_4Cl aq Remove most of residual Na^+	Na-Y ↓ NH_4Cl aq $(NH_4)_{(0.75-0.90)}$ $Na_{(0.25-0.10)}$-Y ↓ Wash; heat to 200°–600°C ↓ Again exchange to remove most residual Na^+. Zeolite metastable at this stage ↓ Heat rapidly in static or steam-containing atmos. to 600°–800°C. 1 to 1·5% unit cell shrinkage.	Na-Y ↓ 0·25 to 0·5 mol H_4EDTA per equivalent of Na^+ added to water slurry over 18 h under reflux. ↓ Al-deficient zeolite ↓ Heat in inert gas purge to 800°C. Unit cell shrinkage ≈1%.

Water vapour must be present during the calcination steps, and there is evidence of removal of framework aluminium in procedures A and B as described earlier in this section. In procedure C a large amount of framework aluminium is removed directly by the acid. The initial and final SiO_2/Al_2O_3 ratios for typical preparations by the three procedures were reported as follows.

$\frac{SiO_2}{Al_2O_3}$	Procedure A	Procedure B	Procedure C
Initial	5·2	5·2	5·2
Final	5·63	6·35	10·3

Especially for procedures A and B part of the Al released from the framework probably remains within the zeolite in the form of fragments such as AlOOH, AlO^+, $Al(OH)^{2+}$, Al^{3+} referred to previously. The great difference

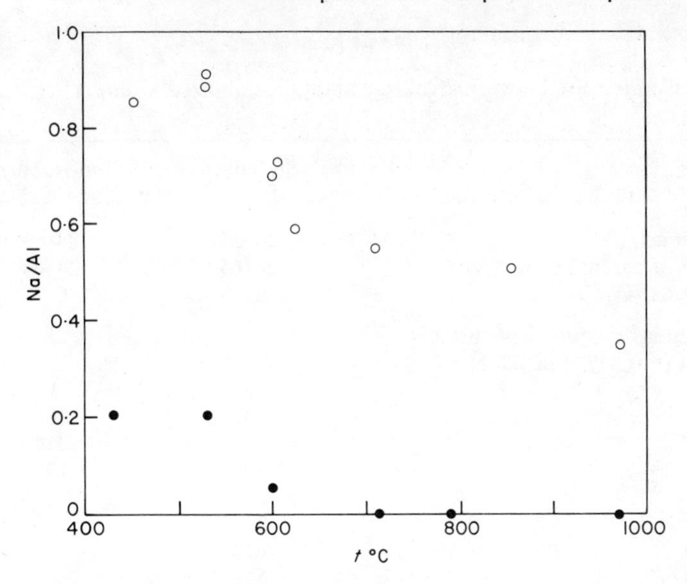

FIG. 4. The Na/Al ratio in NH_4-mordenite heated to various constant temperatures between 455 and 970°C and treated with: (i) 0·01 mol dm^{-3} NaOH solution equivalent to the overall amount of Al in the parent zeolite (⊙); and (ii) 150 cm^3 of 0·01 mol dm^{-3} NaCl solution per 0·575 g of parent zeolite (●).[10]

between stabilized and non-stabilized forms of zeolite Y is exemplified in Table 4. While the chemistry of ultrastabilization is not fully understood, McDaniel and Maher[34] consider that the process comprises several steps, each associated with intermediate structures. The first stage involves generating the H-form from the NH_4-form; the second stage is associated with dealumination of the framework. This creates defects in the structure (cf reaction 8), that in a third stage result, especially for procedure C, in lattice contraction through formation of new $\gtrdot$ Si–O–Si $\lessdot$ bonds:

$$4(-\mathrm{OH}) \rightarrow \left[\begin{matrix} | & | \\ \mathrm{O} & \mathrm{O} \\ | & | \end{matrix}\right] + 2\mathrm{H_2O}. \qquad (17)$$

Such contraction through forming these bonds can also result from reactions 11, 12 and 13, which may occur in all three "ultrastabilization" procedures A, B and C.

At this point it is of interest to summarize general observations which may be made regarding H-zeolites and acidity, as a result of researches described above, and of many others not referred to.[18] Most of these relate to faujasite-type zeolites.

(i) Homoionic non-hydrolysed forms containing cations of group 1A

TABLE 4

Comparison between stabilized and non-stabilized zeolite Y[36]

	Stabilized	Non-stabilized
Unit cell (Å)	24·34	24·69
Cell of parent zeolite (Å)	24·65	24·65
SiO_2/Al_2O_3	6·35	6·35
Monolayer equivalent surface area (m^2g^{-1}) after 2 h at °C		
447	837	1008
815	851	254
844	793	132
900	842	18
925	743	15
940	678	
980	542	

elements are non-acidic, but entrained impurities or cation deficiency due to hydrolysis can lead to acidity.

(ii) Acid sites can be Bronsted or Lewis sites or both. In zeolites X and Y, calcination below 450°C usually produces Bronsted sites and calcination above 600°C yields Lewis site populations. In zeolite H-Y Lewis sites increase as Bronsted sites decrease in numbers, but in H-mordenite both types tend to decline together.

(iii) Bronsted sites whether in cationic or decationized forms of zeolites are similar. In zeolites X and Y they give absorptions at about 3650 cm^{-1} and 3540 cm^{-1}. In zeolites other than X and Y the frequencies differ somewhat among the different zeolites and from those for X and Y, presumably as a result of the various crystallographic environments.

(iv) Multivalent ions can also form Bronsted sites by reactions such as $[M, OH_2]^{2+} \rightleftarrows [MOH]^+ + H^+$ but the $[MOH]^+$ ions do not appear to be strong acid sites.

(v) Faujasite-type and other zeolites all tend to show an –OH frequency at 3740 cm^{-1} which is ascribed to a relatively non-acidic hydroxyl such as is found in silica gel.

(vi) Heat treatments in presence of water vapour result in structural changes and reactions after which acidity in faujasite-type zeolites can no longer be accounted for by the two types of –OH group referred to in (iii).

(vii) Although rehydration of a calcined zeolite restores some of the Bronsted acidity the types of hydroxyl are different.

(viii) The Lewis sites are probably often at tricoordinated aluminium or silicon resulting from dehydroxylation. Exchange cations if accessible can interact with bases and so function as Lewis acid sites, but these are less strongly acid than sites at tricoordinated aluminium.
(ix) Fragments such as AlO^+, shed from the framework during changes referred to in (vi), may function as Lewis sites.

1.3 Silanation

The reactions of NH_3, pyridine or piperidine with acidic groups [$\geqslant$Al $\overset{HO}{\smile}$ Si$\leqslant$] of hydrogen zeolites are reversible chemisorptions. For example, reaction 6 of the previous section can be reversed by the action of dry ammonia on the outgassed hydrogen zeolite. Other chemisorptions are irreversible, including reactions between the acidic hydroxyls of hydrogen zeolites and SiH_4,[37] $SiH(CH_3)_3$[38] and $Si(CH_3)_4$.[39]

Reaction between H-Y and $Si(CH_3)_4$ in the range 250 to 650°C was monitored by measuring pressure increases as a function of time.[39] Two steps were considered to occur as follows:

$$\geqslant SiOH + Si(CH_3)_4 \rightarrow \geqslant Si\text{–}O\text{–}Si(CH_3)_3 + CH_4 \tag{18}$$

and

$$\geqslant SiOH + \geqslant Si\text{–}O\text{–}Si(CH_3)_3 \rightarrow \geqslant Si\text{–}O\text{–}Si(CH_3)_2\text{–}O\text{–}Si \leqslant + CH_4. \tag{19}$$

Reaction 18 causes no pressure change, so that what was monitored was reaction 19, together with any additional secondary processes evolving hydrocarbon gases or hydrogen. Figure 5 shows the pressure increases at 550°C as a function of time for each of a number of doses. The products were gray in colour, but became white on calcination in oxygen at 500°C for six hours. Weight increases and monolayer equivalent areas of the zeolite after exhaustive treatment with $Si(CH_3)_4$ and before and after calcining in oxygen are given in Table 5. The monolayer equivalent area for the parent H-Y was 840 m^2g^{-1}.

The kinetics of the second stage reaction 19 at higher temperatures obeyed the Elovich equation for the initial doses. This equation in its integral form is

$$\theta = \frac{RT}{\alpha} \ln (t + RT/\alpha a) + \text{constant}$$

where θ (proportional to the pressure increase) is the fraction of the –OH's available for secondary reaction which have reacted at time t, and α and a are constants of the differential form $d\theta/dt = a \exp\left(-\frac{\alpha\theta}{RT}\right)$. For higher dose numbers the kinetics changed and there was an approximately linear relation between $1/\Delta p$ and $1/t$ where Δp is the pressure increase. The material treated with $Si(CH_3)_4$ had good ability to accept electrons from perylene and also possessed low cyclopropane and high protoadamantane isomerization capacity. After calcining the treated zeolite the ease of effecting the above two catalyses was reversed.

Silanation with SiH_4 of Na-Y and of its ≈31%, 54% and 73% H-exchanged forms and of Na- and H-mordenite and partially dealuminated H-mordenite

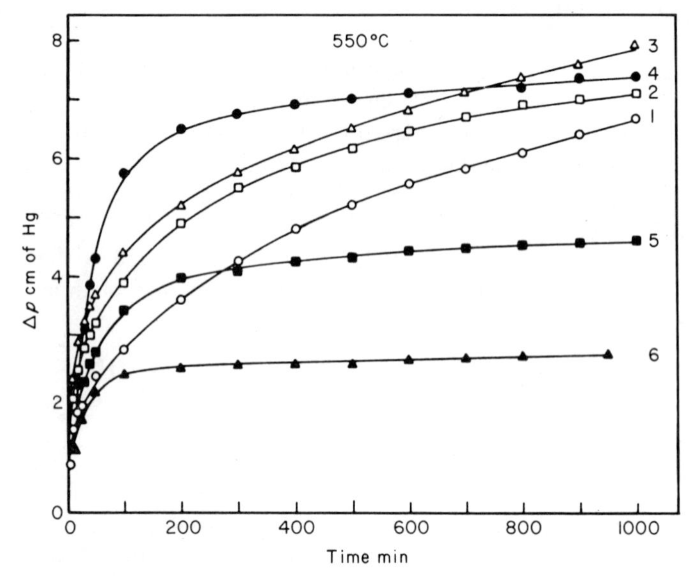

FIG. 5. The change in pressure, Δp, against reaction time, in H-Y at 550°C.[39] ○, dose 1; □, dose 2; △, dose 3; ●, dose 4; ■, dose 5; ▲, dose 6.

TABLE 5

Weight increases of H-Y treated exhaustively with $Si(CH_3)_4$, and monolayer equivalent areas (m^2g^{-1}) before and after heating in O_2[39]

Temp. of reaction with $Si(CH_3)_4$ (°C)	% weight increase	Monolayer equivalent area Before O_2 treatment	After O_2 treatment
250	2·32	711	720
350	3·14	724	754
450	4·22	695	710
550	5·26	714	761
650	4·87	724	757

was studied between 30° and 360°C.[37] Reaction was monitored by measuring the amounts of SiH_4 chemisorbed and physically sorbed and of H_2 evolved, so that all steps were included. In the range 30 to 210°C the primary reaction was considered to be

$$\geqslant SiOH + SiH_4 \rightarrow \geqslant Si\text{–}O\text{–}SiH_3 + H_2. \quad (20)$$

This could be followed by secondary reactions such as

$$\left.\begin{array}{l} 2[\geqslant Si\text{–}O\text{–}SiH_3] \rightarrow \geqslant Si\text{–}O\text{–}SiH_2\text{–}SiH_2\text{–}O\text{–}Si \leqslant + H_2 \\ \geqslant SiOH + \geqslant Si\text{–}O\text{–}SiH_3 \rightarrow \geqslant Si\text{–}O\text{–}SiH_2\text{–}O\text{–}Si \leqslant + H_2. \end{array}\right\} \quad (21)$$

In reaction 20, the ratio H_2 produced : SiH_4 consumed = 1 : 1. For this ratio to exceed unity, secondary processes such as reactions 21 must be involved, and the amount of hydrogen produced in secondary reactions should be equal to the total hydrogen evolved less the silane chemisorbed. Some of the results, for the mordenites and for Na-Y and 0·73HNa-Y are shown in Tables 6 and 7.

The small amounts of reaction of SiH_4 with outgassed parent Na-mordenite and Na-Y represent respectively about 7% and 3·5% replacement of Na by H in these zeolites, and may measure the extent of hydrolysis. Hydrolysis produces surface hydroxyls at external crystal faces (reaction 1), together with internal hydroxyls by intracrystalline hydrolysis (reactions 2 and 3). In Na-mordenite the ratio H_2 produced : SiH_4 used is about unity (Table 6), corresponding with reaction 20, but in Na-faujasite the ratio exceeded unity (Table 7) so that secondary processes must also occur. If terminal surface hydroxyls due to hydrolysis are rather numerous in Na-Y, and so can occur close together, a reasonable explanation is obtained. The results for the parent Na-zeolites compared with those for the hydrogen zeolites show very much greater chemisorptions in the latter, in all cases with ratios H_2 produced : SiH_4 consumed greater than unity, so that secondary reactions are always present.

If all the pendant –O–SiH_3 groups added in reaction 20 then underwent reactions 21, the ratio H_2 produced : SiH_4 consumed would equal two. In some examples recorded in Tables 6 and 7, and in similar experiments with ≈31% and 54% H-exchanged zeolite Y, these ratios are near or less than 2, but there are other instances recorded in the tables where the ratios exceed 2. In such circumstances additional reactions must be possible. These include, perhaps by hydrogen migration, reaction with a third –OH group:

$$\geqslant Si\text{–}O\text{–}SiH_2\text{–}O\text{–}Si \leqslant + HO\text{–}Si \leqslant \rightarrow \geqslant Si\text{–}O\text{–}\underset{\underset{O\text{–}Si \leqslant}{|}}{SiH}\text{–}O\text{–}Si \leqslant + H_2. \quad (22)$$

TABLE 6

Reactions of SiH_4 with mordenites[37]

Sample (dry wt)	Ideal silanol contents of samples (mmol)	Reaction conditions	SiH consumed (mmol) In the stages	SiH consumed (mmol) In total	H_2 produced In the stages	H_2 produced In total	Ratio H_2/SiH_4	Wt % of Si added
Na-mordenite (1·02)	—	100°C for 39 h	—	0·14	—	0·127	0·91	0·39
H-mordenite (1·00)	2·04	Room temp. for 176 h	0·56	0·56	0·69	0·69	1·23	1·57
		+ unreacted SiH_4 pumped off	—	—	—	—	—	
		+ 160°C for 49 h	—	—	+0·35	1·04	1·86	
		+ 210°C for 256 h	—	—	+0·22	1·26	2·25	
H-mordenite (0·955)	1·95	100°C for 88·25 h	0·972	0·972	1·412	1·412	1·45	
		+ 150°C for 65·25 h	+0·18	1·15	+0·488	1·90	1·65	3·40
		+ unreacted SiH_4 pumped off	—	—	—	—	—	
		+ 200°C for 47·5 h	—	—	+0·289	2·19	1·90	
H-mordenite (0·63)	1·29	First dose SiH_4, 300°C for 113 h	0·91	0·91	—	—	—	
		+ second dose, 300°C for 190 h	+0·68	1·59	—	3·02	1·90	7·06
		+ unreacted SiH_4 pumped off	—	—	—	—	—	
		+ reaction with O_2 at 21·7°C consuming 0·637 mmol O_2	—	—	—	—	—	
Dealuminated mordenite (0·6155)	3·15 (in nests of four) + 0·58 (as single silanols)	300°C for 75 h	0·465	0·465	1·06	1·06	2·28	2·48

TABLE 7

Reactions with zeolite Y[37]

Sample (lg dry wt)	Chemisorption of NH_3= silanols (mmol)	High freq. hydroxyls (mmol)	Reaction conditions	SiH_4 consumed (mmol)		H_2 produced (mmol)		H_2/SiH_4	Chemisorption NH_3= silanols[a] (mmol)	% Si added
				In the stages	In total	In the stages	In total			
Na-Y	—	—	60 h at 160°C	0·15	0·15	0·28	0·28	1·89	—	0·42
			+ unreacted SiH_4 evac.	—	—	—	—	—	—	—
			+ 110 h at 160°C	—	—	+0·11	0·39	2·62	—	—
			+ 50 h at 210°C	—	—	+0·04	0·43	2·96	—	—
			+ 96 h at 300°C	—	—	+0·08	0·51	3·44	—	—
(Na, H)-Y (73)	3·17	2·10	150 h at 30°C	0·16	0·16	0·25	0·25	1·52	—	1·12
			+ 150 h at 90°C	0·24	0·40	0·35	0·60	1·50	—	—
			+ unreacted SiH_4 evac.	—	—	—	—	—	—	—
			+ 20 h at 90°C	—	—	0·04	0·64	1·58	—	—
			+ 122 h at 160°C	—	—	0·24	0·88	2·19	2·02	—
(Na, H)-Y (73)	3·17	2·10	First dose SiH_4, 43 h at 210°C	0·89	0·89	1·33	1·33	1·50	—	—
			+ second dose SiH_4, 86 h at 210°C	0·12	1·01	0·20	1·53	1·50	—	2·84
			Unreacted SiH_4 evac.	—	—	—	—	—	—	—
			+ 27 h at 210°C	—	—	0·050	1·58	1·55	—	—
			+ 4 doses of H_2O vap. at 210°C + temp. raised to 300°C at last dose	—	—	2·30	3·88	3·82	1·60	—
(Na, H)-Y (73) first treatment	3·17	2·10	150 h at 210°C, large XS SiH_4	1·23	1·23	2·10	2·10	1·71	—	3·45
			+ unreacted SiH_4 evac.	—	—	—	—	—	—	—
			+ 116 h at 300°C and 170 h at 360°C	—	—	1·02	3·12	2·54	—	—
			+ 2 doses H_2O vap. at 210°C and 1 dose at 210° and then 300°C	—	—	1·62	4·74	3·86	—	—

[a] On modified zeolite equal to 1 g original sample.

Ratios $H_2/SiH_4 > 2$ appear for H-mordenite to require temperatures above 200°C. For partly dealuminated mordenite at 300°C this ratio was also a little above two. However for complete reaction of SiH_4 with nests of 4(–OH), the ratio H_2/SiH_4 would be four:

$$SiH_4 + 4(-OH) \rightarrow \left[\begin{array}{c} | \\ O \\ | \\ -O{-}Si{-}O- \\ | \\ O \\ | \end{array} \right] + 4H_2. \qquad (23)$$

The low ratio $H_2/SiH_4 = 2{\cdot}28$, and the small amount of SiH_4 consumed ($0{\cdot}465$ mmol in a theoretical maximum of $3{\cdot}15 + 0{\cdot}58$) strongly suggests that the nests have already reacted to give two new –Si–O–Si– bonds and water during the outgassing, which was at $\approx$360°C. Thus they were not available to test the interesting possibility of isomorphous replacement of Al by Si via dealumination and silanation:

$$\left[\begin{array}{c} O- \\ |\ominus \\ -O{-}Al{-}O- \\ | \\ -O \end{array} \right] \xrightarrow{\text{dealumination}} \left[\begin{array}{c} O- \\ | \\ H \\ -O{-}H \quad H{-}O- \\ H \\ | \\ -O \end{array} \right] \xrightarrow{\text{silanation}}$$

$$\left[\begin{array}{c} O- \\ | \\ -O{-}Si{-}O- \\ | \\ -O \end{array} \right] + 4H_2. \qquad (24)$$

The total chemisorptions of SiH_4 by the H-mordenites and zeolites Y were always less than the total –OH content, and in the case of zeolite Y, where the –OH groups are of two kinds, high frequency more accessible and low frequency less accessible, the uptake was less than the content of accessible –OH groups. Thus in all cases there were –OH groups available to participate in secondary reactions. At temperatures of 300° and 360°C on two occasions a coloured volatile product stained brown the wall of the apparatus and darkened the zeolite. The zeolite could then be rendered white by calcination in air. It thus seems that at high temperatures, polymerization processes involving SiH_4 and/or chemisorbed SiH_4 may be involved. At temperatures of 210°C or below, such behaviour was not observed.

The zeolite crystals containing only chemisorbed SiH_4 were found to chemisorb oxygen or water. In the latter case more hydrogen was evolved, and when the ultimate ratio total H_2/SiH_4 chemisorbed was evaluated, this ratio was close to 4 (experimental values were 3·82, 3·86, 3·99 and 3·74). The silanated zeolites were investigated as sorbents of oxygen,[37] nitrogen[37] and typical hydrocarbons[40] (*n*-hexane, 3-methyl pentane, 2 : 2 dimethyl butane, benzene and cyclohexane). Very marked effects were observed as illustrated in Fig. 6 for N_2 at −196°C in the mordenites and in

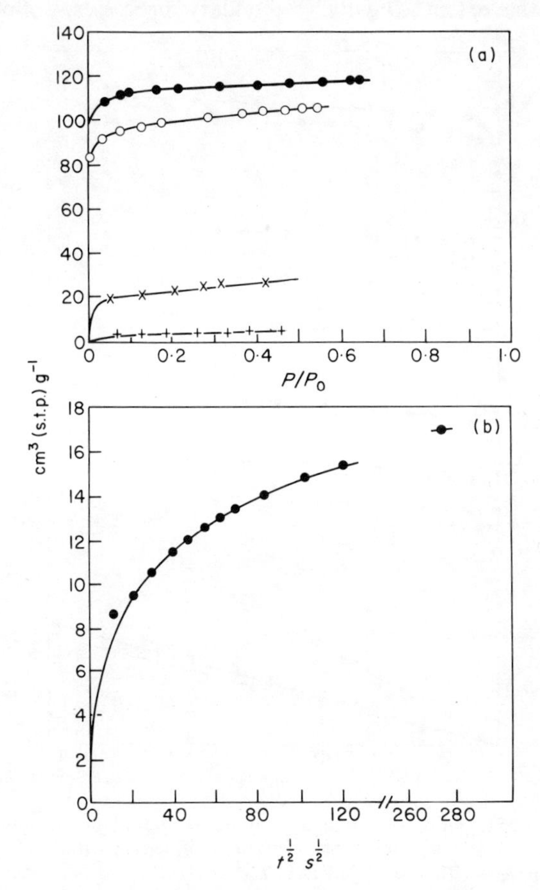

FIG. 6. (a) Sorption of N_2 at −196°C by: •, parent H-mordenite; ○, H-mordenite with 1·5% by weight of added Si; X, H-mordenite with 3·4% by weight of added Si; +, H-mordenite with 7·0₆% by weight of added Si. (b) Rate of sorption of N_2 at −196°C by H-mordenite with 3·40% by weight of added Si.[37]

Fig. 7 for O_2 and N_2 in silanated faujasites. It is noted in Fig. 7 that for 0·31HNa–Y with 2·84% by weight of added Si and for 0·73HNa–Y also with 2·84% added Si the sorption behaviour was extremely different. The ways in which the two lattices have been modified by primary and secondary reactions can thus be made to vary with the extent of hydrogen exchange in the parent material. All the silanated products gave X-ray powder patterns typical of faujasite, with some weakening in the case of 0·73HNa–Y and its silanated forms.

Because the silanated zeolites vary as sorbents according to the amount of SiH_4 chemisorbed, the extent of replacement of Na by H in the parent zeolite and the extent to which secondary processes, following primary

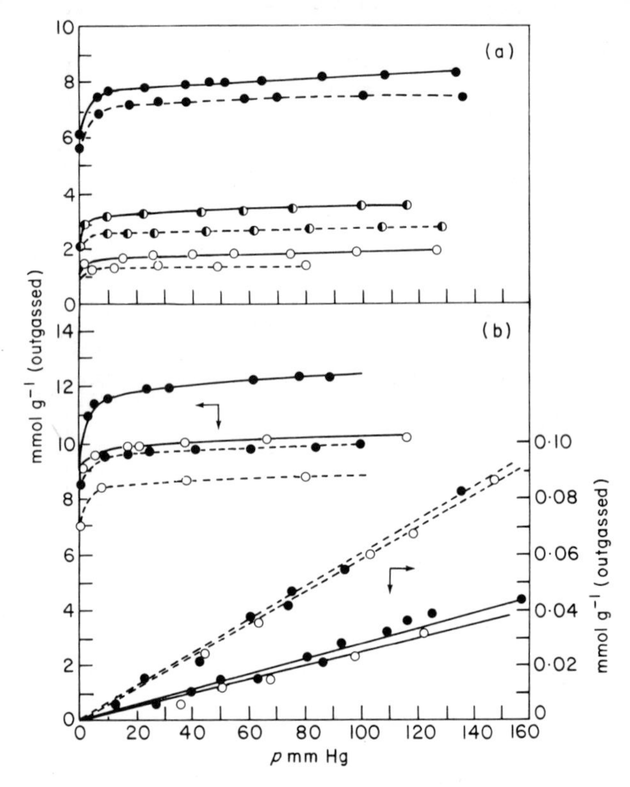

FIG. 7. Sorption of O_2 and N_2 by silanated zeolites Y[37]. Full lines denote O_2, dashed lines denote N_2. (a) ●, parent (Na, H)–Y(73); ◑, parent with 2·84% by wt of added Si; ○, parent with 5·99% added Si. $T = -196$°C. (b) ●, parent (Na, H)–Y(31); ○, parent with 2·84% added Si. In (b) for the left hand ordinate $T = -196$°C and for the right hand ordinate $T = 0$°C.

chemisorption of SiH_4, are allowed to proceed, and because the above three factors are controllable, silanation appears to be a promising method of modifying zeolite sorbents and catalysts. Oxidation of chemisorbed groups by oxygen or water replaces $\geqslant$ Si–H by $\geqslant$ Si–OH and the overall result is to introduce fresh permanent $\geqslant$ Si–O–Si $\leqslant$ bonds and new differently sited –OH groups into zeolite structures to give a controllable range of new compositions of matter. Moreover by using SiH_4 no organic material is introduced into the crystals which could give carbonaceous deposits in subsequent treatments. SiH_4 can be made industrially by the reaction

$$Mg_2Si + 4NH_4Cl \xrightarrow[\text{liq } NH_3]{0^\circ C} SiH_4 + 2MgCl_2 + 4NH_3. \tag{25}$$

From the experience with $Si(CH_3)_4$ and SiH_4 there seems reason to expect analogous reactions between H-zeolites and germane or alkylgermanes, forming —Si–O–Ge— bonds within zeolites. Similarly chemisorption of B_2H_6 might boranate H-zeolites.

1.4 Additional Chemisorptions involving Hydrogen Zeolites

A recent patent[41] has drawn attention to the production of phosphorus-bearing zeolites by refluxing siliceous water-free hydrogen zeolites ($SiO_2 : Al_2O_3 > 12$) with organic solvents containing phosphorus compounds like trimethyl phosphite or phosphorus trichloride. The amounts of phosphorus fixed in the structures were considerable, ranging in different experiments from 1·45 to 4·08% by weight. Phosphation was reported to change the catalytic behaviour. Since H-zeolites were involved it seems possible that the fixation of the phosphorus in part at least involved reactions such as

$$\gg Si\text{–}OH + (CH_3O)_3P \rightarrow \gg Si\text{–}O\text{–}P(CH_3)_2 + CH_3OH \tag{26}$$

$$\gg Si\text{–}OH + PCl_3 \rightarrow \gg Si\text{–}O\text{–}PCl_2 + HCl. \tag{27}$$

The evolution of hydrogen chloride was observed when the phosphorus compound in the reflux solution was PCl_3.

The sorption of methanol was investigated in decationated zeolite X by infra-red absorption spectroscopy[42] and a firmly bound stable form of sorbed methanol was detected which gave a spectrum invariant even after heating up to 400°C. This chemisorbed methanol was considered to have reacted with some of the silanol groups to give $\gg Si\text{–}O\text{–}CH_3$, because the characteristic absorptions due to –OH groups were weakened by the chemisorption process.

In a preliminary study of the sorption of anhydrous $AlCl_3$ vapour by outgassed zeolite Na-X in a closed gravimetric system, Barrer and Wasilewski[43] found that even in the range 200 to 300°C, the $AlCl_3$ was exceptionally strongly sorbed. A small amount of a gas was evolved which remained at room temperature, the crystals being saturated by sorbed $AlCl_3$. It was assumed that this gas was HCl, generated either by residual water or by small amounts of silanol hydroxyl, or by both acting on the $AlCl_3$. In a water-free hydrogen zeolite therefore one could expect a reaction like that given above for PCl_3 and the silanols. Such a reaction would generate new

$\gg$ Si–O–Al $\ll$ bonds and could thus complement silanation as a means of zeolite modification. Barrer, Sikand and Harding[44] in a study of zeolite modification by inclusion of salts from aqueous solution obtained some evidence of a reaction between intracrystalline KCl and outgassed offretite when the salt-bearing zeolite was heated. The reaction was thought to liberate HCl gas:

$$[\gg \mathrm{Al}\overset{\mathrm{HO}}{}\mathrm{Si} \ll] + \mathrm{KCl} \rightarrow [\gg \overset{\ominus}{\mathrm{Al}}\text{–O–Si} \ll]\mathrm{K}^+ + \mathrm{HCl}\uparrow. \qquad (28)$$

As HCl is driven off by heating the reaction tends to proceed from left to right. At low temperatures in outgassed zeolite no such reaction would be expected, although in aqueous systems (§ 1.1) hydronium mordenite and NaCl aq did result in liberation of HCl aq.

Dalla Betta and Boudart[45] studied the kinetics of the two exchange reactions

$$\text{–OH} + \mathrm{D_2} \xrightarrow{k} \text{–OD} + \mathrm{HD} \qquad (29)$$

$$\text{–OD} + \mathrm{H_2} \xrightarrow{k'} \text{–OH} + \mathrm{HD} \qquad (30)$$

in (H, Ca)-Y and in (H, Ca)-Y containing Pt atoms dispersed in the structure. To do this they measured the intensity of the 3640 $\mathrm{cm^{-1}}$ –OH band and of its –OD equivalent at 2690 $\mathrm{cm^{-1}}$, as functions of time. For the reaction in absence of Pt the first order rate equation

$$\ln(1 - A_{\mathrm{OD},t}/A_{\mathrm{OD},\infty}) = kt \qquad (31)$$

was applicable at fixed pressures, p_{H_2}, of H_2 (or D_2) but the values of k were pressure dependent:

$$k \propto p^n_{\mathrm{H_2}} \qquad (32)$$

where $n = 0{\cdot}21$ at 533 K and $0{\cdot}26$ at 610 K. In eqn 31, $A_{\mathrm{OD},t}$ and $A_{\mathrm{OD},\infty}$ are the intensities of the –OD band at time t and when reaction is completed. Figure 8 shows ln k plotted against reciprocal absolute temperature. The activation energy was reported as 24·4 and 23·1 kcal $\mathrm{mol^{-1}}$ for k and k' respectively.

Reaction in the Pt-bearing (H, Ca)-Y was initially very fast, but the rate decreased rapidly with time according to the relation

$$\frac{\mathrm{d}X}{\mathrm{d}t} = k(1 - \alpha X) \qquad (33)$$

where X is the fraction of total hydroxyl and deuteroxyl in the deuteroxyl form at time t. The presence of the Pt increased the exchange rate by as much as 10^5-fold. Water, added to the zeolite in measured doses, further accelerated the exchange as indicated in Table 8. The apparent activation energy was now 18·9 kcal mol^{-1} for k (reaction 29) and 28·9 kcal mol^{-1} for k' (reaction 30).

It is evident from §§ 1.1 to 1.4 that the acidic hydroxyls of outgassed

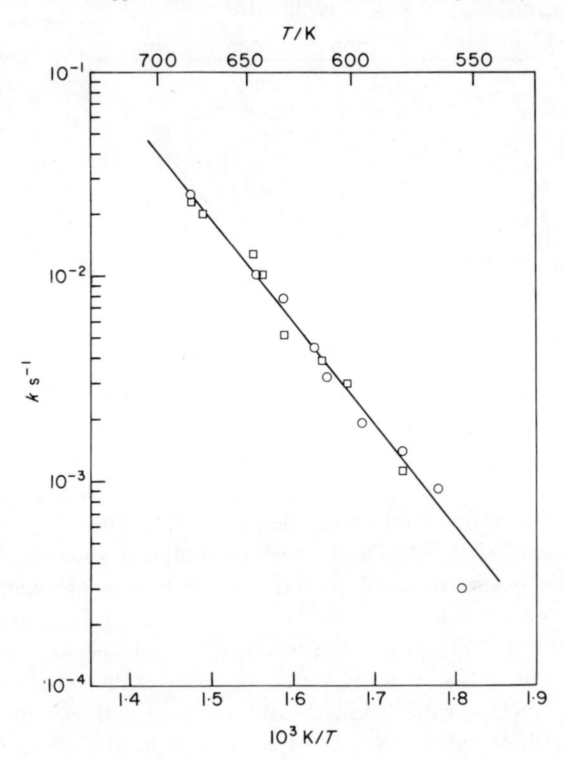

FIG. 8. Hydroxyl exchange rate on (H, Ca)–Y at 107 Pa of D_2 or H_2.[45] ○, $\_OH + D_2 \rightarrow \_OD + HD$; □, $\_OD + H_2 \rightarrow \_OH + HD$.

hydrogen zeolites have a complex chemistry which has been only partially explored and is not yet fully understood.

2. COMPLEXES WITH EXCHANGE IONS

Because of the importance of exchange cations, especially transition metal ions, in zeolite sorbents and catalysts, many investigations have been made of intrazeolitic complexes involving cations and guest molecules. Some examples of the insight which these investigations have given into zeolite chemistry are illustrated in the following sections.

TABLE 8

Acceleration of reaction 29 by H_2O. $T = 100°C$ and $p_{D_2} = 107$ kPa[45]

H_2O content (mg per g)	k ($s^{-1} \times 10^3$)	α
0	8·99	17·5
3·2	6·13	12·9
6·4	6·66	3·38
7·6	7·01	3·18
8·9	13·8	3·01
9·7	26·7	2·91
11·0	22·2	2·76

2.1 Ammonia and other Bases

Bases can react not only with acidic hydroxyls but also with transition metal ions within zeolites, giving complexes often resembling those which they form with the same cations in aqueous solutions. For example ammonia sorbed in Cu^{II}-Y or Cu^{II}-chabazite gives a characteristic cuprammonium blue. Aqueous $[Cu(NH_3)_4]^{2+}$ and other metal ammine ions may be exchanged directly into zeolites.[46,47,48] On the basis of analytical data the NH_3 : cation ratios in Table 9 were obtained.[49] Trivalent cobalt in mordenite carries its full coordination number of six NH_3 molecules. Since for steric reasons, so large an ion is not likely to be able to enter the side pockets these ions are probably present in the wide channels. For the other ions and all three zeolites, $x(x = NH_3 : \text{cation})$ falls below 4. There is thus some competition from framework oxygens or water for coordination of Cu^{II} or Zn^{II}.

Infra-red and e.p.r. spectroscopic studies of Cu^{II}-Y confirm that when dry NH_3 is sorbed there are about $4NH_3$ per Cu^{2+}, the likely configuration being square coplanar.[46, 48, 50] Huang and Vansant[47] made Cu^{II}-Y with

TABLE 9

Compositions of some metal ammine ions[49]

Ion	x for: clinoptilolite	mordenite	phillipsite
$[Cu(NH_3)_x]^{2+}$	3·17	3·24	3 40
$[Zn(NH_3)_x]^{2+}$	3 05	3·08	2·66
$[Co(NH_3)_x]^{3+}$	—	6·11	—

different degrees of exchange and followed the uptake of ammonia gravimetrically. When the NH_3 sorption at 23°C and 20 Torr was plotted against the Cu^{II} content of the zeolite a straight line was obtained having a slope corresponding with 3·3 NH_3 molecules per Cu^{2+}. While fully ammoniated $[Cu(NH_3)_4]^{2+}$ complexes are considered to have square coplanar configurations, desorption of ammonia from Cu^{II}-Y at 100°C gave an e.s.r. spectrum which suggested a distorted tetrahedral symmetry with one NH_3 per Cu^{2+},[46] presumably of the type $[Ox_3Cu–NH_3]^{2+}$ (cf Chapter 2, § 10). When the bases were methylamine and ethylamine the spectra indicated that some of the copper ions did not interact with the amines. For steric reasons Cu^{II}-amine complexes can form only in the 26-hedra of zeolite Y, and those Cu^{2+} ions which did not interact were therefore assumed to be in sodalite cages. Kiselev *et al.*[51] followed the desorption of ammonia by evacuation of Cu^{II}-Y initially charged with the gas, as a function of the temperature of heating. This was expressed as a plot of $\Delta m/M\Delta T$ against mean temperature T where Δm is the loss of mass from sorbent of mass M over the temperature interval ΔT. Figure 9 shows three regions of stability for the co-

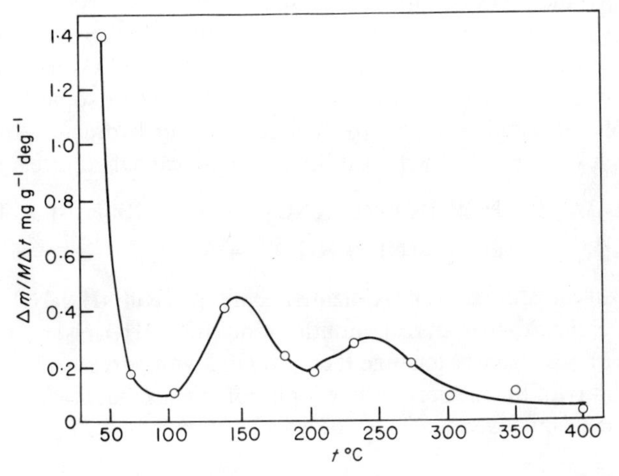

FIG. 9. The differential loss in mass on evacuation of Cu^{II}–Y containing ammonia.[51] M = weight of zeolite; Δm = mass loss over the temperature range Δt.

ordination spheres of Cu^{2+}. The first is in the range 70–110°C and was considered to represent Cu^{2+} coordinated by $2NH_3$ and by framework oxygens. The second, about 200°C, has only one NH_3 per Cu^{2+} and the third at 300°C corresponds with complete loss of ammonia.

Coughlan and coworkers[52] measured reflectance spectra of zeolite L and of mordenite containing Cr^{III}. The zeolites after exchange had respectively the unit cell compositions $Cr_{1\cdot33}K_{5\cdot1}[(AlO_2)_9(SiO_2)_{27}]25H_2O$ and $Cr_{1\cdot56}Na_{3\cdot32}$ $[(A10_2)_8(SiO_2)_{40}]26H_2O$. The ion $[Cr, 6H_2O]^{3+}$ was observed in the hydrated forms. If samples of each were outgassed above 300°C and treated with ammonia gas $[Cr, 6NH_3]^{3+}$ was found in the mordenite, but in zeolite *L* there appeared to be complexes in which Cr^{3+} was bonded both to oxygen and to ammonia.

Gallezot *et al.*[53,54] investigated the influence of a number of guest molecules upon the distributions of Ni^{2+} ions in zeolite Y. Ammonia readily extracted Ni^{2+} from positions in hexagonal prisms (type I sites; Chapter 2 § 10). Pyridine and nitric oxide produced only a slow cation migration, which however was assisted by the presence of residual water. Both NH_3 and NO should penetrate into sodalite cages and so approach Ni^{2+} in sites I, but pyridine cannot. To effect cation migration requires polarity in the guest molecule because the less polar species CO, C_2H_4, C_4H_8 and $C_{10}H_8$ caused no cation migration. This migration represents a perturbation of the zeolite structure which is not then to be regarded as the inert medium of adsorption thermodynamics. On the other hand solution thermodynamics remains appropriate. Water in particular can effect resiting of cations in zeolites, as shown by the differing site populations in hydrated and in water-free zeolite frameworks (Chapter 2, § 10).

An investigation has been made[55] of the stability, reactivity and infra-red spectra of some intracrystalline complex ions of Ru and Os introduced into zeolite Y by exchange of part of the Na^+, followed in some instances by exposure of outgassed products to nitric oxide and hydrazine. In this way ionic complexes were formed as follows, at the exchange levels indicated:

$[Ru(NH_3)_5N_2]^{2+}$, 16%; $[Ru(NH_3)_5NO]^{3+}$, 21%; $[Ru(en)_3]^{2+}$, 14·5%; $[Ru(en)_3]^{3+}$, 19% and $[Os(NH_3)_5N_2]^{2+}$, 14·5%.

In absence of moisture and of oxidizing agents the $[Ru(NH_3)_5N_2]^{2+}$ ion was stable in zeolite Y for several months, and $[Os(NH_3)_5N_2]^{2+}$ for several weeks. It was possible to exchange the N_2 ligand with gaseous N_2, confirming the accessibility of the intracrystalline ionic complexes, but it was not possible to hydrogenate this ligand.

2.2 Oxides of Nitrogen

The sorption of oxides of nitrogen is of particular interest because these gases from automobile exhausts are atmospheric pollutants. Addison and Barrer[56] showed that in Ca- and Na-rich forms of zeolite X and chabazite NO was copiously sorbed and underwent remarkably easy disproportionation:

$$\left.\begin{aligned} 3NO &\rightarrow N_2O + NO_2 \\ 4NO &\rightarrow N_2O + N_2O_3. \end{aligned}\right\} \quad (34)$$

The N_2O could be selectively desorbed at 0°C, by condensation in a trap at

$-183°C$ whereas removal of the other oxides required a higher temperature. The zeolite was heated to 400°C and the evolved brown gas condensed to a blue solid at $-183°C$. Desorption by freezing at $-183°C$ with the zeolite at temperatures not above 150°C liberated only N_2O and NO (the latter mainly from N_2O_3); temperatures above 200°C desorbed the NO_2. The effect of the initial sorption temperature upon the disproportionation is shown in Table 10a. The percentage of NO decomposed is that estimated for the

TABLE 10

(a) Effect of initial sorption temperature upon NO disproportionation in Ca-chabazite[56]

T (°C)	Time (h)	NO sorbed (cm³ at s.t.p.g⁻¹)	N_2O recovered (cm³ at s.t.p.g⁻¹)	% reaction of NO
0	1	16·03	3·18	79·3
0	7	14·51	2·91	80·2
0	24	16·35	3·16	77·3
0	25	16·74	3·31	79·1
0	119	14·45	2·91	80·6
0	1	10·35	2 01	77·7
−22·5	1	24·11	5·29	87·8
−63·5	1	24·10	5·70	94·6
−78	1	34·48	8·47	98·3
−78	31	54·98	13·78	100·2

(b) Disproportionation of NO sorbed at 0°C in several zeolites

Zeolite	NO sorbed (cm³ at s.t.p./g)	N_2O recovered (cm³ at s.t.p./g)	% reaction of NO
Ca-X	21·45	4·82	89·7
Na-X	22·0	4·08	72·8–83·6[a]
Na-chabazite	15·68	3·52	89·8–92·7[a]
Na-*A*	16·31	1·07	26·2–42·7[a]

[a] Results of several runs.

second of reactions 34. At temperatures below $-63 \cdot 5°C$ this stoichiometry corresponds with virtually complete disproportionation (Table 10a, column 5). The zeolites in Table 10b, and mordenite, are also effective for NO disproportionation.

After each sorption of NO in chabazite and thorough outgassing the subsequent uptake of oxygen was found to decrease progressively although the chabazite gave an unimpaired X-ray powder pattern. When a used sample was hydrothermally extracted at 170°C and the crystals filtered off the mother liquor was found to contain Na^+, Ca^{2+}, NO_2^- and NO_3^-. Moreover the extracted crystals had recovered their full capacity for sorbing oxygen. It was suggested that residual water had reacted with the NO_2:

$$H_2O + 2NO_2 \rightarrow HNO_2 + HNO_3. \quad (35)$$

The acids formed nitrate and nitrite and partially hydrogen ion exchanged chabazite. Unlike NO, N_2O was sorbed in chabazite without observable decomposition.

NO is a stable radical which, in spite of its unpaired electron, exhibits no paramagnetism in its $2\Pi_{1/2}$ ground state because the orbital moment exactly cancels the spin moment of the electron. If however the orbital moment is quenched by the intracrystalline environment paramagnetism develops. Kasai and Bishop[57] therefore studied the e.s.r. spectra of NO in Ba–, Na– and Zn–Y. The freshly activated zeolite was sealed with NO for a week at room temperature and then outgassed at this temperature until it no longer exhibited a purple colour or e.s.r. signal when cooled to 77 K. Such zeolites were defined as "NO-treated". When NO was then introduced to freshly outgassed and to NO-treated Ba-, Na- or Zn-Y a broad e.s.r. spectrum on each freshly activated zeolite was dramatically sharpened on each NO-treated zeolite. The modification effected by the NO-treatment survived through exposure to air, moisture and vacuum activation at 300°C overnight. This activation at 400°C finally returned the NO-treated samples to the state giving the broad, ill-defined e.s.r. signal. Heating the NO-treated sample to 400°C caused it to evolve a brown gas, which, after being resorbed at room temperature gave the e.s.r. spectrum of NO_2. The NO-treatment produced strong infra-red bands around 1400 cm^{-1} and, as noted above, yielded NO-treated zeolites which were not paramagnetic and were stable in modified form up to 300°C, capable of sharpening the e.s.r. signal of NO subsequently sorbed. The sharpening was ascribed to improved uniformity of the intracrystalline field consequent upon the NO-treatment, due to ionization of N_2O_3:

$$N_2O_3 \rightleftarrows NO + NO_2 \rightleftarrows NO^+ + NO_2^-. \quad (36)$$

When the NO-treated zeolite was irradiated with ultraviolet light at 77 K the e.s.r. spectrum of NO_2 was photo-induced, a spectrum stable indefinitely at 77 K but disappearing completely at 50°C in 30 minutes. The behaviour with light was taken to indicate a photo-induced reversal of the last stage of reaction 36. Reaction 36 is an alternative to reaction 35 for generating nitrite ion.

Disproportionation of NO sorbed by Mg-, Ca- and Sr-Y was also reported by Lunsford,[50] while Alekseyev and coworkers[59] provided infra-red evidence of the formation of N_2O, NO_2 and NO_3^- when NO was sorbed by zeolite X.

The sorption of NO has also been explored in Cu^{I}-,[60,61] Cu^{II}-,[61,62,63] Ag^{I}-,[64] Cr^{II}-[63] and Ni^{II}-[63] bearing forms of zeolite Y, mainly using e.s.r. spectroscopy. These spectra, and infra-red spectra, were interpreted in terms of the following intrazeolitic cationic complexes:

$$[CuNO]^{+}, [AgNO]^{+}, [Ag_2NO]^{2+}, [Cu^{+}NO^{+}], [FeNO]^{2+}, [Cr^{+}NO^{+}], [Ni^{+}NO^{+}]$$

and a diamagnetic complex formulated as

$$\left[\begin{array}{ccc} & NO & \\ ON\text{-}Cr & & Cr\text{-}NO \\ & ON & \end{array} \right].$$

Nitric oxide may be attached to a single metal ion, or form a bridge between two metal ions as in $[Ag_2NO]^{2+}$ or in the diamagnetic Cr complex. The bridged complexes were considered to occur within the sodalite cages.

NO_2 was also reported to form complexes with Cr^{II} and Ni^{II} in Y, which were formulated as $[Cr^{+}NO_2^{+}]$ and $[Ni^{+}NO_2^{+}]$, according to the e.s.r. spectra. By the charge transfer in such complexes the NO or NO_2 in this type of complex can be regarded as having reduced the metal ions. It has been reported[63] that Cr^{6+} ($3d^0$), Cr^{5+} ($3d^1$), Cr^{3+} ($3d^3$) and Cu^{+} ($3d^{10}$) do not form nitrosyl complexes but that Cr^{2+} ($3d^4$), Co^{2+} ($3d^7$), Ni^{2+} ($3d^8$) and Cu^{2+} ($3d^9$) do. The view that Cu^{+} ($3d^{10}$) does not form a complex with NO appears to be at variance with the occurrence of $[CuNO]^{+}$ reported by Chao and Lunsford.[60] The explanation may lie in the closeness of the energy levels of the $3d^{10}$ and the excited $3d^9 4s^1$ states for Cu^{+}.

2.3 Unsaturated Hydrocarbons

Simple olefines and acetylenes form π-electron complexes with some transition metal ions in zeolites. These complexes have been investigated by reflectance and Raman spectroscopy as well as by standard sorption measurements. In Ni^{II}-*A* the chemisorption on sites S2 (Chapter 2, § 10) containing transition metal ions was interpreted as shown in Fig. 10.[65] The ion in trigonal coordination with three oxygens of a 6-ring is drawn slightly away from the plane of the ring to sites S2* and is in distorted tetrahedral coordination with the same three oxygens and with the hydrocarbon ligand. This model was proposed for cyclopropane, ethylene, propylene and acetylene, the processes of sorption being reported as monocentric and reversible for all the gases save acetylene.

The isosteric heat of sorption of ethylene in Co^{II}-*A* followed the course shown in Fig. 11.[66] A high initial plateau in the heat descends to a second plateau as the number of ethylene molecules per cavity increases. There was one Co^{2+} ion per cavity and the high plateau was attributed to a complex $[Ox_3CoC_2H_4]^{2+}$ involving one C_2H_4 per Co^{2+}, in agreement with Fig. 10. Similar complexes were found for cyclopropane[67] and acetylene, propylene and *cis*- and *trans*-butenes[66] in Co^{II}-*A*, and with acetylene[68] in Mn^{II}-*A*, always in a geometry characteristic of π-bonding. The acetylene molecule

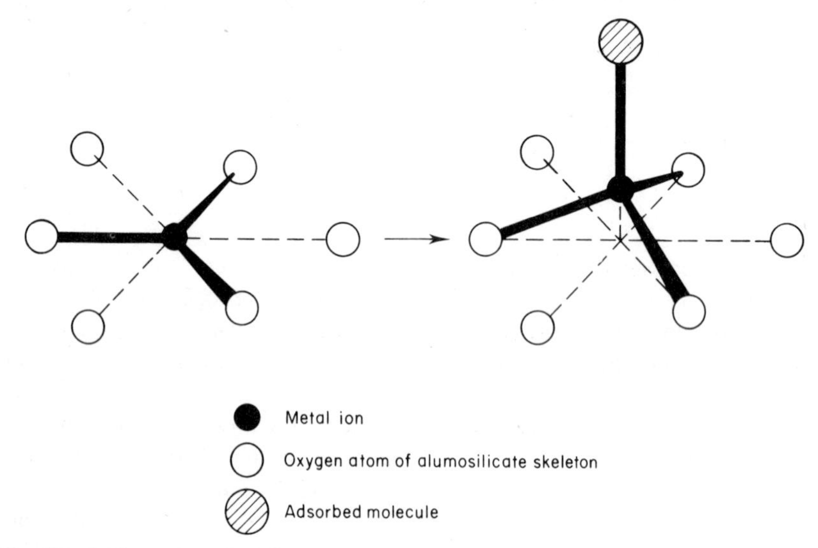

FIG. 10. Configuration of Ni^{II} on the S2 site of zeolite *A* (left); and configuration after chemisorption of a ligand (right).[65]

appeared to rotate freely about the axis joining the centre of the molecule with the Mn^{2+} ion, or else to assume random positions in a plane normal to this axis.

In Cu^{I}-Y Huang and Mainwaring[69] again found a 1 : 1 complex between Cu^{+} and ethylene with characteristic infra-red absorptions at 1428, 1533 and 1920 cm^{-1}. The Cu^{I}-Y was prepared by reducing Cu^{II}-Y with carbon monoxide and the sorption of ethylene was measured gravimetrically on a sample containing Cu^{+} and one containing Cu^{2+}. After sorption at 25°C and 100 Torr, followed by evacuation for 1 hour, 0·85 molecules of C_2H_4 remained per Cu^{+} but less than 0·03 molecules per Cu^{2+}, so that the ethylene -Cu^{+} complex is the more stable of the two. The intense band at 1428 cm^{-1} was assigned to the asymmetric CH_2 deformation. Ag-X, according to Carter *et al.*,[70] also gives a strong interaction with ethylene. Pichat[71] studied the infra-red spectra of acetylene, deuterated acetylene, propyne and but-2-yne, chemisorbed in (Cu, Na)-Y which had been submitted to various treatments yielding –OH groups or Cu^{+} ions. The –OH groups interacted only weakly with sorbed molecules; Cu^{+} were considered not to be involved;

and Cu^{2+}-alkyne bonds to result mainly as π-donation from the unsaturated hydrocarbon to the metallic ion. The results with alkynes appear to be at variance with the observations on C_2H_4 of Huang and Mainwaring.[60] The reason is not clear since π-bonding is involved in each case. Two extremes of the bonding can be represented as

$$\text{(I)}\quad M\begin{matrix}\nearrow \overset{\delta_-}{C}\\ \quad |||\\ \searrow \underset{\delta_-}{C}\end{matrix} \quad \text{and (II)} \quad M \leftarrow \begin{matrix}\overset{\delta+}{C}\\ |||\\ \underset{\delta+}{C}\end{matrix}$$

On the basis of the spectroscopic evidence, II was regarded as the direction

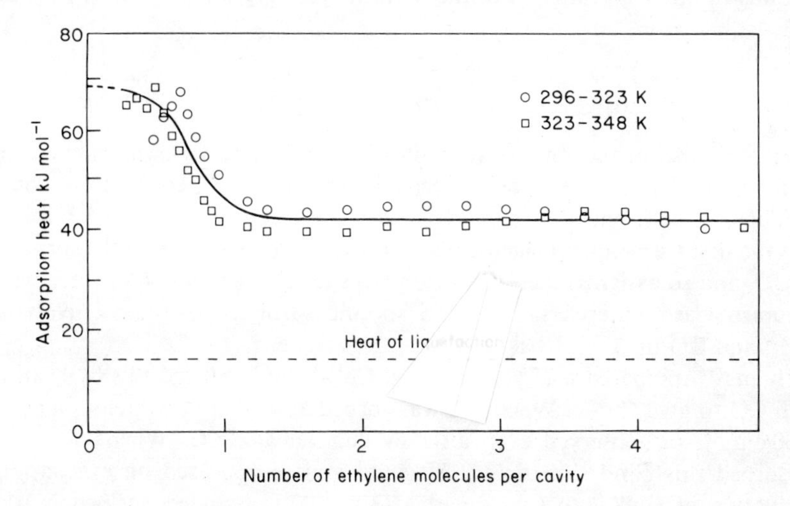

FIG. 11. Isosteric heats of sorption, q_{st}, for C_2H_4 on anhydrous Co^{II}-*A*.[66] There is one Co^{2+} per 26-hedron.

of charge transfer in the Cu^{2+}-alkyne complex. The Cu^+ ion, being less electron deficient than Cu^{2+}, should draw charge less strongly than Cu^{2+}.

Structure determinations of the complexes between ethylene and acetylene and some transition metal ions such as Mn^{II} and Co^{II} in zeolite *A* confirm the arrangement indicated in Fig. 10, the long axis of the hydrocarbon being at right angles to the line joining the mid-point of the molecule to the ion.[68, 72]

2.4 Carbon Monoxide

Angell and Schaffer[73] reported that when carbon monoxide was sorbed by zeolites X and Y containing transition metal ions the stretching frequency shifted by amounts dependent upon the cation to values higher than that for gaseous carbon monoxide. In explanation it was proposed that ions such as Co^{2+} increase the small dipole of CO with donation of electron density from a σ orbital of carbon to a σ-acceptor orbital of the metal. The charge deficiency on the carbon induces charge transfer from oxygen towards carbon and so strengthens the C–O bond. Haber *et al.*[74] subsequently measured isotherms and heats of sorption of CO in (Zn, Na)-X. In the range 0° to 65°C they reported a specific interaction between Zn^{2+} and CO which could be described by Langmuir's isotherm equation, together with a non-specific sorption following Henry's law. The extent of exchange of Na^+ by Zn^{2+} was varied, and from the portion of the sorption governed by Langmuir's isotherm the number N_{CO} of molecules undergoing specific sorption per unit cell was found, as a function of the total number N_{Zn} of Zn^{2+} ions per unit cell.

N_{Zn} =	17	27·5	40·5	42·5
N_{CO} =	0·63	1·42	2·22	2·74

Evidently most of the Zn^{2+} was in sites inaccessible to CO, which could be I and I′, of which there are respectively 16 and 32. Specific sites should include ions at II and II*. Similarly with Ca-Y Egerton and Stone[75] found that the first Ca^{2+} ions replacing Na^+ had very little effect upon the sorption of CO, and so assigned these Ca^{2+} ions to sites I. When all 16 of these were occupied Ca^{2+} entered sites II and specific sorption developed involving Ca^{2+} and CO on a 1 : 1 basis.

Huang[76] prepared a 75% exchanged Cu^{II}-Y and reduced this by heating with CO to give Cu^{I}-Y. When CO was sorbed at 40 Torr an intense band at 2160 cm^{-1} was observed even after evacuation at 25°C. When NH_3 was presorbed this band shifted to 2080 cm^{-1} and disappeared on evacuation. Reduction of Cu^{II}-Y to Cu^{I}-Y with CO at 400°C is indeed helped by the presence of some ammonia, which can draw Cu^{2+} from sites not accessible to CO to sites which are.[77] If ethylenediamine was presorbed and then evacuated at 25°C some was removed, and subsequent addition of CO gave bands at 1916 and 2090 cm^{-1}. That having the lower frequency was ascribed to bridging carbonyls, and disappeared after desorption of more ethylenediamine at 200°C. Finally, when pyridine was presorbed at 25°C and evacuated at 120°C, addition of CO gave a band at 2120 cm^{-1}. In all cases, desorption of the bases at higher temperatures caused increases in the CO stretching frequencies towards the original high value for the sorbed CO in absence of the bases. The shift in frequency when amines are present was attributed to changes in location of the carbonyl complexes.

The structure of a partially exchanged Co^{II}-*A* zeolite containing sorbed carbon monoxide has been determined,[78] there being four Co^{2+} ions per unit cell. The carbon of the CO was attached to the Co^{2+} on a 1 : 1 basis.

The configuration of the complex was similar to that shown in Fig. 10, i.e. $[Ox_3Co\text{-}CO]^{2+}$.

Although concerned with metal carbonyls rather than with ionic complexes, the study by Coudurier *et al.*,[79] which demonstrated the thermal breakdown of $Mo(CO)_6$, $Re_2(CO)_{10}$ and $Ru_3(CO)_{12}$ sorbed into zeolite H–Y is of considerable interest. The release of CO from the first of these carbonyls was complete by about 300°C, and appeared to be a continuous function of the heating temperatures. Both the other carbonyls showed a two stage loss in ill-defined steps. $Re_2(CO)_{10}$ lost three molecules of CO by about 200°C and the second stage loss began around 350°C. With $Ru_3(CO)_{12}$ three molecules had been lost near 200°C and the second stage began near 300°C. Both second stages were complete by 450°C. After heating $Mo(CO)_6$ in H–Y above 250°C the ultraviolet spectrum indicated significant amounts of Mo^V; eventually with all three carbonyls the CO-free metals in some form remain within the zeolite. The method therefore allows metered amounts of these elements into zeolite Y catalysts.

3. ZEOLITE REDOX CHEMISTRY

Chemisorption often involves oxidation and reduction processes, such as the conversion within the zeolite of Cu^{2+} to Cu^+ by heating with carbon monoxide. Many other intrazeolitic transition and noble metal ions can be reduced by H_2 or CO, including those of Fe, Co, Ni, Pd, Pt, Cu, Ag and Zn.[80] Metal atoms are often formed, dispersed initially throughout the zeolite lattice. The ease of reduction depends on the extent of exchange which influences the accessibility of the cations, as well as on the nature of the cation and the Si : Al ratio and structure of the zeolite. The metal atoms can migrate on heating and form crystallites outside the zeolite or small clusters of metal atoms within it. According to Minachev *et al.*[80] the ease of migration decreases in the sequence

$$Ag > Zn > Pd > Cu > Ni > Pt \geq Co.$$

When the metal atoms are sufficiently volatile heating may vaporize them from the crystals.[81] If the conditions of treatment are mild, reduction may proceed, as with $Cu^{2+} \rightarrow Cu^+$, to a lower valence ionic form. X-ray photoelectron spectroscopy has shown that cations of transition metals in dehydrated zeolites interact with the framework by nearly pure ionic bonding.

Quantitative studies of silver ion reduction and reoxidation in (Ag, Na)-Y with various silver ion contents were made by Beyer and coworkers.[82] The chemisorption kinetics of H_2 for 71% exchanged (Ag, Na)-Y are shown in Fig. 12. At 623 K the uptake corresponds with complete reduction of Ag^+

to Ag^0. Similarly Fig. 13 illustrates the kinetics of oxygen chemisorption by the same previously reduced sample of (Ag, Na)-Y, and shows complete reoxidation at 623 K. The reactions involved can be formulated as

$$2\,Ag^+[>\overset{\ominus}{Al}\text{–O–}Si<] + H_2 \rightarrow 2\,Ag + 2[>Al\overset{HO}{\diagdown}Si<] \quad (37)$$

and

$$2\,Ag + 2[>Al\overset{HO}{\diagdown}Si] + \tfrac{1}{2}O_2 \rightarrow 2\,Ag^+[>\overset{\ominus}{Al}\text{–O–}Si<] + H_2O. \quad (38)$$

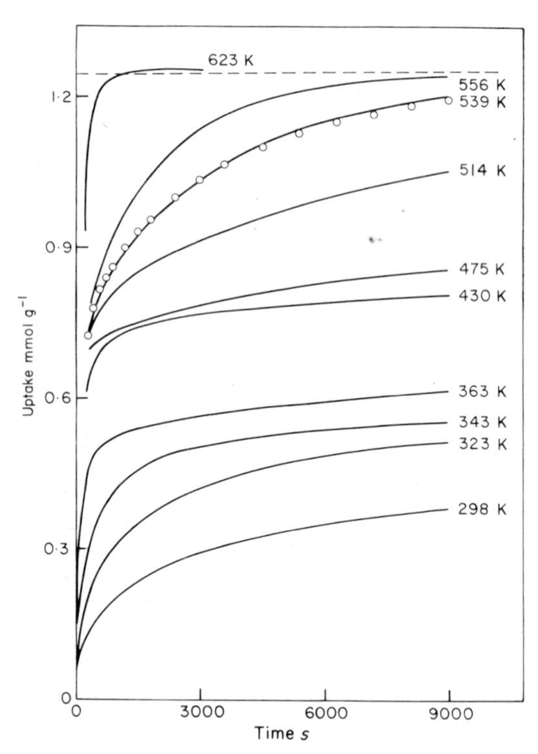

FIG. 12. Curves showing the rates of reduction by H_2 of (Ag, Na)-Y with 71% of the Na^+ exchanged by Ag^+, at each of a series of temperatures.(82)

The affinity of Ag^+ ions for sites I (in hexagonal prisms) may account for incomplete reduction of Ag^+ at lower temperatures (Fig. 12) since these ions may be screened from attack. The reduction and reoxidation are however fully reversible and quantitative processes at high enough temperatures.

Similar investigations have been made of Cu^{2+} in zeolite Y by Herman *et al.*(83) and by Jacobs *et al.*(84) Under mild conditions such as heating in H_2 at 200°C for an hour, Cu^{2+} was converted to Cu^+, this reaction being completely reversible. At 400°C, however, some copper crystallites formed the larger of which could be slowly reoxidized to CuO. Smaller metal clusters and atoms still within the zeolite were readily converted to Cu^{2+} at exchange sites inside the zeolite. When (Cu^{II}, Na)-Y was outgassed at temperatures above about 380°C there was evidence of formation of some Cu^{+}, with

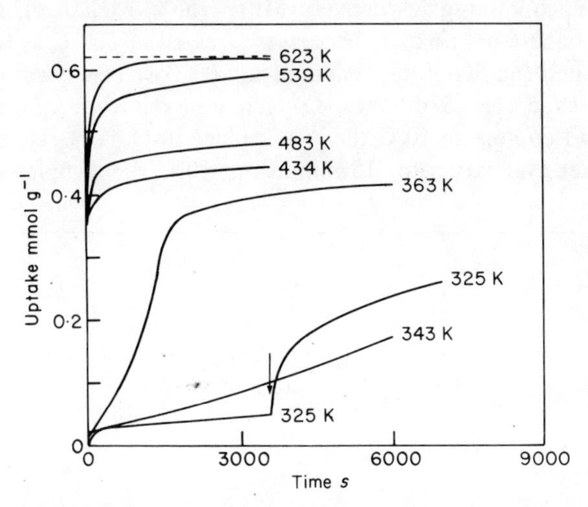

FIG. 13. The re-oxidation kinetic of the fully reduced (Ag, Na)-Y of Fig. 12 by O_2 at each of a series of temperatures.(82) At the point marked by the arrow 0·3 mmol of H_2O was added at 325 K.

evolution of oxygen gas and generation of Lewis acid sites numbering about one for every two Cu^{2+} ions reduced. About 40% of the Cu^{2+} was thus converted to Cu^{+} at 500°C. Reoxidation was rapid but was complete only at 350°C. A stoichiometry which accounted for the observations was proposed:(84)

$$2[>\overset{\ominus}{Al}-O-\underset{\wedge}{Si}-O-\underset{\wedge}{Si}-O-\overset{\ominus}{Al}<]\quad (Cu^{2+})$$

$$\rightleftarrows \tfrac{1}{2}O_2 + [>\overset{\ominus}{Al}-O-\underset{\wedge}{Si}-O-\underset{\wedge}{\overset{\oplus}{Si}}\ \ Al<](Cu^{+}) + [>\overset{\ominus}{Al}-O-\underset{\wedge}{Si}-O-\underset{\wedge}{Si}-O-\overset{\ominus}{Al}<](Cu^{+}). \quad (39)$$

The local charge disbalances and the generation of $>Si^{+}$ may argue against this stoichiometry. An alternative possibility is that residual water has a role:

$$4[>Si-O-\overset{\ominus}{Al}<]2Cu^{2+} + H_2O \rightleftarrows \tfrac{1}{2}O_2 + 2[>Si-O-\overset{\ominus}{Al}<]2Cu^{+} +$$

$$+\,2[>Al\quad \overset{HO}{\diagdown} Si<] \quad (40)$$

and that the appearance of Lewis acid sites is the further result of the heating of the Bronsted sites (cf § 1.2). The reactions 39 and 40 suggest a mechanism for reduction by CO of Cu^{2+} to Cu^{+} within the zeolite since by consuming oxygen the reactions are swung to the right.

Subsequently to this work Iwamoto and his coworkers[85] prepared samples of zeolite Y containing respectively Mn^{2+} (55% exchange), Co^{2+} (70%) Ni^{2+} (60%) and Cu^{2+} (62%). These were treated at 500°C as follows. They were evacuated for two hours, exposed to 100 Torr of oxygen for one hour and reevacuated for 15 minutes. Oxygen was reintroduced at 500°C and 100 Torr and cooling to 10°C then took place in O_2 at a rate of 3·3°C per minute. After evacuation for 15 minutes at 10°C the samples were given a

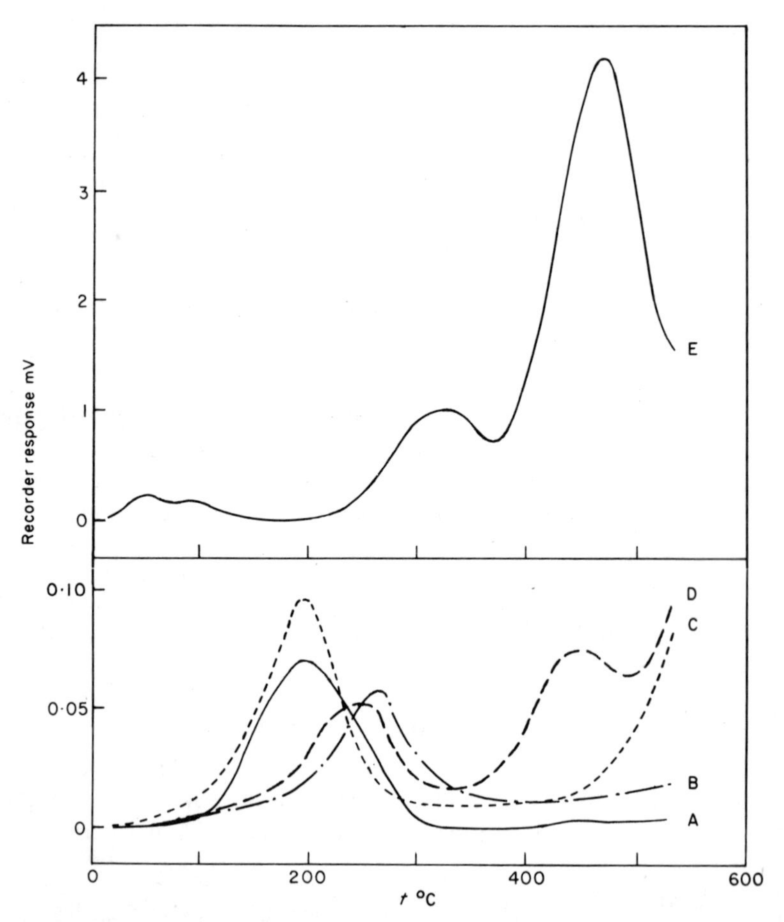

FIG. 14. Temperature programmed desorption chromatograms of oxygen from several transition metal ion-exchanged Y zeolites.[85] A, Na^{I}-Y; B, Ni^{II}-Y; C, Mn^{II}-Y; D, Co^{II}-Y; Y; E, Cu^{II}-Y.

programmed heating at 5°C per minute, with the results shown in Fig. 14. It is notable that, while little oxygen was evolved from Na-, Ni-, Mn- and Co-Y, the Cu-Y released a quantity of the gas. This behaviour may correspond with the reduction processes of reactions 39 or 40.

Reversible bonding of oxygen as transition metal ion complexes is of interest *inter alia* as a model of the oxygen-haemoglobin system.[86] In this connection, Kellerman *et al.*[87] studied the behaviour of zeolite *A* partially exchanged with Cr^{2+} towards oxygen. The pale blue Cr^{II}-*A* contained 1·5 Cr^{2+} ions per unit cell, and became a pale blue-lilac after outgassing at 350°C. Exposure to 760 Torr of dry oxygen then turned it instantly gray. The sorption-desorption isotherm (Fig. 15) shows that uptake of oxygen is

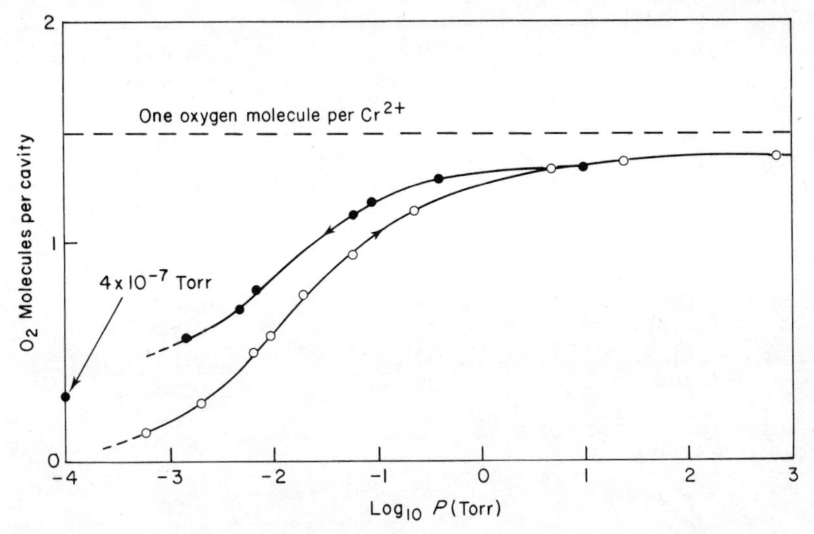

FIG. 15. The isotherm of oxygen at room temperature in anhydrous Cr^{II}-*A*.[87]

partially but not fully reversible; about 0·2 O_2 per Cr remains at 10^{-7} Torr. Near saturation the O_2 : Cr ratio approaches one, and the coordination of the Cr was considered to be that shown in Fig. 10, i.e. $Ox_3Cr^{3+}O_2^-$. Thus the molecular oxygen is converted to an anion radical, π-bonded to the chromium.

Oxygen chemisorption by zeolites X and Y containing transition metal ions was further investigated by Kubo *et al.*[88] The zeolite was outgassed for three hours at 400°C. In case any oxidation had occurred those exchange forms in which the transition metal had been introduced in a lower valence state were reduced in hydrogen at 400°C and again outgassed at this tempera- for three hours, before measuring the chemisorption of oxygen. The oxygen uptakes measured volumetrically at 400°C and 100 mm of Hg pressure are given in Table 11. Tl^+-, Co^{2+}- and Mn^{2+}-forms sorbed little oxygen, as expected from their standard oxidation potentials, and no oxygen was taken up by Cu^{2+}- and Ni^{2}-forms. The Cr^{2+}-, Cu^{+}- and Fe^{2+}-zeolites containing ions with high oxidation potentials, chemisorbed oxygen to give complexes with oxygen to metal atom ratios in the range 0·36 to 0·62, and so not far from 0·5. A ratio of 0·5 is compatible with bridged cations $M^{(n+1)+}–O^{2-}–M^{(n+1)+}$ for ions originally M^{n+}. The amount of oxygen chemisorbed by Fe^{II}-zeolite depended on the extent of exchange of Na^{+} by

TABLE 11

Oxygen chemisorbed at 400°C and 100 mm Hg[88]

(a) Y-type zeolite

Cationic form	% Na^{+} exchanged	Oxygen uptake in atoms per cation
Na^{I}	0	0·000
Co^{II}	48	0·106
Tl^{I}	54	0·006
Fe^{II}	36	0·500
Cu^{I}	34	0·436
Cu^{II}	54	0·000
Cr^{II}	34	0·430
Cr^{III}	38	0·795
Mn^{II}	40	0·100
Ni^{II}	39	0·000

(b) X-type zeolite

Cationic form	% Na^{+} exchanged	Oxygen uptake in atoms per cation
Na^{I}	0	0·000
Co^{II}	41	0·013
Tl^{I}	26	0·000
Fe^{II}	12	0·360
Fe^{II}	35	0·488
Fe^{II}	42	0·499
Cu^{I}	18	0 618
Cr^{II}	28	0 474

Fe^{2+}. It approached the O : Fe ratio of 0·5 closely when the Na^{+} exchange by Fe^{2+} exceeded 35%. At 150°C in anhydrous Cr^{II}-*A*, an unbridged complex results from chemisorption of O_2 or N_2O[87] with one O per Cr, i.e. $Ox_3Cr^{4+}O^{-2}$. This complex can be reduced to the original Cr^{II}-*A* with CO at 300°C, the CO being oxidized to CO_2 in the process. It is thus seen that oxygen may behave differently with Cr^{II}-X and Cr^{II}-Y than it does with Cr^{II}-A. For Cr^{III}-Y the Cr : O ratio in Table 11 is 0·8 and so is approaching unity. Samples oxidized with O_2 at 500°C and then evacuated gave e.p.r. evidence of the presence of Cr^{V}, so that a complex $[Cr^{5+}O^{2-}]$ may have been formed.[63] Reduction in H_2 at 500°C yielded Cr^{2+} ions which with NO give the tetrahedral complex $[Ox_3Cr^{+}NO^{+}]$. As in Chapter 2, Ox_3 means that the cation is associated with the three nearest oxygens of a 6-ring and is in or near sites II. Coughlan and coworkers[52] observed that, when samples of zeolite *L* and of mordenite each containing Cr^{III} were outgassed at 400°C and then exposed to air reflectance spectra characteristic of the CrO_4^{2-} ion were obtained. Thus under various reducing and oxidizing conditions, valence states of 1, 2, 3, 4, 5 and 6 have been indicated for intrazeolitic chromium.

The redox chemistry of intrazeolitic cations is further illustrated in the case of Pd^{II}-Y.[89] The water-free unit cell composition was $Na_{19\cdot5}(NH_4)_{11\cdot5}Pd^{II}_{12\cdot5}[(AlO_2)_{56}(SiO_2)_{136}]$. A bright beige-pink colour characterized the Pd^{2+} ions in the crystals, while an e.p.r. signal at $g_{iso} = 2\cdot223$ was sometimes present and then indicated some Pd^{3+} ions. Oxidation involved heating samples in oxygen at temperatures up to 500°C. Infra-red, e.p.r. and X-ray diffraction spectra served to monitor processes taking place within the zeolite. Interaction with NO was used as a further probe. Reduction of oxidized Pd-zeolite was effected at 500°C in 300 Torr of hydrogen. Following this treatment there was X-ray evidence of Pd crystallites of about 20Å diameter which the authors considered to be inside the zeolite. The dark brown crystals exposed to NO at 25°C quickly became beige pink due to appearance of Pd^{2+} and the Pd crystals disappeared completely in two days. The NO appeared therefore to oxidize Pd atoms present as crystallites or atomically dispersed, giving complexes which redistributed themselves in the zeolite. Evacuation at 200°C next removed NO and Pd^{2+} returned to sites I′ (in sodalite cages near the hexagonal prisms).

4. RADICAL AND RADICAL-ION FORMATION

Several instances were given in §§ 3 and 2.2 of complexes involving radical ions. Further examples of this are of relatively frequent occurrence. If Ti^{III}-Y was evacuated at 500°C, contacted with oxygen at 10 mm for 30 minutes and then evacuated at room temperature, the e.s.r. signal was interpreted as being due to the complex formed in the reaction[88]:

$$Ti^{3+} + O_2 \rightarrow [Ti^{4+}O_2^-]. \tag{41}$$

When both sorption and evacuation temperatures were raised, the relative intensity of signals due to O_2^- and to Ti^{3+} decreased, a result attributed to further oxidation:

$$[Ti^{4+}O_2^-] \rightarrow TiO_2 + \text{trapped electron in solid.} \quad (42)$$

If SO_2 was admitted at 10 mm to Ti^{III}-Y previously outgassed at 500°C, and the system warmed to 70°C for 30 minutes the reaction 43 took place, analogous to reaction 41:

$$Ti^{3+} + SO_2 \rightarrow [Ti^{4+}SO_2^-]. \quad (43)$$

The e.s.r. signal was recorded after evacuation for 15 minutes at room temperature. When oxygen at 20 mm was admitted the SO_2^- signal was replaced by one due to O_2^-. If the $[Ti^{4+}O_2^-]$ complex of reaction 41 was exposed to ethylene, acetylene, 1-butene or carbon monoxide the O_2^- signal disappeared slowly.

Krzyzanowski[91] also found the anion radical O_2^- when oxygen was sorbed by Ce^{III}-X and La^{III}-X, for the best results outgassed at 200°C. The oxygen was sorbed at this temperature and evacuated below it because the signals were found to disappear above 200°C. The complexes are represented as $[Ce^{4+}O_2^-]$ and $[La^{4+}O_2^-]$ also attached to framework oxygens. Sorption of N_2O at room temperature or above, but below 200°C, gave signals attributed to O^- and O_2^-:

$$\begin{aligned} Ce^{3+} + N_2O &\rightarrow [Ce^{4+}O^-] + N_2 \\ 2[Ce^{4+}O^-] &\rightarrow [Ce^{4+}O_2^-] + Ce^{3+} + e. \end{aligned} \quad (44)$$

There was indirect evidence also of the formation of O^{2-}, O_2^{2-} and possibly O_3^- after appropriate treatments. Howe and Lunsford[92] demonstrated the reversible formation of low spin adducts $[Co^{3+}L_xO_2^-]$ within the 26-hedra of (Co^{II}, Na)-Y, where the ligand, L, was NH_3, CH_3NH_2 or n-$C_3H_7NH_2$ and x may have been 5. Dimeric μ-superoxo adducts $[L_xCo^{3+}O_2^-Co^{3+}L_x]$ could also be made with $L = NH_3$ or CH_3NH_2. The e.p.r. signals were similar to those of analogous adducts in solution. The (CO^{II}, Na)-Y zeolites were prepared with 15, 10 and 0·7 Co^{2+} per unit cell and were outgassed at temperatures rising to 400°C and at 10^{-5} Torr. The ligand was then sorbed at room temperature and finally oxygen was sorbed at —78°C and 10 Torr. The formation of the dimeric complex involved prolonged contact of (Co^{II}, Na)-Y with oxygen at 25°C after the ligand had been sorbed. These results suggest stabilization of Co^{III} by the ligands since Kubo *et al.*[88] found only a small uptake of oxygen by Co^{II}-enriched X or Y xeolites (Table 11).

The formation of the radical ion SO_2^- does not necessarily require the presence of transition metal ions (cf reaction 43). Ono and coworkers[93, 94] used H-Y, H-*L* and H-mordenite as sorbents in an e.s.r. study of these radical ions. The H-Y was outgassed at 600°C for five hours, and was then exposed to SO_2 at 13 mm for 2·5 hours. The e.s.r. signal for SO_2^- was recorded, the radical ions being unchanged after seven hours at room temperature. They were also stable at 200°C. The effect of calcination temperature of the zeolite upon the radical ion concentration is shown in Fig. 16. When the SO_2 was sorbed at 200°C on zeolites treated at 600°C for five hours the concentrations of spins were those given in Table 12. Thus Table 12 shows that SO_2^- radical ions form at low concentrations in all the zeolites studied, while Fig. 16 demonstrates that calcination temperature can be an important factor in

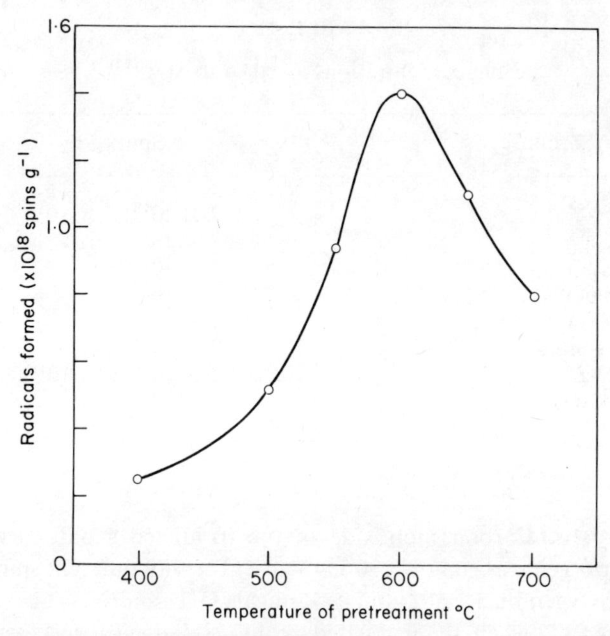

FIG. 16. The effect of calcination temperature upon the concentration of SO_2^- radical anions formed in decationized zeolite Y.[93]

determining the concentration of SO_2^- subsequently obtained. Also it was confirmed that admission of oxygen to the zeolites carrying SO_2^- caused O_2^- radical ions slowly to appear. Donor centres in the outgassed zeolites transfer an electron to the SO_2, which presumably subsequently gives up this electron to sorbed O_2. In further experiments a series of Ce-, La- and Pr-exchanged zeolites X and Y was evacuated for one to four hours at 10^{-5} Torr at temperatures from 200 to 600°C. Sorption of oxygen and carbon monoxide gave signals corresponding with O_2^- and CO_3^-. When mono-olefines, dienes or aromatic hydrocarbons were sorbed into the zeolites containing O_2^- and CO_3^- spectra were observed characteristic of hydrocarbon anion radicals $[RH]^-$ with simultaneous disappearance of O_2^- and CO_3^-.

In the above instances of radical-ion formation, no irradiation was involved. Irradiation by γ-rays, X-rays and ultraviolet light can also yield radical-ions from sorbed materials such as oxygen or carbon dioxide. Thus Coope *et al.*[96] irradiated faujasite and H-mordenite in a ^{60}Co gamma cell for

15 to 20 minutes in order to form Cl_2^- from sorbed Cl_2. Vedrine and co-workers[97] used γ-radiation at $-196°C$ in a successful e.p.r. study of sorption sites and sorbed species in H-Y. Paramagnetic sorption sites appeared to be mainly oxygen nuclei from the lattice, termed *V* centres. Introduction of molecular hydrogen neutralized these sites and molecular oxygen yielded

TABLE 12

Spin concentrations ascribed to SO_2^-[93]

Zeolite	Spins/g
H-Y	$1{\cdot}2 \times 10^{18}$; 2×10^{17}
Na-Y	2×10^{17}
Ca-Y	1×10^{17}
Co-Y	1×10^{17}
Zn-Y	7×10^{17}
H-mordenite	2×10^{17}
H-*L*	2×10^{18} (9×10^{18})[a]
Al_2O_3	3×10^{17}

[a] SO_2 sorbed at 100°C.

O_2^--radical ions. Carbon monoxide in the irradiated zeolite reacted very easily with oxygen sorbed at $-196°C$ to give O_2^-, with different spin densities on the two oxygen nuclei of $\frac{1}{4}$ and $\frac{3}{4}$. Radicals $OH\cdot$ and $NH_2\cdot$ could also be detected but only when the irradiated zeolite contained respectively molecular water or ammonia.

Methyl radicals were formed and trapped inside zeolite Na-*A* when the crystals containing sorbed methane were γ-irradiated at $-196°C$.[98] After an initial preheat in air the zeolite was outgassed for five hours at about 10^{-4} Torr at different temperatures from room temperature to about 550°C. Methane was then sorbed at temperatures up to room temperature before cooling to $-196°C$ and irradiation. The e.s.r. spectra were measured between -196 and $-40°C$. No methyl radicals were stablized at $-196°C$ in zeolite heated below 50°C but zeolite heated at $\approx 65°C$ produced "normal" CH_3, stable on warming up to $-50°C$. Heating at about 80°C gave zeolite which on irradiation gave radicals thought to have additional bonding to a proton as $CH_3 \ldots H^+$, termed "abnormal methyl of type I". The e.s.r. signal due to these began to decay above $-50°C$. In zeolite heated at 250°C "abnormal methyl of type II" was obtained on irradiation at $-196°C$ which began to decay above $-160°C$. The nature of this radical was not made clear. Finally no e.s.r. spectrum was observed for γ-irradiated methane in Na-*A* heated above 500°C. The paper demonstrates a complex behaviour in which the important variable is the pretreatment temperature of the zeolite. In addition to these experiments, Svejda[99] prepared stabilized CF_3 radicals in Na-X by ultraviolet irradiation at $-196°C$ of sorbed hexafluoroacetone. Prior to sorption, the zeolite was outgassed at 400°C for 19 hours. The presence of immobilized radicals was confirmed by the e.s.r. spectra in an amount corresponding with about 10^{16} radicals in 10^{18} intrazeolitic cavities.

TABLE 13

Electron donor and acceptor sites in Zeolites X and Y[100]

Sample	Gu or NH_4 per unit cell	Electron donor centres (spin/g) $\times 10^{-17}$	Electron acceptor centres (spin/g) $\times 10^{-6}$
Na-X	—	29·2	0·8
H(Gu)Na-X	32	1·7	0·7
$H(NH_4)Na$-X	36	11·1	0·8
Na-Y_1	—	6·4	0·8
H(Gu)Na-Y_1	22	3·5	5·1
$H(NH_4)Na$-Y_1	17	11·9	3·7
Na-Y_2	—	2·9	≈0·2
H(Gu)Na-Y_2	21	1·5	2·6
$H(NH_4)Na$-Y_2	21	3·4	1·6
Alumina	—	51·0	0

Radical ions can readily appear when suitable organic molecules are sorbed since such molecules can frequently function as acceptors or donors of electrons. Some illustrations of this behaviour follow. Dudzik *et al.*[100] sorbed tetracyanoethylene and perylene in Na-X, Na-Y and in the mixed (H, Na)-forms of each. The first of these two molecules was reduced to an anion radical by electron donor sites in the outgassed zeolites and the latter was oxidized to a cation radical by electron acceptor sites. The mixed (H, Na)-forms were obtained from the (NH_4, Na)- and (Gu, Na)-zeolites by heating for four hours at 400°C and 10^{-5} Torr (Gu here denotes guanidinium ion). Table 13 gives the numbers of centres able to form the anion radicals and of those able to form the cation radicals. The X-zeolites had 85 and the Y-zeolites 55 or 54 Al atoms per unit cell. The alumina was prepared from aluminium isopropoxide. The symbols (Gu) and (NH_4) in Table 13 indicate whether the (H, Na)-form was obtained by heating guanidinium- or ammonium-enriched zeolite. For zeolite Y the concentration of donor sites decreases in the order

$$H(NH_4)Na\text{-}Y > Na\text{-}Y > H(Gu)Na\text{-}Y$$

and in zeolite X this order is

$$Na\text{-}X > H(NH_4)Na\text{-}X > H(Gu)Na\text{-}Y.$$

In zeolite X electron acceptor sites are at a low, nearly constant level; in Y the sequence is

$$H(NH_4)Na\text{-}Y_1 > H(Gu)Na\text{-}Y_1 > H(NH_4)Na\text{-}Y_2 > Na\text{-}Y.$$

Alumina, while rich in donor centres lacks acceptor sites.

In an earlier investigation, outgassed H-Y was formed from NH_4-Y by heating at 500°C and was allowed to sorb triphenylamine and diphenylethylene.(101) While the parent material had no paramagnetic centres that containing $(C_6H_5)_3N$ returned strong e.s.r. signals, indicating centres at concentrations proportional to the amounts sorbed. There were about three spins per hundred sorbate molecules. The electron acceptor sites were thought to be associated with decationized sites in the zeolite, but this could not be on a 1 : 1 basis because the total number of such sites would provide $\approx 3 \times 10^{20}$ spins/g, about 20-fold larger than experiment indicated for the $(C_6H_5)_3N^+$ radical ions. With diphenylethylene the number of spins remained low for zeolite activated at temperatures up to about 400°C, as shown in Fig. 17. However a large and rapid increase in spin susceptibility then occurs to a plateau starting at about 500°C. There are now numbers of $(Ph)_2C{=}CH_2^+$ radical ions formed and stabilized in the zeolite. The authors considered that a reaction such as that in eqn 7, § 1.2, might provide electron acceptor sites, and that the charge transfer complexes could be of the types

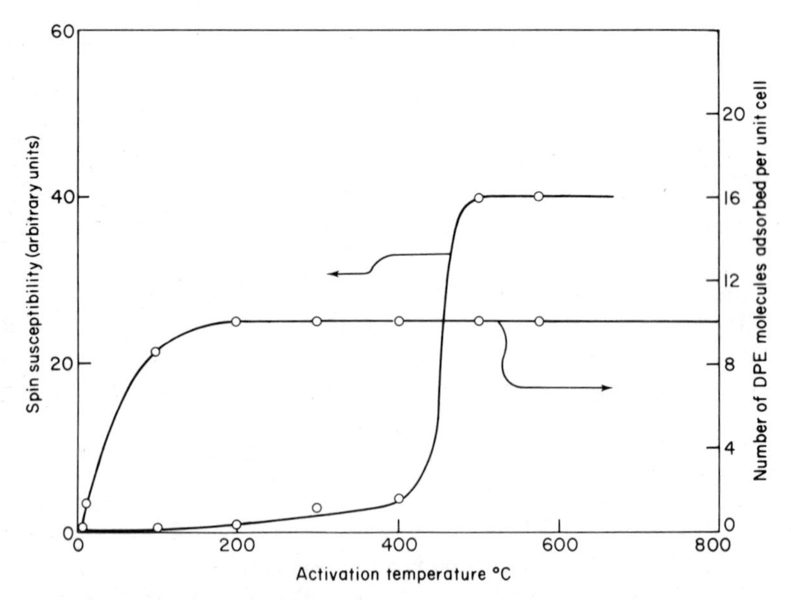

FIG. 17. The influence of calcination temperature of decationised zeolite Y upon the uptake of diphenylethylene and upon the number of spins, for sorption at room temperature.(101)

$$\bar{A}lO_{3/2} \ldots (Ph)_3N^+ \text{ or } \bar{A}lO_{3/2} \ldots [CH_2{=}C(Ph)_2]^+$$

involving the tricoordinated Al. Other cation radicals were reported for tetramethylethylene and cyclopentene for suitably heat-treated H-mordenites which had imbibed these species;[102] and for di-tertiary-butyl nitroxide in zeolites X and Y.[103]

In situations where the sorbed molecule donates negative charge to acceptor centres it is acting as a Lewis base towards Lewis acid centres. It can also be thought of as being oxidized while reducing the acceptor centre. Thus many of the zeolite-sorbate systems considered as generating radical ions can also be considered as redox systems or as Lewis acid or base reactions. There is therefore a common theme in parts of §§ 1.2, 3 and 4.

5. COMPLEXES WITH ELECTRONEGATIVE ELEMENTS

Reference has been made in Chapter 1 to zeolite catalysts containing Se and Te and in Chapters 3 and 4 to the sorption of iodine in several zeolites. Quantitative studies have also been made of sorption of S[104, 105] and P,[104] and X-ray determinations have been made of the structures for intrazeolitic I_2, Br_2, Cl_2 and S complexes.[106, 107, 108, 109]

In the iodine sorption complex of zeolite *A* of water-free composition $Ca_4Na_4[(AlO_2)_{12}(SiO_2)_{12}]$ the Ca^{2+} ions were in sites S2 (in 6-rings) and Na^+ ions in sites S2′ (just within sodalite cages). I_2 molecules were placed in mirror planes of the framework structure near the 8-ring windows and tilted at 32·5° to the window planes. However, the relative orientations of the molecules could not be fully determined and there may be some disorder in the cluster. The I–I bond length was about 2·79Å, compared with 2·68Å in solid iodine. The axial contact distance between I_2 molecules was 3·46Å and the non-axial distance was 4·01Å. These distances are to be compared with 3·54 and 4·06Å respectively in solid iodine. The shortest I . . . O distance was 3·29Å. According to this structure the saturation uptake of a 26-hedron of Type I is six I_2, comparing well with the value of around 5·5 molecules found by Barrer and Wasilewski.[110] In the 26-hedra of Type II of Na-X this figure was $\approx$7·5.[110] For bromine in Na-*A* the maximum uptake, according to the structure of the cluster,[107] was also six Br_2 per 26-hedron. The bromine molecules were so arranged that the individual atoms lay near the vertices of a cubo-octahedron according to one of the two possible configurations of Fig. 18, the maximum point symmetry of an individual bromine cluster being $\bar{3}$. The atomic coordinates of framework atoms and Na^+ ions differed a little between hydrated and bromine-containing forms of zeolite *A*. The structure of the water cluster in hydrated *A* has been described in Chapter 2, § 12.

The structure of the Cl_2 cluster in chabazite has been investigated.[108] Six peaks ascribed to chlorine were reported, at the corners of a distorted octahedron, centred in the 20-hedral cavity. The peaks were non-circular but

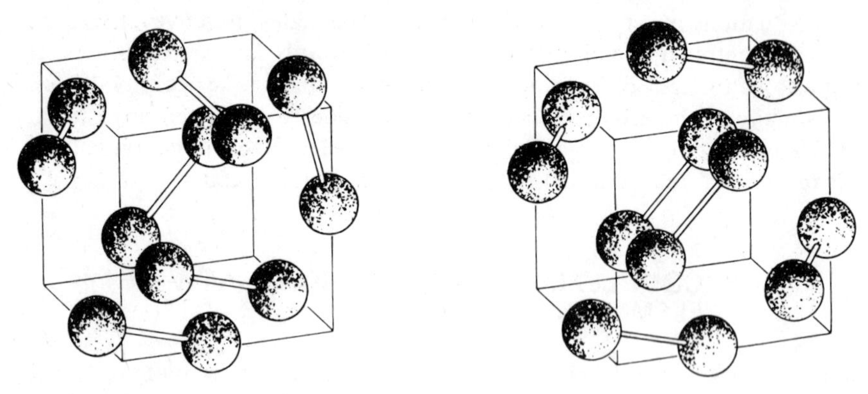

FIG. 18. Two alternative configurations of the six bromine molecules in the Na-*A*, bromine complex.[107]

were not sufficiently elongated to represent oriented Cl_2 molecules. The most probable explanation in the light of the studies on Br_2 and I_2 in zeolite *A*, where molecules of Br_2 and I_2 occurred, is that chlorine molecules were present which were rotating or librating. However six Cl_2 molecules would not be expected to fit into the 20-hedron, which indeed accommodated only two I_2 molecules.[110] Three Cl_2 molecules is about the expected maximum, so it may be that of the six possible positions in each cavity when the appropriate three are occupied by Cl_2 in any one cage the remaining three remain empty for steric reasons. Thus, averaged over all cages, each of the six positions would have 50% occupancy.

Intrazeolitic complexes of S, Se and Te have been studied from several viewpoints. Seff[109] reported that sulphur was present in the 26-hedra of zeolite *A* as two S_8 rings each in a crown configuration (Fig. 19). The planes of the two rings were parallel but 4·96Å apart, considerably more than the van der Waals diameter of sulphur (4·0Å). The planes of the parallel S_8 rings were normal to [001], [010] or [100] and so effectively block the six 8-ring windows of the 26-hedra. Two non-equivalent sulphur atoms, S(1) and S(2), alternate to form the S_8 rings. The S(1)–Na(1) distance was 2·80Å and that of S(2)–O(1) is 3·21Å. The S–S bond length in the ring was 1·94Å, shorter than in rhombic sulphur (2·048Å), and the S–S–S angles of 123° were greater than in rhombic sulphur (107·9°).

The sorption of sulphur vapour by zeolites has been studied quantitatively by Barrer and Whiteman[104] and by Steijns and Mars.[105] Figure 20 shows the reversible isotherm contours in Ca-*A* and Na-X, rectangular even at elevated temperatures. In Ca-*A* close to saturation the isosteric heats were estimated as 24·8 and 25·3 kcal mol^{-1} for uptakes of 0·32 and 0·33 g of sulphur per g of outgassed zeolite; and in Na-X these heats were 32·6 and

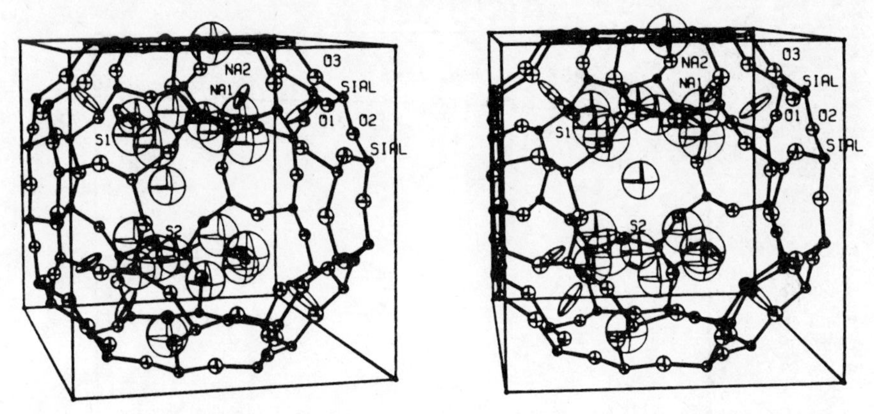

FIG. 19. The structure of the complex of sulphur in zeolite Na-*A*.[109]

30·9 kcal mol^{-1} for sorptions of 0·39 and 0·40 g per g. In zeolite *A*, 0·33 g per g corresponds with $2 \cdot 1_7$ S_8 rings per 26-hedron, in accord with the structure of the complex given by Seff.[109] It is impossible for intact S_8 rings to enter zeolite *A* through the 8-ring openings of only 4·2Å free diameter. Moreover sulphur vapour contains several polymeric forms such as S_2, S_6 and S_8. Sulphur may therefore penetrate zeolite *A* as S_2 molecules which then polymerize to S_8 within the zeolite. The isosteric heats should therefore include components from the shifting with temperature of vapour equilibria such as

$$3S_2 \rightleftarrows S_6; \; 4S_2 \rightleftarrows S_8$$

as well as the heats for processes of sorption and polymerization within the zeolite.

Steijns and Mars[105] measured the uptake of sulphur by Na-X and Ca-*A* between 150 and 350°C from a stream of nitrogen presaturated with sulphur at temperatures between 140 and 160°C. The relative vapour pressure of sulphur was varied between 10^{-4} and 10^{-1} at the temperature of the sorbent, so that the isotherms could be obtained for smaller uptakes than in the experiments of Barrer and Whiteman. The reversibility of sorption under the flow conditions was not established, but in Ca-*A* there is a slight inflexion near 0·05 g of S per g and in Na-X an inflexion around 0·1 g per g. Such curves are usually considered to be the result of sorbate-sorbate interactions as discussed in Chapter 4, § 4.

In Na-X[104] there is no restriction upon entry of S_2, S_6 or S_8 species through the 12-ring apertures. Also, unlike the situation in zeolite *A* the sulphur in one 26-hedron could be in contact with sulphur in another, and indeed chain polymers could form giving filaments extended through a number of cavities. For an uptake of 0·40 g per g there are about 168 S atoms

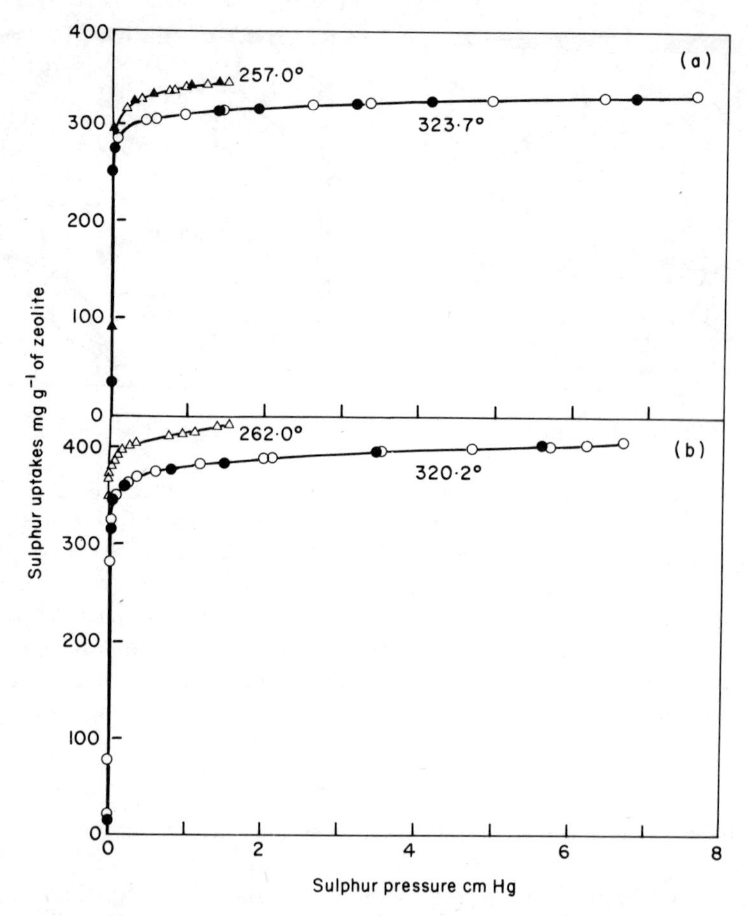

FIG. 20. Isotherms of sulphur in (a) Ca-*A* and (b) Na-X.[104] Temperatures of the isotherms are in °C. △, ○ = adsorption points; ▲, ● = desorption points.

per unit cell or 21 S atoms per large cage, compared with 16 in zeolite A. In chabazite[104] sulphur was taken up more slowly than in Ca-*A* and Na-X with an apparent activation energy around 27·3 kcal mol^{-1} and estimated near saturation uptakes of 0·37 and 0·36 g per g at 261·9° and 320·2°C respectively. It was suggested that the large apparent activation energy could arise if the sulphur diffused as short chains or if chains were immobilized in the 20-hedra until S_2 units dissociated from them as possible diffusing units.

Selenium is considerably less electronegative than sulphur and tellurium is a metalloid. However it is convenient to refer to their complexes with zeolites along with those of sulphur. These elements were initially introduced into zeolite catalysts by ball milling.[111] In a later study[112] tellurium was introduced into zeolites Na-X and Na-Y in a variety of ways: from the element by ball milling, from an aqueous slurry and from the vapour; or from tellurium compounds (TeO_2, Na_2TeO_4, $(NH_4)_2TeO_4$ and K_2TeO_3) as aqueous slurries or by ball milling. Only from equilibrium with the element in the vapour phase can quantitative aspects of sorption be studied, and

TABLE 14

Na^+ and Te distributions in Na-X[113]

Species	Location	Number per unit cell
Na(I)	I	0
Na(I′)	I′	32
Na(II)	II	23
Na(II′)	II′	9
Na(III)	III (undetected)	21
Te(LC)	In 26-hedra	3·7
Te (U)	U	1·3

such measurements are so far lacking. However, Olson and coworkers[113] have made an X-ray crystallographic investigation of Te in Na-X. Dehydrated Na-X crystals were heated with Te powder at 540°C for 116 hours and a single crystal of Na-X was chosen from the batch. This crystal was heated at 475°C in flowing hydrogen for 16·5 hours. Such crystals had a metallic lustre which, by sectioning, was shown to occur throughout their volume. A complete structure determination of the H_2-treated crystals gave the distribution of Na^+ and of tellurium shown in Table 14. The tellurium population corresponds with 4·5 wt %. The coordination of Te at the centre of the sodalite cages (sites U) is to each of four Na(I′) at a distance of 2·59Å. It was inferred that the reduction of intracrystalline elemental Te in H_2 yields Te^{2-} ions and two protons which in turn form OH groups by reaction with the framework, as demonstrated by the development of the infra-red band at 3650 cm^{-1}. Measurement of the amount of hydrogen chemisorbed while heating the Te-bearing Na-X crystals in H_2 was around 0·19 mmol compared with 0·14 mmol of Te taken up. The formal stoichiometry can be represented as

$$Te^\circ + H_2 + 2(Na^+[\gt Al\text{–}O\text{–}Si\lt]) \rightarrow Te^{2-} \ldots 2Na^+ + 2[\gt Al \overset{HO}{\quad} Si\lt]. \tag{45}$$

Usually Na^+ is found in Na-X both on sites I and I′, but the Te^{2-} in the complex appears to have drawn the Na^+ ions out of sites I to I′, the better to neutralize the charge of the Te^{2-}. The configuration wthin the sodalite cages is isostructural with that found in the sodalite-nosean felspathoids where sites U are also occupied by anions (Cl^-, Br^-, SO_4^{2-}) in tetrahedral coordination with Na^+. In the 26-hedra, the Te^{2-} was found to be co-ordinated with Na(II) at a distance of 3·2Å and (by inference only) to an (undetected) Na(III) at a distance of 3·1Å. The Te^{2-} in the 26-hedra are of special importance for the specific dehydrocyclization catalysis effected by reduced Te-bearing Na-X.[111, 112]

Interesting complexes form between phosphorus vapour and zeolites Ca-*A* and Na-X.[104] Sorption and desorption occurred rather slowly, although more rapidly in X than in *A*. The kinetics depend upon temperature and pressure but do not resemble those expected for diffusion control, as can be seen in Fig. 21 for sorption and desorption in Ca-*A*. It therefore appears that chemical processes involving polymerization and depolymerization of phosphorus may be involved. Indications of strong sorbate-sorbate interaction are seen in the reversible sigmoid isotherms of phosphorus in Na-X

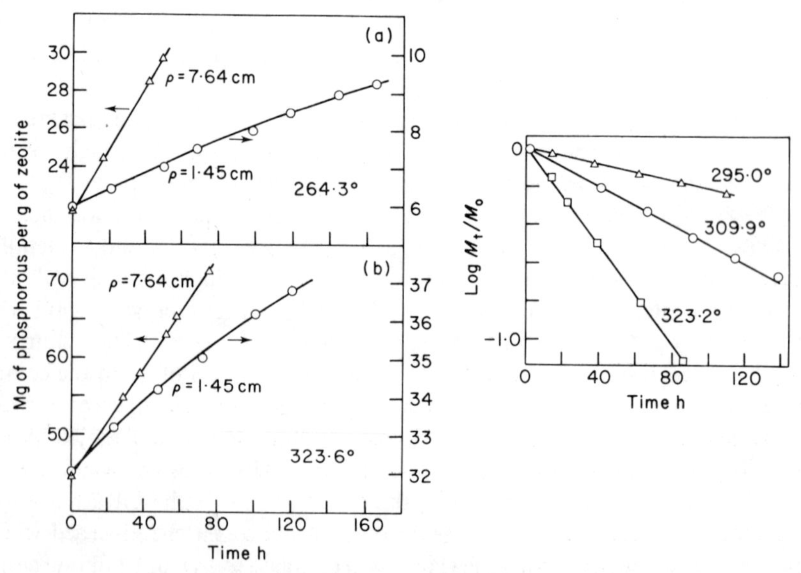

FIG. 21 (a) and (b), sorption kinetics; and, on the right, desorption kinetics, in zeolite Ca-*A*.[104] M_t denotes amount of phosphorus remaining at time t, the amount initially present being M_0. Temperatures are in °C.

(Fig. 22a). This is reflected in the increase in the isosteric heat of sorption shown in Fig. 22b. For sorption with interaction an approximate isotherm (Chapter 3, § 2) is

$$\log \frac{\theta}{p(1-\theta)} = A + B\theta \qquad (46)$$

where θ is the fractional saturation of the zeolite at pressure p, $A = \log K$ where K is the equilibrium constant and $B = -2w/2{\cdot}303\,RT$, with $2w/z$ the

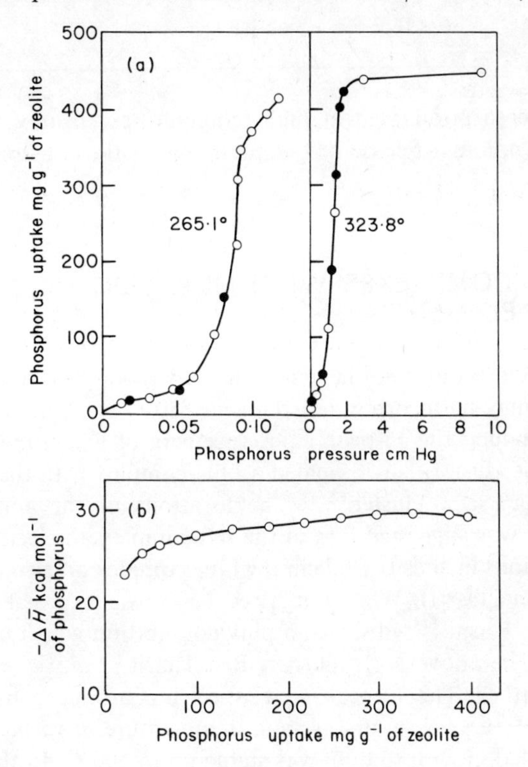

FIG. 22. (a) Isotherms of phosphorus in Na-X. Temperatures are in °C. ○, sorption; ●, desorption. (b) Heat of sorption as a function of amount of phosphorus sorbed, for Na-X.(104)

interaction energy for a pair of molecules. Equation 46 was reasonably well obeyed over a range in θ and some values of relevant quantities derived from the equation are summarized in Table 15. The nature of the interacting pair is not known. The saturation uptake of phosphorus was about 0·44 g per g of outgassed zeolite, representing about 190 phosphorus atoms per unit cell.

An interesting aspect of intracrystalline complexes of elements like iodine, sulphur and phosphorus is the possibility of storing these elements in the crystals for subsequent release in appropriate reactions involving them. The

TABLE 15

Analysis of phosphorus isotherms in Na-X

Temperature (°C)	K (cm^{-1})	w kcal	$\Delta H^{\ominus}$ kcal mol^{-1}	$\Delta S^{\ominus}$ cal mol^{-1}deg^{-1}
265·1	1·30	−2·38	−24·7	−45·5
323·8	0·136	−2·31		−45·6

very energetic sorption even at high temperatures ensures near saturation uptakes with minimal release of the guest species during storage near room temperature.

6. COMPLEXES WITH METALLIC ELEMENTS

When outgassed zeolites are heated with Na vapour considerable quantities of the metal may be incorporated. Thus at 580°C Na-Y took up sodium to give a red product, the 13-peak e.p.r. spectrum of which indicated a Na_4^{3+} intracrystalline cluster. Na-X yielded a blue complex with the 19 peak e.p.r. spectrum of a Na_6^{5+} cluster.[114] The locations of the clusters were not known, but it was suggested that in the red complex an electron was shared by four Na^+ ions in sites II while in the blue complex an electron was shared by six Na^+ on sites III. When outgassed Na-Y was exposed *in vacuo* to X- or γ-radiation Kasai[115] observed a pink colouration which he attributed to formation of the above Na_4^{3+} cluster. Ben Taarit *et al.*[116] examined these complexes further. The 13-peak e.p.r. spectrum produced by γ-irradiation disappeared after a few hours at room temperature *in vacuo*, but that produced by reduction with sodium was stable up to 500°C. In the former case, the spectrum could be restored by further γ-irradiation; in the latter by a further treatment with sodium vapour.

When oxygen was admitted to the Na-reduced zeolite, the pink colour disappeared and the e.p.r. spectrum of O_2^- radical ions was obtained:

$$Na_4^{3+} + O_2 \rightarrow 4Na^+ + O_2^-. \qquad (47)$$

Similar results were obtained when oxygen was sorbed by irradiated Na-Y. However signals corresponding with more than one O_2^- species were present because the g_{zz} peak was resolved into three peaks with $g_{zz} = 2{\cdot}095$, $2{\cdot}073$ and $2{\cdot}058$. This could mean that O_2^- was in three different intracrystalline sites. Sorption of SO_2 caused a slow decay of the O_2^- signal while a new signal due to SO_2^- developed. The simplest picture of the formation of the Na_4^{3+} cluster by sodium reduction is

$$3Na^+ + Na \rightleftarrows Na_4^{3+} \tag{48}$$

and a mechanism for the formation of the same cluster by γ-irradiation is:

$$4Na^+ + \boxed{\uparrow\downarrow} \overset{h\nu}{\rightleftarrows} Na_4^{3+} + \boxed{\uparrow} \tag{49}$$

where a framework component with paired spins ($\boxed{\uparrow\downarrow}$) gives up an electron and yields a defect centre with unpaired spin ($\boxed{\uparrow}$).

Barrer and Cole[117] studied the sorption of sodium vapour in outgassed sodalite hydrate at temperatures between 250° and 500°C. The zeolite rapidly passed through the colour sequence bright blue, blue, deep violet and finally black. The black product exposed to air showed no change and the X-ray pattern of sodalite was retained although there were intensity changes and also some extra diffraction lines. Thermogravimetric analysis of the bright blue product resulted above 500°C in a smooth loss of 5·1 wt % of Na, this loss being complete at 850°C. The white residue still had the essential sodalite structure but there was evidence of some reaction involving the alumino-silicate framework. The weight loss corresponds with the composition for the bright blue phase of $6(NaAlSiO_4)2Na$ for the unit cell, and so with one Na atom per sodalite cage. The e.s.r. spectrum of this phase had 13 lines corresponding with that of the pink Na_4^{3+} complex in zeolite X, and in accord with the above composition. Only sodalite cages occur so that all complexes must be located there. The electron is then trapped in a tetrahedral group of Na^+ ions on sites just within four of the eight 6-ring windows of a given sodalite cage.

The black phase at 20°C gave an e.s.r. spectrum consisting of a single strong absorption while the white product of calcination at 850°C showed no absorption. The black product survived 24 hours of soxhlet extraction with ethanol, 48 hours with methanol and 12 hours with water. Longer extractions with water at 90°C slowly caused the colour to change in the sequence: black, violet, red, pink and eventually white, with rising alkalinity in the water. Similar colour changes occurred over three months when the black phase was left in air. The black phase was at once decomposed by cold dilute HCl with gelation and copious evolution of hydrogen. When it was heated *in vacuo* at 700°C, Na was evolved and the bright blue phase appeared. This colour slowly faded as heating continued. The above properties show that trapped Na in the black phase is protected from reagents such as alcohols too large to penetrate the crystals and is only slowly attacked by water and oxygen, neither of which can diffuse rapidly within crystallites especially when these are blocked by trapped sodium.

Alkali metal vapour is a very good reducing agent for other cationic forms of zeolites in which the cations are lower in the electrochemical series. Rabo *et al.*[114] found that Ni^{II}-Y was reduced by Na vapour at 575°C giving e.s.r. spectra which were interpreted as Ni^+ in octahedral sites I and in trigonal sites II. The Ni^+, protected in site I, was not affected by heating at 400°C or by H_2, NH_3 and CO, although it yielded O_2^- with oxygen. Ni^+ on site II, was not thermally stable and reacted readily with sorbed gases. The reducing

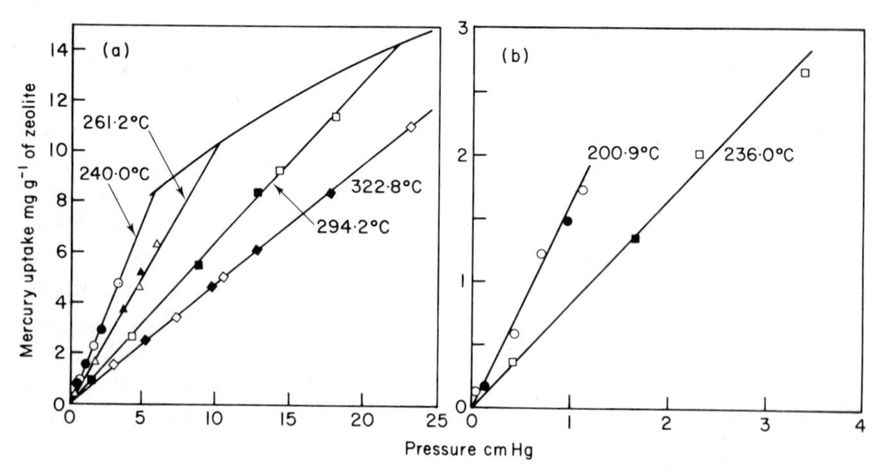

FIG. 23. The sorption of mercury in (a) Na-x and (b) Ph-X.[119] The upper curve in (a) represents the limit to the uptake set by the saturation vapour pressure of liquid mercury at each experimental temperature. ○, Δ, □ and ◇ are for sorption points and ●, ▲, ■ and ◆ for desorption points at the various temperatures.

properties of Na vapour for ions of elements lower in the electrochemical series should be duplicated by any other metal vapour, and indeed this behaviour was reported when suitable exchange forms of zeolites were exposed to mercury vapour. In Na- and Ca-forms of zeolites such as chabazite and zeolite X and even in Pb-X mercury vapour sorption was slight and isotherms obeyed Henry's law[118, 119] and resembled isotherms of the heavier inert gases at similar temperatures (Fig. 23). On the other hand in the silver exchanged forms of zeolite *A*, chabazite, zeolite X and gmelinite copious uptakes of mercury were observed, as was also the case with Hg^{II}-X. Fig. 24 shows the observed isotherms with zeolites Hg- and Ag-X. The uptakes are extensive and in Ag-X are of Type IV in Brunauer's classification.[120]

In the Hg-X at 235·7°C the sorption shows an upward inflexion after about 0·28 g of mercury/g had been sorbed. There appeared thereafter to be two steps which were not however reflected in the desorption branch, which rejoined the adsorption branch once more when about 0·28 g/g of mercury remained in the crystals. The first stage in which 0·28 g/g was sorbed without inflexion was considered to represent reduction of Hg^{2+} in the zeolite by chemisorption of mercury:

$$Hg^{2+} + Hg \rightarrow Hg_2^{2+}. \tag{50}$$

Thereafter more mercury was thought to be sorbed by forming intracrystalline clusters:

$$Hg_2^{++} + xHg \rightarrow Hg_{x+2}^{2+}. \tag{51}$$

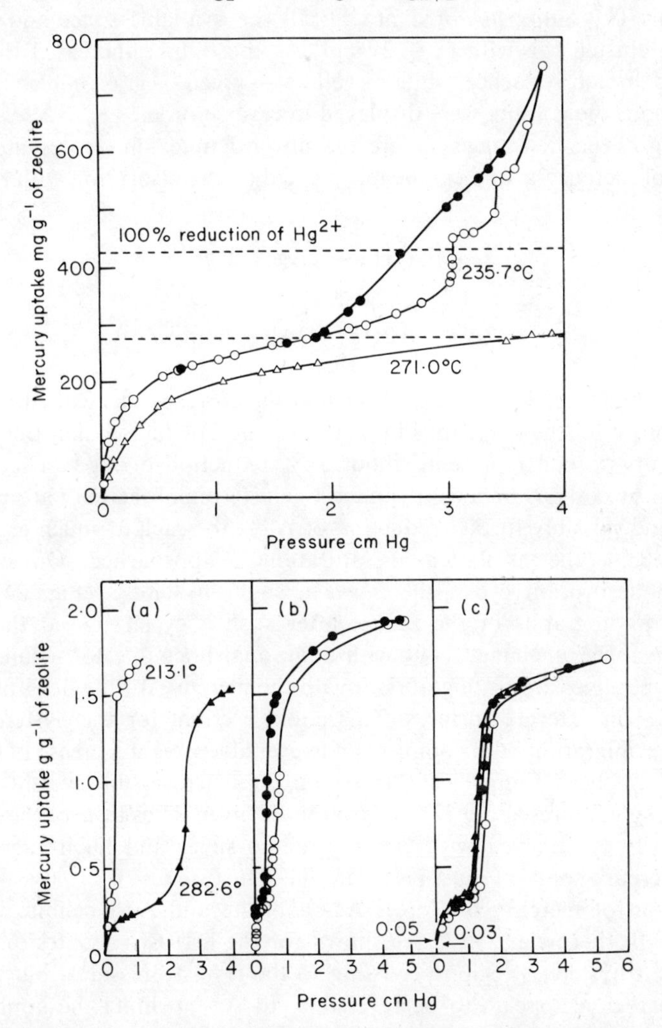

FIG. 24. Sorption of mercury in Hg-X (upper half of figure) and in Ag-X ((a), (b) and (c) in the lower half, for which temperatures are in °C).[119] ○, Δ are sorption and ● denotes desorption points in the upper half. (a) gives sorption branches at two temperatures; (b) shows a sorption-desorption cycle at 235·2°C; (c) shows two successive sorption-desorption cycles at 270°C.

The Na-X contained 86 Na^+ ions per unit cell which for 100% exchange by mercury would contain 43 Hg^{2+} ions, of diameter 2·24Å and so able to penetrate to all exchange sites. The uptake of 0·28 g/g represents reduction of about 27Hg^{2+} ions per unit cell. This could either be the extent of exchange of Na^+ by Hg^{2+}, or some Hg^{2+} ions might be protected from reduction, for example if they were in the hexagonal prisms (site I). The maximum uptake in Fig. 24 for Hg-X corresponds with 76 extra mercury atoms per unit cell. This number would not fill all the available space and gives an average cluster size with $(x + 2) \approx 4$. As the uptake increased the colour changed in the sequence white → yellow → green → gray → black. During desorption, the colours were displayed in reverse order.

In Ag-X the first stage, before the upward inflexion, represented about 0·26 g of mercury/g, and can be represented as chemisorption with reduction of silver:

$$2Ag^+ + Hg \rightarrow 2Ag + Hg^{2+} \tag{52}$$

or

$$2Ag^+ + 2Hg \rightarrow 2Ag + Hg_2^{2+}. \tag{53}$$

For the first reaction 0·26 g of mercury/g corresponds with about 63% reduction of all the silver ions initially present; for the second, this amount of mercury sorbed represents about 31% reduction of Ag^+. This stage is followed by copious uptake attributed to clustering of mercury around silver atoms and possibly further reduction of Ag^+, to reach as much as $1{\cdot}9_2$ g/g before the isotherms flatten as saturation is approached. On successive sorption–desorption cycles there was a slight hysteresis (Fig. 24). X-ray powder photographs of the zeolite after such a cycle showed the zeolite structure to be unchanged but with additional lines due to metallic silver. Thus some silver atoms migrate out with the mercury during desorption and aggregate on external surfaces. This could account for the hysteresis, but had silver migration been complete an irreversible weight increase of 0·41 g/g would have been found. The largest observed irreversible weight increase was 0·05 g/g, representing 12% migration of silver. Thus little of the sorption of mercury is involved with extracrystalline silver and an intracrystalline silver-mercury solution must permeate the structure.

Sorption of mercury by Ag-*A*, Ag-chabazite and Ag-gmelinite was also copious. Isotherms at 235°C are shown for the first two zeolites in Fig. 25. There is a hysteresis region persisting to the lowest pressures, but the high pressure region appears to be reversible. In Ag-gmelinite the approach to final equilibrium was very slow. Irreversible weight increases on thorough outgassing following mercury sorption, which measure the extent of migration of silver to external surfaces, are given in Table 16. The numbers of atoms in the last column of the table refer to the maximum uptakes on the

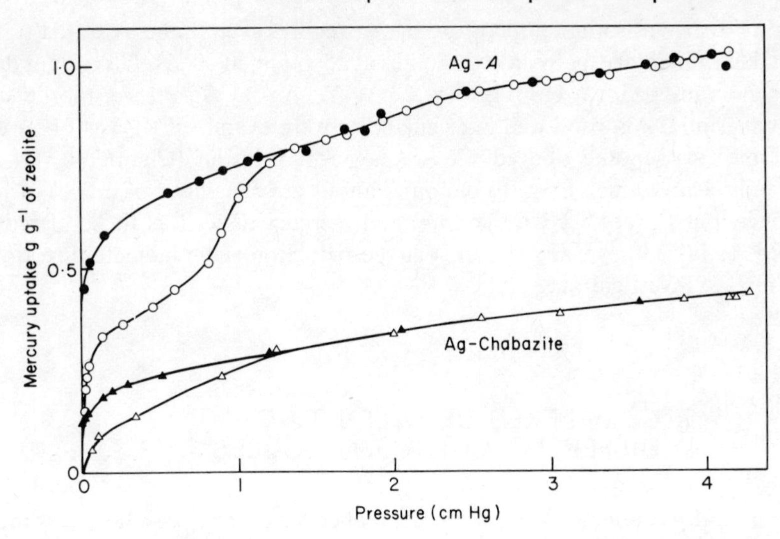

FIG. 25. Sorption of mercury in Ag-*A* and Ag-chabazite at 235°C.[119] Sorption points ○, △. Desorption points ●, ▲.

first sorption before the desorption of mercury has swept any silver to the outer surfaces. The third column of the table shows a variable tendency for silver migration which is most extensive in Ag-*A* and Ag-gmelinite.

The copious uptake of mercury in silver zeolites could be of interest in dealing with pollution by mercury vapour, or by liquid mercury droplets, especially since Bukowiecki and Meier[121] demonstrated that mercury is

TABLE 16

Irreversible weight increases in silver zeolites after sorbing mercury[119]

Zeolite	Maximum sorption Hg at *ca* 235°C (g/g)	Irreversible wt increase g/g	Wt increase for formation Hg^{2+}-zeolite +Ag	Maximum (Ag + Hg) per unit cell assuming Hg^{2+} zeolite
Ag-X	1·92	0·03 (1st cycle)	0·41	86 Ag + 156 Hg
		0·05 (2nd cycle)		
Ag-*A*	1·03	0·45 (1st cycle)	0·44	12 Ag + 8 Hg
Ag-gmelinite	0·67	0·20 (1st cycle)	0·35[a]	8 Ag + 3·7 Hg
		0·28 (2nd cycle)		
Ag-chabazite	0·44	0·13 (1st cycle)	0·35[a]	6 Ag + 0·8 Hg[b]

[a] Assuming Ag_2O, Al_2O_3, $4SiO_2$ as anhydrous oxide formula of parent zeolite.
[b] In the rhombohedral unit cell.

readily sorbed at room temperature by Ag-zeolites even in the hydrated state. Uptakes of mercury by hydrated zeolites at 25°C and at 50% relative humidity ranged from 0·3 to 1·3 g/g. Ag-X, Ag-Y, Ag-*A*, Ag-Pl, Ag-mordenite, Ag-cancrinite, Ag-losod and Ag-heulandite were examined. It was observed that mercury uptake obeyed the $\sqrt{t}$ law for diffusion (Chapter 6, § 4.1). There is clearly much more to be found about the behaviour of metals inside zeolites, but the studies with sodium and mercury as well as those of redox processes in zeolites using sodium for the reduction stage indicate directions for further investigation.

7. COMPLEXES BETWEEN TWO DIFFERENT GUEST MOLECULES

Barrer and Woodhead[118] found that when chabazite was heated simultaneously in mercury vapour and air there was an increase in weight of the chabazite, accompanied by the development of a yellow colour deepening to bronze as the uptake increased. The simultaneous consumption of oxygen was demonstrated. When the bronze crystals were allowed to sorb water they became a lustrous black and on heating the colour sequence black → orange → yellow → white was followed, water being first evolved and then mercury. Similarly H_2S gave a lustrous dark gray colour with the Hg-O-chabazite complex rather like that of lump graphite; SO_2 gave a lustrous black with a brownish tendency; and NH_3 yielded silvery gray crystals, not so dark as with H_2S and resembling galena. These observations suggest that first an intracrystalline mercury-oxygen complex formed; and that this complex could react further with polar molecules both acidic and basic.

A quantitative investigation was subsequently made of the interaction between NH_3 and HCl gases within zeolites.[122] These gases were obtained as a 1 : 1 mixture by vaporizing dry NH_4Cl which is known to dissociate virtually completely in the gas phase[123] into such a mixture. The zeolite sorbents were Na-mordenite, H-mordenite, H-Y, Na-Y and K-*L*, chosen for their acid stability. Figure 26 gives results for ammonia alone (curve B), for HCl alone (curve C) and for the 1 : 1 mixture (NH_3 + HCl) (curve A) in H-mordenite at 230°C. As with phosphorus in zeolite Na-X (§ 5) and mercury in Ag-X (§ 6) the isotherms for the mixture are of Type IV in Brunauer's classification. Ammonia alone and hydrogen chloride alone give Type I isotherms with much smaller uptakes. The isotherms for the mixture therefore indicate a strong interaction between intracrystalline NH_3 and HCl:

$$NH_3 + HCl \leftrightarrows NH_3 \ldots HCl. \quad (54)$$

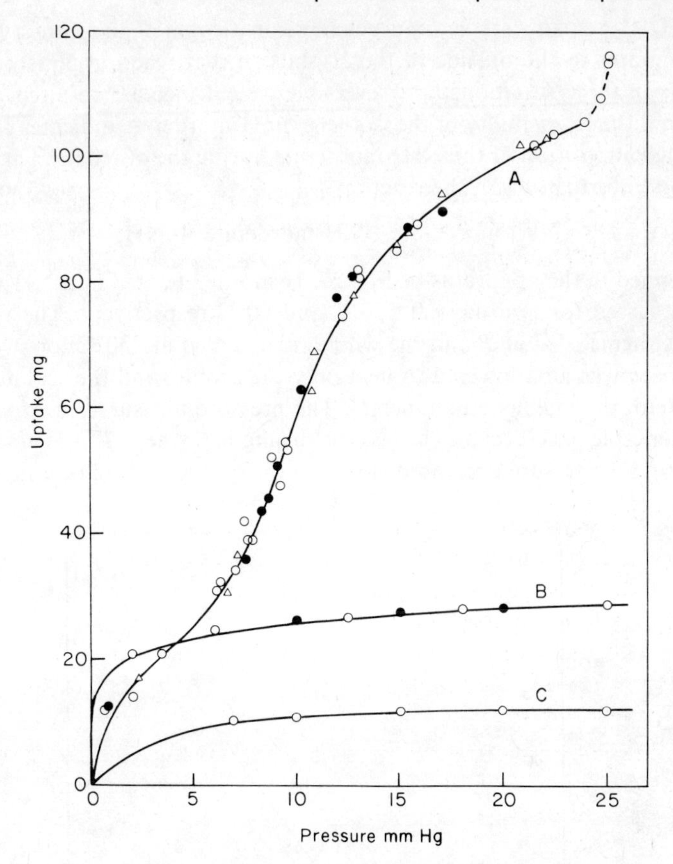

FIG. 26. Sorption of A, NH_3 + HCl; B, NH_3 alone; and C, HCL alone, in H-mordenite at 230°C. ○ denote sorption points; ● are desorption points. △ are points for final pressures and uptakes.[122]

In the hydrogen zeolites some of the ammonia also reacts to give the ammonium form, for example,

$$NH_3 + \text{H-Y} \rightleftarrows NH_4 - \text{Y} \tag{55}$$

while in the Na-Y or K-*L* a reaction such as

$$n\text{HCl} + n\text{NH}_3 + \text{Na-Y} \rightleftarrows [\text{Na}_{(1-n)}(\text{NH}_4)_n] - \text{Y} + n\text{NaCl} \tag{56}$$

is expected. Reactions 55 and 56 account for the irreversibility of sorption and desorption branches of the isotherms observed in H-Y and Na-Y, provided, in the case of H-Y, that the ammonium form does not decompose to NH_3 and H-Y on evacuation at 246°C; and, in the case of Na-Y, if the

(Na, NH_4)-Y remains stable on evacuation or if some of the intracrystalline NaCl migrates to the outside of the crystals so that reaction 56 is not fully reversed. In these situations an irreversible weight increase resulted. On the other hand the reversibility of the isotherm in H-mordenite indicates eventual if slow decomposition of the NH_4-mordenite during the evacuation at 230°C.

Further information on the reaction

$$NH_4\text{-mordenite} \rightleftarrows \text{H-mordenite} + NH_3 \qquad (57)$$

was obtained in the apparatus of Fig. 27. H-mordenite at Z and NH_4Cl at A were outgassed for two days at 350°C and 90°C respectively. The reaction tube was then sealed at P and the whole tube heated at 230°C for two days. The furnace was next lowered to heat only the zeolite, and the seal at B was broken with the magnetic hammer H. The pressure measured in M was due to condensable gas because it became negligible when T was cooled to −196°C. This pressure decreased slowly with time according to a first order

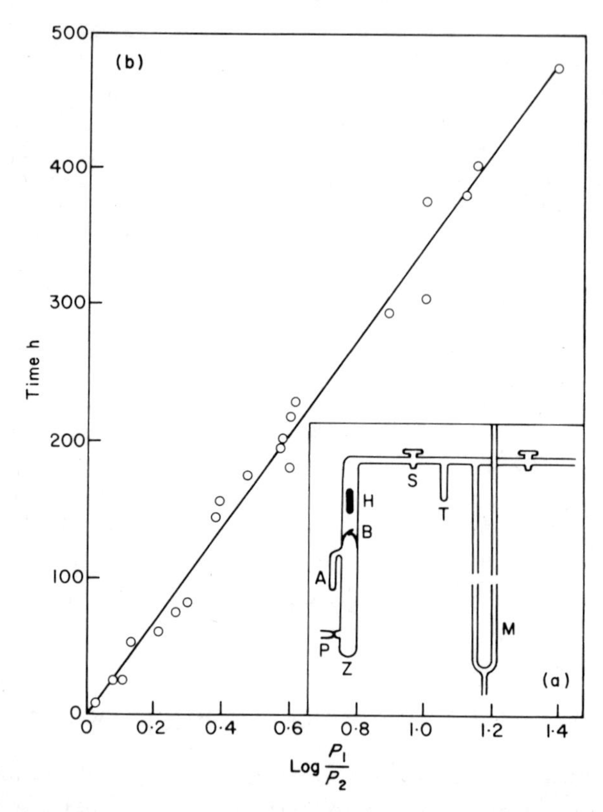

FIG. 27. (a) Apparatus for measuring dissociation of NH_4-mordenite in presence of HCl gas. (b) The first order reaction plot for dissociation of NH_4-mordenite in presence of HCl gas at 230°C. p_1 and p_2 are pressures at times t_1 and t_2.[122]

law, as seen in Fig. 27, and approached zero after one month. The velocity constant k at $\approx 230°C$ was $1{\cdot}94 \times 10^{-6} s.^{-1}$ This behaviour is understood as follows. When the 1 : 1 mixture of $HCl + NH_3$ is initially sorbed, the reaction 57 occurs from right to left and in addition inside the crystals reaction 54 takes place. There results an excess of HCl gas in the system,

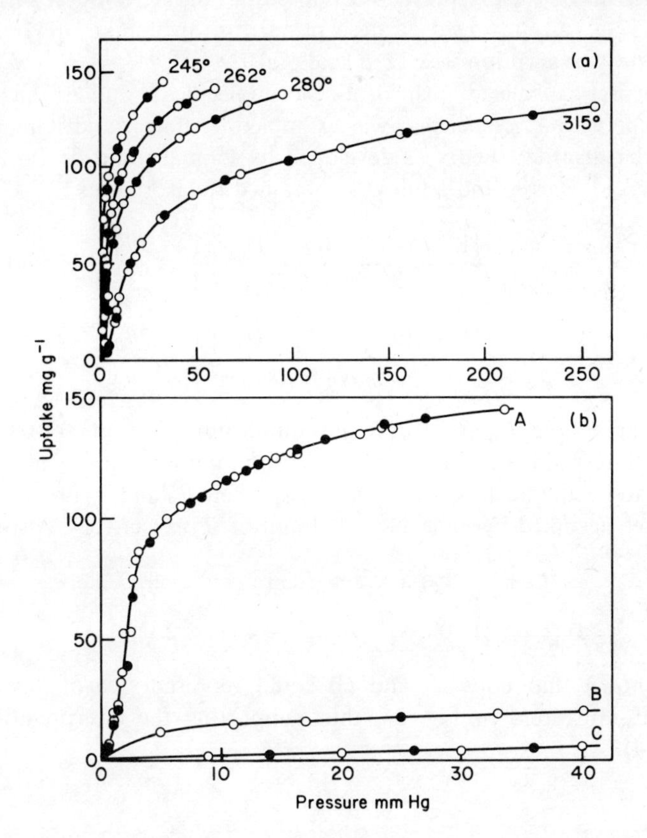

FIG. 28. (a) Sorption isotherms of vaporized NH_4Cl (i.e. NH_3 + HCl) in zeolite K-*L*.(122) ○ are sorption and ● are desorption points. All temperatures are in °C. (b) Sorption isotherms at 245°C in K-*L*.(122) A, vaporized NH_3Cl (i.e. NH_3 + HCl); B, Hcl alone; C, NH_3 alone.

which was the condensable gas referred to above. However, during the subsequent heating at 230°C in absence of the 1 : 1 mixture of $HCl + NH_3$ generated by vaporizing NH_4Cl, reaction 57 proceeds from left to right, releasing NH_3 which condenses with the HCl as NH_4Cl on the cold surfaces outside the furnace. Thus the HCl is consumed at a rate equal to the rate of dissociation of the NH_4-mordenite.

The sorption of $NH_3 + HCl$ within zeolite K-*L* also proved to be reversible as shown by the group of isotherms in Fig. 28. The reversibility indicates that if on sorption the reaction

$$m\mathrm{NH_3} + m\mathrm{HCl} + \mathrm{K}\text{-}L \rightleftarrows [\mathrm{K}_{(1-m)}(\mathrm{NH_4})_m] - L + m\mathrm{KCl} \qquad (58)$$

takes place inside the crystal the KCl does not migrate to outer surfaces and reaction 57 is fully reversed on desorption. For an uptake of 120 mg/g the isosteric heat of sorption was 22·5 kcal mol^{-1}.

The theoretical basis of coadsorption of molecules A and B which interact with an energy $2\omega_{AB}/\nu$ per pair was examined by Barrer and Klinowski,[124] using order-disorder theory as developed by Hijmans and de Boer[125] and by Domb.[126] Barrer and Klinowski obtained the isotherms

$$K_A = \frac{\theta_A}{p_A\theta_0}\left\{\frac{[1 + t(\theta_A - \theta_B) - D_{AB}](1 - 2\theta_A)}{2\theta_A(t\theta_0 + D_{AB} - 1)}\right\}^z \quad \text{and} \qquad (59)$$

$$K_B = \frac{\theta_B}{p_B\theta_0}\left\{\frac{[1 + t(\theta_A - \theta_B) - D_{AB}](1 - 2\theta_B)}{2\theta_B(t\theta_0 + D_{AB} - 1)}\right\}^z . \qquad (60)$$

In these expressions K_A and K_B are the equilibrium constants for the sorption of A and B respectively, z is the coordination number of a site, all sites being assumed to be identical. $\theta_0 = 1 - \theta_A - \theta_B$ where θ_A and θ_B are the fractions of the sites occupied by molecules of A and of B respectively. Also

$$t = 1 - \exp(-2\omega_{AB}/zkT) \qquad (61)$$

$$D_{AB} = \{[1 + t(\theta_A - \theta_B)]^2 - 4t(1 - 2\theta_B)\theta_A\}^{1/2}. \qquad (62)$$

It was shown that eqns 59 and 60 could represent reasonably well the behaviour illustrated in Fig. 26, thus supporting the interpretation given (cf eqn 54).

8. CONCLUSION

The survey given in this chapter provides an introduction to chemical aspects of sorption by molecular sieve zeolites. These sorbents participate in a rich variety of intracrystalline chemisorptions, any account of which suggests additional areas requiring investigation. There is much in this aspect of zeolite chemistry which is important for the understanding of zeolite catalysts and catalysis and much that is relevant to modification of molecule sieving and selective sorption. Spectroscopic methods have proved notably successful in furthering these investigations and can strongly complement classical procedures. The unique intracrystalline environments can stabilize unusual valence states of ions, radicals, radical ions and chemisorption complexes; certain ligands promote cation migration and framework modification; unusual coordination numbers and configurations arise due to competition for the exchange ions between anionic oxygens of the framework and ligand guest molecules. As regards the defect chemistry of zeolites there is much yet to be learned, especially concerning the role of defects in catalysis.

REFFERENCES

1. "Zeolite Chemistry and Catalysis", ed. by J. A. Rabo, A.C.S. Monograph 171, Amer. Chem. Soc. (1976) Chapters 8 to 13.
2. P. Venuto and P. S. Landis, *Adv. Catal.* (1968) **18**, 259.
3. P. B. Weisz, *Ann. Rev. Phys. Chem.* (1970) **21**, 175.
4. J. Turkevich and Y. Ono, *Adv. Catal.* (1969) **20**, 135.
5. R. M. Barrer, W. Buser and W. F. Grutter, *Helv. Chem. Acta* (1956) **39**, 518.
6. J. L. Carter, P. J. Luccesi and D. J. C. Yates, *J. Phys. Chem.* (1964) **68**, 1385.
7. H. W. Habgood, *J. Phys. Chem.* (1965) **69**, 1764.
8. H. Hattori and T. Shiba, *J. Catal.* (1968) **12**, 111.
9. C. L. Angell and P. C. Schaffer, *J. Phys. Chem.* (1965) **69**, 3463.
10. R. M. Barrer and J. Klinowski, *J. Chem. Soc., Faraday* I (1975) **71**, 690.
11. R. M. Barrer, *Nature* (1949) **164**, 112.
12. J. W. Ward, *J. Catal* (1967) **9**, 225.
13. B. I. Shikunov, L. I. Lafer, V. I. Yakerson and A. M. Rubenshtein, Izvst. Akad. Nauk SSSR, Ser. Khim. (1973) 449.
14. E. J. Detrekoy, P. A. Jacobs, D. Kallo and J. B. Uytterhoeven, *J. Catal.* (1974) **32**, 442.
15. D. F. Best, R. W. Larson and C. L. Angell, *J. Phys. Chem.* (1973) **77**, 2183.
16. D. J. C. Yates, *Catal. Rev.* (1968) **2**, 113.
17. Y. Ono, M. Kaneko, K. Kogo, H. Takayanagi and T. Keii, *J. Chem. Soc., Faraday* I (1976) **72**, 2150.
18. "Zeolite Chemistry and Catalysis", ed. by J. A. Rabo, A.C.S. Monograph 171, Amer. Chem. Soc. (1976) Chapter 3.
19. J. W. Ward, *J. Catal.* (1968) **10**, 34.
20. P. E. Eberly Jr, *J. Phys. Chem.* (1968) **72**, 1042.
21. W. Kladnig, *J. Phys. Chem.* (1976) **80**, 262.
22. B. V. Liengme and W. K. Hall, *Trans. Faraday Soc.* (1966) **62**, 3229.
23. J. W. Ward, *J. Catal.* (1969) **14**, 365.
24. T. Nishizawa, H. Hattori, T. Uematsu and T. Shiba, 4th International Congress Catal., Moscow (1968) Paper 55.
25. F. R. Cannings, *J. Phys. Chem.* (1968) **72**, 4691.
26. M. Lefrancois and G. Malbois, *J. Catal.* (1971) **20**, 350.
27. H. Karge, *Zeit. Phys. Chem., Neue Folge* (1971) **76**, 133.
28. K.-H. Steinberg, H. Bremer and P. Falke, *Zeit. Phys. Chem., Leipzig* (1976) **257**, 151.
29. G. T. Kerr, *J. Phys. Chem.* (1967) **71**, 4155.
30. G. H. Kuhl, in Proc. 3rd International Conf. on Molecular Sieves, ed. by J. B. Uytterhoeven, Zurich, Sept. 3–7, 1973, p. 227.
31. A. P. Bolton and M. A. Lanawela, *J. Catal.* (1970) **18**, 154.

32. G. T. Kerr, *J. Catal.* (1969) **15**, 200.
33. B. A. Morrow and I. A. Cody, *J. Phys. Chem.* (1976) **80**, 1995 and 1998.
34. C. V. McDaniel and P. K. Maher, in Ref. 1, Chapter 4, for a review.
35. "Zeolite Chemistry and Catalysis", ed. by J. A. Rabo, A.C.S. Monograph 171, Amer. Chem. Soc. (1976) p. 320.
36. "Zeolite Chemistry and Catalysis", ed. by J. A. Rabo, A.C.S. Monograph 171, Amer. Chem. Soc. (1976) p. 323.
37. R. M. Barrer, R. G. Jenkins and G. Peeters, *in* 4th International Conf. on Molecular Sieve Zeolites, April 1977, Chicago.
38. G. T. Kerr, in Proc. 3rd International Conf. on Molecular Sieves, ed. by J. B. Uytterhoeven, Sept. 3–7, 1973, p. 38.
39. J. C. McAteer and J. J. Rooney, in "Molecular Sieves", ed. by W. M. Meier and J. B. Utterhoeven, Advances in Chemistry Series No. 121, Amer. Chem. Soc. (1973) p. 258.
40. R. M. Barrer and A. Sikand, in preparation.
41. S. A. Butter and W. W. Kaeding, to Mobil Oil Corp., U.S.P. 3,972,832, 1976.
42. V. Bosacek and Z. Tvaruzkova, Coll. Czech. Chem. Comm. (1971) **36**, 551.
43. R. M. Barrer and S. Wasilewski, unpublished.
44. R. M. Barrer, A. Sikand and D. Harding, in preparation.
45. R. A. Dalla Betta and M. Boudart, *J. Chem. Soc., Faraday* I (1976) **72**, 1723.
46. E. F. Vansant and J. H. Lunsford, *J. Phys. Chem.* (1972) **76**, 2860.
47. Y. Huang and E. F. Vansant, *J. Phys. Chem.* (1973) **77**, 663.
48. D. R. Flengte, J. H. Lunsford, P. A. Jacobs and J. B. Uytterhoeven, *J. Phys. Chem.* (1975) **79**, 354.
49. R. M. Barrer and R. P. Townsend, *J. Chem. Soc., Faraday* I (1976) **72**, 2650.
50. J. Turkevich, Y. Ono and J. Soria, *J. Catal.* (1972) **25**, 44.
51. A. V. Kiselev, N. M. Kuz'menko and V. I. Lygin, *Russ. J. of Phys. Chem.* (1973) **47**, 88.
52. B. Coughlan, W. A. McCann and W. M. Carroll, *Chem. and Ind.* (1977) 7th May, 358.
53. P. Gallezot, Y. Ben Taarit and B. Imelik, *J. Phys. Chem.* (1973) **77**, 2556.
54. P. Gallezot, Y. Ben Taarit and B. Imelik, *J. Catal.* (1972) **26**, 481.
55. K. R. Laing, R. L. Leubner and J. H. Lunsford, *Inorg. Chem.* (1975) **14**, 1400.
56. W. E. Addison and R. M. Barrer, *J. Chem. Soc.* (1955) 757.
57. P. H. Kasai and R. J. Bishop Jr, *J. Amer. Chem. Soc.* (1972) **94**, 5560.
58. J. H. Lunsford, J. Phys. Chem. (1970) **74**, 1518.
59. A. V. Alekseyev, V. N. Filimonov and A. N. Terenin, Dokl. Akad. Nauk SSSR (1962) **147**, 1392.
60. C. C. Chao and J. H. Lunsford, *J. Phys. Chem.* (1972) **76**, 1546.
61. Yu. A. Lokhov and A. A. Davydov, *React. Kin. and Catal. Letters* (1975) **3**, 39.
62. C. Naccache, M. Che and Y. Ben Taarit, *Chem. Phys. Letters* (1972) **13**, 109.
63. C. Naccache and Y. Ben Taarit, *J. Chem. Soc., Faraday* I (1973) **69**, 1475.
64. J. W. Jermyn, T. J. Johnson, E. F. Vansant and J. H. Lunsford, *J. Phys. Chem.* (1973) **77**, 2964.
65. K. Klier and M. Ralek, *J. Phys. Chem. Solids* (1968) **29**, 951.
66. K. Klier, R. Kellerman and P. J. Hutta, *J. Chem. Phys.* (1974) **61**, 4224.
67. K. Klier in "Molecular Sieve Zeolites", *Advances in Chemistry Series No. 121*, Amer. Chem. Soc. (1973) p. 480.
68. P. E. Riley and K. Seff, *J. Amer. Chem. Soc.* (1973) **95**, 8180.
69. Y. Huang and D. E. Mainwaring, *Chem. Comm.* (1974) 584.

70. J. L. Carter, D. J. C. Yates, P. J. Lucchesi, J. J. Elliott and V. Kevorkian, *J. Phys. Chem.* (1966) **70**, 1126.
71. P. Pichat, *J. Phys. Chem.* (1975) **79**, 2127.
72. P. E. Riley, K. B. Kunz and K. Seff, *J. Amer. Chem. Soc.* (1975) **97**, 537.
73. C. L. Angell and P. C. Schaffer, *J. Phys. Chem.* (1966) **70**, 1413.
74. J. Haber, J. Ptaszynski and J. Sloczynski, *Bull. Acad. Pol. Sci., Sci. Chim.* (1975) **23**, 709.
75. T. A. Egerton and F. S. Stone, *Trans. Faraday Soc.* (1970) **66**, 2364.
76. Y. Huang, *J. Amer. Chem. Soc.* (1973) **95**, 6636.
77. Y. Huang, *J. Catal.* (1973) **30**, 187.
78. P. E. Riley and K. Seff, *Inorg. Chem.* (1974) **13**, 1355.
79. G. Coudurier, P. Gallezot, H. Praliand, M. Primet and B. Imelik, *C.R. Acad. Sci., Paris* (1976) **282**, Série C, 311.
80. Kh. M. Minachev, G. C. Antoshin, E. S. Shpiro and Yu. A. Yusifov, 6th International Congr. on Catal., London, 12–16th July, 1976, Paper B2.
81. D. C. Yates, *J. Phys. Chem.* (1965) **69**, 1676.
82. H. Beyer, P. A. Jacobs and J. B. Uytterhoeven, *J. Chem. Soc., Faraday* I (1976) **72**, 674.
83. R. G. Herman, J. H. Lunsford, H. Beyer, P. A. Jacobs and J. B. Uytterhoeven, *J. Phys. Chem.* (1975) **79**, 2388.
84. P. A. Jacobs, W. de Wilde, R. A. Schoonheydt, J. B. Uytterhoeven and H. Beyer, *J. Chem. Soc., Faraday* I (1976) **72**, 1221.
85. M. Iwamoto, K. Maruyama, N. Yamazoe and T. Sieyama, *Chem. Comm.* (1976) 615.
86. J. Valentine, *Chem. Rev.* (1973) **73**, 235.
87. R. Kellerman, P. J. Hutta and K. Klier, *J. Amer. Chem. Soc.* (1974) **96**, 5946.
88. T. Kubo, H. Tominaga and T. Kunugi, *Bull. Chem. Soc. Jap.* (1973) **46**, 3549.
89. M. Che, J. F. Dutel, P. Gallezot and M. Primet, *J. Phys. Chem.* (1976) **80**, 2371.
90. Y. Ono, K. Suzuki and T. Keii, *J. Phys. Chem.* (1975) **79**, 752.
91. S. Krzyzanowski, *J. Chem. Soc., Faraday* I (1976) **72**, 1573.
92. R. F. Howe and J. H. Lunsford, *J. Amer. Chem. Soc.* (1975) **97**, 5156.
93. Y. Ono, H. Tokunaga and T. Keii, *J. Phys. Chem.* (1975) **79**, 752.
94. H. Tokunaga, Y. Ono and T. Keii, *Bull. Chem. Soc. Jap.* (1973) **45**, 3362.
95. S. Krzyzanowski, *Chem. Comm.* (1974) 1036.
96. J. A. R. Coope, C. L. Gardner, C. A. McDowell and A. J. Pelman, *Mol. Phys.* (1971) **21**, 1043.
97. J. C. Vedrine, J. Massardier and A. Abou–Kais, *Canadian J. Chem.* (1976) **54**, 1678.
98. M. Shiotani, F. Yuasa and J. Sohma, *J. Phys. Chem.* (1975) **79**, 2669.
99. P. Svejda, *J. Phys. Chem.* (1972) **76**, 2690.
100. Z. Dudzik, R. Fiederow and A. Wiechowski, *Bull. Acad. Pol. Sci., Sci. chim.* (1975) **23**, 955.
101. D. N. Stamires and J. Turkevich, *J. Amer. Chem. Soc.* (1964) **86**, 749.
102. P. L. Corio and S. Shih, *J. Phys. Chem.* (1971) **75**, 3475.
103. G. P. Lozos and B. M. Hoffman, *J. Phys. Chem.* (1974) **78**, 2110.
104. R. M. Barrer and J. L. Whiteman, *J. Chem. Soc., A* (1967) 13.
105. M. Steijns and P. Mars, *J. Coll. Interface Sci.* (1976) **57**, 175.
106. K. Seff and D. P. Shoemaker, *Acta Cryst.* (1967) **22**, 162.
107. W. M. Meier and D. P. Shoemaker, *Zeit. Krist.* (1966) **123**, 5.
108. J. H. Fang and J. V. Smith, *J. Chem. Soc.* (1964) 3749.

109. K. Seff, *J. Phys. Chem.* (1972) **76**, 2601.
110. R. M. Barrer and S. Wasilewski, *Trans. Faraday Soc.* (1961) **57**, 1140.
111. J. N. Miale and P. B. Weisz, *J. Catal.* (1971) **20**, 288.
112. W. H. Lang, R. J. Mikovsky and A. J. Silvestri, *J. Catal.* (1971) **20**, 293.
113. D. H. Olson, R. J. Mikovsky, G. F. Shipman and E. Dempsey, *J. Catal.* (1972) **24**, 161.
114. J. A. Rabo, C. L. Angell, P. H. Kasai and V. Schomaker, Faraday Soc. Discussion No. 41 (1966) 328.
115. P. H. Kasai, *J. Chem. Phys.* (1965) **43**, 3322.
116. Y. Ben Taarit, C. Naccache, M. Che and A. J. Tench, *Chem. Phys. Letters* (1974) **24**, 41.
117. R. M. Barrer and J. F. Cole, *J. Phys. and Chem. of Solids* (1968) **29**, 1755.
118. R. M. Barrer and M. Woodhead, *Trans. Faraday Soc.* (1948) **44**, 1001.
119. R. M. Barrer and J. L. Whiteman, *J. Chem. Soc., A* (1967) 19.
120. S. Brunauer, "The Adsorption of Gases and Vapours", Oxford (1944) p. 150.
121. S. Bubowiecki and W. M. Meier, in Proc. 3rd International Conf. on Molecular Sieves, ed. by J. B. Uytterhoeven, Sept. 3–7 (1973) Zurich, p. 250.
122. R. M. Barrer and A. G. Kanellopoulos, *J. Chem. Soc., A* (1970) 775.
123. P. Goldfinger and G. Verhaegen, *J. Chem. Phys.* (1969) **50**, 1467.
124. R. M. Barrer and J. Klinowski, *J. Chem. Soc. Faraday II*, to be published.
125. J. Hijmans and J. de Boer, *Physica* (1955) **21**, 471, 485, 499.
126. C. Domb, *Advan. Phys.* (1960) **9**, 149, 245.

8

Sorption and Molecule Sieving by Layer Silicates

1. INTRODUCTION

Among the clay minerals, especially smectites, there are some which imbibe guest molecules as prolifically as do the zeolites. For the most part intercalation between the anionic siliceous sheets of these minerals causes their crystals to swell, in order to make room for the sorbed species. In this respect sorption differs from that in zeolites where the porous but rigid frameworks suffer only minor or negligible dimensional changes. Swelling of clay minerals is possible because the parallel lamellae in these structures are bonded to each other not covalently but by van der Waals and electrostatic forces. Electrostatic bonding arises if there is anionic charge on the sheets neutralized by interlayer cations. These cations provide a "cement" which helps to hold together the anionic sheets on either side of them. The strength of the bond increases with the anionic charge density on the sheets and so is at a maximum in the micas. Here the sheets come together as closely as the size of the interlayer cations permits and there is little tendency of the crystals to swell in order to allow intercalation of neutral molecules. On the other hand in smectites such as montmorillonite or hectorite the charge density on the siliceous sheets is low and so therefore is the concentration of interlayer counter-ions. Such crystals swell readily by penetration of polar species between the sheets. When they have been outgassed, interlayer pore space does not normally exist in the minerals; it is created only by penetration of guest molecules and the consequent swelling. If the potential guest species are non-polar they are not usually able to intercalate in anhydrous smectites or vermiculites. However, as discussed later (§§ 9 and 10) it is possible to open the interlayer regions permanently and so to allow the expanded clay minerals to sorb non-polar guest molecules and to function like the zeolites as molecular sieves.

2. STRUCTURAL FEATURES OF LAYER SILICATES

In all the groups of minerals comprising talc, pyrophyllite, micas, hydrous micas and illites, chlorites, vermiculites, smectites, kandites, serpentines and palygorskites a common structural feature is a hexagonal sheet of linked SiO_4 tetrahedra. In all these minerals save the palygorskites and members of the serpentine group the vertices of the tetrahedra in each sheet point in one way, as in Fig. 10, 1 of Chapter 2. If such a sheet is envisaged as being formed by condensation-polymerization of $Si(OH)_4$ molecules it will be seen that all the vertices are hydroxyl groups and that the composition of the sheet is $Si_2O_3(OH)_2$. The hydroxyl groups can be thought to undergo further condensation-polymerization with $Al(OH)_6$ octahedra. If one layer of octahedra reacts with one sheet $(Si_2O_3(OH)_2)_n$ the two-layer sheet typical of kandites (kaolinite, dickite, nacrite and halloysite) results. This two-layer sheet has the composition $Al_2SiO_5(OH)_4$. If one layer of octahedra reacts with two sheets $(Si_2O_3(OH)_2)_n$ a triple layer sheet with the octahedral layer at its centre is the result. It has the composition $Al_2Si_4O_{10}(OH)_2$ and is found with various isomorphous substitutions in talc, pyrophyllite, micas, vermiculites and smectites.

2.1 Silicates with two-layer Sheets

The condensation-polymerization of $Al(OH)_6$ octahedra with the $(Si_2O_3(OH)_2)_n$ layer is represented in Fig. 1a[1], and a portion of the resultant double sheet with repeat distance 7·2 Å normal to the plane of the sheet is

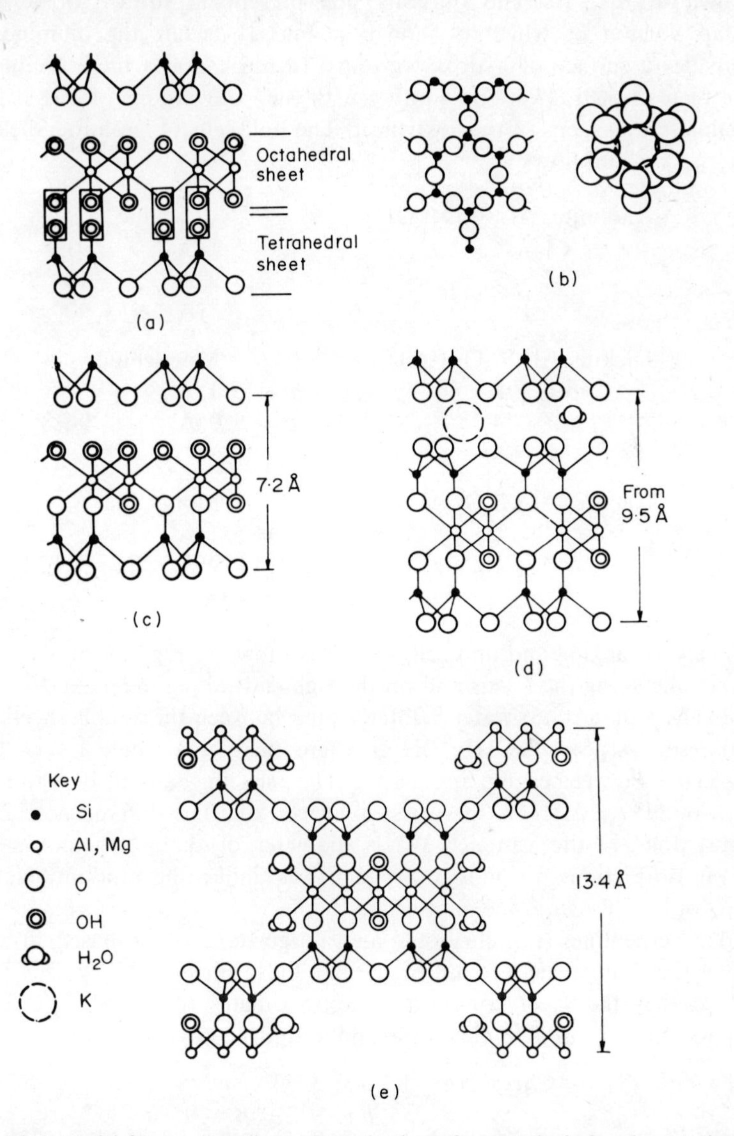

FIG. 1. (a) A model for condensation-polymerization of an octahedral and a tetrahedra layer. (b) A portion of the hexagonal sheet of linked SiO_4 tetrahedra, and on the right a 6-ring in this sheet with a partially recessed K^+ ion. (c) The two-layer sheet found in kandites. (*d*) The three-layer sheet found in micas, vermiculites, smectites, talc and pyrophyllite. (e) A section in the *ab* plane of attapulgite.[1]

shown in Fig. 1c. The siliceous side presents a surface of oxygens, a small portion of which is shown in Fig. 1b, while the aluminous side provides a surface of hydroxyl groups. In the kandites these double layers are then stacked upon one another with the –OH groups of one such sheet against the oxygens of the next sheet. The unit cells of kaolinite, dickite and nacrite are as follows[2]:

Kaolinite, $Al_2(Si_2O_5)(OH)_4$	Triclinic
$C_1^1 = C1$	$a = 5{\cdot}14$Å $\alpha = 91{\cdot}8°$
$Z = 2$	$b = 8{\cdot}93$Å $\beta = 104{\cdot}7°$
	$c = 7{\cdot}37$Å $\gamma = 90°$
Dickite, $Al_2(Si_2O_5)(OH)_4$	Monoclinic
$C_8^4 = Cc$	$a = 5{\cdot}15$Å
$Z = 4$	$b = 8{\cdot}95$Å $\beta = 96{\cdot}8°$
	$c = 14{\cdot}42$Å
Nacrite, $Al_2(Si_2O_5)(OH)_4$	Monoclinic
$C_8^4 = Cc$	$a = 5{\cdot}15$Å
$Z = 12$	$b = 8{\cdot}95$Å $\beta = 90°\ 20'$
	$c = 43$Å

The layer stacking and unit cell of each is shown in Fig. 2 viewed on the left hand side along the b axis and on the right side along the a axis.[3]

In the kandites, no water is intercalated between the double sheets save in halloysite, $Al_2(Si_2O_5)(OH)_4$, $2H_2O$ where there is a single layer of water molecules between each pair of sheets. The c spacing is about 10 Å and shrinks to around 7·2 Å when the water is driven off. The difference of 2·8 Å is approximately the van der Waals diameter of the water molecule. The X-ray diffractions are mostly broad bands, indicating random stacking of the double sheets on one another.

The serpentines (e.g. chrysotile and antigorite) are also based structurally upon the double sheet found in the kandites, but now Mg largely replaces Al so that the sheet composition approximates to $Mg_3Si_2O_5(OH)_4$. The monoclinic unit cells of chrysotile and antigorite are as follows:

Chrysotile	$a = 5{\cdot}33$Å
	$b = 9{\cdot}2$Å, $\beta = 93°\ 7'$
	$c = 14{\cdot}66$Å ($2 \times 7{\cdot}33$)
Antigorite	$a = 43{\cdot}5$Å ($8 \times 5{\cdot}44$)
	$b = 9{\cdot}26$Å, $\beta = 91°\ 24'$
	$c = 7{\cdot}28$Å

The unit cell of chrysotile has a, b and c dimensions very close to those of dickite, but the substitution of Mg_3 for Al_2 has an interesting consequence.

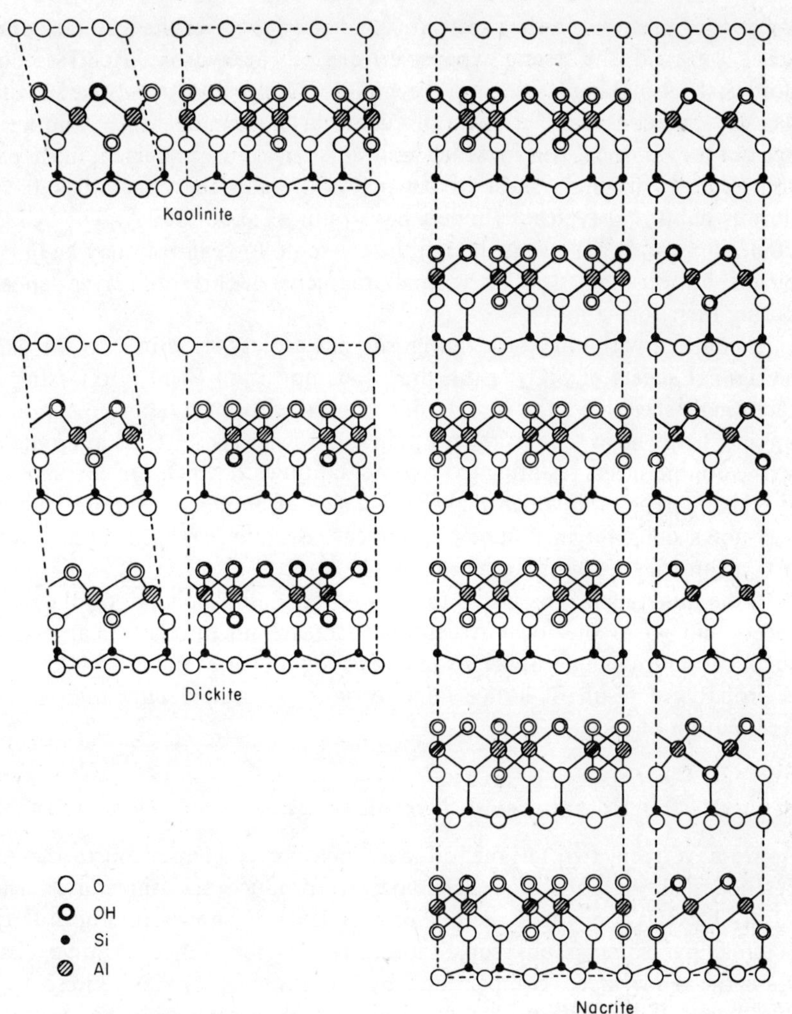

FIG. 2. The structures of kaolinite, dickite and nacrite,[3] viewed on the left hand side along *b* and on the right hand side along *a*. The *c*-axis is always vertical.

Whereas AlO_6 octahedra and SiO_4 tetrahedra fit well together in the kaolinite double sheet, with little distortion of natural bond angles and distances, there is in contrast considerable distortion when MgO_6 and SiO_4 form the same kind of double sheet. The octahedral layer resembles that in brucite (crystalline $Mg(OH)_2$) and if the *a* and *b* dimensions of each type of sheet are compared for kaolinite and brucite one finds

Si-O sheet (kaolinite) $a = 5\cdot16$Å, $b = 8\cdot90$Å
Mg-OH sheet (brucite) $a = 5\cdot39$Å, $b = 9\cdot33$Å.

Because of misfit when the sheets are condensed together the resultant double sheet is curved, the strong expansive force on the hydroxyl-rich side of the double sheet making it curl into a cylinder with the fibre axis parallel with *a*. The crystals consists of a series of curved layers one on top of another, or even of a cylindrical roll in which a sheet is curled around itself, like a paper roll made of a single sheet of paper. As a result the chrysotiles have a fibrous habit and typical cylinders have radii of about 100 Å. The cylinders do not appear to behave as though they were hollow, and so may be blocked by adventitious impurities or by small fragments of chrysotile layer deposited during their formation.

In contrast with chrysotile, antigorite has a platy structure. In the parent hexagonal sheets of SiO_4 tetrahedra of composition $Si_2O_3(OH)_2$ strips of each single sheet have the vertices of the tetrahedra (the OH groups) pointing up and strips have the tetrahedra pointing down. These strips alternate and so the double sheets resulting from combination with MgO_6 are corrugated[4]. It is this corrugation which makes the *a* axis of antigorite so long, although *a* is not a constant in different specimens. Brindley and Zussman[5] found that *a* can vary between about 30 and 45 Å.

In the minerals of this group there is no net anionic charge on the double sheets and so in the ideal structures there are no interlayer cations. The small cation exchange capacity sometimes found, for example with kaolinite, is probably a result of lattice imperfections, associated *inter alia* with the termination of crystal faces.

2.2 Silicates with three-layer Sheets

As we have seen, two tetrahedral layers may be condensed on to the AlO_6 octahedral layer, one on each side, to produce the three-layer sheet, $Al_2Si_4O_{10}(OH)_2$. A small section of this layer is shown in Fig. 1d. It is possible by isomorphous substitutions to develop a net negative charge upon the triple layer, compensated by interlayer cations as represented in this figure. If the cations are small enough they can, as in the micas, be recessed into the 6-rings of the tetrahedral sheets (as illustrated at the right of Fig. 1b). Hydroxyls arising from the octahedral layer are located on axes through the centre of, and normal to, the plane of each 6-ring, at the height of the two planes through the vertices of the SiO_4 tetrahedra forming the two hexagonal networks in each triple sheet. In the upper network, the vertices point down; in the lower they point up. Some minerals in which the triple sheets occur are given in Table 1, together with idealized formulae and interlayer cation densities.

The Si_4O_{10} unit in pyrophyllite and talc represents the composition of the tetrahedral sheets in the triple layer. The octahedral layer contains the Al (pyrophyllite) or Mg (talc) given in the formulae. The Al occupies only two

TABLE 1

Some minerals having the three-fold sheets found in micas

Mineral	Ideal formula	Interlayer cations (meq/100 g)
Pyrophyllite	$Al_2(Si_4O_{10})(OH)_2$	0
Talc	$Mg_3(Si_4O_{10})(OH)_2$	0
Micas		
Muscovite	$KAl_2(AlSi_3O_{10})(OH)_2$	252
Paragonite	$NaAl_2(AlSi_3O_{10})(OH)_2$	262
Phlogopite	$KMg_3(AlSi_3O_{10})(OH)_2$	240
Biotite	$K(Mg, Fe)_3(AlSi_3O_{10})(OH)_2$	240 to 204
Lepidolite	$K(Li_2, Al)(Si_4O_{10})(F, OH)_2$	279 to 277
Zinnwaldite	$K(Li,Fe, Al)(AlSi_3O_{10})(F, OH)_2$	≈229
Brittle Micas		
Margarite	$CaAl_2(Al_2Si_2O_{10})(OH)_2$	502
Chloritoid	$(Mg, Fe^{II})Al_2(Al_2Si_2O_{10})(OH)_2$	524 to 440
Seyberite	$Ca(Mg, Al)_3((Si, Al)_4O_{10})(OH)_2$	
Vermiculites[a]		
General formula	$(Ca, Mg)_{x/2}(Mg, Fe, Al)_3((Al, Si)_4O_{10})(OH)_2, mH_2O$	≈100 to 165
Smectites[a]		
Montmorillonite	$Na_x(Al_{(2-x)})Mg_x)(Si_4O_{10})(OH)_2, mH_2O$	≈ 60 to 100
Saponite	$Ca_{x/2}Mg_3(Al_xS_{(4-x)}O_{10})(OH)_2, mH_2O$	
Nontronite	$M_x(Fe^{III}, Al)_2 (Al_xSi_{(4-x)}O_{10}) (OH)_2, mH_2O$	
Beidellite	$M_xAl_2(Al_xSi_{(4-x)}O_{10})(OH)_2, mH_2O$	
Sauconite	$M_x(Zn, Mg)_3(Al_xSi_{(4-x)}O_{10})(OH)_2, mH_2O$	
Hectorite	$(Na_2, Ca)_{x/2}(Li, Al)_2(Si_4O_{10})(OH)_2, mH_2O$	
Fluorhectorite	$(Na, Ca)_{x/2}(Li, Al)_2(Si_4O_{10})(F, OH)_2, mH_2O$	

[a] M denotes one equivalent of an exchangeable cation such as Na, Ca and Mg, and so x is the number of equivalents of exchangeable cations present. In a typical smectite x in the above formulae is about 0·33, representing around 90 meq per 100 g of clay mineral. The water contents vary with relative humidity and with the cations present. A typical value of m is ≈3·5.

out of every three six-fold positions (a dioctahedral mineral) whereas the Mg in talc occupies all three of these positions (a trioctahedral mineral). In both cases the triple sheet is electrically neutral. The sheets are stacked in an *ABAB* . . . sequence to give the unit cells

Talc	Pyrophyllite
$a = 5{\cdot}16Å$	$a = 5{\cdot}15Å$
$b = 9{\cdot}10Å, \beta = 100°$	$b = 8{\cdot}90Å, \beta = 90° 55'$
$c = 18{\cdot}81Å$	$c = 18{\cdot}55Å$

Thus the thickness of the triple sheets is 9·3 to 9·4 Å. Both minerals show the type of diffuse scattering in the X-ray powder photographs which indicates frequent stacking faults.

In the micas, various isomorphous replacements are found. Very important is the substitution of Si by Al in the tetrahedral layers, which introduces one negative charge on the framework for each Al. In addition, Li, Mg and Fe may replace Al in the octahedral layer with no alteration in framework charge, to change dioctahedral micas (e.g. muscovite and paragonite) to trioctahedral ones (e.g. phlogopite and biotite). Lepidolite and zinnwaldite are interesting in that fluorine replaces hydroxyl, and Li is found in the octahedral layer. In lepidolite this causes the development of negative charge in that layer, balanced by potassium ions between pairs of triple sheets. Thus the triple sheets can develop negative charge by substitution either in tetrahedral or in octahedral layers, or in both. Such substitutions are further exemplified in vermiculites where Mg and Fe, and in smectites where Li, Mg and Fe can replace Al in octahedral layers and Al can replace Si in tetrahedral layers (Table 1). In synthetic smectites and micas Ga can replace Al and Ge can substitute for Si.[(6)]. In synthesis also, ammonium or alkylammonium ions can replace inorganic interlamellar cations.[(7)] Among the smectites given in the table montmorillonite, beidellite and nontronite are dioctahedral, and saponite, sauconite and hectorite are trioctahedral.[(8)]

In the brittle micas the extent of isomorphous replacement of Si by Al in the tetrahedral layers is doubled as compared with that in muscovite. The resultant high negative charge of the triple sheets is now neutralized by divalent instead of monovalent cations in number equal to that of potassium ions in muscovite.

The area of a single 6-ring in the hexagonal sheets of the micas, vermiculites and smectites is 24 Å^2. Each 6-ring is large enough to allow cations to be partly or wholly keyed into them in the anhydrous outgassed crystals (Fig. 1b). Thus each 6-ring can provide a cation site allowing the triple sheets to come into close contact. It is therefore of interest that in margarite, chloritoid, muscovite, paragonite, phlogopite, biotite, lepidolite and zinnwaldite the area available to each cation is 24 Å^2, so that there are no vacant 6-ring sites. On the other hand with vermiculites and smectites much larger areas are available per interlamellar cation so that in the anhydrous crystals only some 6-ring sites need be occupied. In the hydrated swollen states exchange ions such as Li, Na or Ca may not be associated with 6-rings, but may as hydrated ions be "floating" in the interlamellar water layers. Here the idea of fixed cation sites is much less appropriate.

Another consequence of the above considerations is that none of the micas could accommodate more interlamellar cations without a significant expansion of *d*(001). The tendency to a fixed cation stoichiometry suggests that 6-rings provide trapping sites which stabilize the cation composition and so determine negative charge on the triple sheets. The variable exchange capacities reported for vermiculites and smectites may, especially for smectites, arise in part from impurities, but may also be influenced by the absence of fixed exchange sites in these hydrated crystals, which allows more variable, if smaller, anionic charge on triple sheets during crystal growth.

If one considers the two sheets of SiO_4 tetrahedra facing one another across the interlayer cations, because of their hexagonal symmetry the pair may be stacked with angles of 0°, 60°, 120°, 180°, 240° or 300° between their directions of stagger. These ways of staggering are not independent because each sheet has a plane of symmetry, but combinations of them give rise to a number of stacking schemes.[9,10] The simplest gives a $d(001)$ of ≈ 10 Å; others give c repeat distances of ≈ 20, 30 and 60 Å. The $\approx$ 10 Å spacing was observed in a biotite and is common in phlogopites. The ≈ 20 Å spacing occurs in muscovite and in lepidolite (as two different sequences) and a 30 Å spacing has been observed in a muscovite. Also, as in talc and pyrophyllite there may be stacking faults in layer sequences. In the kandites likewise different layer sequences can occur as exemplified by kaolinite, dickite and nacrite.

The chlorites represent sequences of talc triple layers alternating with the layers which occur in brucite (crystalline $Mg(OH)_2$), and which are represented in the formula as $3Mg(OH)_2$. In vermiculites and smectites the brucite layers are replaced by layers of water. Indeed these water layers so "lubricate" the lamellae that they may no longer be stacked in ordered sequences but form turbostratic arrays.

2.3 The Palygorskites

The fibrous clay minerals of the palygorskite group are represented by attapulgite and sepiolite. A section of the attapulgite structure is shown in Fig. 1e, and sections of both attapulgite and sepiolite are compared with additional detail in Fig. 3[11] which, like Fig. 1e, is a projection on the (001) plane. Hexagonal networks of linked SiO_4 again appear but with the vertices of the tetrahedra lying in strips. In one strip all vertices point up, in the next all point down, and so on in alternation. In attapulgite the strips are four tetrahedra and in sepiolite they are six tetrahedra wide. MgO_6 octahedra then join a strip in an upper layer with down-pointing vertices to a strip in a lower layer with up-pointing vertices, so that three-fold strips are obtained. The cross-sections shown in Fig. 3 can be likened to a brick wall with each alternate brick removed. The resultant parallel channels contain water molecules which have been divided into zeolitic water and water of crystallization (or bound water). An idealized attapulgite composition is $Mg_5Si_8O_{20}(OH)_2(H_2O)_4$, $4H_2O$ and for sepiolite is $Mg_8Si_{12}O_{30}(OH)_4(H_2O)_4$, $8H_2O$. Complete removal of the water by heat and outgassing can cause

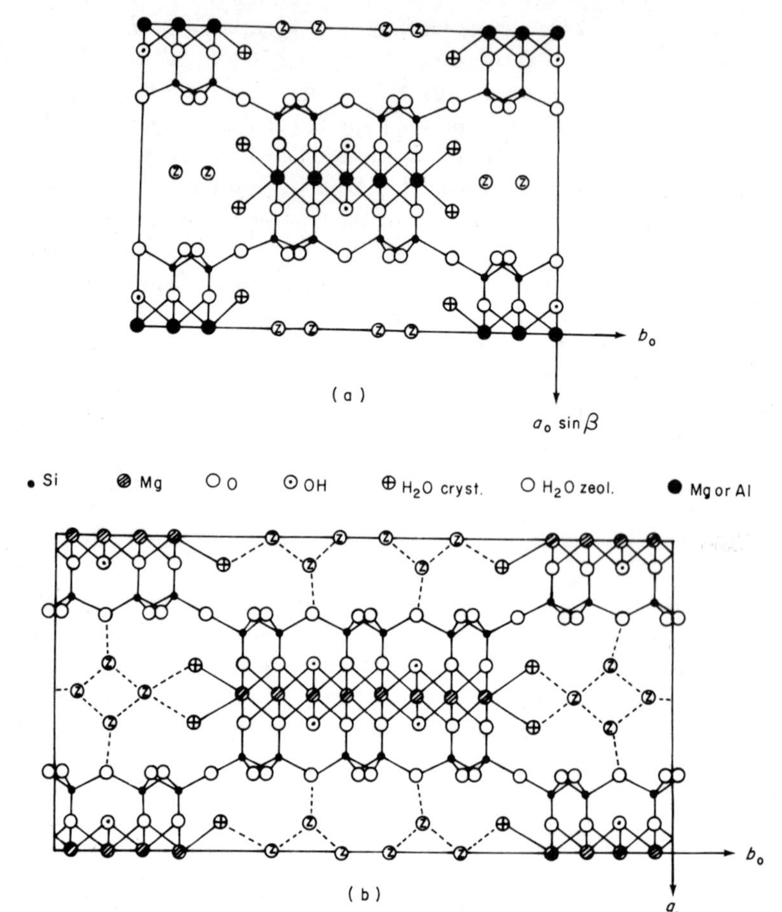

FIG. 3. A projection on the (001) plane of the cross-sections of the linked strips in (a) attapulgite and (b) sepiolite.[11]

irreversible changes in the lattices.[12] The tunnels in the outgassed structures appear to be impeded so that only small polar molecules such as NH_3 can replace the water, but the fine fibrous habit of the crystals ensures a large surface.

3. INTERCALATION IN KANDITES

As noted in § 2.1, there is an absence of charge on the aluminosilicate sheets of kandites and of water layers between the sheets (except for halloysite). This does not, however, prevent the direct intercalation of a range of guest species.[13] These include hydrazine, urea, formamide and acetamide

TABLE 2

Some kaolinite intercalation complexes[13]

Guest Species	Intercalation conditions	Temp. (°C)	*c*-repeat of intercalate (Å)	Expansion on intercalation (Å)
Urea	Aqueous or alc. solution >10M. Can remove by washing crystals	20, 65	$10{\cdot}7_2$	$3{\cdot}5_6$
Formamide	From pure formamide or conc. aqueous solution	25, 40	$10{\cdot}0_3$	$2{\cdot}8_7$
Hydrazine	From conc. aqueous solutions	0, 25, 40	$10{\cdot}4_2$	$3{\cdot}2_6$
Acetamide	From melts or more readily by treatment of urea or formamide intercalates with acetamide melts	0, 25, 40, 65	$10{\cdot}9_2$	$3{\cdot}7_6$
Acetates	From saturated aqueous solutions. Intercalation pH-sensitive. Intercalated materials washed out		$9{\cdot}5_5(Li^+)$	$2{\cdot}3_9$
			$10{\cdot}0_6(Na^+)$	$2{\cdot}9_0$
			$10{\cdot}4_2(Ca^{2+})$	$3{\cdot}2_6$
			$10{\cdot}7_6(N_2H_5{}^+)$	$3{\cdot}6_0$
			$14{\cdot}0_1(K^+)$	$6{\cdot}8$
			$14{\cdot}4_2(Rb^+)$	$7{\cdot}2_6$
			$14{\cdot}6_8(Cs^+)$	$7{\cdot}5_2$
			$14{\cdot}0_5$ (NH_4^+)	$6{\cdot}8$
			$17{\cdot}1_5$ (NH_4^+)	$9{\cdot}9_9$
K, Rb and Cs	From saturated aqueous solutions			
Propionates	Intercalation pH sensitive (7 to 13 range is favourable)			
K-cyanoacetate	From saturated aqueous solutions in pH range 7 to 10·5		$12{\cdot}9_8$	$5{\cdot}8_2$

as well as the K, Rb, Cs and NH_4 salts of lower fatty acids (acetates, propionates, cyanoacetates). The intercalation conditions and lattice expansions of kaolinite are given in Table 2. The pH dependence of intercalation referred to in the table is illustrated for K-acetate at several temperatures in Fig. 4b, and was observed with other acetates and propionates. The kinetics of uptake were characteristic: an induction period of a duration dependent upon temperature was followed by a period of acceleration and then an asymptotic approach to the final uptake.[13] These sigmoid kinetic curves are illustrated in Fig. 4a for formamide at 25° and 40°C. There is a large temperature coefficient in the rate of intercalation at a given uptake of the guest. However this does not denote a clear-cut activated process in the usual chemica

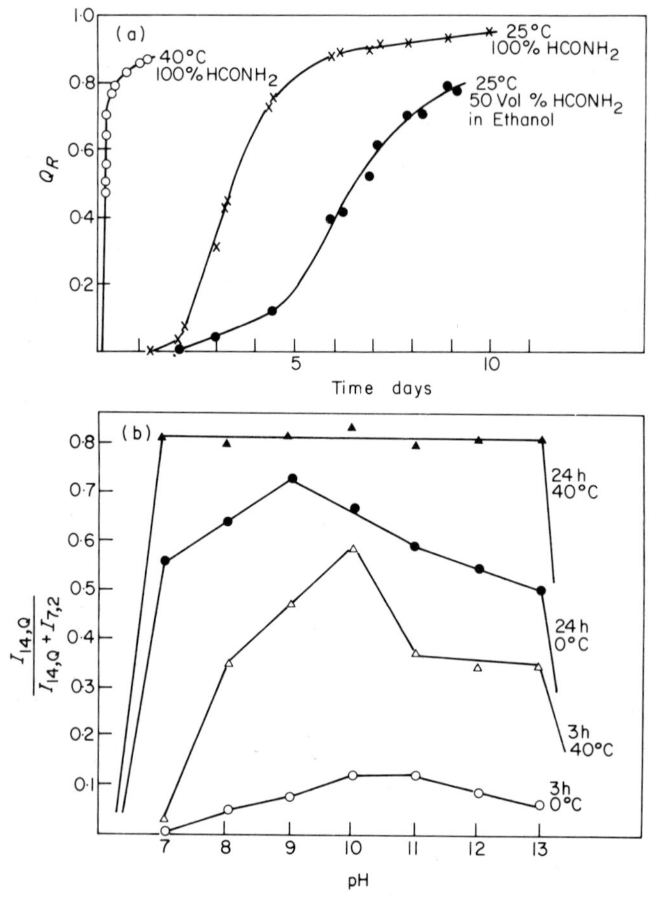

FIG. 4. (a) Degree of intercalation, Q_R, as a function of time (in days) and of temperature, for formamide in kaolinite. (b) pH and temperature dependence of inclusion of K-acetate in kaolinite. $I_{14\cdot0}$ and $I_{7\cdot2}$ are the intensities of the 14·0 and 7·2Å spacings.[13]

sense, as is illustrated by the behaviour with hydrazine. From its 20M and $10\cdot9_5$M aqueous solutions at 2, 25 and 40°C intercalation rates had a strong positive temperature coefficient, but with $4\cdot6_8$M solution there was a comparably large negative temperature coefficient.

The behaviour with NH_4-acetate was also unusually complex. Intercalation occurred only in the pH range 7 to 10·5. Between 0 and 25°C the rate of intercalation increased rapidly, but intercalation did not occur above

TABLE 3

c-repeat distances of some intercalation complexes formed from solutions of guests in 24%hydrazine solutions in water

Compound intercalated	Repeat distance (Å) In presence of hydrazine solution	After removal of hydrazine solution
K-acetate	10·4	14·0
Na-acetate	10·4	$10\cdot0_6$
K-oxalate	10·4	$10\cdot2_8$
K-glycollate	10·7	$12\cdot4_4$
K-alaninate	10·7	$12\cdot4_5$
K-lysinate	10·6	$14\cdot8_8$
K-lactate	10·6	$11\cdot1_8$
Glycerine	10·4	10·52
n-octylamine	$31\cdot7_1$	$31\cdot7_1$
Benzidine[a]	10·6	$10\cdot_8$

[a] In this case the benzidine was dissolved in 24% hydrazine in aqueous alcoholic solution.

40°C. If however the $14\cdot0_5$ Å complex was first formed at low temperatures and was then heated to 65°C still in the acetate solution the *c* repeat distance increased to $17\cdot1_5$ Å. This $17\cdot1_5$ Å phase was obtained much more rapidly when the $14\cdot0_5$ Å complex was dried in air.[13]

There have been one-dimensional Fourier analyses of the structures of the intercalation complexes of kaolinite with urea, formamide and hydrazine[13,14] and of dickite with formamide.[15] In kaolinite and dickite the ⟩CO of the formamide approaches the hydroxyl-rich side of the alumino-silicate two-layer sheet. In urea-kaolinite one -NH_2 and the ⟩CO are associated with this surface and the remaining -NH_2 with the oxygen surface of the next sheet.

The number of compounds which can be intercalated can be increased considerably by having in the aqueous phase not only the potential new guest species but also a molecule such as hydrazine which can intercalate directly and independently. This appears to act as a lubricant ("schlepper") for promoting penetration by other guests. Examples of the introduction of various species dissolved in 24% solutions of hydrazine are given in Table 3.[13] In the resultant complex with benzidine these molecules must lie parallel with the aluminosilicate sheets because the $10 \cdot 2_8$ Å c-spacing is too small to permit any alternative configuration. However when this complex was heated in a benzidine melt at 130°C the $20 \cdot 7_8$ Å phase was formed, which incorporates considerably more benzidine. The c-spacing of $20 \cdot 7_8$ Å is compatible with an arrangement in which the guest molecules are perpendicular to the aluminosilicate sheets. When the $17 \cdot 1_5$ Å phase of NH_4-acetate in kaolinite was introduced into appropriate liquid amines further swelling occurred. With *n*-decylamine and trimethylcetylamine the spacings were respectively $34 \cdot 9_9$ Å and $41 \cdot 1_4$ Å and with *n*-octylamine it was 31·7 Å.

Dimethylsulphoxide (DMSO) and *N*-methylformamide (NMFA) are also able to swell kandites (kaolinite, dickite, nacrite, halloysite and metahalloysite) as well as smectites and vermiculites.[16] The $d(001)$ spacings are 11·18Å for DMSO-kandites and 10·7Å for NMFA-kandites, which represent unimolecular layers of the guest. In natural halloysite these spacings were 10·98 and 10·5Å, but the larger spacings were obtained when the halloysite was first dehydrated at 140°C.

These various examples demonstrate that kandites under special conditions can intercalate both polar and ionic guest species, and that in part the problem of intercalation is the problem of initiating the process at the periphery of the host crystal. While some guest species can initiate nucleation others require the assistance of a second guest before penetration begins.

4. SORPTION BY FIBROUS CLAYS

In attapulgite and sepiolite, slit-shaped channels run parallel with the fibre axes (Fig. 2). If they could be freed of water molecules and of any adventitious inclusions these channels would have free dimensions of about $3 \cdot 7 \times 6 \cdot 3$Å (attapulgite) and $3 \cdot 7 \times 9 \cdot 3$Å (sepiolite). It was therefore of interest that Nederbragt and de Jong[17, 18] reported that when *n*-paraffins such as *n*-hexadecane and *n*-tetracosane were dissolved in pentane together with analogous branched chain paraffins, and passed through a column of attapulgite, the *n*-paraffins were preferentially retained by the crystals. It was suggested that the *n*-paraffins might be entrained in the channels referred to above, which the branched chains could not enter. Migeon[19] also suggested that zeolitic sorption of oils and other species might occur in sepiolite, and McCarter *et al.*[20] considered that sorption of nitrogen in the channels of attapulgite was possible. On the other hand Granquist and Amero[21] questioned the availability of these channels for such sorption.

Following this work several systematic investigations were made to find out whether, and how readily, intracrystalline sorption by palygorskites could occur.[22,23,24,25] Sorption of O_2, N_2, CO_2, H_2O, NH_3, CH_3OH, C_2H_5OH, CH_3NH_2, $C_2H_5NH_2$ and of various C_5, C_7 and C_8 straight and branched chain paraffins was investigated on attapulgite or sepiolite under systematically altered conditions of outgassing. The conclusions from this work were that water and ammonia could diffuse into the channels after appropriate initial outgassing, with much less penetration by simple alcohols. Sorption of most species, including nitrogen and other non-polar molecules, was confined to the external surfaces of the crystallites or to sites

TABLE 4

Effect of pretreatment on BET surface areas of palygorskites

Attapulgite[22]		Sepiolite[24]	
Outgassing at T°C[a]	Area (m^2g^{-1})	Outgassing at T°C	Area (m^2g^{-1})
20	195	25 (48 h)	354
70	195	70 (24 h)	380
95	192	96 (1 h)	375
115	128	96 (3 h)	378
120	128	300 (1 h)	195
150	127	400 (1 h)	196
200	127	300 (1 h)[b]	158
250	125	500 (1 h)[b]	158
350	123	700 (1 h)[b]	157

[a] For times up to 70 h.

[b] The heating was carried out in these cases in air. Subsequent outgassing was for 5 min at 25°C.

just at the channel entrances. However, the sorption was very substantial because of the extremely fine fibrous habit of the crystals and the correspondingly large external areas. The effect upon these areas of the outgassing temperatures is illustrated in Table 4 for an attapulgite[22] and a sepiolite.[24]

When air-dried fibrous clays were outgassed there was about 9% weight loss with attapulgite and 15 to 16% loss in sepiolite at 100°C. This represented easily removed zeolitic water, rather than bound water (water of crystallization, §2·3), and for these conditions maximum surface areas were developed. Isotherms of non-polar sorbates are of Type II in Brunauer's classification[26] and equilibrium is established rapidly. The sorption results support the view that much if not all the uptake is on external surfaces, because true intracrystalline sorption isotherms of N_2 are of Type I (Chapter 3). The rapid equilibration also supports this view since diffusion into individual cyrstal fibres can occur only through cross-sections normal to the fibre axes. These cross-sections are very small compared with the length of the crystals, so that for intracrystalline diffusion the uptake would be expected to occur rather slowly even for perfect unblocked channels.

On outgassing above 100°C there tends to be a rapid decrease in the surface areas of the fibrous clays to lower values, which thereafter decline very little with further outgassing at considerably higher temperatures

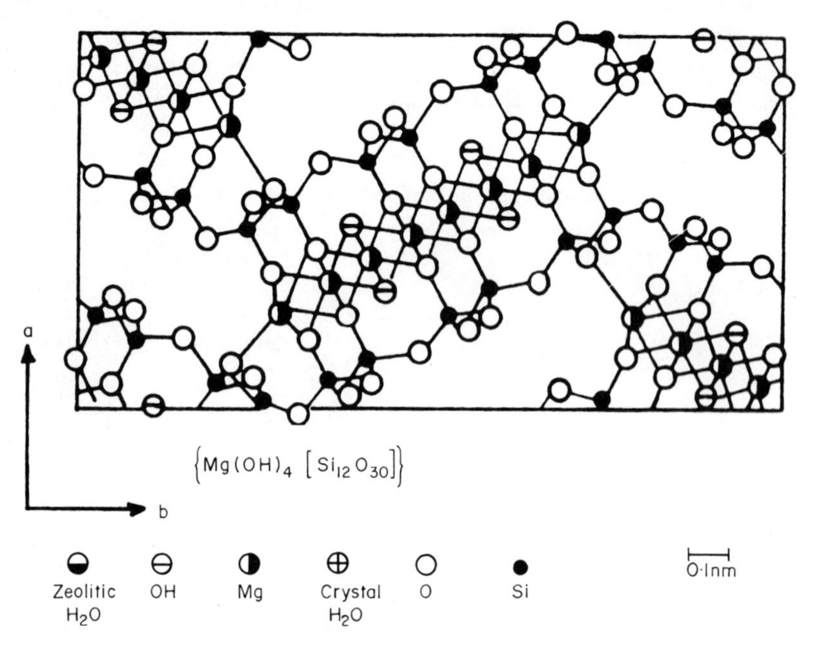

FIG. 5. The buckling of the hexagonal sheets of sepoilite on dehydration, shown as the projection on the (001) plane of the cross-sections of the linked strips (after Preisinger.)[12]

(Table 4). The removal of bound water at these higher temperatures constitutes a second stage in the dehydration of palygorskites which is certainly not reflected in higher surface areas. According to Preisinger[12] the formation of water-free sepiolite "anhydride" produces a corrugation of the hexagonal sheets of oxygen atoms and a mutual reorientation of the three-fold strips. This is shown in Fig. 5, and is to be compared with the parent structure in Fig. 3.

Barrer *et al.*[23] investigated the selectivity shown by attapulgite for *n*-paraffins relative to iso- and neo-paraffins. For attapulgite outgassed at 70°C there is a clear selectivity sequence *n*- > iso- > neo-pentane, but when the attapulgite was outgassed at 215°C this selectivity was greatly reduced and the sorption of each hydrocarbon was also lower. Figure 6 shows the same sequence of selectivities for sorption of the three pentane isomers on a montmorillonite outgassed at 50°C. It was possible with attapulgite to correlate the rate of water loss on outgassing, the affinity coefficient, C, of the BET isotherm equation, the surface area, and the amounts sorbed at a relative pressure of $0 \cdot 1$ with the outgassing temperature. The inter-relations, shown in Fig. 7, are striking. The differential weight loss curve shows a deep minimum at a point where zeolitic and adsorbed water has been removed,

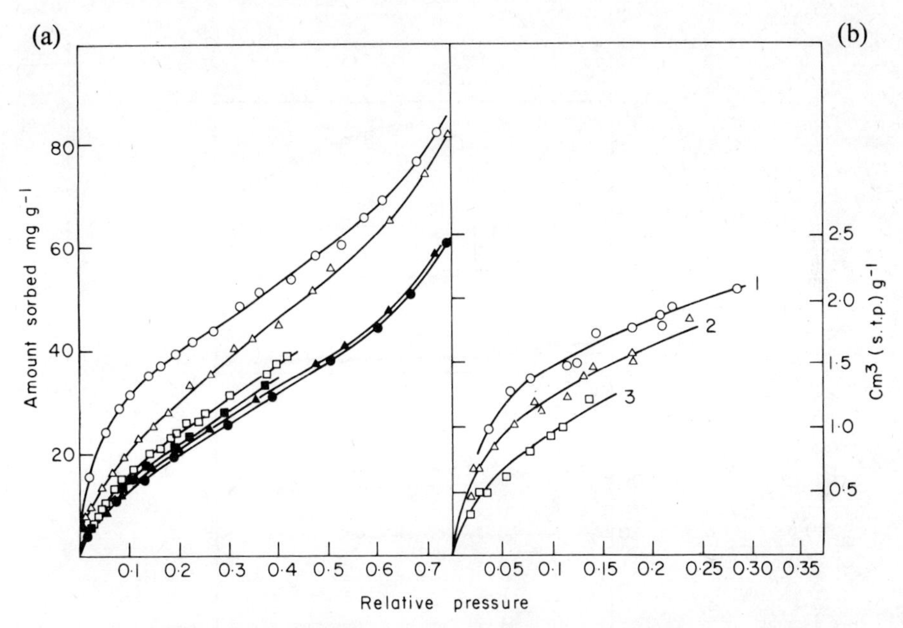

FIG. 6. (a) n-C_5H_{12} (○), iso-C_5H_{12} (△) and neo-C_5H_{12} (□) at 20°C on attapulgite outgassed at 70°C. Also n-C_5H_{12} (●), iso-C_5H_{12} (▲) and neo-C_5H_{12} (■) at 20°C on attapulgite outgassed at 215°C, showing loss of selectivity. (b) Sorption of the isomeric pentanes at 50°C on montmorillonite outgassed at 50°C. Curves 1, 2 and 3 refer to n-C_5H_{12}, iso-C_5H_{12} and neo-C_5H_{12} respectively.[23]

followed by a rapid increase again as water of crystallization is evacuated. Near the temperature of this minimum the affinity constants, C, of all three isomers also come together, and the surface areas shown towards n- and iso-pentane have undergone a rapid decline, which is however, much less pronounced for neo-pentane. The sorptions at $0 \cdot 1$ relative pressure behave in the same way as the surface areas. An explanation was sought in terms of the corrugations expected on the surface of attapulgite, as a consequence of its structure (§ 2.3). It was suggested that n-paraffins could fit well into such corrugations, but neo-, and to some extent iso-paraffins were not so well accommodated. The effective fit according to this explanation must become much less selective for n-paraffins once the water of crystallization begins to be moved. This may be a result of the onset of structural changes analogous to those observed when sepiolite (Fig. 3) changes to sepiolite "anhydride" (Fig. 5).

The selectivity sequence *n*- > iso- > neo-pentane found with montmorillonite, where sorption is mainly on the basal surfaces that provide most of the area, probably reflects how close to the surface the centre of gravity of the molecules may approach. With *n*-pentane, as one extreme, all CH_3

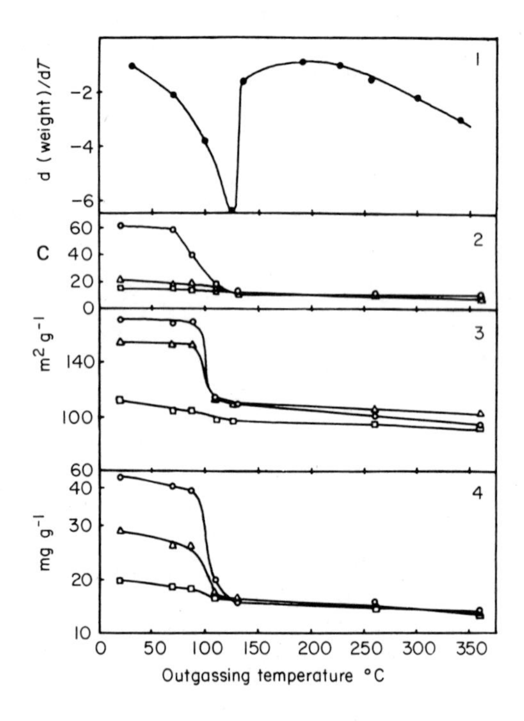

FIG. 7. Correlations of various properties, in sorption of the pentane isomers on attapulgite, with outgassing temperature. ○, *n*-C_5H_{12}; △, iso-C_5H_{12}; ⊡, neo-C_5H_{12}. 1, water loss; 2, BET affinity constant, C 3, surface area; 4, amounts sorbed at relative pressure 0·1.[23]

and CH_2 groups can lie on the surface; with neo-pentane as the other, three CH_3 groups only can lie on the surface. Such an explanation is not fully adequate for attapulgite because the above selectivity sequence is not maintained if higher outgassing temperatures than 100°C are used (Fig. 7).

From these and other studies it is concluded that, despite good sorption capacities and interesting selectivities the development of a true molecular sieve function is not characteristic of palygorskites.

5. SWELLING OF VERMICULITES AND SMECTITES IN WATER

By far the largest amount of research on the sorption properties of clay minerals has been devoted to smectites and vermiculites. Since water is a natural component of these crystals and is one of the most important sorbates the behaviour of such crystals with water will first be considered. In smectites and vermiculites the amounts of interlayer water can be changed according to the cations present[27] and to the physical conditions, with parallel alterations in the *d*(001) spacing. This is well illustrated in a study by Walker[28,29] of dehydration of Mg-vermiculite where the stages of Fig. 8[29,30] were observed. The 14·81Å phase resulted from several days

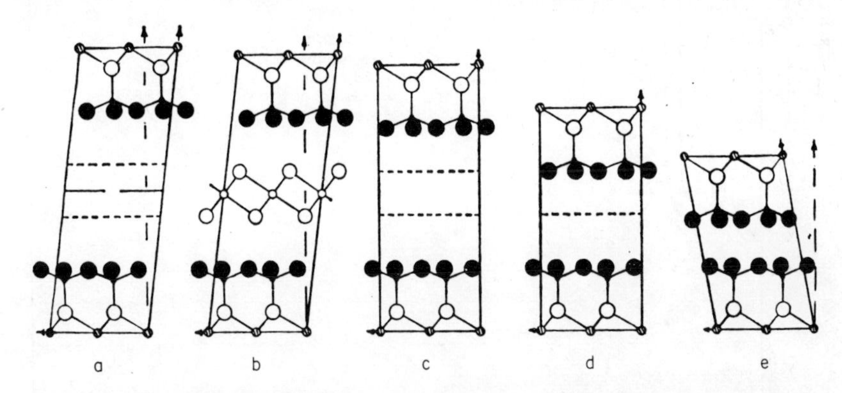

FIG. 8. A series of projections of Mg-vermiculite on (010), showing the silicate layer relationships at various stages of dehydration.[30] (a) 14·81Å complex; (b) 14·36Å complex; (c) 13·82Å complex; (d) 11·59Å complex; and (e) 9·02Å phase. In (a), (c) and (d) the dashed lines are formal representations of water layers.

immersion in water of the normal 14·36Å phase. That having *d*(001) equal to 13·82Å involved a double water layer between silicate lamellae which was considered to have a different arrangement from that in the 14·36Å phase. With only a monolayer of water remaining *d*(001) fell to 11·59Å, while for the anhydrous crystals this spacing was only 9·02Å. Water isotherms at 50°C reflect some of these stages, in that they show a clear step at the uptake corresponding with the transition from the monolayer to a thicker layer complex[31] (Fig. 9). Heats of sorption of water determined by direct calorimetry from heats of immersion are given in Table 5. The first mg per g is sorbed with an exceptionally high heat, presumbly upon a few special sites which are not characteristic of the whole crystal. The heat involved in forming the completed monolayer complex (up to about 63 mg per g) is approximately twice that for forming the completed bilayer complex or

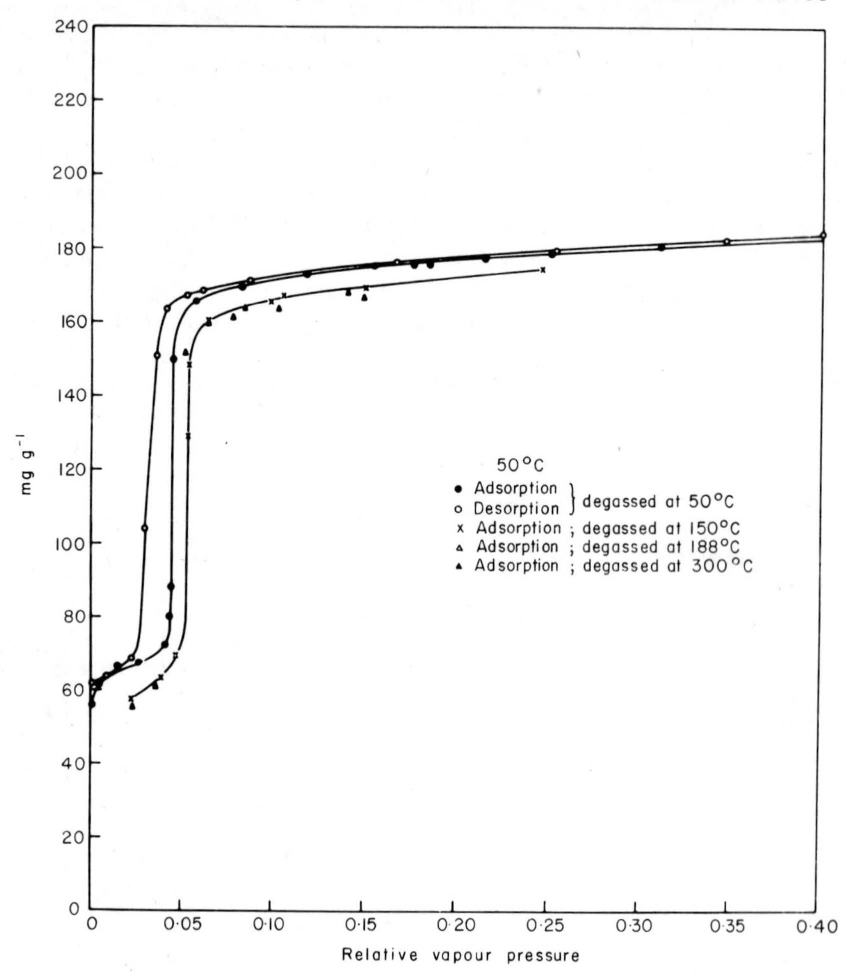

FIG. 9. The step-wise sorption of water in Mg-vermiculite outgassed at various temperatures.[31]

beyond (up to around 188 mg per g). The heat then declines further towards zero, as expected for heats of immersion of water-saturated crystals.

Gillery[32] measured the *d*(001) spacings of a number of synthetic Na- and Ca-beidellites and saponites as functions of the relative humidity. Some of the relationships obtained are illustrated in Fig. 10, *a*, *c* and *f* for Na-beidellites, *e* for a natural Na-montmorillonite and *b* and *d* for two Na-saponites. As with vermiculite the *d*(001) spacings are step functions of the vapour pressure. This is also the case for the uptake of water in a natural Na-rich montmorillonite.[33]

A feature of hydrated vermiculites and smectites is the readiness with which exchange of interlayer cations can occur. The interlayer water solvates ions such as Li, Na, Mg and Ca, and makes them mobile within the crystals. In contrast with the behaviour of hydrated layer silicates exchange of interlayer cations in micas is difficult and slow. The triple mica sheets are very close together, with the potential exchange ions recessed into and trapped by the 6-rings of the hexagonal silica sheets. The cations are thus fixed unless and until some entering species separates the sheets and releases

TABLE 5

Average heats of sorption of water in Mg-vermiculite[31]

Coverage ($mg\ g^{-1}$)	Average heat ($kcal\ mol^{-1}$)
0 to 0·96	32·31
0·96 to 62·58	7·99
62·58 to 83·8	4·25
83·8 to 188·6	3·50
188·6 to 207·25	0·69

and solvates the trapped cations. In the micas aqueous alkylammonium ions do very slowly swell the crystals and replace interlayer potassium.[34,35,36] Also heating muscovite mica in molten lithium nitrate replaced much of the potassium by lithium ions, which could not then be readily exchanged.[37,38] Some Li appeared to enter the octahedral layer and thus reduced the negative charge on the lamellae. The reduction of charge was progressive with time as the figures in Table 6 show. The layer charge of the mica was estimated as the sum of the potassium and cation-exchange capacity of the treated Delamica. Once the layer charge had been reduced to about $6{\cdot}7 \times 10^4$ e.s.u. cm^{-2} or less it became possible to swell the crystals by water vapour from air to give an 11·8Å *c* repeat distance (a single layer of intersheet water), and in glycerol to give 17·8Å or 14Å repeat distances. These experiments demonstrate a relation between charge density and ability of the crystals to swell. The connection between charge density and this ability is however not simple because the talc and pyrophyllite structures which have no interlayer cations are non-swelling, just like margarite or certain other micas which have the maximum charge densities. It is the vermiculites and smectites with intermediate charge densities that swell most easily, particularly the smectites.

Further quantitative information on the swelling in distilled water of various layer silicates is contained in Table 7. While talc and pyrophyllite do not imbibe water, in the Na-forms of some montmorillonites the individual triple sheets can be dispersed altogether. There must be a balance between the increase in free energy of the system when the sheets are

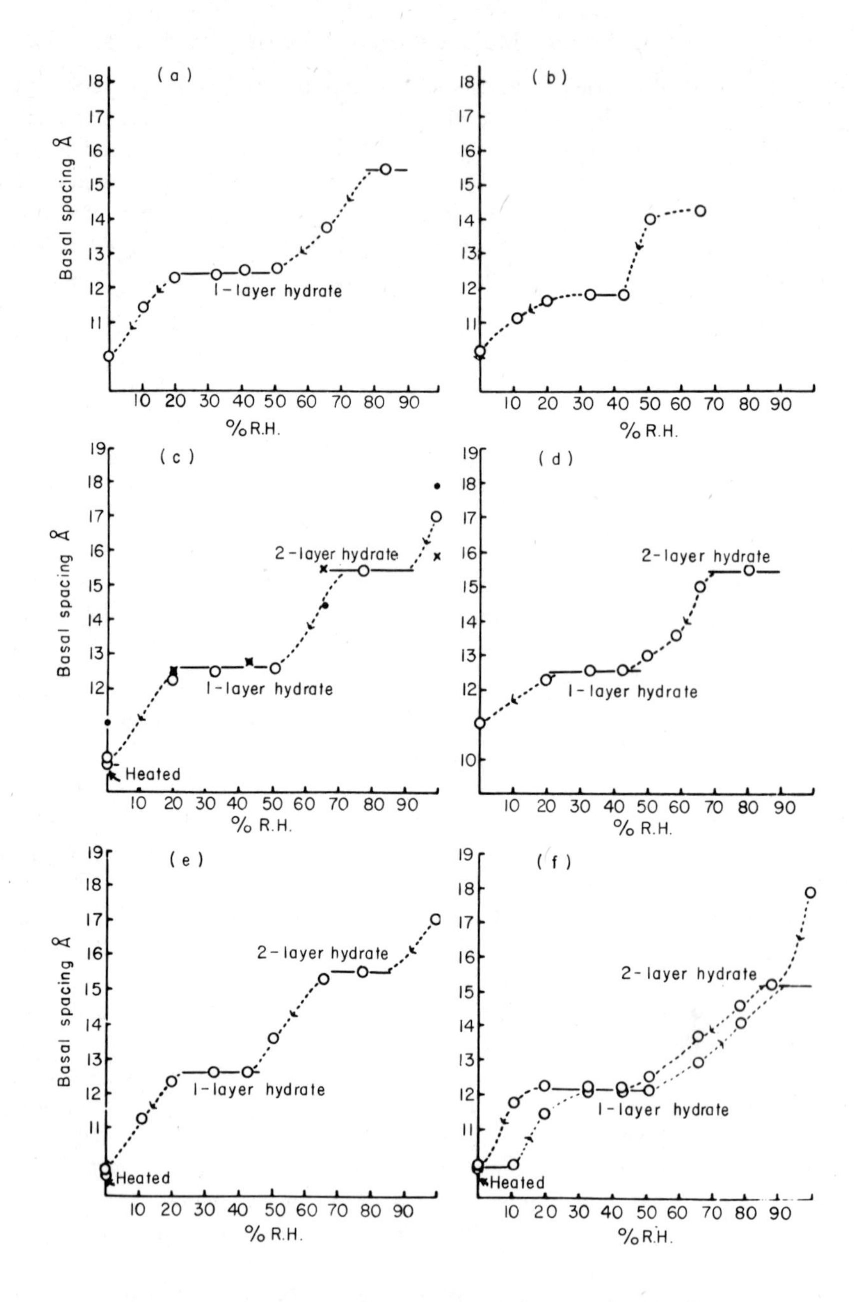

FIG. 10. Relations between *d*(001) spacings and relative humidity.[32] (a) Beidellite, (b) Na-saponite, (c) Na-beidellite, (d) Na-saponite, (e) natural Na-montmorillonite (f) Na-beidellite. The saponite and beidellite samples differ *inter se* in the extent of isomorphous replacements of Si by Al and hence in their cation exchange capacities.

separated against the forces holding them together, and the decrease in free energy when guest molecules such as water invade the interlayer regions. An important part of the decrease is the change in free energy when interlayer cations are hydrated rather than being solvated by lattice oxygens. If there are too many cations, as in margarite, the positive free energy of sheet separation exceeds the decrease due to penetration of water, so that no swelling occurs. Also in pyrophyllite and talc there are no cations to solvate

TABLE 6
Surface density of charge (σ) in Delamica ($1 - 0{\cdot}2\mu$) as a function of time of heating at 300°C with $LiNO_3$ melt (1 : 60 ratio)[38]

Reaction time (h)	σ (e.s.u. cm^{-2})	meq per 100 g
0	$9{\cdot}8 \times 10^4$	238
0·3	$9{\cdot}1 \times 10^4$	221
1·0	$9{\cdot}1 \times 10^4$	221
3·0	$8{\cdot}2 \times 10^4$	199
6·0	$6{\cdot}7 \times 10^4$	163
18·0	$6{\cdot}3 \times 10^4$	153
36·0	$6{\cdot}3 \times 10^4$	153

and so penetration by water decreases the free energy less than sheet separation increases it. In these two minerals the dispersion energy of interaction per unit area, and for atoms all of one kind, in two adjacent triple sheets will have the form[40]

$$E = -\frac{\pi N^2 A}{48}\left[\frac{1}{d_0^2} + \frac{1}{(d_0 + d)^2} - \frac{2}{(d_0 + d/2)^2}\right] \quad (1)$$

where d is the sheet thickness, $2d_0$ is the sheet separation, N is the number of the atoms per unit volume of the sheet and A is the dispersion energy constant for the interaction energy between a pair of isolated atoms. Since oxygens are the most numerous and polarizable of the atoms a reasonable estimate of E per unit area of a pair of sheets should follow by considering only these atoms. Because the dispersion energy is expressed in terms of the centre to centre distance between each interacting pair of atoms, for two sheets with direct contact between pairs of the hexagonal oxygen layers $2d_0$ is the van der Waals diameter of an oxygen atom, which is about 2·8Å. Accordingly d is the c repeat distance less 2·8Å. Thus for pyrophyllite, $d = 9{\cdot}3 - 2{\cdot}8 = 6{\cdot}5$Å. Where there are interlayer cations keeping the three layer sheets apart $2d_0$ is appropriately increased since the repeat distance is increased, unless the cations are small enough to be recessed into the 6-rings of the hexagonal oxygen layers. Equation 1 is an approximation in which integration through the volume of each sheet replaces pair summation. As discussed in Chapter 4, § 6 the value to be taken for the dispersion energy constant A also varies according as the London, Slater–Kirkwood or Kirkwood–Muller approximations are used.

TABLE 7

The relation between swelling in distilled water and charge density of a number of silicates having the triple layers found in mica[39]

Silicate	Interlayer surface per unit charge (Å^2)	Extent of swelling in Å	
		Na^+-form	Ca^{2+}-form
Margarite	12	0	0
Muscovite	24	1·9	2·8
Biotite	24	1·9	2·8
Lepidolite	24	1·9	2·8
Zinnwaldite	24	1·9	2·8
Seladonite	27	2·4	2·8
Glauconite	31	3·8	2·8
Dioctahedral illite	32	4·2	2·8
Trioctahedral illite	36	5·1	4·3
Vermiculite	36	5·1	4·3
Vermiculite	36	5·0	4·2
Vermiculite	37	5·1	4·2
Batavite	36	5·1	4·3
Beidellite I	41	5·4	4·9
Saponite	42	4·9	4·9
Nontronite	46	∞	9·2
Beidellite II	57	∞	9·2
Montmorillonite	60	∞	9·2
Montmorillonite	75	∞	9·6
Hectorite	100	∞	10·6
Pyrophyllite	∞	0	0
Talc	∞	0	0

A simplified estimate[41] of electrostatic cohesion energy per m^2 of interlamellar area with cation density σ, cationic charge q and an effective permittivity, ϵ, of the medium between the charge gives

$$E = \frac{\sigma q}{16\pi\epsilon}\left[\frac{1}{d_1} + \frac{1}{d_2}\right] \qquad (2)$$

where d_1 and d_2 are respectively the distances between the plane containing the cations and the planes containing centres of negative charge in each of two adjacent sheets. Thus the coulomb energy should increase approximately as the charge density. In a series of montmorillonites the coulombic energy according to eqn 2 was estimated to be from 162 J g^{-1} to 124 J g^{-1}. Because of the uncertainties in such quantities as ϵ, as well as assumptions in the derivation of eqn 2 such estimates are very approximate. However, for micas with much higher charge densities (Table 7) the electrostatic component of the binding energy between sheets should be several fold larger than in smectites, whereas eqn 1 indicates a dispersion energy component which will not be very different in pyrophyllite, talc, micas and anhydrous smectites.

Equations 1 and 2 provide a basis for explaining the difficulty of intercalating guests such as water between mica sheets and the comparative ease with which this may be done with water-free smectites and vermiculites. However a third factor must operate which makes intercalation difficult also in talc and pyrophyllite which carry no sheet charge. This factor may be the kinetic, non-equilibrium one of nucleation. The formation of crystal swollen by intercalation involves initiation of a new phase at peripheries of crystals of a parent phase, with substantial positive free energy terms arising from strain and interfacial surface tension. The larger these are the greater the induction time for nucleation, and if they are very large nucleation may not occur in any practicable period. Nucleation is a function of the nature of the guest, its concentration, the nature, size and degree of perfection of the host crystals and the temperature. The difficulty of nucleating the intercalation process is also seen with the kandites (§ 3). These, like talc and pyrophyllite, carry no sheet charge. However they can intercalate appropriate guests, and the kinetic curves of imbibition of guest against time are sigmoid, with induction periods which demonstrate the role of nucleation.

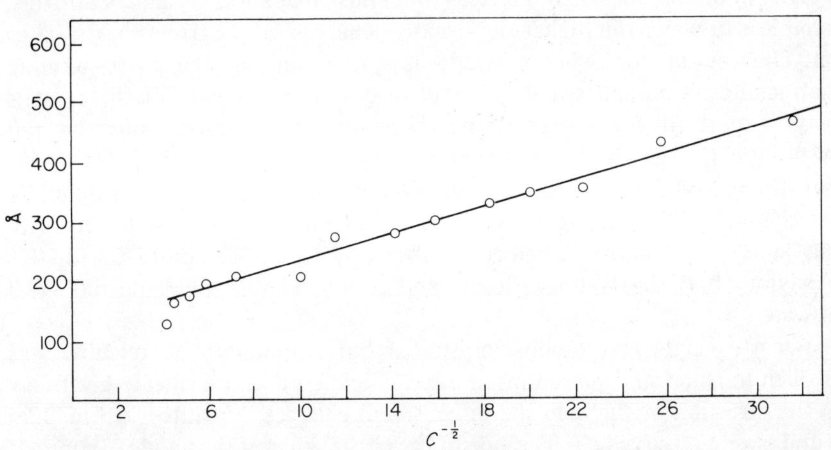

FIG. 11. The interlayer spacing in Å of butylammonium-vermiculite as a linear function of $C^{-1/2}$ where C is the concentration of butylammonium chloride in the external solution.[42] C is in mol dm^{-3}.

The swellings recorded in Table 7 are those obtained after immersion in distilled water. If salt solutions replace water the extent of swelling is reduced. The manner in which the concentration of *n*-butylammonium chloride modified the swelling of *n*-butylammonium-vermiculite is shown in Fig. 11.[42] As dilution increases, the sheet separation increases, and over a considerable range in concentration, C, this separation varies linearly with $C^{-1/2}$.[42,43] Similar results were obtained with Li-vermiculite in LiCl solutions.[44] In the linear region of Fig. 11 the swelling was reversible if water-swollen crystals were allowed to dry out again at room temperature.

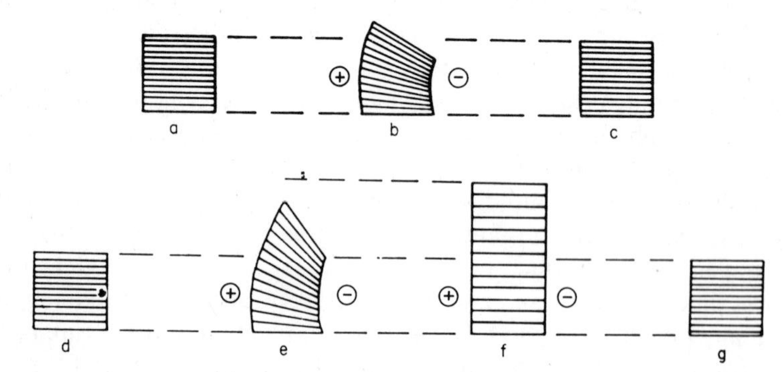

FIG. 12. Dimensional changes in water-swollen butylammonium-vermiculite placed in an electric field (schematic).42 a to c, in distilled water or dilute butylammonium-chloride. a, before, b, during and c, after application of field. d to g, in dilute aqueous butylamine. d, before field applied; e, in field for ≈ 2 min; f, in field for ≈ 10 min; and g, after removal of field.

Butylammonium forms of the clay minerals, like the Na- and Ca-forms, swell less in water the higher the negative charges on the triple sheets. Thus butylammonium phlogopite swells less than any of the corresponding vermiculites, although swelling is still in evidence. The size of the crystals plays a part since fine-grained butylammonium montmorillonite did not swell even though the layer charge was less than that of vermiculite. Also powdering vermiculite and mica to give particles less than 50 μ diameter largely inhibited swelling irrespective of whether they were treated with butylammonium chloride before or after powdering. In contrast with this behaviour both Li-montmorillonite and powdered Li-vermiculite did swell in water.

When a single macroscopic crystal of butylammonium vermiculite was swollen in water and placed in an electric field so that the silicate layers lay in the direction of the field the dimensional changes illustrated in Fig. 12a, b and c were observed.[42] The side of the crystal nearest the anode expanded; that nearest the cathode shrank. When the field was removed the crystal recovered almost completely (Fig. 12c). The same behaviour was observed when dilute butylammonium chloride replaced water. With dilute aqueous butylamine, first the anode side swelled and then the cathode side to give parallel planes once more, the sequence being shown in Fig. 12 d, e and f. On removing the field the crystal reverted slowly to its pristine state. Although the behaviour was less reproducible Li-vermiculite behaved in distilled water

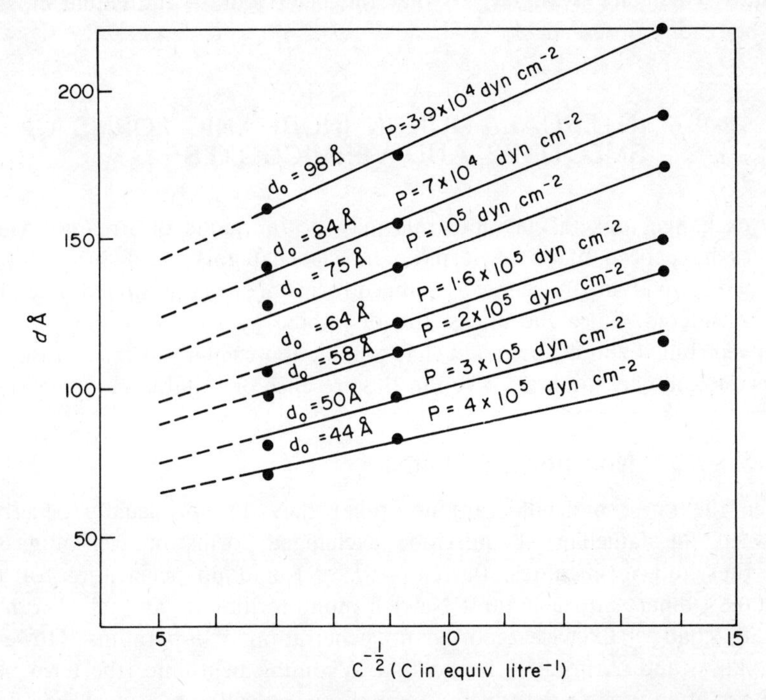

FIG. 13. Swelling of butylammonium-vermiculite in aqueous butylammonium-chloride as a linear function of $C^{-1/2}$ at each of a series of hydrostatic pressures.[43] C denotes the concentration of the aqueous solution in mol dm^{-3}.

or dilute lithium chloride or hydroxide in a similar manner to butylammonium-vermiculite.

Another variable which modifies the interlayer distance is the pressure. This is illustrated in Fig. 13 where $d(001)$ is plotted against $C^{-1/2}$ at each of a series of presures, for butylammonium vermiculite in butylammonium chloride solutions. Both the slopes of the linear plots and their intercepts, d_0, on the $d(001)$ axis are very sensitive to the pressure, and the behaviour was fully reversible. The sheets are thought to be kept apart on account of the interaction of diffuse double layers and this repulsion is to some extent overcome by the applied pressure.

The foregoing observations reval that swelling of layer structures such as the vermiculites in water and in electrolyte solutions is sensitive to the nature of the interlamellar exchange ions, external electrolyte concentration, electric fields, pressure and the density of negative charge carried by the triple sheets. Sometimes swelling is affected by the extent of grinding and hence by possible physical and chemical damage to or the size of crystals. An advantage in using vermiculites for swelling studies lies in the large crystals which are available, so that the behaviour of individual crystals can be readily observed.

6. INTERCALATION IN INORGANIC FORMS OF SMECTITES AND VERMICULITES

The exchange ions in clay minerals may be inorganic or organic. Many studies have been made of sorption in each category, and in this part complexes involving inorganic exchanged forms of smectites and vermiculites are considered. Since the micas do not imbibe guest species they are not involved, but it should be noted that organic ion-exchanged micas can, like other organo-clay minerals, swell in the presence of suitable guests (§ 8).

6.1 Non-polar Sorbates

When the guest molecules are non-polar they do not usually penetrate between the lamellae of inorganic exchanged forms of well-outgassed smectites and vermiculites. Barrer *et al.*[23] found no penetration of the pentane isomers into a natural Na-rich montmorillonite, and MacEwan[45] and Barshad[46] likewise reported no penetration of *n*-paraffins. However Eltantawy and Arnold[47] found that Wyoming bentonite (the term bentonite is often used to describe an impure montmorillonite), which had been merely air-dried rather than outgassed, swelled in *n*-hexane vapour and after three hours at room temperature gave a monolayer complex with *d*(001) equal to 14·01Å. More prolonged exposure gave a system with mixed mono- and bilayers of interlamellar hexane, with *d*(001) of 17·96Å and 14·03Å. However infra-red absorption showed water to be present in addition to the hydrocarbon, and it could be that water is needed to initiate penetration by the hexane. Ca-montmorillonite oven-dried at 200°C did not readily intercalate *n*-hexane, although after two weeks the monolayer complex was detected. The same oven-dried clay also very slowly intercalated *n*-dodecane. It seems therefore that for rigorously outgassed Na-, Ca- and Mg-smectites and water-free *n*-paraffins these hydrocarbons would intercalate only very slowly if at all.

Such inorganic forms of montmorillonite also do not intercalate permanent gases, but Barrer and Reay[48] examined the NH_4- and and Cs-forms to see whether these ions, by virtue of their greater size, keep the *d*(001) spacing large enough to allow small molecules to penetrate. The isotherms on the well-outgassed sorbents were of Type II in Brunauer's classification,[26] and gave hysteresis loops which closed at relative pressures around 0·25 and 0·5 for CH_4 and O_2 respectively, but which persisted to the lowest pressures

TABLE 8

Equivalent monolayer areas of some ion-exchanged montmorillonites[48]

Sorbate	T/K	Areas (m^2g^{-1}) Na-rich	NH_4	Cs	NH_3CH_3
N_2	90	21·7	40·6	141	198
O_2	90	21·9	37·0	122	166
Ar	90	—	34·6	103	142
CH_4	90	—	45·1	—	140
C_6H_6	323	—	38·0	143	~180

for benzene and water. In Table 8 a comparison is made of the monolayer equivalent areas of Na-rich, NH_4-, Cs- and NH_3CH_3-forms of montmorillonite. In the outgassed NH_3CH_3-form *d*(001) is 12·0Å as compared with 9·6Å for Na-montmorillonite, and it is known that in this organo-clay all the sorbates of Table 8 penetrate between the silicate layers (§§ 9 and 10). The table shows a progressive increase in sorption capacity in the order $Na < NH_4 < Cs < NH_3CH_3$. The hysteresis between sorption and desorption branches of the isotherms of benzene, persisting to the lowest pressure, indicates some interlamellar uptake in NH_4- and Cs-montmorillonites, and the comparison of monolayer equivalent areas of benzene and the other sorbates suggests limited penetration by O_2, N_2, Ar and CH_4. However with benzene the X-ray powder pattern of the sorption complex showed no substantial change in the diffuse first order *d*(001) line for either NH_4- or Cs-form so that intercalation of benzene may have occurred in some layers but certainly not in all.

Even allowing for a small amount of interlamellar penetration by permanent gases the sigmoid shapes of all isotherms indicates that much of the uptake was upon external surfaces. Much of the increase in surface could be accounted for if there were on average fewer lamellae per crystal in NH_4- and Cs- than in Na-montmorillonite. If during exchange a limited number of single lamellae were to peel off the parent crystals, enhanced areas would result. Mooney *et al.*[49] also found an increased monolayer equivalent area of Cs-montmorillonite as compared with the parent Na-form.

The conclusion from these and other experiments is that interlamellar penetration of non-polar molecules into water-free smectites containing small ions like Na^+ and Ca^{2+} is difficult to achieve if it occurs at all. If the cations are larger (e.g. Cs^+) some intercalation may take place, although it is unlikely to be complete. Water in small amounts may however promote penetration of *n*-paraffins.

6.2 Polar Sorbates

When the sorbate molecules are polar interlamellar uptake by smectites containing exchange ions such as Na^+, Ca^{2+} or Mg^{2+} usually takes place readily to give a great variety of complexes in which layers of guest molecules of different thicknesses may be present. In this connection attention has already been drawn to the particular case of water in vermiculites and smectites (§ 5).

6.2.1 *Isotherm Contours, Hysteresis and Sorption Kinetics*

Figure 14[33] shows isotherm contours obtained on a Na-rich natural montmorillonite for several polar guest molecules. Wide hysteresis loops, persisting to the lowest pressures, differentiate adsorption and desorption branches. With each sorbate, except *t*-butyl alcohol, a threshold pressure on the adsorption branch precedes the onset of copious uptake. This is associated with an increase in the *d*(001) spacings, and produces a single step in isotherms of ammonia and an ill-defined double step in isotherms of water. The double step is also clearly seen in the isotherm of water in Llano vermiculite, obtained by van Olphen.[31,50] *t*-Butyl alcohol, which has a bulky, globular molecule, gives a much reduced uptake, but still gives hysteresis persisting down to the lowest pressures.

Such hysteresis contrasts with that observed in many inert porous sorbents for which the hysteresis loops, if they occur, close at intermediate pressures. It appears as though, once the lattice of the clay mineral has expanded, some or all of it remains expanded until almost all the sorbate has drained out. This behaviour may be formulated in thermodynamic terms by regarding the sorbent (1) and sorbate (2) as a two-component system.[51] At constant temperature and external pressure, the Gibbs–Duhem relation is

$$d\mu_1 = -(n_2/n_1)d\mu_2 \tag{3}$$

where μ_1, μ_2 and n_1, n_2 are the chemical potentials and the moles of the respective components. For the sorbate in equilibrium with its vapour at pressure P_2, $d\mu_2 = RT\,d\ln P_2$ and so

$$\mu_1 = \mu_1^0 - RT\int_0^{P_2}(n_2/n_1)\frac{dP}{P} \tag{4}$$

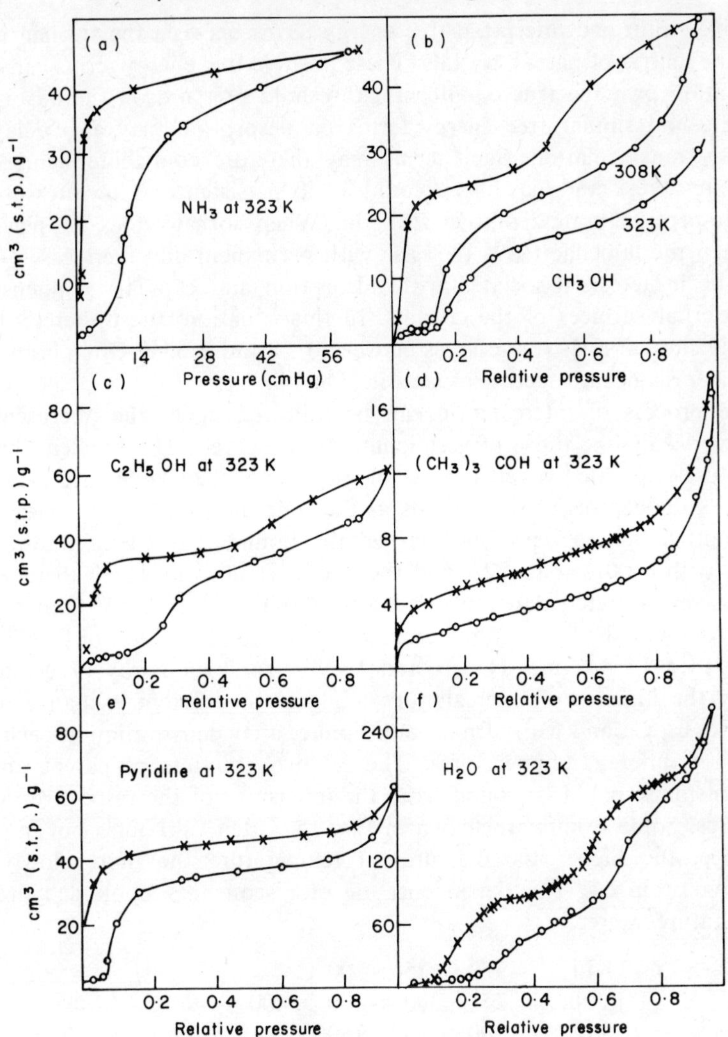

FIG. 14. Isotherms for some polar molecules in a natural Na-rich montmorillonite at 308 or 323 K.[33] ⊙, adsorption and ×, desorption points.

where μ_1^0 is the chemical potential of the sorbent free of sorbate. Low pressure hysteresis occurs if, at some point on the adsorption branch, the structure of the solid changes to a new configuration (the expanded form of the clay mineral) which then persists to a low sorbate content on the desorption branch and has a different value of μ_1^0 from that for the unexpanded mineral.

The threshold pressures of Fig. 14 suggest that the onset of swelling involves a nucleation process which, as noted earlier, introduces extra, positive strain and interfacial free energy terms between the swollen nuclei and the matrix of parent crystals. These positive free energy terms can delay nucleation past the true equilibrium threshold pressure on the adsorption branch, and similar free energy terms on desorption may also delay the reverse transformation. Such terms may therefore contribute significantly to the hysteresis and may also account for its persistence to low pressures on the desorption branch of the isotherm. When sorbate does not penetrate between the lamellae (as is the case with permanent and inert gases) there may be hysteresis associated with adsorption and capillary condensation on external surfaces of the crystals. In this situation, the hysteresis loops usually close at relative pressures between 0·25 and 0·5.[33] Often both types of hysteresis occur together as seen in Fig. 14b.

The process of intercalation can be followed under the microscope if larger crystals like those of vermiculite are observed. The swollen phase is seen to be initiated at crystal peripheries. A swollen zone then advances towards the centres of the crystals as the time increases.[52] In reflected or transmitted light a dark line marked the boundary between the swollen phase with $d(001) = 13{\cdot}82$Å and the parent crystals in which there was a monolayer of intercalated water with $d(001) = 11{\cdot}59$Å. A fainter dark band followed the first and represented the further expansion of the 13·82Å phase to a 14·4Å one. Hazart and Wey[53] used an X-ray procedure to follow the kinetics of intercalation of glycol in different exchange forms (Na, K, Li, Ca and Mg) of montmorillonite, after dehydration at each of a series of different temperatures. The relative amounts of parent and of swollen phase could be found from the intensities of the respective $d(001)$ spacings. Some results are shown in Fig. 15. Often, although not in every instance, the higher the dehydration temperature the more slowly the swelling occurred. The dependence of rate sequences upon dehydration temperatures were:

Li	$150 > 25 > 300$°C
Na	$25 > 300 > 150 > 400 > 490$°C
K	$300 > 25 > 150$°C
Mg	$25 > 300 > 150 > 400$°C
Ca	$25 > 150 > 300$°C.

The curves of uptake against time tended to be sigmoid in shape (Fig. 15). This, as with the threshold pressures in Fig. 14, indicates once more that initially nuclei of swollen phase must form and then grow at the expense of the parent phase. On the other hand the desorption rate curves of glycol from Ca-, Mg-, K- and Na-forms of montmorillonite were not sigmoid in shape,[54] suggesting that when the glycol drained out of the crystals nucleation of the unswollen form of the crystals was not involved in the observable pressure range.

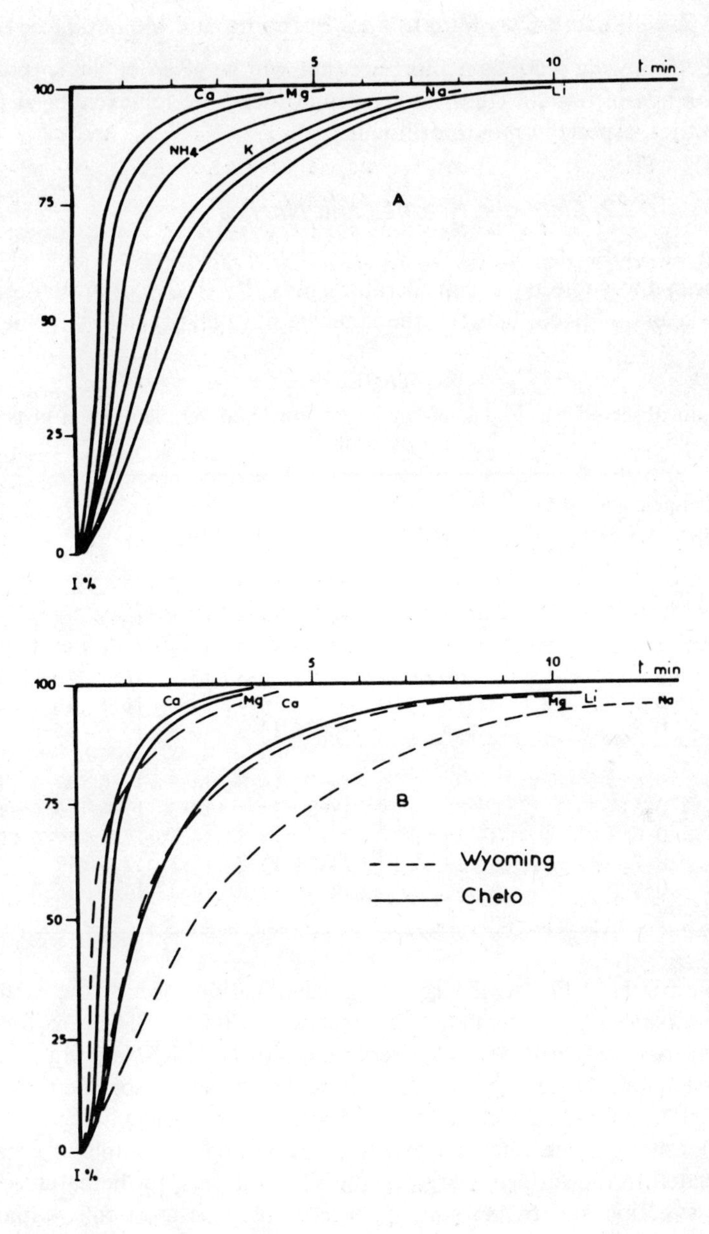

FIG. 15. The growth in intensity with time of the X-ray powder line corresponding with the formation of the glycol-montmorillonite complex for samples cation-exchanged as shown in the figure. A, montmorillonite of Camp Bertaux; B, montmorillonites of Wyoming and of Cheto.(53)

In the following sections a short account will be given of the sorption of typical kinds of organic guest molecules in inorganic ion-exchanged forms of smectites, especially montmorillonite.

6.2.2 *Alcohols, Ketones and Nitriles*

Among alcohols, the complexes formed by ethylene glycol and glycerol have been the subjects of considerable study.[54, 55] Glycol smectites with double layers of glycol between the lamellae have *d*(001) ≈17Å, while with

TABLE 9

Maximum observed basal spacings in Å, of powdered vermiculites in glycol and glycerol[55]

Exchange capacity in meq per $O_{10}(OH)_2$ unit of structure	Guest	Li	Na	Mg	Ca	Sr	Ba
0·6	Glycol	16·1	16·3	16·3	16·2	16·1	16·2
0·6		16·1	16·2	15·5	16·1	16·1	16·2
0·65		16·2	16·3	15·2	16·1	16·2	16·2
0·7		16·1	16·1	14·3	16·1	16·1	16·2
0·8		16·0	14·8	14·3	15·6	15·6	16·0
0·6	Glycerol	14·2	14·8	14·3	17·6	17·6	17·6
0·6		14·2	14·8	14·3	17·6	17·6	17·6
0·65		14·3	14·8	14·3	17·6	17·6	17·6
0·7		14·3	14·8	14·3	17·6	17·6	17·6
0·8		14·3	14·8	14·3	14·3	15·8	14·3

glycerol, *d*(001) ≈ 17·7Å. If large monovalent cations occupy the exchange sites monolayers of intercalated glycerol with *d*(001) ≈ 14Å are formed. Monolayers of glycerol were also reported with H-, Li-, Na-, Mg-, Ca- or Ba-vermiculites, but more expanded complexes were also often formed. The *d*(001) spacings are recorded in Table 9 for vermiculite powders of <300 mesh size. The rate and extent of expansion in complex formation were related to the surface charge on the silicate layers, to the nature of the interlayer cation and to the size of the crystals. The most fully expanded glycol-vermiculites have slightly smaller *d*(001) spacings than their montmorillonite counterparts, an effect which may also be due to the increased layer charge on the vermiculites. However, the most fully expanded glycerol-vermiculites have *d*(001) very close indeed to that found in smectites. Basal spacings near to those for a monolayer of intercalate are also apparent for certain of the ions and charge densities on the vermiculite sheets, both for glycerol and glycol, and also several which lie between the unimolecular and bimolecular extremes. The latter may represent randomly interstratified mixtures of the two extremes. The glycol and glycerol complexes were prepared from vermiculite powders already containing unimolecular or bimolecular layers of intercalated water.

Complexes in which the organic molecules lie between the clay lamellae to form layers one and two molecules thick are found for polar species of

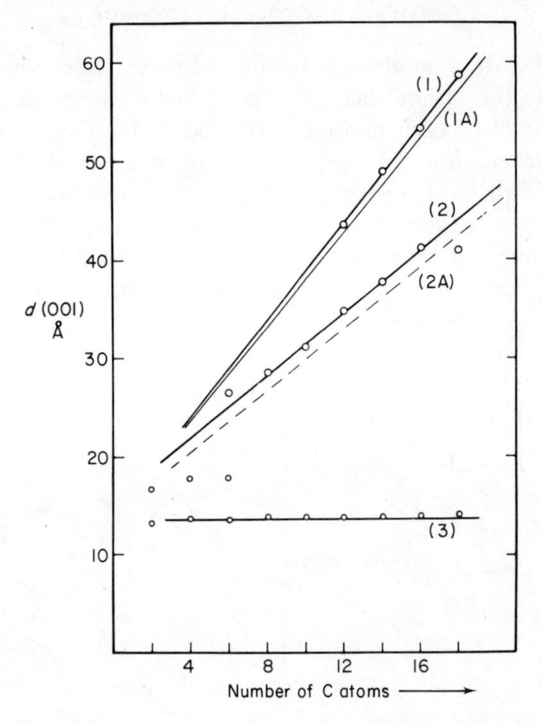

FIG. 16. Observed basal spacings (open circles) and calculated spacings (lines) for a series of primary monohydric *n*-alcohols in Ca-montmorillonite.[56]

considerable chemical variety and molecular weight ranges. In some circumstances the layers of organic molecules may adopt different orientations. Complexes of primary monohydric alcohols of even carbon number with Ca-montmorillonite illustrate these situations[56] (Fig. 16). In one kind of complex—the single layer series (line 3 in Fig. 16)—the values of $d(001)$ lay between 13·2Å (for C_2) and 14·1Å (for C_{18}). For the C_2, C_4 and C_6 alcohols two-layer complexes with $d(001)$ = 16·7, 17·8 and 17·9Å also formed, the points being shown on the figure. Two series of long-spacing complexes are also seen, in which $d(001)$ is a linear function of the chain length. That of lower slope was formed from C_6 onwards with the alcohols above their melting points in contact with the mineral. The second long-spacing series was formed with the C_{12} to C_{18} alcohols in contact with the mineral at room temperature.

In the single and double layer series, the chains lie parallel with basal surfaces. For the most expanded long-spacing series the lines 1 and 1A in the figure represent the two expressions

$$d(001) = 19{\cdot}16 + (n - 2)2{\cdot}48\text{Å}$$
$$d(001) = 18{\cdot}62 + (n - 2)2{\cdot}40\text{Å}$$

where n is the carbon number. In the first of these, the chain axes are at an angle of 77° to the silicate sheets, the layer being two molecules thick with the terminal -OH of each molecule attached to the oxygen surface of the clay.[56] The second line, 1A, is calculated for an angle of chain orientation of 70·5°.[57] With the long-spacing complexes formed with alcohols above their melting points (line 2) it was suggested that each chain began with a steep orientation, but then by a rotation around the bound OC_1 or C_2C_3 the rest of the chain takes another opposing orientation of 38·9°. For double layers of alcohol molecules

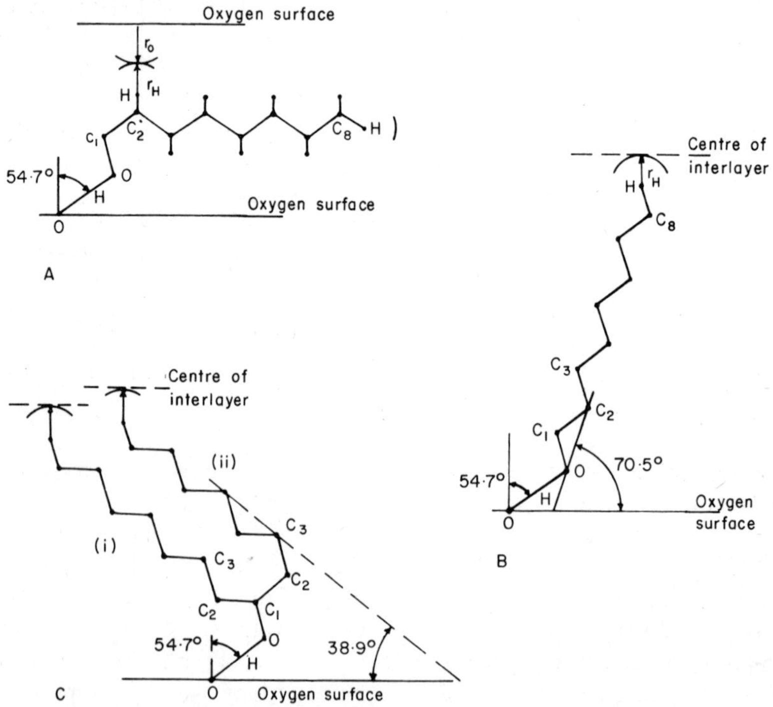

FIG. 17. Some chain orientations proposed for n-alcohols relative to the silicate sheets of the clay mineral.[56]

$$d(001) = 21 \cdot 76 + (n - 4)1 \cdot 58$$

gives the full line 2 in Fig. 16. The line 2A represents the $d(001)$ spacings expected for rotation around the OC_1 bond, which are slightly too small. Several of these chain orientations are illustrated in Fig. 17.

One and two layer complexes having the organic molecules parallel with the basal surfaces were reported for the polyethylene glycol ester of oleic acid,

$$\underset{\displaystyle \mathrm{OH}}{\overset{|}{\mathrm{CH_2}}}\text{–}\mathrm{CH_2(OCH_2\text{-}CH_2)_xO\text{–}CH_2\text{–}}\underset{\displaystyle \mathrm{O}}{\overset{\|}{\mathrm{C}}}\text{–}\mathrm{(CH_2)_7CH = CH(CH_2)_7CH_3}$$

in Na-, Ca- and Mg-montmorillonites.[58] The complexes were prepared from mixtures of water, organic compound and clay mineral. The thickness of the organic monolayer was 4·2Å and of the bilayer 7·7Å. It thus appears that certain high molecular weight materials can readily intercalate from aqueous solutions. Examples of other mono- and bilayer complexes formed from aqueous solutions are given in Table 10,[59] for which the $d(001)$ spacings refer to samples first throughly dried. The chains lie parallel with the surface as in Fig. 17 A, and the symbols "||" and "⊥" in the final column refer to the two possible orientations of the zig-zag of the carbon chain.

TABLE 10

Some one and two layer complexes of organic guests in Ca-montmorillonite[56]

Guest	$d(001)$/Å one layer	$d(001)$/Å two layer	One layer spacing −9·50Å	Probable orientation of guest to basal surfaces
Acetylacetone	$13 \cdot 0_0$	$17 \cdot 1$	$3 \cdot 5_0$	||
α-methoxyacetylacetone	$13 \cdot 0_0$	$16 \cdot 9_5$	$3 \cdot 5_0$	||
Acetoaceticethylester	$13 \cdot 1_0$	$17 \cdot 1_5$	$3 \cdot 6_0$	||
Nonanetrione-2 : 5 : 8	$13 \cdot 1_0$	$17 \cdot 3_5$	$3 \cdot 6_0$	||
Hexanedione-2 : 5	$13 \cdot 0_5$	$16 \cdot 8_5$	$3 \cdot 5_5$	||
β : β′-oxydipropionitrile	$13 \cdot 1_5$	$15 \cdot 7_5$	$3 \cdot 6_5$	||
β-Ethoxypropionitrile	$13 \cdot 2_5$	—	$3 \cdot 7_5$	|| (?)
Bis-(2-ethoxyethyl)-ether	$13 \cdot 4_0$	$16 \cdot 7_5$	$3 \cdot 9_0$	⊥
Bis-(2-methoxyethyl)-ether	$13 \cdot 2_5$	$17 \cdot 0_0$	$3 \cdot 7_5$	⊥
Ethyleneglycoldiglycid ether	$13 \cdot 4_0$	$17 \cdot 6_0$	$3 \cdot 9_0$	—
Triethyleneglycol	$13 \cdot 3_0$	$17 \cdot 3_5$	$3 \cdot 8_0$	⊥
Diethyleneglycol	$13 \cdot 3_0$	$15 \cdot 7_5$	$3 \cdot 8_0$	⊥
Triethyleneglycoldiacetate	$13 \cdot 3_5$	$16 \cdot 5_0$	$3 \cdot 8_5$	⊥
Diethyleneglycoldiacetate	$13 \cdot 1_5$	$16 \cdot 5_5$	$3 \cdot 6_5$	|| (?)
Hexanediol-1 : 6	$13 \cdot 5_5$	$17 \cdot 4_5$	$4 \cdot 0_5$	⊥
Pentanediol-1 : 5	$13 \cdot 6_5$	$17 \cdot 4_5$	$4 \cdot 1_5$	⊥
2 : 4-Hexadiynediol-1 : 6	$13 \cdot 0_5$	$16 \cdot 1_0$	$3 \cdot 5_5$	||

The "||" sign signifies that the plane of the zig-zag is parallel with the silicate sheets and "⊥" that this plane is perpendicular to the sheets. The zig-zag of most straight chain aliphatic compounds was reported to adopt the ⊥ configuration in interlamellar complexes, but with a contraction of the contact distances of about 1·0Å for two clay-organic contacts.[60] The ring compounds tetrahydropyran, 1,4-dioxan[61] and pyridine[62] (see Fig. 18)

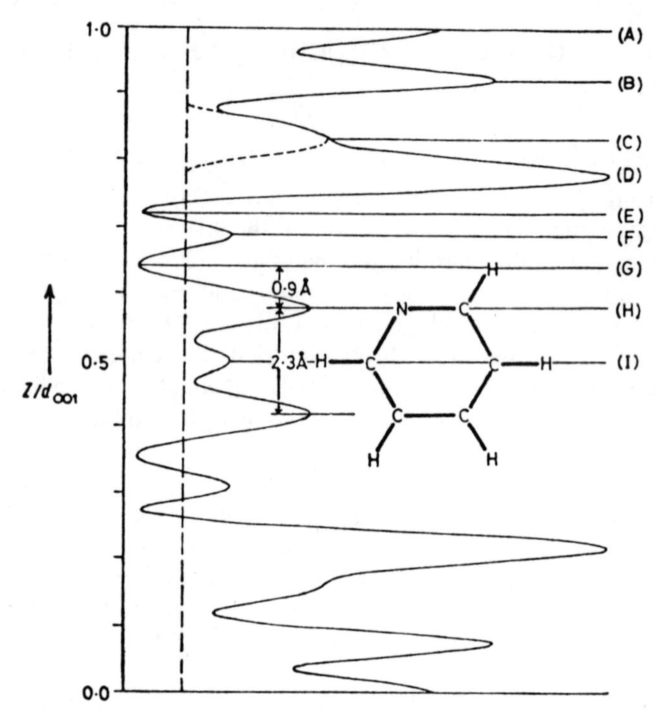

FIG. 18. One-dimensional projection of neutron scattering density normal to (001) for pyridine intercalated in Na-montmorillonite.[62]

likewise adopt configurations in which the plane of the ring is steeply oriented to the silicate sheets. The $d(001)$ spacings were respectively for Na- and Sr-montmorillonite 14·99 and 14·81Å (tetrahydropyran); 14·94 and 15·00Å (1,4-dioxan); and 14·8Å (pyridine in the Na-form). In the presence of traces of moisture pyridine also forms a regular complex with $d(001) =$ 23·3Å.

The simplest nitrile is acetonitrile. It forms complexes in which three $d(001)$ spacings are observed, according to the kind of interlayer cation and the amount of co-sorbed water.[63] These spacings were 13·0–13·2, 17·8–17·9 and 16·3Å, the latter being obtained only with NH_4-montmorillonite. The 17·8–17·9Å complexes formed most readily when the exchange ions possessed high polarizing power (Ca^{2+}, Co^{2+}, Ni^{2+} and Mg^{2+}). Where both water and organic molecules are co-sorbed the interlamellar layers are effectively two-dimensional solutions.

6.2.3 Organic Bases

Heats of immersion of Na-montmorillonite, first dried in air at 200°C and then outgassed at the same temperature, have been measured for various liquid amines and give the integral heats of interaction in the last column of

TABLE 11

Heats of interaction of liquid amines with Na-montmorillonite

Guest	Dipole moment (Debye)	K_b	$d(001)$ (in Å)	Amount sorbed (meq g $^{-1}$ clay)	ΔH (Kcal mol^{-1})
n-propylamine	1·17	$3{\cdot}8 \times 10^{-4}$	13·4	0·88	−14·5
n-butylamine	0·92	$4{\cdot}4 \times 10^{-4}$	13·4	0·90	−16·1
n-hexylamine	1·32	$4{\cdot}4 \times 10^{-4}$	13·8	0·89	−17·7
Diethylamine	0·92	$9{\cdot}5 \times 10^{-4}$	13·3	0·52	−16·2
Triethylamine	—	$5{\cdot}9 \times 10^{-4}$	14·5	0·46	−14·0
Benzylamine	—	$0{\cdot}23 \times 10^{-4}$	16·7	1·15	−11·6
Aniline	1·48	$4{\cdot}4 \times 10^{-10}$	15·1	0·42	− 6·4
o-Toluidine	—	$2{\cdot}5 \times 10^{-10}$	16·7	0·48	− 5·4

Table 11 when the amount of amine taken up is that in the second to last column.[64] The heats correlate better with the magnitudes of the base dissociation constants K_b than with the dipole moments (columns 3 and 2). For the three aliphatic primary *n*-amines with similar K_b the heats also increase with chain length and so indicate significant contributions from dispersion energy. The $d(001)$ values are consistent with monolayer complexes. The heats of immersion in *n*-amines were about four times greater than for immersion in *n*-propyl and *n*-butyl alcohols and l-nitropropane. In the immersion process no Na^+ was displaced into the pure liquid amines, but if the sorption of amine occurred from aqueous solutions some displacement was observed. The uptakes of the three primary amines from their liquids were nearly equal to the ion exchange capacity of the clay mineral, suggesting an approximate 1 : 1 association between amine and interlamellar Na^+.

The displacement of Na^+ into the aqueous solutions of the amines, referred to above, indicates ion exchange. In general the formation of cations from amines may occur in several ways.[65] If the bases, *A*, are in donor solvents such as water there is some proton transfer in the aqueous phase:

$$A + HOH \rightleftarrows AH^+ + OH^-. \quad (5)$$

TABLE 12

Colours of various montmorillonite-organic complexes[67]

Guest	Colours developed Wet	Dry	Aged	Comments
Benzidine	Blue	Yellow	Yellow form stable	Yellow form easily produced by dehydration, easily reversed
NNN′N′-tetramethylbenzidine	Green	Brown	Green form stable	Brown form only produced at 80°C and 10^{-3} Torr
Triphenylamine	Green	Purple	Purple form stable	Not easily reversed; density of colour dependent on nature of montmorillonite
p-Phenylenediamine	Brown	Purple	Purple form stable	Easily reversed
NN′-Dimethyl-*p*-phenylenediamine	Green	Purple	Purple form stable	Easily reversed
NNN′N′-Tetramethyl-*p*-phenylenediamine	Purple	Purple	Faded to pale blue after 1 month	Not easily reversed
trans-4,4′-Diaminostilbene dihydrochloride	Green	Brown	Brown form stable	
7,7,8,8-Tetracyanoquinomethane	Brown	Brown	Stable	
Tetracyanoethylene	No change	No change		Solution turns green

The AH^+ cations then compete with the ions initially in the mineral for the exchange sites. It may also happen in absence of any donor solvent that there are protonated species, DH^+, already at the surfaces which can transfer protons:

$$A + DH^+ \rightleftarrows AH^+ + D. \tag{6}$$

Thus in acid-washed smectites, protons may be supplied by the anhydrous hydrogen forms of the clays:

$$RNH_2 + \text{H-clay} \rightleftarrows RNH_3\text{-clay}. \tag{7}$$

The base may be ammonia, amine, diamine or other compounds such as pyridine. Reaction 7 has a parallel in the reactions between such bases and hydrogen zeolites (Chapter 7). Finally, at low water contents of the clay mineral residual water, polarized by the exchangeable cation, may transfer protons:

$$[M(H_2O)_m]^{n+} + RNH_2 \rightleftarrows RN^+H_3 + [M(OH)(H_2O)_{(m-1)}]^{(n-1)+} \tag{8}$$

This behaviour also has its counterpart among zeolites.

Smectites form coloured complexes with a number of bases. Pyridine sorbed into well outgassed natural montmorillonite gave a blue complex, the colour of which became gray after access to water vapour. The blue colour was restored when the crystals were again outgassed.[66] The benzidine blue reaction has been attributed to the formation of a positively charged

TABLE 13

Glycine and its peptides in montmorillonites

Organic Sorbate	Complex in	Amount Sorbed (mmol/100g)	*d*(001) Å
None	Moist Ca-form	0	19·7
Glycine		>10	22·4
Glycylglycine		17	22·5
Diglycylglycine		25	24·0
Triglycylglycine		7	21·8
Glycine	Dried H-form	47	12·7
Glycylglycine		53	13·00
Diglycylglycine		48	13·00
		85	16·33
Triglycylglycine		54	13·00
		78	16·00
		90	16·75
Glycylglycine	Dried Na-form	14	12·84
Diglycylglycine		20	12·75
Triglycylglycine		40	13·00

cation radical of benzidine,[68] and a Mossbauer spectroscopic study has identified structurally incorporated Fe^{3+} ions as electron acceptor sites involved in the formation of the blue complex.[69] The origin of the yellow colour is not certain.[70,71] The sorption of benzidine released some Na^+, but there was no 1 : 1 correspondence between Na^+ released and benzidine taken up. Ultraviolet and visible transmission and reflectance studies supported the view that a large part of the benzidine was present as the blue monovalent cation radical. When montmorillonite-amine complexes were immersed in a solution of benzidine in benzene the blue colour did not develop, but when the presorbed species were *n*-propyl- and butyl-alcohols or l-nitropropane the blue complex formed.[64]

The structure of the Na-montmorillonite-pyridine complex has been investigated by means of neutron diffraction.[62] The one-dimensional projection of nuclear scattering density normal to (001) is shown for the 14·8Å complex in Fig. 18 together with the proposed interpretation of the scattering peaks in terms of the positions of the atoms of pyridine. The guest is orientated with the long C–C axis parallel with the silicate sheets, and the plane of the ring perpendicular to these sheets, as previously noted. The assignment of the peaks in Fig. 18 was:

A	Al, Mg, Fe	*D*	O
B	O, OH	*E*	H of H_2O
C	Si	*F*	O of H_2O

The kinetics of interconversion of the 23·3Å and the 14·8Å complexes of pyridine with vermiculite were monitored by observations of changes with time of appropriate peak areas, the crystals after immersion in pyridine being maintained at 295 K and 59% relative humidity. After an initial delay period the curves could be represented by apparent first order rate equations.

6.2.4 *Long-chain alkylammonium Salts*

l-*n*-alkyl pyridinium bromides[72] and cetyltrimethyl ammonium bromide[73] in aqueous solutions are sorbed into Na- and Ca-montmorillonites. For the pyridinium salts with alkyl chains up to eight carbon atoms long the uptakes reached limits close to the exchange capacity of the clay mineral. With longer alkyl chains sorption of salt molecules occurred, so that the interlayer region contained some Br^- ions. The interlameller organic cations normally lie parallel with the silicate sheets, but cetylpyridinium and cetyltrimethylammonium ions were reported to be steeply oriented to these sheets, as shown in Fig. 19 for the 20·5Å complex of cetyltrimethylammonium. In general, sorption from aqueous solutions of the salts is partially by ion exchange and partly by intercalation of whole salt molecules.

6.2.5 *Amino-acids and Peptides*

Glycine and its peptides[74] as well as a variety of other amino-acids[75,76,77] are sorbed from aqueous solution by Na-, Ca- and H-montmorillonites. In the moist state of the treated mineral glycine and its peptides increased the sheet separation beyond the values found with water alone (Table 13). The dried complexes of the Na- and H-forms shrink to the values expected either for a monolayer or for a bilayer of the guest, with chains parallel to the

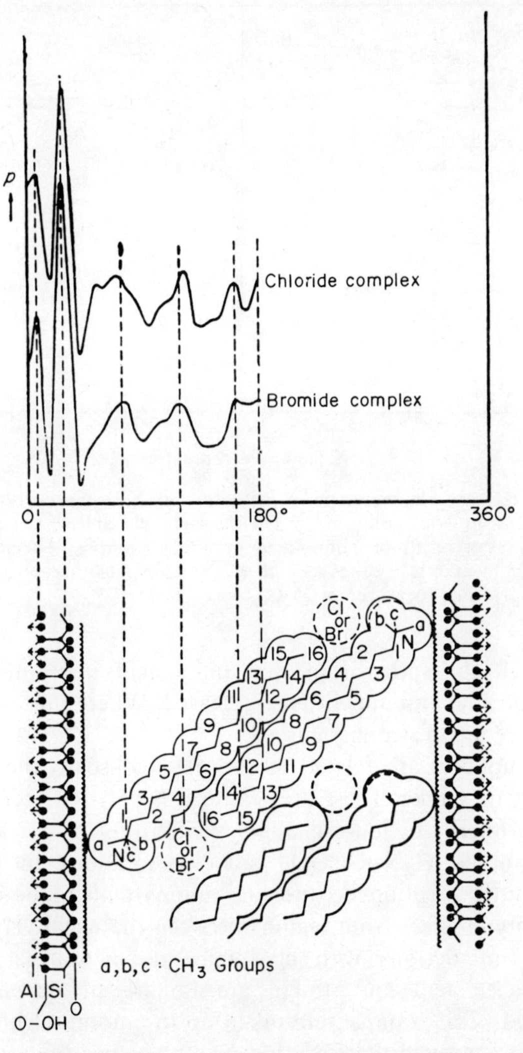

FIG. 19. One-dimensional Fourier syntheses for X-ray scattering of cetyltrimethylammonium bromide and chloride complexes in montmorillonite.[73]

silicate surfaces. Thus, the moist complexes comprise both organic molecule and water. Isotherms of solutions of glycine and its peptides at their natural pH's approximated to Henry's law in Ca-montmorillonite and Ca-illite. In H-montmorillonite, however, the isotherms were strongly curved and approached limiting values. This was also found for other amino acids (α- and β-alanine, leucine and serine) in the H-clay, although the total titratable hydrogen content exceeded the amounts sorbed. Figure 20[74]

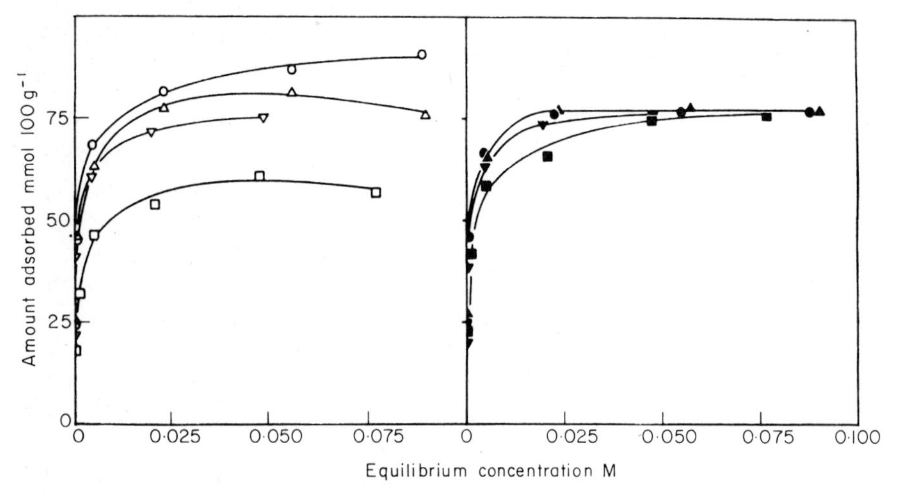

FIG. 20. On the left are shown sorption isotherms in Na-montmorillonite for amino-acids from aqueous solution.[74] On the right are shown the amounts of Na^+ liberated as a result of amino-acid uptake. △ and ▲, *l*-arginine; ○ and ●, histidine; □ and ■, lysine; ▽ and ▼, carnosine. Isotherms at 20°C except for that of carnosine measured at 25°C.

shows, on the left, the isotherms of some amino acids in Na-montmorillonite, and, on the right, the amounts of Na^+ released. Where this exchange occurs the isotherms are again strongly curved.

Amino-acid uptakes are related to the pH and so to the several forms which the acids may take. These are: cations, N^+H_3-R-COOH (acid conditions; uptake primarily by ion-exchange); zwitterions, N^+H_3-R-COO^- (near neutral); and anions, H_2N-R-COO^- (alkaline solution). In alkaline conditions there is little sorption because the anions tend to be excluded by a Donnan membrane effect from regions between the negatively charged clay lamellae. The interactions with clay minerals of purines, pyrimidines, nucleosides, nucleic acids and proteins are similarly pH dependent. Proteins and nucleic acids, for example, are taken up in amounts which decrease as pH increases. It is argued that sorption may be helped by a gain in entropy because one protein or nucleic acid molecule displaces many water molecules.

7. INTERLAYER COMPLEXES OF LIGANDS WITH INORGANIC CATIONS

As with the zeolites (Chapter 7) transition metal ions in clay minerals may complex with a variety of ligands. Desiccated Cu^{II}-montmorillonite formed complexes even with benzene vapour,[78] the original pale blue changing to dark red in a few minutes. In room conditions, the red phase turned green or gray but mild desiccation restored the red colour. Toluene, xylene and chlorobenzene also formed complexes, respectively bright green, dark purple and dark red to green. Benzene complexes were observed only with the Cu^{II}-forms of those clay minerals in which the source of negative charge on the three layer sheets was in the octahedral layer. The arenes are associated with Cu^{II} through their π-electrons. Two kinds of benzene complex exist,[79,80] respectively green (type I) and red (type II), and can be interconverted by adding or removing controlled amounts of water. In type I, the benzene ring is planar and retains its aromatic character; in type II the ring is greatly distorted. Type II complexes did not form with methyl-substituted benzenes. No Cu^{II}-arene complexes have been observed with dissolved Cu^{II} salts so that the interlayer environment contributes in a remarkable way to their stability. Arene complexes were not observed when the exchange ions in the clay were Na, Ca, Al, Cr^{III}, Fe^{II}, Fe^{III}, Ag^{I}, Mn^{II}, Ni^{II} or Co^{II}.

Cu^{II} in montmorillonites also forms complexes with pyridine,[81] diamines,[82, 83] ethylamine,[84] amino acids[85] and rubeanic acid,[86] the

TABLE 14

Complexes of organic ligands and Cu^{II}-montmorillonite

Ligand	Complex	Colour
Ethylenediamine	$[Cu\ en_2]^{2+}$	mauve
1 : 2-Propylenediamine	$[Cu\ pn_2]^{2+}$	mauve
1 : 3-Propylenediamine	$[Cu\ pn_2]^{2+}$	mauve
Pentamethylenediamine	$[Cu\ pen_2]^{2+}$	light blue
Diethylenetriamine	$[Cu\ dien]^{2+}$	blue
Triethylenetetramine	$[Cu\ trien]^{2+}$	mauve
Tetraethylenepentamine	$[Cu\ tepa]^{2+}$	blue
Ethylamine	—	blue
Pyridine	$[Cu\ (H_2O, py)_6]$	pale blue ($d(001) = 14{\cdot}7$Å)
Rubeanic acid[a]	$[Cu\ ra_2]^{2+}$	black ($d(001) = 12{\cdot}6$Å)

[a] Molecular formula $H_2N{-}C({=}S){-}C({=}S){-}NH_2$

colours and some typical complexes being indicated in Table 14. With pyridine, the pale blue 14·7Å complex formed rapidly. The infra-red absorptions at 1595 and 1607 cm^{-1} indicated two kinds of pyridine, the first hydrogen-bonded to water and the second coordinated to Cu^{II}. The 1595 band was lost on evacuation, but reappeared in air. The analysis of the pale blue complex indicated about four pyridines per Cu^{II}. On longer standing in pyridine Cu^{II}-montmorillonite expanded further, giving a broad diffraction peak corresponding with higher spacings interstratified with the 14·7Å spacing. With hot pyridine, irreversible changes occurred in the infra-red spectrum together with colour changes. Cu^{II}-montmorillonite films in cold pyridine became brown and then blue-black; in heated pyridine the films became intensely blue. All the films turned russet-brown when heated overnight in air at 100°C. None of these changes took place when the pale blue complex was heated in air. Pyridine molecules also form strong hydrogen bonds with pyridinium ions in the interlayer spaces of pyridinium montmorillonite, resulting in perturbation of vibrations involving the $\geqslant N^+H$ group, and progressively displace water molecules directly coordinated to interlayer Ca^{2+} and Mg^{2+}. These complexes have also the 14·7Å spacing of the pale blue Cu^{II}-pyridine complex, and are thought to be $M^{2+}(H_2O, Py)_6$.[81]

The Cu^{II}-rubeanic complex was formed, with $d(001) = 12{\cdot}6$Å, when Cu^{II}-montmorillonite was soaked in a solution of the acid in acetone.[86] The ratio of cupric ion to rubeanic acid was 1 : 2. The e.s.r. spectra showed that the plane of the complex was parallel to the interlamellar surfaces and the structure was deduced as

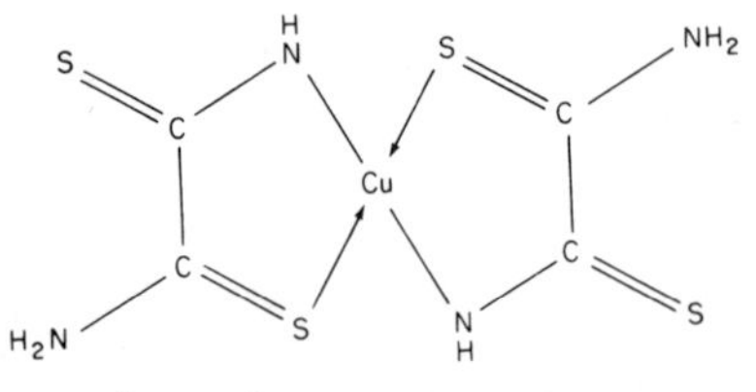

Naturally occurring amino acids can be divided into basic, acidic, neutral and sulphur-bearing, examples being respectively lysine, glutamic acid, glycine and methionine. Aqueous solutions of these four amino acids were shaken with Cu^{II}-montmorillonite, and with natural montmorillonite respectively. In the latter, glutamic acid, glycine and methionine were not significantly sorbed at their natural pH values. With the Cu^{II}-form of the mineral some Cu^{II} entered the solution even when no amino acid was sorbed. This contrasts with the behaviour found using amines with which complex formation occurred preferentially in the interlayers so that Cu^{II} was liberated into solution only when excess amine was present. Except with glutamic acid the presence of interlayer copper increased the sorption of the amino acid, but the distribution of copper and amino acid between the mineral and the solution varied among the different acids.[85]

Although most of these examples of interlayer complex formation have involved copper ions, this property is shared by other transition metal ions. Examples are $Co,^{III}$[87] which is stabilized with ethylenediamine as $[Co\,en_3]^{3+}$, and Ni^{II}, Cd^{II} and Zn^{II}.[82] As a final illustration of an organometallic interlayer complex one may refer to that of silver-urea.[88] This complex invades the interlayer region from solution with extremely high selectivity, and with complete displacement of uni- di- or trivalent ions (Na^+, Ca^{2+}, Al^{3-}) using only minimal excess of the complex. High selectivities of this kind provide means of collecting and concentrating small amounts of ions from solutions. The selectivity depends on the organic ligand because the exchange $Na^+ \rightleftarrows Ag^+$ was not particularly selective for the Ag^+.

8. INTERCALATION IN ORGANO-CLAYS

Organic cations in great variety have been introduced into the interlayer regions of smectites, vermiculites and, to some extent, micas by ion exchange, and even by direct synthesis of organo-clays.[7] Theng[1] has summarized some general conclusions about organo-clays:

(i) Sorption by ion exchange approaches a maximum equal or close to the exchange capacity of the mineral. However, for longer chains ($>C_8$) sorption may exceed the exchange capacity, the excess being taken up as free amine or as amine salt (cf. § 6.2.4).

(ii) For montmorillonite the selectivity for the organic ions has been found to increase with chain length, probably as a result of enhanced (because largely additive) dispersion energy contributions to the overall bonding energy.

(iii) The $d(001)$ spacings indicate that the organic ion is often sorbed with the chains parallel with the basal surfaces. It is usual to consider the region between each pair of silicate layers in terms of the differences, Δ, between the observed $d(001)$ spacing of the complex and the thickness of an individual three layer siliceous sheet (often taken as 9·5Å). In many examples Δ is less than the minimum van der Waals thickness of the sheet. This behaviour is attributed to limited keying of the organic molecules into the 6-rings of the hexagonal sheets (cf. § 6.2.2).

(iv) Monolayer complexes with $d(001)$ in the range $\approx$12·5 to 13·5Å are formed with montmorillonite provided the area of the cation (up to C_{10}) is less than the area per exchange site, because there is then room in the interlayer region for a monolayer of cations. When the cation area exceeds the area per exchange site monolayers are replaced by bilayer complexes, with $d(001)$ of $\approx$16·5 to 17·5Å.[89,90,91,92]

(v) After uptake of a sufficient amount of organic ion the chains often pack with the charged ends on the interlayer surfaces and the alkyl chains oriented steeply away from the surfaces (cf. the alcohols, § 6.2.2.). In this conformation, greater van der Waals interactions become possible and the interlayer space is increased permitting larger amounts of the organic component to be accommodated.

(vi) The tilted configuration referred to in (v) is favoured by increasing exchange capacity. With vermiculites it can therefore be observed with shorter chains than with montmorillonite, and with mica the tilted configuration is the only one which can be expected, because there is only 24Å^2 of surface available per cation (§ 2.2). Double layer complexes of steeply oriented chains have been observed in vermiculite at high solute concentrations (cf. the alcohols, § 6.2.2).

(vii) In partially ion-exchanged crystals there is evidence of interstratified layers respectively rich in the organic ion and in the inorganic ion.[93,94]

Certain of the long chain ion organo-clays have found application as thickeners in paints, inks, lubricants, ointments and even lipstick. Organo-clays have also proved useful as the stationary phase in gas-solid chromatography and as sorbents (§§ 9 and 10). The organophilic nature of long-chain alkylammonium clays introduces new properties. They imbibe not only polar but also some non-polar guest molecules and can produce thixotropic organogels.[90,95] Several illustrative studies of sorption equilibria and energetics are given below.

Isotherms of various organic vapours in dimethyldioctadecylammonium bentonite (Bentone 34) have the contours shown in Fig. 21.[96] For the aromatic hydrocarbons these contours recall isotherms of hydrocarbons sorbed in elastomers. There is however a slight hysteresis between sorption and desorption branches, persisting to low relative pressures. With *n*-heptane, iso-octane and cyclohexane the sorption is smaller and the hysteresis loops more marked, with partial closure at $p/p_0 \approx 0{\cdot}3$ to $0{\cdot}4$. On the other hand dioxane and pyridine were strongly sorbed and the isotherms changed in shape from type III to type II in Brunauer's classification.[26] The selectivity sequence was

pyridine > dioxane > benzene, toluene, ethylbenzene > *n*-heptane, iso-octane, cyclohexane >*n*- and iso-butane.

For aromatic hydrocarbons the *d*(001) spacing increased continually with p/p_0, but for the paraffins and cyclohexane this spacing changed only a little (Table 15). By assuming molecular volumes for the guest molecules equal to those in their bulk liquids the observed expansions of *d*(001) can be compared with those calculated for additivity of volumes (columns 6 and 7). For the *n*-, iso- and cycloparaffin the calculated expansions exceed those actually

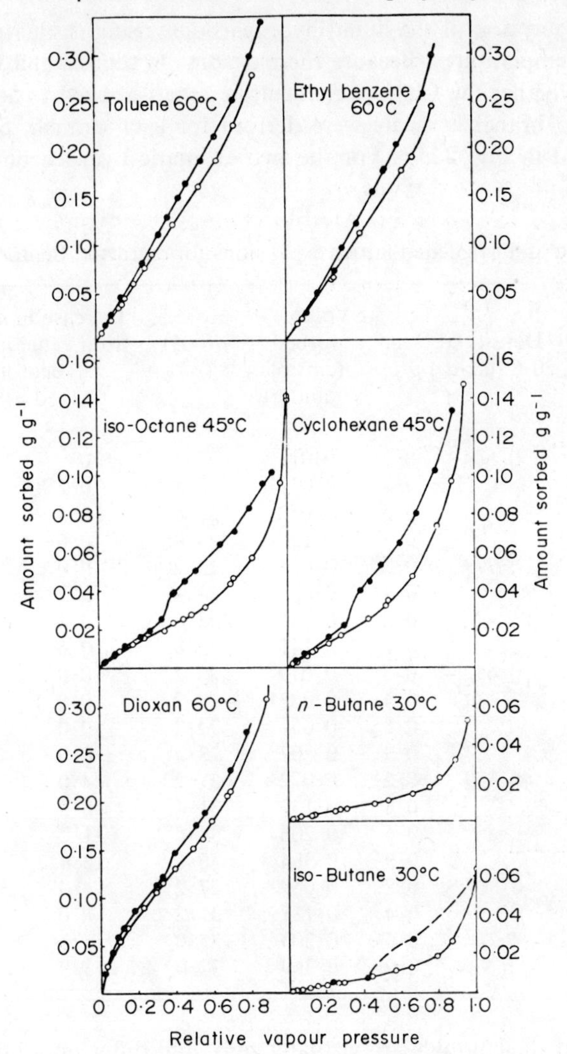

FIG. 21. Typical isotherms obtained with extracted Bentone 34 (dimethyldioctadecylammonium-bentonite) for hydrocarbons of different structural types and dioxane.[96] ○, sorption; ●, desorption.

observed. This indicates either that most of the sorption of paraffins occurs on external surfaces or that there is some free interlamellar space, allowing insertion of paraffins between the long chain ions. For the aromatic hydrocarbons the calculated and observed swellings are more nearly comparable and are very much larger than for the paraffins. In their case it is concluded that most of the uptake is in the interlamellar region.

Estimates were made via the Clapeyron equation of isosteric heats and entropies of sorption of the liquid hydrocarbons, using isotherms measured at different temperatures. Because there is some hysteresis and because it is not certain whether the Clapeyron equation is best applied to adsorption or to desorption branches, heats were derived for each branch. Some results are illustrated in Fig. 22.[96] For the two aromatic hydrocarbons the heats

TABLE 15

Observed and calculated lattice expansions for extracted Bentone 34[96]

Sorbate	Density at 20°C(g/cm³)	p/p_0	Volume sorbed (cm³ of liquid/g)	$d(001)$ (Å)	Increase in $d(001)$ in Å from value in outgassed Bentone 34 observed	calculated
n-heptane	0·684	0·2	0·018	23·2	0	0·8
		0·4	0·031	23·4	0·2	1·3
		0·6	0·050	23·6	0·4	2·2
		0·9	0·126	23·8	0·6	5·5
cyclohexane	0·779	0·2	0·015	23·2	0·0	0·6
		0·4	0·030	23·4	0·2	1·3
		0·6	0·051	23·6	0·4	2·2
		0·9	0·141	23·8	0·6	6·1
isooctane	0·692	0·2	0·019	23·2	0·0	0·8
		0·4	0·033	23·2	0·0	1·4
		0·6	0·051	23·2	0·0	2.2
		0·9	0·107	23·2	0·0	4·6
benzene	0·879	0·2	0·072	27·2	4·0	3·1
		0·4	0·139	31·4	8·2	6·0
		0·6	0·206	34·6	11·2	8·9
		0·9	0·364	36·6	13·4	15·7
toluene	0·867	0·2	0·069	27·2	4·0	3·0
		0·4	0·137	31·2	8·0	5·9
		0·6	0·205	35·0	11·8	8·8
		0·9	0·369	37·0	13·8	16·0

of sorption of their liquids are virtually zero, and differ only minimally for the two branches. The large uptakes are therefore the result of the considerable positive entropies which, as discussed later in this section, are expected if sorption involves mixing of hydrocarbon with the paraffin chains of the dimethyldioctadecylammonium ions. For *n*-heptane and cyclohexane the heats, according to the interpretation of the small degree of swelling given in the preceding paragraph, refer primarily to sorption on external surfaces of crystallites as, especially at larger relative pressures, the main component of sorption. The heats tend initially to be exothermic.

For the small polar molecules, H_2O and CH_3OH, heats of sorption from the vapours were derived from isotherms in Bentone 18-C (coconut fatty amine bentonite), Bentone 34 and Bentone 38 (dimethyl-dioctadecylammonium hectorite).[97] Sorption of water vapour was initially strongly exothermic and even at larger uptakes the heat was always larger than the heat of condensation. For methanol, the exothermal heats declined from high initial values to values less than the heat of liquefaction. Thus from its

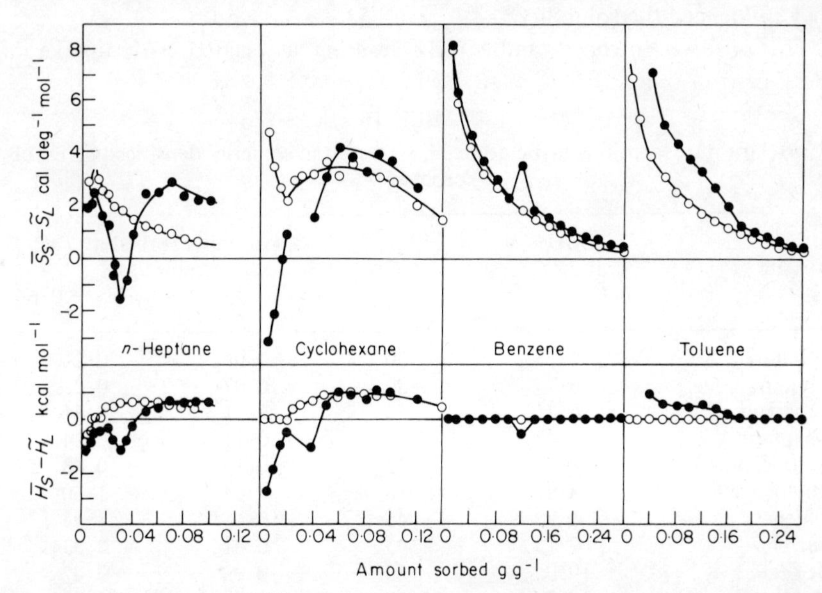

FIG. 22. Apparent heats and entropies of sorption by Bentone 34.[96] ○, derived from sorption branches of isotherms; ●, derived from desorption branches. $\bar{H}_s$, $\bar{S}_s$ denote differential molar enthalpy and entropy of sorbed hydrocarbon and $\tilde{H}_L$, $\tilde{S}_L$ are the integral molar enthalpy and entropy of bulk liquid hydrocarbon.

liquid an initial exothermal uptake becomes slightly endothermal at larger uptakes.

The behaviour of Bentone 34 led Barrer and Kelsey[96] to suggest that interlamellar sorption in this material was a function of the cohesive energy density (CED) of the sorbates. At a given relative pressure among a series of guest molecules the uptakes should increase in the order of the CED to a maximum for a molecule having a CED equal to that of the interlamellar region and should then decrease. The results in Table 16 suggest that the decrease may have begun with nitromethane. At the anionic silicate surfaces strong additional binding may arise with polar molecules. This was suggested as a cause of the change in isotherm shape from type III for aromatic hydrocarbons to type II for dioxane and pyridine.

The correlation between uptake and cohesive energy density of the sorbate is supplemented by a second which was established for sorption of various polar molecules from their aqueous solutions into trimethyloctadecylammonium montmorillonite.[101] The equilbrium concentration in solution for a fixed uptake of 1·1 mmol g^{-1} was a linear function of the logarithm of the solubility for resorcinol, phenol, aniline, valeric acid, *m*-cresol *p*-dinitrophenol, picric acid, phenylacetic acid and benzoic acid. Such physical properties as dipole moment and dielectric constant were important only insofar as they influenced the solubility.

The positive entropies and strong swelling associated with uptake of

TABLE 16

Correlation of sorptions in Bentone 34 with cohesive energy densities (CED) of sorbates[96]

Sorbate	CED[a] at 25°C $(cals/cm^3)^{1/2}$	T°(C)	Amounts sorbed at T°C (mmol/g)	
			$p/p_0 = 0{\cdot}2$	$p/p_0 = 0{\cdot}6$
isobutane	6·25	−30	$0{\cdot}06_2$	0·21
n-butane	6·7	−30	$0{\cdot}07_7$	0·22
isooctane	6·85	45	0·11	0·31
n-heptane	7·45	45	0·11	0·34
cyclohexane	8·20	45	0·16	0·48
ethylbenzene	8·8	60	0·52	1·60
toluene	8·90	45	0·65	1·93
benzene	9·15	45	0·81	2·33
dioxane	10·0	60	0·92	2·07
pyridine	10·7	60	1·79	3·08
nitromethane	12·6	45	1·38	2·54

[a] CED from Hildebrand and Scott[98]

benzene in Bentone 34, were considered by Barrer and Kelsey[96] in terms of disorientation of the alkyl chains of the organic ions and mixing of these chains with molecules of aromatic hydrocarbon. Each dimethyldioctadecylammonium ion was considered as a single chain with the nitrogen atom at its centre attached to a site on the silicate sheet. As in the treatment of monomer-polymer mixtures the available space, i.e. the interlamellar space, was visualized as divided into cells, each able to accommodate a benzene molecule or a segment of the long chain equal in volume to the benzene molecule. The chains were distributed at random among the cells as far as the restraint on the central nitrogen and the continuity of chains permitted. The benzene molecules were then placed in the unoccupied cells. The total number of cells was taken as the sum of the number of benzene molecules and of chain segments, and increases as the crystals swell. The estimated differential entropy of this system was compared with the differential entropy of sorption from liquid benzene into the crystals, and a reasonable correspondence was found. This supports the view that benzene uptake in Bentone 34 is associated with disordering of the long alkyl chains and their solvation with benzene.

Slabaugh and Hiltner[92] investigated sorption from the vapour phase of water, methanol, ethanol, l-propanol and *n*-heptane in *n*-alkylammonium bentonites with even numbers of carbon atoms per chain, and up to C_{18} in chain length. When methanol was sorbed in the bentonites with C_{12}, C_{14}, C_{16} and C_{18} chains the $d(001)$ spacing remained nearly constant for $0 < p/p_0 < 0{\cdot}8$ but thereafter rose rapidly to values dependent on the carbon number, *n*, of the chain, according to the expression (in Å)

$$d(001) = 12{\cdot}6 + n(1{\cdot}20).$$

It was considered therefore that the chains lying parallel with the surface were, for $p/p_0 > 0{\cdot}8$, lifted off the surface into steeply oriented configurations, presumably as a result of methanol uptake. An ordered steeply oriented arrangement would correspond with a lower entropy of sorption than calculated by Barrer and Kelsey[96] since the orientation does not lead to such random configurations as those assumed in the statistical treatment. When *n*-heptane was sorbed by dimethyldioctadecyl ammonium montmorillonite $d(001)$ remained nearly constant over the range $0 < p/p_0 < 0{\cdot}9$. However, when increasing amounts of ethanol were presorbed the *n*-heptane swelled the crystals at lower and lower relative pressures. It is probable that the polar ethanol was sorbed on to the silicate surfaces, thereby displacing the non-polar alkyl chains of the organic ions. These reoriented chains are then able to solvate and mix with heptane. The charged ends of the organic ions are of course expected to remain anchored to the silicate lamellae.

A further investigation was made by Barrer and Millington[102] with *n*-alkylammonium montmorillonites and α-ω-alkyldiammonium hectorites and montmorillonites as sorbents. The outgassed parent organo-clays had the $d(001)$ spacings given in Table 17. The exchange capacities in column 3 are derived from the nitrogen contents determined experimentally (column 2). The consistency and numerical values of the exchange capacities indicate that exchange must have been nearly complete in each preparation. The $d(001)$ spacings typify monolayers of chains lying parallel with the silicate layers.

In agreement with previous researches[92,96] *n*-heptane gave only 0·1 to 0·2Å swelling in any of the organo-clays; *n*-primary amines and alcohols did not swell the alkyldiammonium clays after equilibration at 80°C for three days. With alkylammonium montmorillonites some swelling was observed in forms containing longer chains. Benzene, dioxane and cyanobenzene gave spacings between 14·6 and 15·1Å, values compatible with a single layer of penetrant with the planes of the rings nearly perpendicular to the sheets and intermixed with the organic cations.

Although acetonitrile swelled all the clays to the extent shown in Table 18, ethyl and propyl cyanides were not sorbed sufficiently for *d*(001) to exceed 13·6Å after three days at 80°C. With acetonitrile, two complexes were formed with *d*(001) ≈ 16·3 and 19·9Å. 1,3-diaminopropane expanded the 13·3Å spacing to 14·3Å, a value consistent with vertical orientation of the long axis of the amine with the two -NH_2 groups embedded in oxygen 6-rings.

TABLE 17

d(001) for exchanged forms of outgassed montmorillonite and hectorite[102]

Cation	N content (%by wt in dry clay)	Cation exchange capacity (meq/100g)	*d*(001) (Å)
(a) Alkylammonium montmorillonites			
N^+H_4	1·16	83	10·5
$N^+H_3CH_3$	1·10	80	11·6
$N^+H_3C_2H_5$	1·09	81	12·8
$N^+H_3C_3H_7$	1·16	88	13·1
$N^+H_3C_4H_9$	—	—	13·3
$N^+H_3C_5H_{11}$	1·13	86	13·4
$N^+H_3C_6H_{13}$	1·06	82	13·4
$N^+H_3C_{12}H_{25}$	1·05	87	14·6
(b) α-ω-Alkyldiammonium montmorillonites			
$N^+H_3NH_3^+$	—	—	10·7
$N^+H_3(CH_2)_2N^+H_3$	1·18	86	12·2
$N^+H_3(CH_2)_3N^+H_3$	1·18	87	12·9
$N^+H_3(CH_2)_4N^+H_3$	1·19	89	13·4
$N^+H_3(CH_2)_5N^+H_3$	1·10	82	13·3
$N^+H_3(CH_2)_6N^+H_3$	1·12	85	13·2
$N^+H_3(CH_2)_7N^+H_3$	1·10	84	13·3
$N^+H_3(CH_2)_8N^+H_3$	1·12	86	13·3
$N^+H_3(CH_2)_9N^+H_3$	1·14	87	13·3
$N^+H_3(CH_2)_{10}N^+H_3$	1·17	90	13·3
$N^+H_3(CH_2)_{12}N^+H_3$	1·08	85	13·4
(c) α-ω-Alkyldiammonium hectorites			
N^+H_4	1·24	89	—
$N^+H_3(CH_2)_2N^+H_3$	1·25	90	12·2
$N^+H_3(CH_2)_3N^+H_3$	1·25	91	13·0
$N^+H_3(CH_2)_4N^+H_3$	1·25	94	—
$N^+H_3(CH_2)_5N^+H_3$	1·29	94	13·3
$N^+H_3(CH_2)_9N^+H_3$	1·21	91	13·3
$N^+H_3(CH_2)_{12}N^+H_3$	1·20	91	—

Adsorption branches of some isotherms for benzene are shown in Fig. 23.[102] For exchange ions having the same carbon number the alkylammonium montmorillonite sorbed less than the alkyldiammonium form, and as the carbon numbers increased, the uptakes, whether of benzene, *n*-heptane, 1,3-diaminopropane or acetonitrile, all decreased. Hysteresis was marked and persisted to the lowest pressures (Fig. 24).

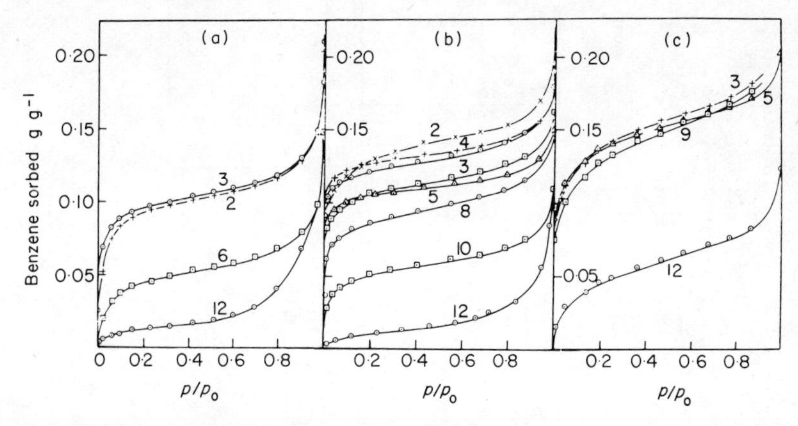

FIG. 23. Isotherms for benzene sorbed by: (a) alkylammonium montmorillonites at 50°C; (b) alkyldiammonium montmorillonites at 40°C; and (c) alkyldiammonium hectorites at 40°C.[102] The number on each curve gives the number of carbon atoms in the organic exchange ion.

The above examples of intercalation in organo-clays typify several kinds of behaviour. In one, swelling is limited by the length of the alkyl chains steeply oriented to the silicate sheet surfaces.[92] In another (Table 18), the limit to expansion is determined not by the alkyl chains but by the guest molecule. A third behaviour is exhibited by butylammonium vermiculite in butylammonium chloride solutions where swelling over a range of concentrations, C, varied linearly with $C^{-1/2}$ (Fig. 11).

9. PERMANENT INTRACRYSTALLINE POROSITY

In zeolite molecular sieves, the tectosilicate frameworks are permanently porous whether guest molecules are present or not. In smectites and vermiculites on the other hand removal of guest molecules normally causes the silicate layers to come close together so that inorganic ion exchanged forms of the outgassed minerals possess little permanent porosity. However in 1955 Barrer and McLeod[103] showed that if the exchange ions were $N^+(CH_3)_4$ or $N^+(C_2H_5)_4$ the silicate sheets were kept permanently apart and thereby provided intracrystalline free volume readily accessible to

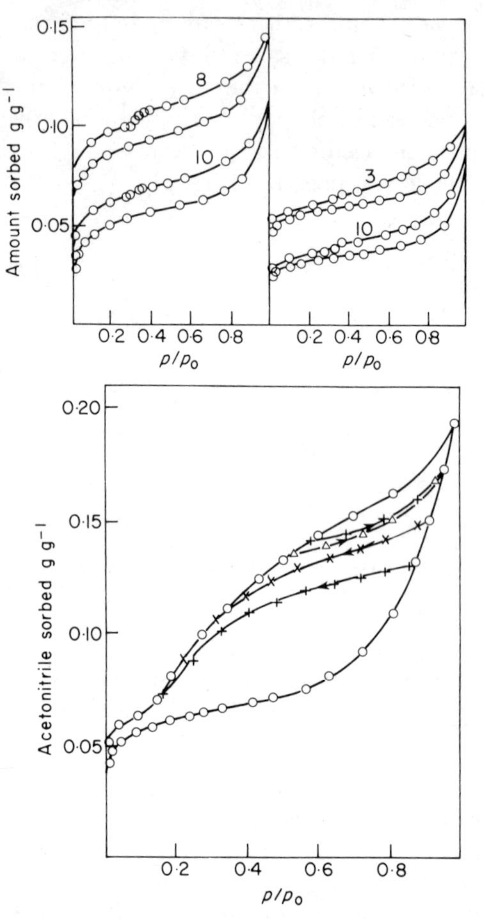

FIG. 24. Hysteresis at 40°C for sorption of benzene (top left) and *n*-heptane (top right) in alkyldiammonium montmorillonites. The numbers on the curves give the numbers of carbon atoms in the organic exchange ion. The bottom gives the hysteresis loop with scanning curves for sorption of acetonitrile at 50°C by $N^+H_3(CH_2)_3N^+H_3$-montmorillonite.(102)

various non-polar as well as to polar molecules. Thus interlamellar sorption of O_2, N_2 and Ar occurred freely at 78 K and 90 K; also *n*- and iso-paraffins and aromatic hydrocarbons were readily taken up, and at room temperature, the selectivity sequence for the isomeric pentanes was $n\text{-}C_5 > \text{iso-}C_5 > \text{neo-}C_5$. Polar molecules, H_2O, NH_3, CH_3OH, C_2H_5OH and pyridine were sorbed without evidence of the threshold pressures in the isotherms which were required before these molecules penetrate outgassed Na-montmorillonite crystals, as shown in Fig. 14 and discussed in § 6.2.1. Moreover the hysteresis loops no longer persisted to the lowest pressures (Fig. 14) but, as

TABLE 18

Swelling in acetonitrile

Form	d(001) in Å with complex Not outgassed	Outgassed at 60°C	Outgassed at 110°C
(a) Alkylammonium montmorillonites			
$CH_3N^+H_3$	16·4	—	—
$C_2H_5N^+H_3$	16·3	—	—
$C_3H_7N^+H_3$	20·6	—	—
$C_6H_{13}N^+H_3$	21·0	—	—
$C_{12}H_{25}N^+H_3$	16·3	—	—
(b) α-ω-dialkylammonium montmorillonites			
$N^+H_3N^+H_3$	20·0	—	16·1
$N^+H_3(CH_2)_2N^+H_3$	19·6	19·6	16·4
$N^+H_3(CH_2)_3N^+H_3$	19·7	15·9	15·9
$N^+H_3(CH_2)_4N^+H_3$	16·2	16·4	—
$N^+H_3(CH_2)_5N^+H_3$	16·4	16·3	—
$N^+H_3(CH_2)_6N^+H_3$	16·9	16·8	—
$N^+H_3(CH_2)_7N^+H_3$	19·8	—	—
$N^+H_3(CH_2)_8N^+H_3$	19·6	19·8	—
$N^+H_3(CH_2)_9N^+H_3$	—	19·8	19·7
$N^+H_3(CH_2)_{10}N^+H_3$	—	20·0	—
$N^+H_3(CH_2)_{12}N^+H_3$	20·6	20·4	20·4
(c) α-ω-dialkylammonium hectorites			
$N^+H_3(CH_2)_2N^+H_3$	19·7	19·6	—
$N^+H_3(CH_2)_3N^+H_3$	19·4	—	—
$N^+H_3(CH_2)_4N^+H_3$	16·2	—	—
$N^+H_3(CH_2)_5N^+H_3$	16·3	—	—
$N^+H_3(CH_2)_9N^+H_3$	19·5	19·7	—
$N^+H_3(CH_2)_{12}N^+H_3$	19·9	19·9	—

seen in Fig. 25, closed at relative pressures p/p_0 between about 0·25 and 0·5, according to the sorbate. The hysteresis is now associated primarily with that part of the sorption which occurs upon external surfaces of the crystals. The interlayer free distances of ≈4·1 and 4·5Å allow penetration of the crystals by the sorbates with the minimum of additional lattice expansion.

The amount of the interlamellar component of sorption depends *inter alia* upon the size of the organic cation, because the larger the ion the greater the fraction of interlamellar surface it covers. Accordingly the $N(C_2H_5)_4$-montmorillonite sorbed less copiously than the $N(CH_3)_4$-form. Barrer and Reay[104] demonstrated permanent intracrystalline porosity when the quaternary ammonium ions were replaced by mono-, di- and trimethyl-

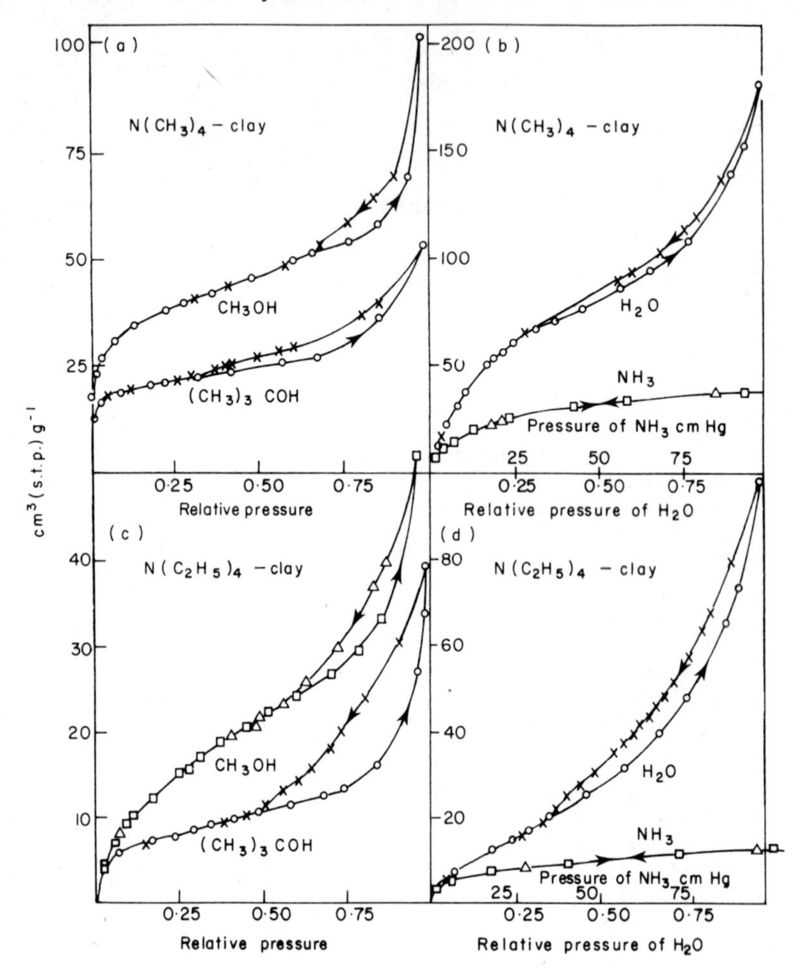

FIG. 25. Isotherms and hysteresis loops observed in the sorption of typical polar molecules by tetra-alkylammonium-montmorillonites at 50°C[102] The hysteresis loops close at intermediate relative pressures.

ammonium. The $d(001)$ spacings and free distances, $d(001)$—9·4, between the silicate sheets are, in order

$N^{+}(C_2H_5)_4$-montmorillonite	13·9$_5$Å and 4·5$_5$Å
$N^{+}(CH_3)_4$-montmorillonite	13·5Å and 4·1Å
$N^{+}H(CH_3)_3$-montmorillonite	13·0Å and 3·6Å
$N^{+}H_2(CH_3)_2$-montmorillonite	12·2Å and 2·8Å
$N^{+}H_3CH_3$-montmorillonite	12·0Å and 2·6Å

The free distances are rather smaller than the minimum heights of the ions, the difference being ascribable to keying of the ions into the 6-rings of the silicate sheets. The $N^{+}(C_2H_5)_4$ can assume a flattened configuration in which its height is little different from that of $N^{+}(CH_3)_4$.[104] The free distances are big enough for penetration of small molecules such as oxygen with little or no further swelling. For larger molecules, where additional swelling may be appreciable, the extent of the sheet separation already imposed by the exchange ions makes intercalation relatively easy.

The greater the further swelling needed to accommodate the guest, the

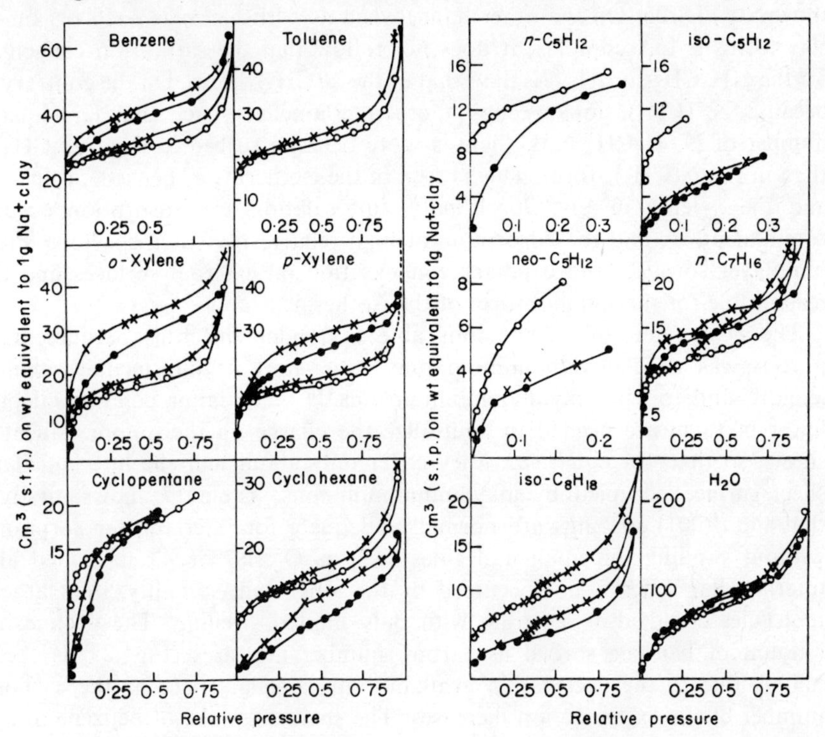

FIG. 26. A comparison of isotherms of various sorbates in $N(CH_3)_4$- and NH_3CH_3-montmorillonites at 50°C[104] ○, sorption by the $N(CH_3)_4$-form; ●, sorption by the NH_3CH_3-form; ×, desorption.

more marked the hysteresis between sorption and desorption branches tends to be. If the guest molecules are too bulky, little or no interlamellar penetration may occur. Isotherms of some guest species in tetra- and monomethylammonium montmorillonites are compared in Fig. 26. The molar integral free energy of interlamellar sorption, ΔG, at pressure p is the sum of two terms:

$$\Delta G = \Delta G_s + \Delta G_I \tag{9}$$

where ΔG_s is the free energy change associated with separating the lamellae and ΔG_I is this change for insertion of guest molecules into the expanded interlamellar pore volume. ΔG_s is expected to be positive and to be bigger the larger the separation required to insert the guest; ΔG_I is negative. For example to insert cyclohexane into NH_3CH_3-montmorillonite will require a much more positive ΔG_s than its insertion into the $N(CH_3)_4$-form because the free distance between the silicate sheets is 1·5Å smaller in the NH_3CH_3-clay and therefore the needed additional swelling is correspondingly greater. This can account for the large difference in the two cyclohexane isotherms shown in Fig. 26. On the other hand, when as with benzene, each organo-clay forms a 15Å complex, it does not follow that the saturation capacity for the NH_3CH_3-form is less than that of the $N(CH_3)_4$-form. On the contrary, because $N^+(CH_3)_4$ ions occupy more interlamellar space than an equal number of $N^+H_3CH_3$ ions there is more benzene sorbed in the NH_3CH_3- than in the $N(CH_3)_4$-form. This is seen in the isotherms of benzene, toluene and the xylenes in Fig. 26. Finally, intercalation and adsorption occur simultaneously, and the adsorption at high relative pressures develops into multilayer sorption and capillary condensation on external surfaces and is responsible for the sigmoid form of the isotherms.

The next extension of the ions able to render smectites permanently porous was to *n*-alkylammonium ions with short or intermediate chain lengths, and to α-ω-alkyldiammonium ions.[(102)] The latter being divalent, fewer of them are needed to neutralize the charge on the anionic silicate sheets, so that, for equal size, they cover only about half the interlamellar sheet surface occupied by alkylammonium ions. Table 17 shows, firstly, that the $d(001)$ spacings are normally adequate for interlamellar sorption without swelling for small molecules such as O_2, N_2 or Ar, provided all interlamellar space is not occupied by the ions; and, secondly, that larger molecules could also penetrate with only limited swelling. The decline in amount of benzene sorbed as carbon number increases (Fig. 23) is then mainly due to the decrease in available interlamellar space as the carbon number of the exchange ion increases. The smaller uptake of benzene in an alkylammonium-montmorillonite than in an alkyldiammonium-montmorillonite in which the cation has the same carbon number (Fig. 23) received a similar explanation since the interlamellar divalent ions are only half as numerous as the monovalent ions.

As noted above, sorption by clay minerals is the sum of interlamellar uptake and uptake on external surfaces. Barrer and Millington[(102)] attempted to separate the two components by subtracting the amounts of nitrogen or argon sorbed on a given amount of the non-permeable Na-montmorillonite or Na-hectorite from the total uptakes on the amounts of their various alkylammonium forms equivalent to 1 g of their Na-forms. This assumes that for equivalent amounts the external areas of the Na-form and its organic ion exchanged forms are the same. The resultant estimates of pure intra-crystalline sorption of N_2 and Ar at 78 K are then shown in Fig. 27. As expected, at this temperature all isotherms are extremely rectangular and so of Type I in Brunauer's classification. The effect of chain length upon the uptake is apparently greater for mono- than for divalent ions and appears less pronounced for divalent ions in the hectorite than in the montmorillonite. The saturation capacities for interlamellar sorption of nitrogen are summarized, with the monolayer equivalent areas, in Table 19. With divalent ions in particular good capacities are achieved, and the areas compare

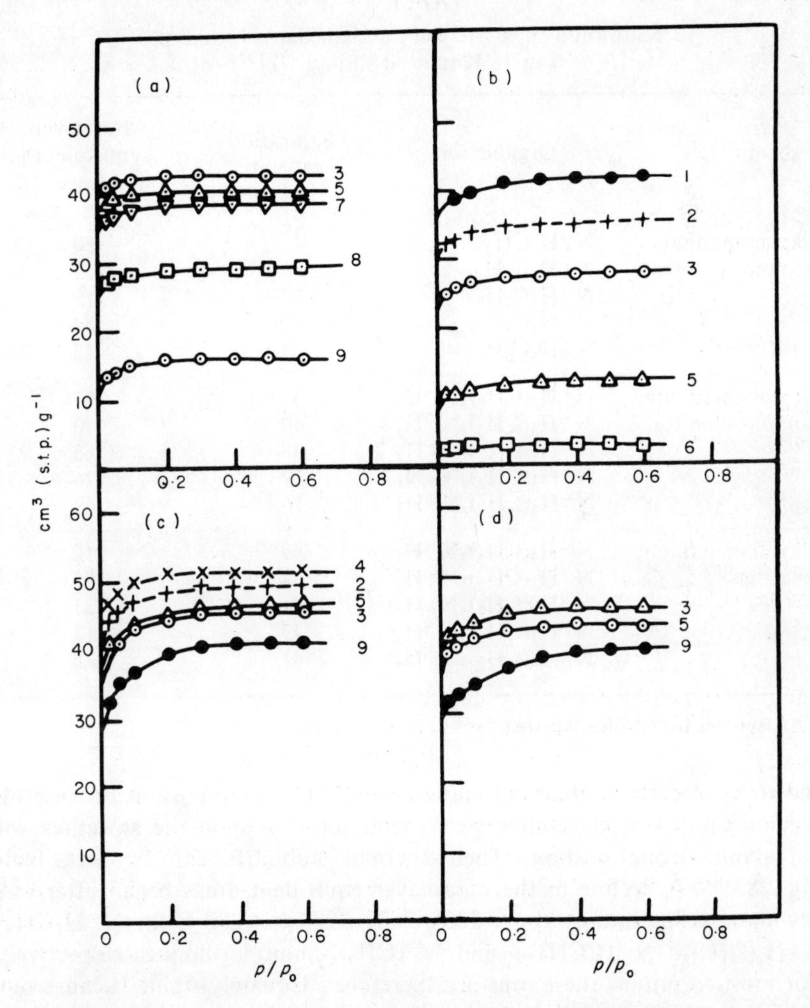

FIG. 27. Estimated sorptions of N_2 and Ar at 78 K in the interlamellar free volumes only, for:[102] (a) N_2 in alkyldiammonium montmorillonites; (b) N_2 in alkylammonium montmorillonites; (c) N_2 in alkyldiammonium hectorites; (d) Ar in alkyldiammonium hectorites. The number on each curve is the number of carbon atoms in each organic exchange ion.

with those of Vycor porous glass[105] or fibrous clays (Table 4), although they are smaller than those of some zeolites, carbons and silica gels.

In the interlamellar regions of the clay minerals there should be reasonable energetic homogeneity for monolayer complexes, unlike amorphous microporous sorbents or, indeed, many zeolites (Chapter 4). Also the clay minerals are more easily outgassed both with respect to the original zeolitic water

TABLE 19

Saturation capacities for interlamellar sorption of N_2 at 78 K (cm^3 at s.t.p. g^{-1})[102]

Sorbent	Organic ion	Saturation capacity	Monolayer equivalent[a] area (m^2g^{-1})
Alkylammonium montmorillonites	$N^+H_3CH_3$	41·5	180
	$N^+H_3C_2H_5$	35	152
	$N^+H_3C_3H_7$	27·5	119
	$N^+H_3C_5H_{11}$	12·5	54
	$N^+H_3C_6H_{13}$	3	13
Alkyldiammonium montmorillonites	$N^+H_3(CH_2)_3N^+H_3$	43	187
	$N^+H_3(CH_2)_5N^+H_3$	40	173
	$N^+H_3(CH_2)_7N^+H_3$	38	165
	$N^+H_3(CH_2)_8N^+H_3$	29	126
	$N^+H_3(CH_2)_9N^+H_3$	16	69
Alkyldiammonium hectorites	$N^+H_3(CH_2)_2N^+H_3$	49	212
	$N^+H_3(CH_2)_3N^+H_3$	44	191
	$N^+H_3(CH_2)_4N^+H_3$	51	221
	$N^+H_3(CH_2)_5N^+H_3$	49	212
	$N^+H_3(CH_2)_9N^+H_3$	41	178

[a] Assuming 16·2 Å² for the area per nitrogen molecule.

and to hydrocarbons than are many zeolites. This is important because the organic cations which confer permanent porosity upon the smectites will not stand strong heating. Their thermal stabilities can be seen from Fig. 28.[104] A decline in the monolayer equivalent areas began after outgassing at temperatures above 150°, 200°, 220° and 280°C for $N^+H_3CH_3$-, $N^+H_2(CH_3)_2$-, $N^+H(CH_3)_3$– and $N^+(CH_3)_4$-montmorillonites respectively. For most sorptions, these ions are therefore adequately stable because outgassing is usually readily effected at temperatures not above 100°C. The interlamellar part of the uptake of gases like Ar, N_2, O_2 or CH_4 can serve as a model for true monolayer sorption, and the interlamellar region could prove to be energetically very uniform.

Other developments[106] followed the synthesis of two series of fluorhectorites. Those made by sintering reactions had cation exchange capacities of about 90 meq/100 g, typical of smectites; and those made from LiF melts had capacities of about 150 meq/100 g, typical of vermiculite. Because F in these materials replaced the OH in hectorites the phases were thermally stable to at least 800°C. They swelled readily in water and easily exchanged interlamellar Li^+ present in the parent crystals for various other cations, including Na^+, Cs^+, $[Co(en)_3]^{3+}$, $N^+H_3CH_3$, $N^+(CH_3)_4$ and $N^+H_3(CH_2)_2N^+H_3$. The crystals were bigger than those of natural hectorites so that sorption

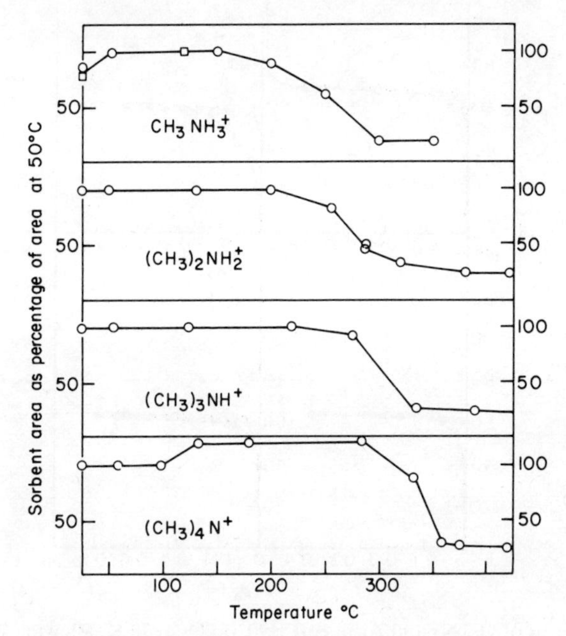

FIG. 28. The variation with outgassing temperature of the relative monolayer equivalent areas measured with N_2. This area after outgassing at 50°C is taken as 100.[104]

upon external surfaces was reduced, and the role of interlamellar sorption was correspondingly emphasized. The *d*(001) spacings of several cationic forms of the outgassed fluorhectorites were, in Å: Cs^+, 10·5; $N^+H_3CH_3$, 11·6; $N^+H_3(CH_2)_2N^+H_3$, 12·2; $N^+(CH_3)_4$, 13·4; and $[Co(en)_3]^{3+}$, 13.6.

On the Cs^+- and $N^+H_3CH_3$-forms of the fluorhectorites of 90 and 150 meq/100 g exchange capacities (respectively termed FH90 and FH150) there was no evidence of interlamellar penetration of O_2 or N_2 at 78 K, in contrast with the behaviour of the $N^+H_3CH_3$-forms of natural hectorite[102] and montmorillonite.[103,104] The difference may be the result of the larger and more perfect crystals of FH90 and FH150, together with the slightly smaller value of $[d(001) - 9{\cdot}41] = 2{\cdot}2$Å as compared with 2·6Å. With

$N(CH_3)_4$-FH90 O_2, N_2, and Ar at 78 K all penetrated the crystals but there was little if any penetration of this form of FH150 by N_2. This indicates a new factor which governs molecule sieving by and penetration of layer silicates. The greater the charge density on the siliceous layers the smaller the lateral distance between adjacent pairs of ions. For a two-dimensional packing of tetramethylammonium ions independent of the

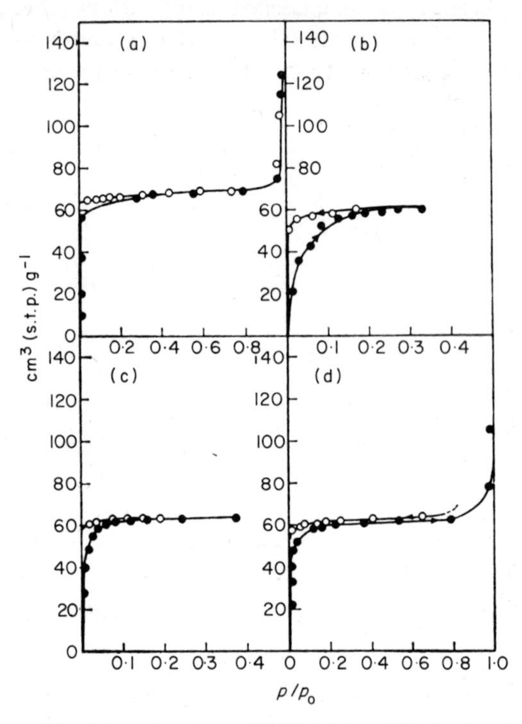

FIG. 29. Isotherms of O_2, N_2 and Ar in en-FH90. (a) O_2 at 78 K, allowing 20 min per isotherm point; (b) N_2 at 78 K, allowing 15 min per point; (c) O_2 at 90 K, 20 min per point; (d) Ar at 78 K, 30 min per point.[106] ●, sorption points; ○, desorption points.

hexagonal oxygen sheets the centre to centre distances for FH150 and FH90 would be $6{\cdot}8_6$ and $8{\cdot}8_2$Å respectively. The free distances were then estimated as $\approx 0{\cdot}8$ and $\approx 2{\cdot}7$Å respectively, in accordance with the experimental finding that FH150 did not but FH90 did sorb permanent gases at 78 K. The $N^+(CH_3)_4$ ions are however not spheres but tetrahedra, and it is expected that 50% would have their triangular bases against lower and 50% against upper lamellae. While this modifies the free distances somewhat[107] it is not likely to alter the above conclusion.

The number of $N^+H_3(CH_2)_2N^+H_3$ (en) ions between the silicate layers is only half the number of $N^+(CH_3)_4$ ions, with corresponding enhancement of the average free distances between adjacent pairs of ions in FH90 and FH150. Centre to centre distances were estimated respectively as 12·4 and 9·7Å and for en ions of length ≈7·4$_6$Å and cross-sectional diameter ≈4·9Å the free distances for FH90 could vary from ≈4·9 to ≈7·5Å. If the $-N^+H_3$ groups of the en ions were embedded in 6-rings of oxygens with a 56° orientation of the axis of the molecule to the oxygen surfaces of the clay

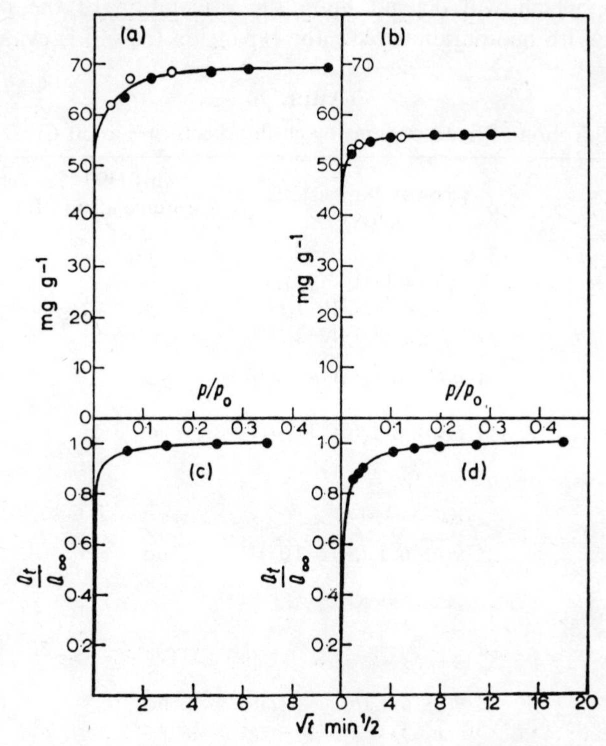

FIG. 30. Isotherms of *n*-octane at 50°C.[106] (a) Isotherm in en-FH90; (b) Isotherm in en-FH150; (c) Sorption kinetics in en-FH90; (d) Sorption kinetics in en-FH150. ●, sorption points; ○, desorption points. Q_t denotes amount sorbed at time t and Q_∞ this amount at equilibrium.

mineral the effective cross-section is ellipsoidal with major and minor axes about 5·9 and 4·4Å. Then even for FH150 the free distances between adjacent ions can be from ≈3·8 to ≈5·3Å. In confirmation of this argument it was found that en-forms of both FH90 and FH150 freely sorbed permanent gases and certain hydrocarbons. Isotherms for permanent gases in FH90 are illustrated in Fig. 29 and for *n*-octane in both FH90 and FH150 in Fig. 30. Uptake of *n*-octane in each sorbent is rapid. Results obtained with some hydrocarbons are summarized in Table 20. Because of the larger crystal size and consequently reduced external surface the isotherms in

Figs 29 and 30 are much more rectangular than isotherms obtained in organo-montmorillonites and hectorites. Sorption is dominated by its interlamellar component.

Since the vertical free distance between lamellae in the en-fluorhectorites is 2·8Å, interlamellar penetration of hydrocarbons having the van der Waals dimensions given in column 2 is expected to result in additional expansions, which will depend upon the orientation of the intercalated molecules. With neo-paraffins, ΔG_s for expansion (eqn 9) is evidently more

TABLE 20

Sorption of hydrocarbons by en-fluorhectorites at 50°C[102]

Guest	Critical Dimensions in (Å)	en-FH90 Intercalation and wt %	en-FH150 Intercalation and wt %
n-C_4	$4\cdot9(b)$, $4\cdot0(t)$, $7\cdot7_7$ (l)	yes	yes
n-C_6	$4\cdot9(b)$, $4\cdot0(t)$, $10\cdot3(l)$	yes	yes, 6·7
n-C_8	$4\cdot9(b)$, $4\cdot0(t)$, $12\cdot3(l)$	yes, 6·9	yes, 6·4
iso-C_4	$4\cdot6_5(h)$, $6\cdot1_8(b)$, $6\cdot1_8(l)$	yes	yes

C–C–C–C
  | |
  C C
	$4\cdot6_5(h)$, $6\cdot1_8(b)$, $7\cdot7_7(l)$	yes, 5·1	yes, 4·5

C
  |
C–C–C
  |
  C
	$5\cdot9(h)$, $6\cdot1_8(b)$, $6\cdot1_8(l)$	no	no

C
  |
C–C–C–C
  |
  C
	$5\cdot9(h)$, $6\cdot1_8(b)$, $7\cdot7_7(l)$	no	no

C
      |
C–C–C–C–C
  | |
  C   C
	$5\cdot9(h)$, $6\cdot1_8(b)$, $9\cdot0_3(l)$	no	no

cyclo-C_5	$6\cdot4(b)$, $4\cdot0(t)$, $6\cdot5(l)$	yes, 7·2	yes, 7·6
cyclo-C6	$6\cdot4(b)$, $4\cdot9(t)$, $7\cdot2(l)$	yes, 5·9	no
Benzene	$6\cdot7(b)$, $3\cdot7(t)$, $7\cdot4(l)$	yes, 12·9	yes, 8·1
Toluene	$6\cdot7(b)$, $4\cdot0(t)$, $8\cdot6(l)$	yes, 9·6	no
m-Xylene	$7\cdot9(b)$, $4\cdot0(t)$, $8\cdot6(l)$	yes, 5·6	no
Mesitylene	$9\cdot0(b)$, $4\cdot0(t)$, $8\cdot6(l)$	yes, 5·4	no

[a] The dimensions given are in three directions at right angles. h = height; t = thickness; b = breadth; l = length.

positive than ΔG_I, the free energy of insertion of guest into the expanded lattice, is negative, because no penetration was found. Benzene, which is considered to be oriented with the plane of the ring steeply oriented to the plane of the silicate layer, is particularly strongly sorbed. Especially in en-FH90 the packing seems very economical of space and the sorption capacity is high. However substitution of one, two or three substituent methyl groups in benzene gave progressively reduced uptakes in en-FH90 and intercalation was not found in en-FH150. Thus very interesting selectivity differences of a molecular sieve character are apparent.

In subsequent chromatographic separations of hydrocarbons a short fore-column of en-fluorhectorite was used before main columns of Celite coated with dinonylphthalate.[87] The behaviour under kinetic conditions did not always duplicate those in the static sorption systems leading to Table 20. For example the fore-column of en-FH150 did not remove cyclopentane, benzene, 2,3-dimethylbutane or 3-methylpentane from the nitrogen carrier gas. These differences may relate to the extra expansion of the crystals needed to accommodate the hydrocarbons, and the consequent need for a threshold pressure and an induction time to initiate this expansion (cf Fig. 14 and § 6.2.1). In the chromatographic runs these requirements may not have been met.

Another globular organic ion which permanently expands montmorillonite[108,109] and vermiculite[108] is the divalent cage-like ion

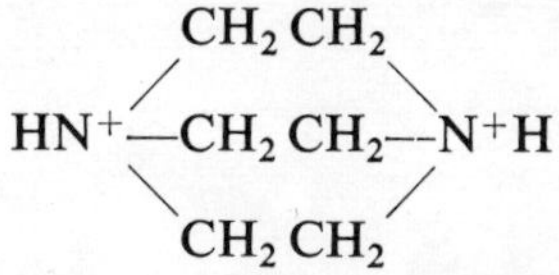

```
          CH2 CH2
         /       \
HN+—CH2 CH2—N+H
         \       /
          CH2 CH2
```

(from triethylenediamine, i.e. 1,4-diazabicyclo[2,2,2]octane or DABCO). The $d(001)$ spacing is 14·8Å so that the vertical free distance is 5·4Å. In montmorillonite the free distance between two adjacent cations was estimated as ≈6Å. The triethylenediammonium forms of montmorillonite and vermiculite sorbed nitrogen, ethane and 2,4-dimethylpentane.[108] The monolayer equivalent areas for nitrogen in the montmorillonite and vermiculite were 280 m^2g^{-1} and 144 m^2g^{-1} respectively, most of which can be ascribed to interlamellar sorption. The lower value in vermiculite corresponds with its higher exchange capacity, larger cation population, and consequently smaller interlamellar free surface.

Another exchange cation which permanently opens vermiculite and smectite structures[87] is $[Co(en)_3]^{3+}$. The vertical free distance between silicate sheets was 4·2Å and the average free distances between adjacent pairs of cations were estimated to be around 6·7Å and 3·4Å for the FH90 and the FH150 respectively. The FH90 readily sorbed permanent and inert gases, methane, isobutane and even neopentane. The FH150 readily sorbed

H_2, D_2, Ne and O_2 at 77 K but sorbed very little N_2, Ar and CH_4, a selectivity difference of particular interest (Fig. 31).[100] Kr and CH_4 were however intercalated at 195 K and at higher temperatures.

It is possible to calculate the total interlamellar area for the clay minerals per gram and to subtract from this area the area covered by the interlamellar cations. The area per $[Co(en)_3]^{3+}$ ion was taken as 63Å², and the free

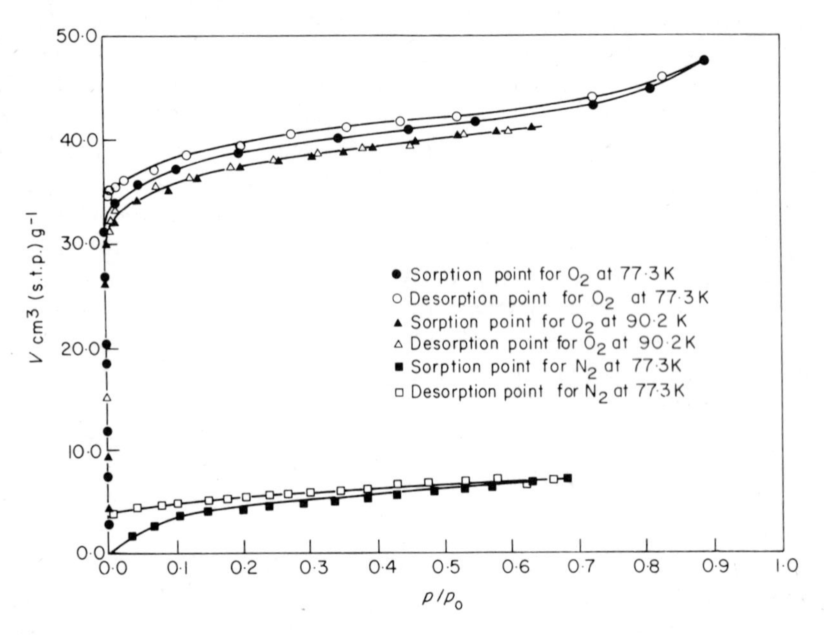

FIG. 31. Isotherms for O_2 and N_2 at about 78 and 90 K on $Co^{III}(en)_3$-FH150,[110] showing very great differences in uptake.[110]

areas per gram were estimated as 242 m² and 165 m² for FH90 and GH150 respectively. From the isotherms, total monolayer equivalent areas were found, and the "experimental" interlamellar free area was obtained by subtracting the external surface area taken as $8 \cdot 1$ m^2g^{-1}. The calculated and experimental free areas can then be compared, as exemplified in Table 21.[87] The ratios in the last two columns are close to unity, which provides strong support for the interpretation, already given on other grounds, of the factors controlling interlamellar sorption. Similar calculations were made by Barrer and Millington[102] for sorption of nitrogen into alkylammonium montmorillonites and alkyldiammonium montmorillonites and hectorites. The ratios of observed to calculated interlamellar saturation capacities, though somewhat more variable than those in Table 21, were, for shorter alkyl chains in particular, rather near unity.

TABLE 21
Calculated and experimental interlamellar free areas for $Co(en)_3$-FH90[87]

Sorbate and area per molecule ($Å^2$)		T/K	Interlamellar free area, by experiment FH90	FH150	Experimental free area ÷ calculated free area FH90	FH150
O_2	13·7	77	242	202	1·0	1·2
	14·1	90	244	182	1·0	1·1
N_2	16·3	77	217	—	0·90	—
	17·1	90	202	—	0·83	—
Ar	14·2	77	242	—	1·0	—
	14·5	90	240	—	0·99	—
CH_4	17·5	90	227	—	0·94	—
$C(CH_3)_4$	38·6	273	238	—	0·98	—

10. SMECTITE MOLECULAR SIEVES AND SELECTIVE SORPTION

The picture which has emerged from the studies referred to in § 9 is an interesting one, showing a number of features parallel with those of the molecular sieve zeolites. The openings into, and throughout the interlamellar free volumes of clay minerals permanently expanded by relatively globular exchange ions tend to be slit-shaped. The dimensions of the slits can readily be varied according to the shape and size of the exchange ions. Thus good sorptions were possible in clay minerals where the vertical free distances between silicate lamellae were varied from 2·2 to 5·4Å. If necessary this distance may increase slightly to accommodate the guest. Secondly, the lateral free distances between pairs of adjacent cations has also been regulated by varying the numbers of cations and their shapes and sizes. The numbers have been changed by altering the layer charge on the silicate sheets (e.g. smectites and vermiculites) and by altering the valence of the exchange cations (e.g. $N^+(CH_3)_4$, $N^+H_3(CH_2)_2N^+H_3$ and $[Co(en)_3]^{3+}$). Thus the slits giving access to the interlamellar free volumes may be tailored to modify this access. The diffusion within the lamellae is always two-dimensional.

A further attempt was made to modify the mean lateral distance between cations by partial exchange of Na^+ in montmorillonite by $N^+H_3CH_3$ and by $N^+(CH_3)_4$[93]. If the Na^+ and the organic ions in each interlamellar region mix in partially exchanged forms, the expanded molecular sieve smectites could be made with regulated free distances between the organic ions. The sorption capacity should be thereby enhanced. However it was found that the saturation sorption capacity of the partially exchanged forms increased from a low value in a nearly linear dependence upon the extent of exchange. This, and X-ray evidence, led to the conclusion that interlamellar regions with mixed cations did not occur, but that when organic ions invaded a given region the inorganic cations were swept out. A partially exchanged form was randomly interstratified with layers rich either in the organic ions or in the inorganic ions, but not in both.

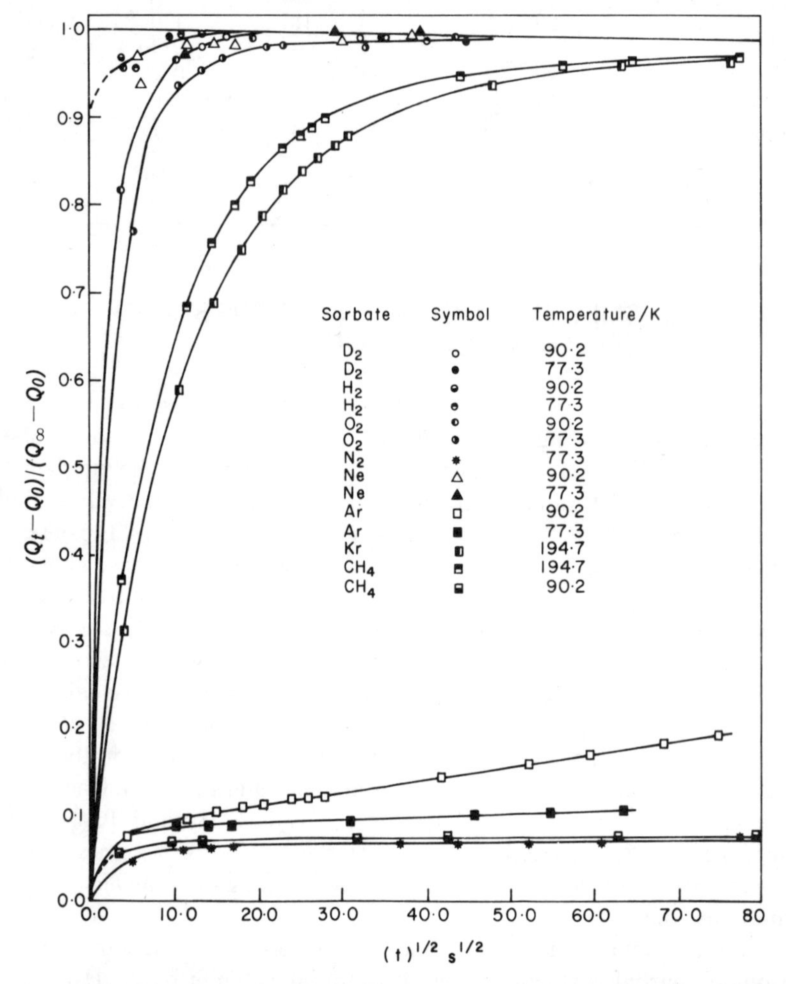

FIG. 32. Sorption kinetics for H_2, D_2, O_2, N_2, Ne, Ar, Kr and CH_4 At $\approx$78 and 90 K the group N_2, Ar, Kr and CH_4 is strongly differentiated from H_2, D_2, O_2 and Ne, which are very rapidly sorbed.[111] Q_t = amount sorbed at time t. Q_0 = amount sorbed at time zero. Normally $Q_0 = 0$ in these experiments. Q_∞ = amount sorbed at equilibrium.

In § 6.1 of Chapter 6 reference was made to striking differences in the rates of sorption in certain zeolites of permanent and inert gases based upon the different van der Waals dimensions of these molecules. Figure 12 in Chapter 6 illustrates these differences for O_2, N_2 and Ar in levynite at 89 K and in Ca-mordenite at 195 K. Similar differences arise for these gases in zeolite Na-*A*, the sequence again being $O_2 > N_2 > Ar$. Figure 32[111] shows that very great differences also occur in rates of sorption of permanent and inert gases by $Co^{III}(en)_3$-FH150 molecular sieve. Extremely rapid uptakes of H_2, D_2, Ne and O_2 and remarkably low rates of Ar, N_2 and CH_4 were observed at 78 K or 90 K. The differentiation is based in the same way upon differences in molecular dimensions as observed among the above molecular sieve zeolites. Table 20 also illustrates molecular sieve selectivities shown for hydrocarbons in the ethylenediammonium fluorhectorites, and chromato-

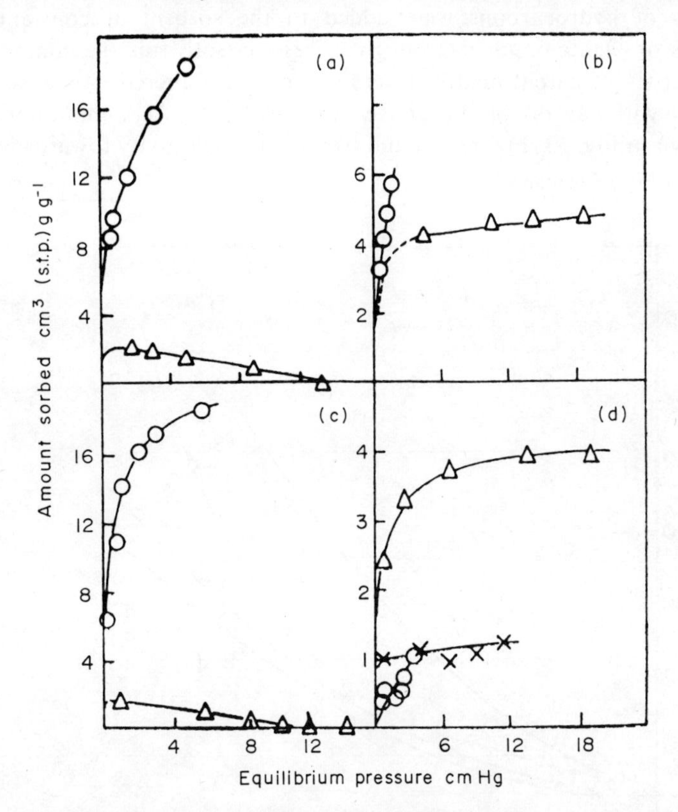

FIG. 33. (a) and (b) Isotherms in $N(CH_3)_4$-montmorillonite for benzene (○) and *n*-heptane (△) from two mixtures with (a) 0·67 and (b) 0·30 mole fraction of benzene. (c) Isotherms for benzene (○) and cyclohexane (△) from a mixture with 0·67 mole fraction of benzene. (d) Cyclohexane isotherms from several mixtures with *n*-heptane with mole fractions 0·85 (○), 0·56 (×) and 0·16 (△). All isotherms were obtained at 80°C.[107]

graphic experiments demonstrated that *n*-paraffins could be separated from branched chain paraffins again recalling the same separations with zeolite Ca-*A* or with chabazite. Various other possibilities for specific separations present themselves from Table 20, in which the shape and size of the guest molecules is the cause of differences in sorbability.

As with the zeolites, selectivity in smectite porous crystals can depend upon molecule sieving or upon differences in free energy of sorption (eqn 10) for two guest species each of which has access to the interlamellar free volume. Mixture isotherms have been determined in one or both of methylammonium- and tetramethylammonium-montmorillonite for the pairs:[107,112] $C_6H_6 + n\text{-}C_7H_{16}$; $C_6H_6 + CH_3OH$; $CH_3OH + CCl_4$; C_6H_6 + cyclohexane; cyclohexane + isooctane;* $n\text{-}C_7H_{16}$ + cyclohexane; $n\text{-}C_7H_{16}$ + isooctane; and $n\text{-}C_6H_{14}$ + isooctane. The doses of parent mixtures of hydrocarbons were added to the sorbent at constant mole fractions of each component; but in successive isotherms the compositions of the doses of parent mixtures were progressively altered. Some isotherms obtained with several of the above mixtures in $N(CH_3)_4$-montmorillonite are shown in Fig. 33. Figure 33a and b shows the selectivity towards benzene

* 2,2,4-trimethylpentane

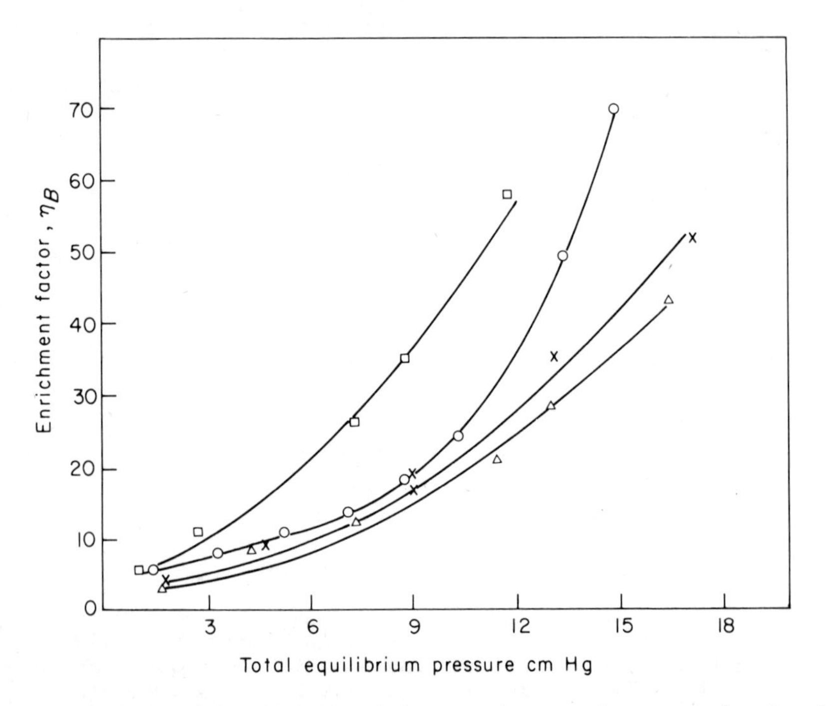

FIG. 34. Enrichment factors for benzene in benzene-*n*-heptane mixtures, as a function of temperature and pressure.[107] ⊡, 80°; ○, 85°; ×, 90°; and △, 95°C.

in two mixtures of this compound with n-heptane; Fig. 33c shows the same strong selectivity in a mixture of benzene with cyclohexane and Fig. 33d shows cyclohexane isotherms in several mixtures with n-heptane. Enrichment factors for benzene over n-heptane are illustrated as functions of total pressure and of temperature in Fig. 34; both hydrocarbons are intercalated but there is a strong preference for benzene which increases with falling temperature or rising total pressure. The doses of parent mixtures all had mole fractions of benzene of 0·67.

The enrichment factor, η_A, for a component A in a mixture of $(A + B)$ is defined by

$$\eta_A = \frac{N_A^s p_B}{N_B^s p_A} = \frac{n_A^s p_B}{n_B^s p_A} \tag{11}$$

where the N are mole fractions and the n are numbers of moles. The superscript "s" denotes in the sorbed phase. Sorption may occur in the interlamellar region and on external surfaces so that two enrichment factors may contribute to the overall selectivity:

$$\eta_A^i = \frac{n_A^i p_B}{n_B^i p_A} = ; \quad \eta_A^a = \frac{n_A^a p_B}{n_B^a p_A} \tag{12}$$

where superscripts "i" and "a" denote "intercalated" and "adsorbed" respectively. For the total sorptions one has

$$n_A^s = n_A^i + n_A^a; \; n_B^s = n_B^i + n_B^a \tag{13}$$

and for the measured separation factor

$$\eta_A = \frac{p_B(n_A^i + n_A^a)}{p_A(n_B^i + n_B^a)}. \tag{14}$$

By rearranging and making use of eqns 12 one obtains

$$n_B^s \eta_A = n_B^i \eta_A^i + n_B^a \eta_A^a \tag{15}$$

which gives the general relation between the overall enrichment factor and that associated with interlamellar sorption and with adsorption on external surfaces respectively. For the special case when component A is intercalated but not component B

$$\eta_A = \eta_A^a \left(1 + \frac{n_A^i}{n_A^a}\right) \tag{16}$$

so that $\eta_A \rightarrow \infty$, as required, when $n_A^i/n_A^a \rightarrow \infty$. Equation 16, which expresses the behaviour for true molecule sieving modified by any adsorption on external surfaces, is appropriate in NH_3CH_3-montmorillonite for the first named of the pairs n-heptane + cyclohexane, benzene + cyclohexane and hexane + isooctane.[107] In this sorbent and also in $N(CH_3)_3$-montmorillonite the selectivity sequence based upon the sorption equilibria was

benzene > *n*-heptane, *n*-hexane > cyclohexane > isooctane.

The sequence is related to the structural type of the hydrocarbons.

Equilibrium studies of selectivity have been supplemented by chromatographic measurements. With nitrogen as carrier gas, Barrer and Hampton[112] used 2·7 g of $N(CH_3)_4$- and 2·5 g of NH_3CH_3-montmorillonite in a column

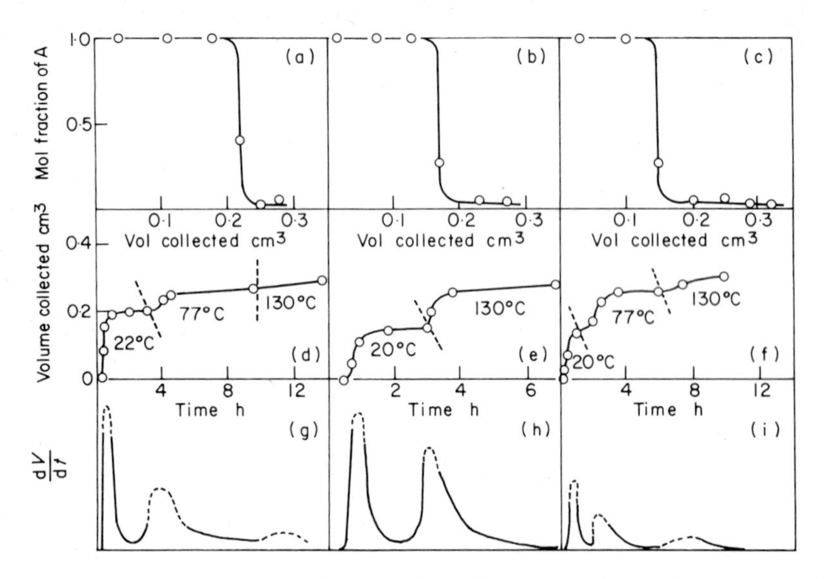

FIG. 35. Gas-solid chromatographic separations of mixtures using $N(CH_3)_4$-montmorillonite.[112]
a, d, g : component A is cyclohexane (0·78 mole fraction in parent mixture); B is *n*-heptane.
b, e, h : A is isooctane (0·45 mole fraction); B is *n*-heptane.
c, f, i : A is cyclohexane (0·45 mole fraction); B is *n*-heptane.

50 cm long and 3·5 mm internal diameter. Liquid volumes of mixtures injected into the carrier gas were between 0·293 and 0·362 cm³ for the former and between 0·120 and 0·221 cm³ for the latter sorbent, so that the sorbents could become rather heavily loaded with sorbates. Experiments were usually conducted at 77°C, but some were made at room temperature and at 130°C. Good separations were obtained using $N(CH_3)_4$-montmorillonite with benzene or toluene + cyclohexane, benzene + *n*-heptane, CCl_4, toluene or methanol, and with *n*-heptane or *n*-hexane + cyclohexane or isooctane. Figure 35 illustrates compositions of effluent as functions of the volumes collected, temperature and time. Good separations were found with NH_3CH_3-montmorillonite for benzene + cyclohexane or isooctane, for

TABLE 22

Some mixture separations with $N(CH_3)_4$-montmorillonite.[112]

Wt of sorbent (g)	Mixture A	Mixture B	Volume used (cm^3)	Mole fraction of A in parent mixture	Volume of sample recovered (cm^3)	Composition of sample recovered
7·5	benzene	isooctane	1·5	0·5	0·2	~100% isooctane
6·5	cyclohexane	benzene, thiophene, toluene	2·0	0·7 (+0·1 benzene, 0·1 thiophene, 0·1 toluene)	0·6	97% cyclohexane
8·5	methanol	carbon tetrachloride	1·5	0·4	~0·25	99% carbon tetrachloride
7·5	benzene	*n*-heptane	1·7	0·6	0·2	98% *n*-heptane
7	*n*-heptane	isooctane	1·5	0·67	0·2	75–80% isooctane
6	benzene	cyclohexane	4·5	0·25	1·0	92% cyclohexane

methanol + *n*-heptane or isooctane, and for methanol + toluene. The established selectivity sequences were

$N(CH_3)_4$-form: pyridine > thiophene > benzene > toluene > methanol > *n*-heptane > *n*-hexane > carbon tetrachloride, cyclohexane, cyclopentane and isooctane.

NH_3CH_3-form: methanol > benzene, toluene, *n*-heptane and thiophene > cyclo-hexane and isooctane.

Additional separations were effected in which 6 to 8·5 g of the sorbents were contacted with mixtures and samples of the non-sorbed liquid were then obtained and analysed refractometrically. Some results are given in Table 22 which show how readily molecules of the wrong shape and size may be removed from others which have the right form to penetrate smectite molecular sieves.

Packing materials for gas chromatographic columns have been employed in the form of dimethyldioctadecylammonium bentonite (Bentone 34), which gave good separations of *o*- *m*- and *p*-xylenes and ethylbenzene.[113] The packing of the column was Celite +8% Bentone 34 and the carrier gas was argon. A chromatogram is shown in Fig. 36. Bentone 34 was also used in

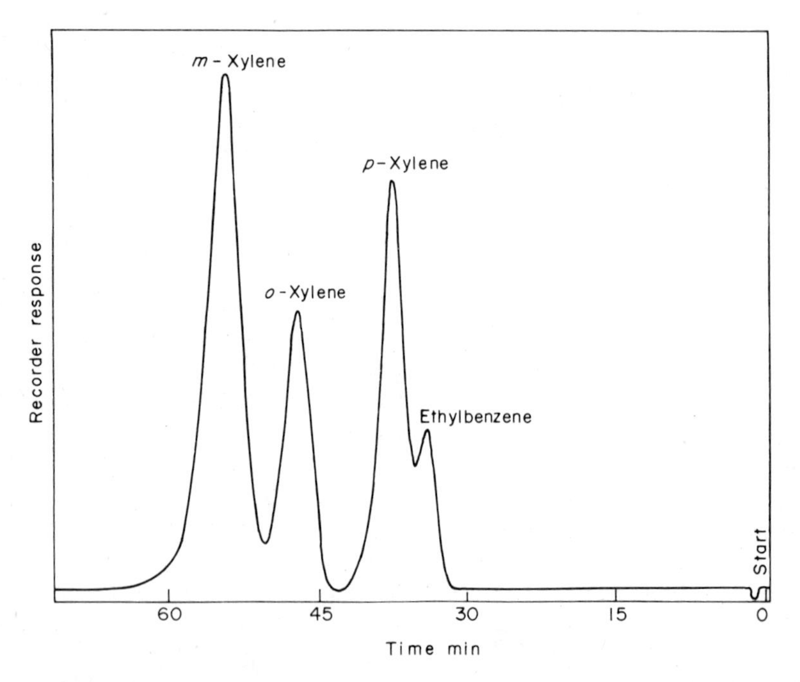

FIG. 36. Gas-solid chromatogram for substituted benzenes in a column of Celite +8% Bentone 34.[113]

chromatographic separations of aromatic and aliphatic compounds,[114] of toluidines and cresols,[115] and of benzene and its di- and trichloro derivatives.[116] Kiselev *et al.*[117] had success in separations of *o*-, *m*- and *p*-isomers of terphenyl, phenoxyphenol and nitrophenol with the sorbent as packing in liquid chromatography. Using Bentone 34 modified with silicone oil and dispersed in Celite, Mortimer and Gent[118] measured retention times for twenty three different hydrocarbons. However the separations obtainable are not due only to the Bentone 34 since the silicone oil in particular plays a part in the mixed columns.

Molecular sieve smectites in the form of the tris(ethylenediamine) complexes of Cr^{-III} and Co^{-III} montmorillonites gave good separations of C_1 to C_5 light hydrocarbons on very short columns at relatively high temperatures and carrier gas flow rates.[119] The retention times of these hydrocarbons and of oxides of nitrogen increased with % exchange of Na^+ by the complex cations up to 100%, but decreased if, through uptake of the salts of the complex cations, the apparent extent of exchange exceeded 100%.

11. CONCLUSION

This account of the sorbent properties of smectites, vermiculites, kandites and palygorskites draws attention to some of their special features and shows that some have potentialities, not yet fully explored, which in various directions parallel those of the zeolites. Molecular sieve smectites have been prepared in which the apertures giving entry to interlamellar regions rendered permanently porous have been regulated in many ways, so that their selectivity can rival that shown by zeolite sieves, as illustrated in § 10. It is interesting that the synthetic fluorhectorite molecular sieves are magnesium silicates rather than aluminosilicates.

Because of their layer structure and ability to swell the clay minerals can sorb bulky and long chain molecules with a facility not apparent in many of the zeolites. The packing geometry of intercalated guest molecules is varied and is subject to rather subtle influences. The chains can imitate in packing and orientation that of the chains in monolayers of amphipathic long chain compounds on water surfaces, but other arrangements, including bilayers, are also of common occurrence in the intercalates. The same capacity of ligands to attach themselves to transition metal ions is found in zeolites and in clay minerals, with notable changes in cation exchange selectivities when organometallic ionic complexes form. In some cases, as with benzene in Cu^{-II}smectite, complexes are stabilized which do not form in aqueous solutions containing Cu^{2+}. The special environment can thus serve, as with zeolites, to bring about unusual chemisorption as well as physisorption processes.

REFERENCES

1. B. K. G. Theng, *J. Roy. Soc. N.Z.* (1972) **2**, 437.
2. Sir L. Bragg and G. F. Claringbull, "Crystal Structures of Minerals", Bell and Sons (1965) p. 279.
3. Sir L. Bragg and G. F. Claringbull, "Crystal Structures of Minerals", Bell and Sons (1965) p. 281.
4. Sir L. Bragg and G. F. Claringbull, "Crystal Structures of Minerals", Bell and Sons (1965) p. 287.
5. G. W. Brindley and J. Zussman, *Amer. Mineralog.* (1957) **42**, 666.
6. R. M. Barrer and L. W. R. Dicks, *J. Chem. Soc.* A (1966) 1379.
7. R. M. Barrer and L. W. R. Dicks, *J. Chem. Soc.* A (1967) 1523.
8. G. E. Marshall, "Physical Chemistry and Mineralogy of Soils", Vol. I, Wiley (1964) p. 133.
9. S. B. Hendricks and M. E. Jefferson, *Amer. Mineralog* (1939) **24**, 729.
10. J. V. Smith and H. S. Yoder, *Mineralog. Mag.* (1956) **31**, 209.
11. R. M. Barrer *in* "Non-stoichiometric Compounds", ed. L. Mandelcorn Academic Press (1964) p. 329.
12. A. Preisinger, *Clays and Clay Minerals* (1957) **6**, 61.
13. A. Weiss, W. Thielpape, G. Goring, W. Ritter and H. Schafer, Internat. Clay Conference, Vol. 14, Pergamon (1963) p. 287.
14. A. Weiss, W. Thielepape, W. Ritter, H. Schafer and G. Goring, *Zeit. anorg. u. allg. Chemie* (1963) **320**, 183.
15. J. M. Adams and D. A. Jefferson, *Acta Cryst.* (1976) B **32**, 1180.
16. J. L. M. Vivaldi, A. Pozzuoli, P. Mathias and E. Galan–Huertos, International Clay Conference, Madrid (1972) p. 455.
17. G. W. Nederbragt and J. J. de Jong, *Rec. Trav. Chim.* (1946) **65**, 831.
18. G. W. Nederbragt, *Clay Min. Bull.* (1949) **3**, 72.
19. G. Migeon, *Bull. Soc. franc. Mineralog.* (1936) **59**, 6.
20. W. S. McCarter, K. A. Krieger and H. Heinemann, *Ind. Eng. Chem.* (1950) **42**, 529.
21. W. T. Granquist and R. C. Amero, *J. Amer. Chem. Soc.* (1948) **70**, 3265.
22. R. M. Barrer and N. Mackenzie, *J. Phys. Chem.* (1954) **58**, 560.
23. R. M. Barrer, N. Mackenzie and D. M. MacLeod, *J. Phys. Chem.* (1954) **58**, 568.
24. A. J. Dandy, *J. Phys. Chem.* (1968) **72**, 334.
25. A. J. Dandy, *J. Chem. Soc.*, A (1971) 2383.
26. S. Brunauer, "The Adsorption of Gases and Vapours" Oxford University Press (1944) p. 150.
27. G. F. Walker and A. Milne, *in* Trans. Int. Cong. of Soil Sci., Amsterdam (1950) Vol. II, p. 62.
28. G. F. Walker, *Clays and Clay Minerals* (1956) **4**, 101.
29. G. F. Walker, *in* "The X-ray Identification and Crystal Structures of Clay Minerals", ed. G. Brown, The Mineralogical Society, London (1961) p. 304.
30. Sir L. Bragg and G. F. Claringbull, "Crystal Structures of Minerals", Bell and Sons (1965) p. 274.
31. H. van Olphen *in* Proc. Internat. Clay Conf., Tokyo, Vol. I (1969) p. 649.
32. F. H. Gillery, *Amer. Mineralog.* (1959) **44**, 806.
33. R. M. Barrer and D. M. MacLeod, *Trans. Faraday Soc.* (1954) **50**, 980.

34. A. Weiss, *Zeit. anorg. u. allg. Chemie* (1958) **297**, 17.
35. A. Weiss, A. Mehler and U. Hofmann, *Zeit. Naturforsch.* (1956) **11** B, 431.
36. G. F. Walker, *Clay Minerals* (1967) **7**, 129.
37. J. White, *Clays and Clay Minerals* (1956) **4**, 133.
38. J. White, *Clays and Clay Minerals* (1958) **5**, 289.
39. A. Weiss, *Chem. Ber.* (1958) **91**, 487.
40. E. J. W. Verwey and J.Th.G. Overbeek, "Theory of Stability of Lyophobic Colloids", Elsevier (1948) p. 101.
41. M. S. Stuhl and J. B. Uytterhoeven, *J. Chem. Soc. Faraday* I (1975) **71**, 1396.
42. W. G. Garrett and G. F. Walker, *Clays and Clay Minerals* (1962) **9**, 557.
43. J. A. Rausel–Colom, *Trans. Faraday Soc.* (1964) **60**, 190.
44. K. Norrish and J. A. Rausel–Colom, *Clays and Clay Minerals* (1963) **10**, 123.
45. D. C. MacEwan, *Trans. Faraday Soc.* (1948) **44**, 349.
46. I. Barshad, *Soil Sci. Soc. Amer. Proceedings* (1952) **16**, 176.
47. I. M. Eltantawy and P. W. Arnold, *Nature*, *Phys. Sci.* (1972) **237**, 123.
48. R. M. Barrer and J. S. S. Reay, *J. Chem. Soc.* (1958) 3824.
49. R. W. Mooney, A. G. Keenan and L. A. Wood, *J. Amer. Chem. Soc.* (1952) **74**, 1367 and 1371.
50. H. van Olphen, *J. Coll. Sci.* (1965) **20**, 882.
51. S. J. Gregg and K. S. W. Sing *in* "Surface and Colloid Science", Vol. 9, ed. E. Matijevic, John Wiley, (1976) p. 342.
52. G. F. Walker, *Nature* (1956) **177**, 239.
53. J.-P. Hazart and R. Wey, *Bull. Serv. Carte géol. Als. Lorr* (1965) **18**, 191.
54. I. M. Eltantawy and P. W. Arnold, *J. Soil Sci.* (1974) **25**, 99.
55. G. F. Walker, *Clay Miner. Bull.* (1958) **3**, 302.
56. G. W. Brindley and S. Ray, *Amer. Mineralog.* (1964) **49**, 106.
57. W. W. Emerson, *Nature* (1957) **180**, 48.
58. G. W. Brindley and M. Rustom, *Amer. Mineralog.* (1958) **43**, 627.
59. R. W. Hoffmann and G. W. Brindley, *Geochim. Cosmochim. Acta* (1960) **20**, 15.
60. R. Greene-Kelley, *Trans. Faraday Soc.* (1956) **52**, 1281.
61. J. M. Adams, *J. Chem. Soc. Dalton* (1974) 2286.
62. J. M. Adams, J. M. Thomas and M. J. Walters, *J. Chem. Soc. Dalton* (1975) 1459.
63. S. Yamanaka, F. Kanamuru and M. Koizumi, *J. Phys. Chem.* (1975) **79**, 1285.
64. T. Masuda and H. Takahashi, *Bull. Inst. Chem. Res. Kyoto Univ.* (1975) **53**, 147.
65. M. M. Mortland, "Clay Organic Complexes and Interactions" *in* Advances in Agronomy (1970) **22**, 75.
66. R. M. Barrer and J. S. S. Reay, unpublished observation.
67. D. T. B. Tennakoon, J. M. Thomas, M. J. Tricker and J. O. Williams, *J. Chem. Soc. Dalton* (1974) 2207.
68. B. K. G. Theng, *Clays and Clay Minerals* (1971) **19**, 383.
69. D. J. B. Tennakoon, J. M. Thomas and J. Tricker, *J. Chem. Soc. Dalton*, (1974) 2211.
70. C. G. Dodd and S. Ray, *Clays and Clay Minerals* (1960) **8**, 237.
71. Y. Matsunaga, *Bull. Chem. Soc. Japan* (1972) **45**, 770.
72. D. J. Greenland and J. P. Quirk, *Clays and Clay Minerals* (1962) **9**, 484.
73. P. Franzen, *Clay Minerals Bulletin* (1954) **2**, 223.
74. D. J. Greenland, R. H. Laby and J. P. Quirk, *Trans. Faraday Soc.* (1962) **58**, 829; (1965) **61**, 2013; (1965) **61**, 2024.

75. O. Sieskind, *C.R.* (1960) **250**, 2228, 2392.
76. O. Sieskind and R. Wey, *C.R.* (1959) **248**, 1652.
77. O. Talibudeen, *Trans. Faraday Soc.* (1955) **51**, 582.
78. H. E. Doner and M. M. Mortland, *Science* (1969) **166**, 1406.
79. T. J. Pinnavia and M. M. Mortland, *J. Phys. Chem.* (1971) **75**, 3957.
80. T. J. Pinnavia and M. M. Mortland, *Nature* (1971) **229**, 75.
81. V. C. Farmer and M. M. Mortland, *J. Chem. Soc.* A (1966) 344.
82. W. Bodenheimer, B. Kirson and Sh. Yariv, *Israel J. Chem.* (1963) **1**, 69.
83. W. Bodenheimer, B. Kirson and Sh. Yariv, *Anal. Chem. Acta* (1963) **29**, 582.
84. V. C. Farmer and M. M. Mortland, *J. Phys. Chem.* (1965), **69**, 683.
85. W. Bodenheimer and L. Heller, *Clay Minerals* (1967) **7**, 167.
86. S. Son, S. Ueda, F. Kanamaru and M. Koizumi, *J. Phys. Chem.* (1976), **80**, 1780.
87. R. M. Barrer and R. J. B. Carven, in preparation.
88. J. Pleysier and A. Cremer, *J. Chem. Soc. Faraday* I (1975) **71**, 256.
89. J. E. Gieseking, *Soil Sci.* (1939) **47**, 1.
90. J. W. Jordan, *J. Phys. Coll. Chem.* (1949) **53**, 294.
91. E. Suito, M. Arakawa and S. Kanjo, *Bull. Inst. Chem. Res. Kyoto Univ.* (1966) **44**, 316.
92. W. Slabaugh and P. A. Hiltner, *J. Phys. Chem.* (1968) **72**, 4295.
93. R. M. Barrer and K. Brummer, *Trans. Faraday Soc.* (1963) **59**, 959.
94. B. K. G. Theng, D. J. Greenland and J. P. Quirk, *Clay Minerals* (1967) **7**, 1.
95. J. W. Jordan, B. J. Hook and C. M. Finlayson, *J. Phys. Coll. Chem.* (1950) **54**, 1196.
96. R. M. Barrer and K. Kelsey, *Trans. Faraday Soc.* (1961) **57**, 625.
97. W. H. Slabaugh and G. H. Kennedy, *J. Coll. Sci* (1963) **18**, 337.
98. J. H. Hildebrand and R. L. Scott, "The Solubility of Non-electrolytes", 3rd Edition, Rheinhold (1950) p. 435.
99. P. Flory, *J. Chem. Phys.* (1942) **10**, 51.
100. M. Huggins, *J. Chem. Phys.* (1941) **9**, 440.
101. G. B. Street and D. White, *J. Appl. Chem.* (1963) **13**, 288.
102. R. M. Barrer and A. D. Millington, *J. Coll. Interface Sci.* (1967) **25**, 359.
103. R. M. Barrer and D. M. MacLeod, *Trans. Faraday Soc.* (1955) **51**, 1290.
104. R. M. Barrer and J. S. S. Reay, *Trans. Faraday Soc.* (1957) **53**, 1253.
105. R. M. Barrer and J. A. Barrie, *Proc. Roy. Soc.* (1952) A **213**, 250.
106. R. M. Barrer and D. L. Jones, *J. Chem. Soc.* A (1971) 2595.
107. R. M. Barrer and G. S. Perry, *J. Chem. Soc.* (1961) 842 and 850.
108. M. M. Mortland and V. Berkheiser, *Clays and Clay Minerals* (1976) **24**, 60.
109. J. Shabtai, N. Frydman and R. Lazar, 6th Internat. Cong. on Catalysis, London, 12–16th July (1976) Paper B5.
110. R. J. B. Craven, Ph.D. Thesis, London University (1976) p. 164.
111. R. J. B. Craven, Ph.D. Thesis, London University (1976) p. 231.
112. R. M. Barrer and M. G. Hampton, *Trans. Faraday Soc.* (1957) **53**, 1462.
113. J. van Rysselberge and M. van der Stricht, *Nature* (1962) **193**, 1281.
114. D. White, *Nature* (1957) **179**, 1075.
115. D. White, *Nature* (1959) **184**, 1795.
116. C. T. Cowan and J. M. Hartwell, *Nature* (1961) **190**, 712.
117. A. V. Kiselev, N. P. Lebedeva, I. I. Frolov and Ya. I. Yashin, *Chromatographia* (1972) **5**, 341.
118. J. V. Mortimer and P. L. Gent, *Nature* (1963) **197**, 789.
119. V. J. Thielmann and J. L. McAtee, *J. Chromatog.* (1975) **105**, 115.

INDEX

Italics indicates that some or all of the reference is in a figure. **Bold face** indicates that a major section of the text is devoted to the topic. Note: prefixes (such as cyclo and iso) have been ignored in the alphabetization of organic compounds.

D

E

F

N